Ľubica Holá, Dušan Holý, and Warren Moors
USCO and Quasicontinuous Mappings

De Gruyter Studies in Mathematics

Volume 81

Ľubica Holá, Dušan Holý, and Warren Moors

USCO and Quasicontinuous Mappings

DE GRUYTER

Mathematics Subject Classification 2020
Primary: 54C08, 54C60, 46B20; Secondary: 49J50, 49J52, 49J53

Authors
Ľubica Holá
Slovak Academy of Sciences
Mathematical Institute
Štefánikova ul. 49
814 73 Bratislava
Slovakia
lubica.hola@mat.savba.sk

Warren Moors
The University of Auckland
Department of Mathematics
Private Bag 92019
Auckland 1142
New Zealand
w.moors@auckland.ac.nz

Dušan Holý
Trnava University
Faculty of Education
Priemyselná 4
918 43 Trnava
Slovakia
dusan.holy@truni.sk

ISBN 978-3-11-075015-7
e-ISBN (PDF) 978-3-11-075018-8
e-ISBN (EPUB) 978-3-11-075022-5
ISSN 0179-0986

Library of Congress Control Number: 2021940877

Bibliographic information published by the Deutsche Nationalbibliothek
The Deutsche Nationalbibliothek lists this publication in the Deutsche Nationalbibliografie; detailed bibliographic data are available on the Internet at http://dnb.dnb.de.

Typesetting: VTeX UAB, Lithuania
Printing and binding: CPI books GmbH, Leck

MIX
Papier aus verantwortungsvollen Quellen
FSC® C083411

www.degruyter.com

Preface

The purpose of this monograph is threefold. Firstly, to introduce to nonspecialists (in general topology) and graduate students alike, the classes of usco and quasicontinuous mappings, which both generalise, in natural ways, the classical notion of a single-valued continuous function. Secondly, to present the most fundamental properties of these mappings and, thirdly, to highlight their utility across many areas of mathematics.

In fact, we hope that this monograph will serve as a pathway to the study of both usco and quasicontinuous mappings as well as to their applications.

The first generalisation of the classical notion of a continuous function that we consider is that of an usco mapping. Here the restriction of single-valuedness is relaxed while still retaining a natural notion of continuity.

In Chapter 1, we explore the basic properties of usco mappings. Some of their best known applications are also presented.

Chapter 2 concerns a different generalisation of continuity, a quasicontinuous mapping. This time, we retain the single-valuedness of a mapping, but we slightly relax the requirement that the inverse image of each open set is open. In this chapter, we present the fundamental properties of quasicontinuous mappings and provide a wide range of applications of this notion. We also explore topological properties of the space of all quasicontinuous mappings acting between two given topological spaces.

In Chapter 3, we consider six areas of applications of usco mappings. This chapter is designed to highlight the vast utility of these mappings. The reader should note that the applications given here are merely a reflection upon the backgrounds of the authors. Such applications undoubtedly exist in many other areas of mathematics as well.

Finally, in Chapter 4, we consider the topological structure of the space of all usco mappings acting between two fixed topological spaces. We answer many questions concerning the countability, metrisability, and completeness of these spaces.

Ľ. Holá and D. Holý are grateful to Professor Neubrunn for showing them this area of research and also to colleagues J. Činčura, B. Novotný, M. Sleziak, and V. Toma, who read various parts of Chapter 2.

https://doi.org/10.1515/9783110750188-201

Contents

1 Usco mappings

In many situations that occur in mathematics, there arise correspondences that are not in the form of a function. For example, partial orderings, equivalence relations, incidence relations, best approximations, and so on. Even in the case where we have a function $f : X \to Y$ acting between sets X and Y, it may be the case that $f^{-1} : Y \to X$ does not exist, possibly because f is not onto or not one-to-one. However, even in these cases, it may be desirable to consider the "generalized inverse" $f^{-1} : Y \to 2^X$ defined by

$$f^{-1}(y) := \{x \in X : f(x) = y\}.$$

In optimization, we are confronted with situations where we must find

$$\operatorname{argmax}(f) := \left\{x \in X : f(x) = \sup_{y \in X} f(y)\right\}$$

for some function $f : X \to [-\infty, \infty)$. Of course, it may be that $\operatorname{argmax}(f) = \varnothing$, in which case the function f does not attain a maximum value. On the other hand, it may that f attains its maximum value at many points. Hence, in this case, $\operatorname{argmax}(f)$ will not be a singleton. Thus, if we want to consider the variational problem $f \mapsto \operatorname{argmax}(f)$, then we are inextricably led to consider set-valued mappings. Beyond these few examples, there is a whole plethora of examples where the use of set-valued mappings is unavoidable.

In this chapter, we consider the class of "usco" mappings/correspondences/multifunctions. These are a robust, well-behaved class of set-valued mappings that act between topological spaces. They generalize, in a natural way, the notion of continuity, of a single-valued mapping. However, their definition is general enough to encompass a wide range of applications, including, but not restricted to: approximation theory, control theory, convex analysis, differential inclusions, functional analysis, game theory, mathematical economics, and optimization theory.

We will begin by giving the basic definitions and fundamental properties of usco mappings. Then we will present an array of examples that will help demonstrate the wide utility of usco mappings. After that, we present some fundamental constructions of usco mappings. Following this, we will provide a proof of the famous Kakutani–Glicksberg–Fan fixed-point theorem and then give some its applications.

The next topic considered is that of "minimal uscos". These provide a class of mappings that behave somewhat like continuous functions (or at least like quasicontinuous functions). This topic naturally leads to the study of selections. In this direction, we prove some general "selection theorems", most notably, Michael's selection theorem. Then we present some applications of these results.

Finally, we end this chapter by analysing the general behaviour of usco mappings that map into metric spaces. Among the results, we present a proof of "Fort's theorem".

https://doi.org/10.1515/9783110750188-001

1.1 Introduction

In this section, we provide a gentle introduction to the theory of usco mappings.

1.1.1 Basic properties

In this subsection, we present the most basic and fundamental results concerning general usco mappings utilized in this chapter.

We begin with a notion of continuity applied to set-valued mappings. Let (X, τ) and (Y, τ') be topological spaces, and let $\Phi : X \to 2^Y$, that is, for each $x \in X$, $\Phi(x)$ is a (possibly empty) subset of Y. If $x_0 \in X$, then we say that Φ is *τ'-upper semicontinuous at x_0* if for every $W \in \tau'$ containing $\Phi(x_0)$, there exists a neighborhood U of x_0 such that $\Phi(U) \subseteq W$. Here we use the notation $\Phi(U) := \bigcup\{\Phi(u) : u \in U\}$. If Φ is τ'-upper semicontinuous at every point of X, then we say that Φ *is τ'-upper semicontinuous on* (X, τ). When there is no ambiguity concerning the topology τ', we will simply say that Φ is *upper semicontinuous on* (X, τ).

From this definition we can immediately deduce some simple facts.

Exercise 1.1.1. Let (X, τ) and (Y, τ') be topological spaces, and let $\Phi : X \to 2^Y$. Show that $\Phi : X \to 2^Y$ is upper semicontinuous on (X, τ) if and only if for each open subset W of (Y, τ'), $\{x \in X : \Phi(x) \subseteq W\}$ is an open subset of (X, τ).

Exercise 1.1.2. Let (X, τ) and (Y, τ') be topological spaces, and let $\Phi : X \to 2^Y$. Show that $\Phi : X \to 2^Y$ is upper semicontinuous on (X, τ) if and only if for each closed subset C of (Y, τ'), $\{x \in X : \Phi(x) \cap C \neq \varnothing\}$ is a closed subset of (X, τ).

The following exercise provides an important source of upper semicontinuous set-valued mappings.

Exercise 1.1.3. Let (X, τ) and (Y, τ') be topological spaces, and let $f : X \to Y$ be a closed mapping (i. e., it maps closed sets in X to closed sets in Y). Then $f^{-1} : Y \to 2^X$ defined by $f^{-1}(y) := \{x \in X : f(x) = y\}$ is an upper semicontinuous mapping on (Y, τ').

Upper semicontinuity has topological consequences for the effective domain of a set-valued mapping.

Exercise 1.1.4. Let (X, τ) and (Y, τ') be topological spaces, and let $\Phi : X \to 2^Y$. Show that if $\Phi : X \to 2^Y$ is upper semicontinuous on (X, τ), then

$$\mathrm{Dom}(\Phi) := \{x \in X : \Phi(x) \neq \varnothing\}$$

is a closed subset of (X, τ).

Note: $\mathrm{Dom}(\Phi)$ is called the *effective domain* of Φ or simply the *domain* of Φ.

Next, suppose that (X,τ) and (Y,τ') are topological spaces and $\Phi : X \to 2^Y$. If $x_0 \in X$, then we say that Φ is a *τ'-usco at x_0* if $\Phi(x_0)$ is a nonempty compact subset of (Y,τ') and Φ is τ'-upper semicontinuous at x_0. If Φ is a τ'-usco at every point of X, then we say that Φ is a *τ'-usco on (X,τ)*. When there is no ambiguity concerning the topology τ', we will simply say that Φ is an usco on (X,τ).

In the subsequent proposition, we will consider the restriction of a set-valued mapping. Suppose that A is a subset of a set X and $\Phi : X \to 2^Y$ is a set-valued mapping from the set X into subsets of a set Y. Then $\Phi|_A : A \to 2^Y$ is defined by

$$\Phi|_A(a) := \Phi(a) \quad \text{for all } a \in A.$$

Proposition 1.1.5. *Let (X,τ) and (Y,τ') be topological spaces, and let $\Phi : X \to 2^Y$ be an usco on (X,τ). If A is a nonempty subset of X, then $\Phi|_A$ is an usco on (A,τ''), where τ'' denotes the relative τ-topology on the set A.*

Proof. The proof of this fact follows directly from the relevant definitions. □

The subsequent proposition demonstrates that the usconess (if such a word exists) is locally determined.

Proposition 1.1.6. *Let (X,τ) and (Y,τ') be topological spaces, and let $\Phi : X \to 2^Y$. Then Φ is an usco on (X,τ) if and only if for every $x \in X$, there exists a neighbourhood U_x of x such that $\Phi|_{U_x}$ is an usco on (U_x,τ''), where τ'' denotes the relative τ-topology on the set U_x.*

Proof. As before, the proof of this fact follows directly from the relevant definitions. □

We now prove that usco mappings map compact sets to compact sets.

Proposition 1.1.7 ([24, Proposition 6.2.11]). *Let (X,τ) and (Y,τ') be topological spaces, and let $\Phi : X \to 2^Y$ be an usco on (X,τ). If K is a compact subset of (X,τ), then $\Phi(K)$ is a compact subset of (Y,τ').*

Proof. Let K be a compact subset of (X,τ), and let $\{O_\alpha : \alpha \in A\}$ be a τ'-open cover of $\Phi(K)$.

For each $k \in K$, let U_k be a τ-open neighborhood of k in (X,τ), and let F_k be a finite subset of A such that $\Phi(U_k) \subseteq \bigcup_{\alpha\in F_k} O_\alpha$. Note that this is possible since $\Phi(k)$ is τ'-compact and Φ is τ'-upper semicontinuous at k. Now let $\{U_{k_j} : 1 \le j \le n\}$ be a finite subcover of the τ-open cover $\{U_k : k \in K\}$ of K, and let $F := \bigcup_{1\le j\le n} F_{k_j}$. Then F is a finite subset of A, and

$$\Phi(K) \subseteq \Phi\Big(\bigcup_{1\le j\le n} U_{k_j}\Big) = \bigcup_{1\le j\le n} \Phi(U_{k_j}) \subseteq \bigcup_{1\le j\le n}\Big(\bigcup_{\alpha\in F_{k_j}} O_\alpha\Big) = \bigcup_{\alpha\in F} O_\alpha.$$

Thus $\{O_\alpha : \alpha \in F\}$ is a finite subcover of $\{O_\alpha : \alpha \in A\}$. □

The following proposition shows that usco mappings, like continuous functions, possess closed graphs (when the range space is Hausdorff).

Proposition 1.1.8 ([59]). *Let (X,τ) and (Y,τ') be topological spaces, and let $\Phi : X \to 2^Y$ is an usco on (X,τ). If (Y,τ') is a Hausdorff space, then the set*

$$Gr(\Phi) := \{(x,y) \in X \times Y : y \in \Phi(x)\},$$

called the graph of Φ, is a closed subset of $X \times Y$ when the latter is endowed with the product topology of (X,τ) and (Y,τ).

Proof. We will show that $(X \times Y) \setminus \mathrm{Gr}(\Phi)$ is an open subset of $X \times Y$, endowed with the product topology of (X,τ) and (Y,τ'). To this end, let $(x,y) \in (X \times Y) \setminus \mathrm{Gr}(\Phi)$. Then $y \notin \Phi(x)$. Choose disjoint τ'-open sets V and W of (Y,τ') such that $y \in V$ and $\Phi(x) \subseteq W$. Note that this is possible since (Y,τ') is Hausdorff and $\Phi(x)$ is compact. By the upper semicontinuity of Φ at x there exists a τ-open neighbourhood U of x such that $\Phi(U) \subseteq W$. In particular, this means that $\Phi(U) \cap V = \varnothing$, which in turn means that $(U \times V) \cap \mathrm{Gr}(\Phi) = \varnothing$. Thus $U \times V$ is an open neighbourhood of (x,y), contained in $(X \times Y) \setminus \mathrm{Gr}(\Phi)$, which shows that $(X \times Y) \setminus \mathrm{Gr}(\Phi)$ is open. □

In the ensuing text, we will show that there is a partial converse to Theorem 1.1.8.

Exercise 1.1.9. Let (X,τ) and (Y,τ') be topological spaces, and let C be a closed subset of $X \times Y$ endowed with the product topology. Suppose also that $x \in X$ and K is a nonempty compact subset of (Y,τ'). If $(\{x\} \times K) \cap C = \varnothing$, then there exist open subsets U of (X,τ) and V of (Y,τ') such that $x \in U$, $K \subseteq V$, and $(U \times V) \cap C = \varnothing$.

In fact, there is a more general result, which states that if K_1 is a nonempty compact subset of (X,τ), K_2 is a nonempty compact subset of (Y,τ'), and $(K_1 \times K_2) \cap C = \varnothing$, then there exist open subsets U of (X,τ) and V of (Y,τ') such that $K_1 \subseteq U$, $K_2 \subseteq V$ and $(U \times V) \cap C = \varnothing$.

Proposition 1.1.10. *Let (X,τ) and (Y,τ') be topological spaces, and let $\Phi : X \to 2^Y$ be an usco on (X,τ). If C is a closed subset of $X \times Y$ endowed with the product topology and $\pi : X \times Y \to X$ is defined by $\pi(x,y) := x$ for all $(x,y) \in X \times Y$, then $\pi(Gr(\Phi) \cap C)$ is a closed subset of (X,τ).*

Proof. We will show that $X \setminus \pi(\mathrm{Gr}(\Phi) \cap C)$ is open in (X,τ). To achieve this, consider an arbitrary element $x \notin \pi(\mathrm{Gr}(\Phi) \cap C)$. Then $(\{x\} \times \Phi(x)) \cap C = \varnothing$. Therefore by Exercise 1.1.9 there exist open subsets U of (X,τ) and V of (Y,τ') such that $x \in U$, $\Phi(x) \subseteq V$, and $(U \times V) \cap C = \varnothing$. Since Φ is upper semicontinuous, there exists an open neighbourhood N of x, contained in U, such that $\Phi(N) \subseteq V$. Therefore

$$\varnothing \subseteq [N \times \Phi(N)] \cap C \subseteq [U \times V] \cap C = \varnothing, \quad \text{and so} \quad [N \times \Phi(N)] \cap C = \varnothing.$$

Hence $N \cap \pi(\mathrm{Gr}(\Phi) \cap C) = \varnothing$. Thus $x \in N \subseteq X \setminus \pi(\mathrm{Gr}(\Phi) \cap C)$, which completes the proof. □

Theorem 1.1.11 ([36, 37]). *Let (X,τ) and (Y,τ') be topological spaces, and let $\Phi : X \to 2^Y$ be an usco on (X,τ). If $\Omega : X \to 2^Y$ has a closed graph and $\{x \in X : \Phi(x) \cap \Omega(x) \neq \varnothing\}$ is dense in (X,τ), then $\Psi : X \to 2^Y$ defined by*

$$\Psi(x) := \Phi(x) \cap \Omega(x) \quad \text{for all } x \in X$$

is an usco on (X,τ).

Proof. We will first show that Ψ has compact (possibly, empty) images on X. Let $x \in X$. Since Ω has a closed graph, it follows that $\Omega(x)$ is a closed subset of (Y,τ'). Therefore $\Psi(x) = \Phi(x) \cap \Omega(x)$ is a closed subset of the compact set $\Phi(x)$ and hence is compact itself. Next, we will show that Ψ is upper semicontinuous on (X,τ). To do this, we will employ Exercise 1.1.2. To this end, let F be a closed subset of Y, and let $C := \mathrm{Gr}(\Omega) \cap (X \times F)$. Then C is closed in $X \times Y$, as it is the intersection of two closed sets and

$$\begin{aligned}\{x \in X : \Psi(x) \cap F \neq \varnothing\} &= \pi(\mathrm{Gr}(\Psi) \cap (X \times F)) \\ &= \pi(\mathrm{Gr}(\Phi) \cap \mathrm{Gr}(\Omega) \cap (X \times F)) = \pi(\mathrm{Gr}(\Phi) \cap C),\end{aligned}$$

which is closed by Proposition 1.1.10. This shows that Ψ is upper semicontinuous on (X,τ). Finally, to show that Ψ has nonempty images, we appeal to Exercise 1.1.4. □

Because of the fundamental importance of this result and also because the proof above relies upon some unproven (in these notes) exercises, we will now give a more naive self-contained proof of this theorem.

Proposition 1.1.12. *Let (X,τ) and (Y,τ') be topological spaces, and let $\Phi : X \to 2^Y$ be an usco on (X,τ). If $\Omega : X \to 2^Y$ has a closed graph, $x \in X$, and $\Phi(x) \cap \Omega(x) = \varnothing$, then there exists a τ-open neighbourhood U of x such that $\Phi(y) \cap \Omega(y) = \varnothing$ for every $y \in U$.*

Proof. Let $x \in X$. Since Ω has a closed graph, for each $y \in \Phi(x)$, there exist a τ'-open neighbourhood V_y of y and a τ-open neighbourhood U_y of x such that $(U_y \times V_y) \cap \mathrm{Gr}(\Omega) = \varnothing$, that is, $\Omega(U_y) \cap V_y = \varnothing$. Since $\Phi(x)$ is compact and $\Phi(x) \subseteq \bigcup\{V_y : y \in \Phi(x)\}$, there exists a finite subcover $\{V_{y_j} : 1 \le j \le n\}$ of $\{V_y : y \in \Phi(x)\}$. Let $U_1 := \bigcap_{1 \le j \le n} U_{y_j}$ and observe that $\Omega(U_1) \cap \bigcup_{1 \le j \le n} V_{y_j} = \varnothing$. Now since Φ is upper semicontinuous and $\Phi(x) \subseteq \bigcup_{1 \le j \le n} V_{y_j}$, there exists a τ-open neighbourhood U_2 of x such that $\Phi(U_2) \subseteq \bigcup_{1 \le j \le n} V_{y_j}$. Let $U := U_1 \cap U_2$. Then U is a τ-open neighbourhood of x, and

$$\varnothing \subseteq \Phi(U) \cap \Omega(U) \subseteq \Phi(U_2) \cap \Omega(U) \subseteq \bigcup_{1 \le j \le n} V_{y_j} \cap \Omega(U) \subseteq \bigcup_{1 \le j \le n} V_{y_j} \cap \Omega(U_1) = \varnothing.$$

Thus $\Phi(y) \cap \Omega(y) = \varnothing$ for every $y \in U$. □

Proof. (Second proof of Theorem 1.1.11) It follows from Proposition 1.1.12 that $\mathrm{Dom}(\Psi)$ is a closed subset of (X,τ). Therefore $\mathrm{Dom}(\Psi) = X$. Next, we show that Ψ has compact images on X. Let $x \in X$. Since Ω has a closed graph, it follows that $\Omega(x)$ is a closed subset of (Y,τ'). Therefore $\Psi(x) = \Phi(x) \cap \Omega(x)$ is a τ'-closed subset of the compact set $\Phi(x)$ and hence is compact itself.

So it remains to show that Ψ is upper semicontinuous at each point of X. To this end, let $x \in X$, and let W be an open subset of (Y,τ') such that $\Psi(x) \subseteq W$. We consider two cases:

(i) If $\Phi(x) \subseteq W$, then from the upper semicontinuity of Φ it follows that there exists a τ-open neighbourhood U of x such that $\Psi(U) \subseteq \Phi(U) \subseteq W$.

(ii) On the other hand, suppose that $K := \Phi(x) \setminus W \neq \varnothing$. Then for each $y \in K$, choose open subsets U_y of (X,τ) and V_y of (Y,τ') such that $(x,y) \in U_y \times V_y$ and $(U_y \times V_y) \cap \mathrm{Gr}(\Omega) = \varnothing$, that is, $\Omega(U_y) \cap V_y = \varnothing$. Since K is compact and $K \subseteq \bigcup\{V_y : y \in K\}$, there exists a finite subcover $\{V_{y_j} : 1 \le j \le n\}$ of $\{V_y : y \in K\}$. Let $U_1 := \bigcap\{U_{y_j} : 1 \le j \le n\}$ and observe that $\Omega(U_1) \cap \bigcup\{V_{y_j} : 1 \le j \le n\} = \varnothing$. Now, since Φ is upper semicontinuous and

$$\Phi(x) \subseteq W \cup K \subseteq W \cup \bigcup\{V_{y_j} : 1 \le j \le n\},$$

there exists an open neighbourhood U_2 of x such that

$$\Psi(U_2) \subseteq \Phi(U_2) \subseteq \bigcup\{V_{y_j} : 1 \le j \le n\} \cup W.$$

Let $U := U_1 \cap U_2$. Then since

$$\varnothing \subseteq \left[\Psi(U) \cap \bigcup_{1\le j\le n} V_{y_j}\right] \subseteq \left[\Psi(U_1) \cap \bigcup_{1\le j\le n} V_{y_j}\right] \subseteq \left[\Omega(U_1) \cap \bigcup_{1\le j\le n} V_{y_j}\right] = \varnothing,$$

we must have that

$$\Psi(U) \subseteq \left(\bigcup\{V_{y_j} : 1 \le j \le n\} \cup W\right) \setminus \left(\bigcup\{V_{y_j} : 1 \le j \le n\}\right) \subseteq W.$$

Hence, in both cases, there exists a τ-open neighbourhood U of x such that $\Psi(U) \subseteq W$. □

From this result we may obtain the following useful corollaries.

Corollary 1.1.13. *Let (X,τ) and (Y,τ') be topological spaces, and let $\Phi : X \to 2^Y$ be an usco on (X,τ). If C is a closed subset of (Y,τ') and $\{x \in X : \Phi(x) \cap C \neq \varnothing\}$ is dense in (X,τ), then $\Psi : X \to 2^Y$ defined by $\Psi(x) := \Phi(x) \cap C$ for $x \in X$ is an usco on (X,τ).*

The next result concerns densely defined set-valued mappings. Recall that a set-valued mapping $\Phi : X \to 2^Y$ acting from a topological space (X,τ) into subsets of a set Y is said to be *densely defined* if $\{x \in X : \Phi(x) \neq \varnothing\}$ is dense in (X,τ), that is, if $\mathrm{Dom}(\Phi)$ is dense in (X,τ).

Corollary 1.1.14 ([59]). *Let (X,τ) and (Y,τ') be topological spaces, and let $\Phi : X \to 2^Y$ be an usco on (X,τ). If $\Omega : X \to 2^Y$ is densely defined and has a closed graph and if $\Omega(x) \subseteq \Phi(x)$ for all $x \in X$, then Ω is an usco on (X,τ).*

Corollary 1.1.15. *Let (X,τ) and (Y,τ') be topological spaces, and let $\Omega : X \to 2^Y$. If (Y,τ') is compact and if Ω is densely defined and has a closed graph, then Ω is an usco on (X,τ).*

Thus we see that there is a close relationship between mappings with closed graph and usco mappings.

Next we will see how to construct a mapping with a closed graph from any given set-valued mapping acting between topological spaces. Essentially, the graph of the new mapping is just the closure of the graph of the original mapping.

Let (X,τ) and (Y,τ') be topological spaces, and let $\Phi : X \to 2^Y$. Then we define $\mathrm{USC}(\Phi) : X \to 2^Y$ by

$$\mathrm{USC}(\Phi)(x) := \bigcap\{\overline{\Phi(U)} : U \text{ is an open neighbourhood of } x\}.$$

Exercise 1.1.16. Let (X,τ) and (Y,τ') be topological spaces, and let $\Phi : X \to 2^Y$. Show that $\mathrm{Gr}(\mathrm{USC}(\Phi)) = \overline{\mathrm{Gr}(\Phi)}$. Moreover, show that if (Y,τ') is Hausdorff and Φ is an usco at $x \in X$, then $\mathrm{USC}(\Phi)(x) = \Phi(x)$.

Exercise 1.1.17. Let (X,τ) and (Y,τ') be topological spaces, and let $\Phi : X \to 2^Y$. Suppose also that Φ is densely defined and has a closed graph. Show that if for each $x \in X$, there exists an open neighbourhood U_x of x such that $K_x := \overline{\Phi(U_x)}$ is compact, then Φ is an usco on (X,τ). *Hint:* Use (i) Proposition 1.1.6; (ii) the fact that $\Phi|_{U_x} : U_x \to 2^{K_x}$ has a closed graph, and (iii) Corollary 1.1.15.

1.1.2 Examples

In this subsection, we present an array of examples of usco mappings. Let (X,τ) and (Y,τ') be topological spaces.

(i) If $f : X \to Y$ is a continuous function, then $\Phi : X \to 2^Y$ defined by $\Phi(x) := \{f(x)\}$ for $x \in X$ is an usco on (X,τ).

(ii) If B is a nonempty compact subset of (Y,τ'), then $\Phi : X \to 2^Y$ defined by $\Phi(x) := B$ for $x \in X$ is an usco on (X,τ).

(iii) If $f : X \to Y$ is a *perfect function* (i. e., a continuous closed surjection onto Y such that $f^{-1}(y)$ is a compact subset of (X,τ) for each $y \in Y$), then $\Phi : Y \to 2^X$ defined by

$$\Phi(y) := \{x \in X : f(x) = y\} \quad \text{for } y \in Y$$

is an usco on (Y,τ').

(iv) If $f : X \to Y$ is a continuous surjection and (X,τ) is compact, then $\Phi : Y \to 2^X$ defined by

$$\Phi(y) := \{x \in X : f(x) = y\} \quad \text{for } y \in Y$$

is an usco on (Y,τ').

(v) Let (X, τ) be a topological space, and let $n \in \mathbb{N}$. Then the mapping $\Phi : X^n \to 2^X$ defined by $\Phi(x_1, x_2, \dots, x_n) := \{x_1, x_2, \dots, x_n\}$ for all $(x_1, x_2, \dots, x_n) \in X^n$ is an usco on X^n (here X^n is endowed with the product topology).

(vi) If $(A, +, \cdot, \|\cdot\|)$ is a Banach algebra (with multiplicative identity, denoted by $\mathbf{1}$), then $\sigma_A : A \to 2^{\mathbb{C}}$ defined by

$$\sigma_A(x) := \{\lambda \in \mathbb{C} : x - \lambda\mathbf{1} \text{ is singular}\} \quad \text{for } x \in A$$

is an usco on $(A, \|\cdot\|)$.

(vii) If $(X, \|\cdot\|)$ is a finite-dimensional normed linear space and C is a nonempty closed subset of $(X, \|\cdot\|)$, then $P_C : X \to 2^C$ defined by

$$P_C(x) := \left\{y \in C : \|x - y\| = \inf_{c \in C} \|x - c\|\right\} \quad \text{for } x \in X$$

is an usco on $(X, \|\cdot\|)$; see Corollary 3.2.6.

(viii) If $(X, \|\cdot\|)$ is a finite-dimensional normed linear space and C is a closed and bounded nonempty subset of $(X, \|\cdot\|)$, then $F_C : X \to 2^C$ defined by

$$F_C(x) := \left\{y \in C : \|x - y\| = \sup_{c \in C} \|x - c\|\right\} \quad \text{for } x \in X$$

is an usco on $(X, \|\cdot\|)$; see Corollary 3.2.28.

(ix) If (K, τ'') is a nonempty compact Hausdorff space and L is a nonempty closed subset of (K, τ''), then $M_L : C(K) \to 2^L$ defined by

$$M_L(f) := \left\{x \in L : f(x) = \max_{l \in L} f(l)\right\} \quad \text{for } f \in C(K)$$

is an usco on $(C(K), \|\cdot\|_\infty)$, see Proposition 3.4.15.

(x) If f is a continuous convex function defined on a nonempty open convex subset K of a normed linear space $(X, \|\cdot\|)$ and $x \in K$, then we define the *subdifferential of f at x* as the set $\partial f(x)$ of all $x^* \in X^*$ satisfying

$$x^*(y - x) \le f(y) - f(x) \quad \text{for all } y \in K.$$

The mapping ∂f acting from K, endowed with the relative norm topology, into subsets of (X^*, weak^*) is a weak*-usco on K; see Proposition 3.3.12.

(xi) If f is a real-valued locally Lipschitz function defined on a nonempty open subset K of a normed linear space $(X, \|\cdot\|)$ and $x \in K$, then we define the *Clarke generalized directional derivative of f at $x \in K$* as

$$f^0(x; y) := \limsup_{\substack{z \to x \\ \lambda \to 0^+}} \frac{f(z + \lambda y) - f(z)}{\lambda}.$$

Furthermore, we define the *Clarke subdifferential mapping of f at x* as

$$\partial_C f(x) := \{x^* \in X^* : x^*(y) \le f^0(x; y) \text{ for all } y \in X\}.$$

The mapping $\partial_C f$ acting from K, endowed with the relative norm topology, into subsets of (X^*, weak^*) is a weak*-usco on K; see Theorem 3.6.8.

(xii) If $(X, \|\cdot\|)$ is a normed linear space, then the *duality mapping* $D : S_X \to 2^{X^*}$ defined by $D(x) := \{x^* \in S_{X^*} : x^*(x) = 1\}$ for $x \in S_X$ is a weak*-usco on S_X endowed with the relatively topology inherited from $(X, \|\cdot\|)$.

(xiii) If Y is a subspace of a normed linear space $(X, \|\cdot\|)$, then the *Hahn–Banach extension operator* $H_Y : Y^* \to 2^{X^*}$ defined by

$$H_Y(y^*) := \{x^* \in X^* : \|x^*\| = \|y^*\| \text{ and } x^*|_Y = y^*\} \quad \text{for } y^* \in Y^*$$

is a weak*-usco on $(Y^*, \|\cdot\|^*)$, where $\|\cdot\|^*$ denotes the dual norm on Y^*.

1.1.3 Construction of usco mappings

In this subsection, we present some basic constructions of usco mappings. Let (X, τ) and (Y, τ') be topological spaces.

(i) Let $\Phi : X \to 2^Y$ be an usco on (X, τ), and let U be a nonempty open subset of (X, τ). If $\Omega : U \to 2^Y$ is an usco on (U, τ'') (where τ'' denotes the relative topology on U) and $\Omega(u) \subseteq \Phi(u)$ for all $u \in U$, then $\Psi : X \to 2^Y$ defined by

$$\Psi(x) := \begin{cases} \Omega(x) & \text{for } x \in U, \\ \Phi(x) & \text{for } x \notin U, \end{cases}$$

is an usco on (X, τ).

(ii) If F is a nonempty finite set and for each $i \in F$, $\Phi_i : X \to 2^Y$ is an usco on (X, τ), then $(\bigvee_{i\in F} \Phi_i) : X \to 2^Y$ defined by $(\bigvee_{i\in F} \Phi_i)(x) := \bigcup_{i\in F} \Phi_i(x)$ for $x \in X$ is an usco on (X, τ).

(iii) If $\Phi : X \to 2^Y$ is an usco on (X, τ) and $f : (Y, \tau') \to (Z, \tau'')$ is a continuous mapping, then $(f \circ \Phi) : X \to 2^Z$ defined by $(f \circ \Phi)(x) := f(\Phi(x))$ for $x \in X$ is an usco on (X, τ).

(iv) If $f : X \to Y$ is a continuous function, (Z, τ'') is a topological space, and $\Phi : Y \to 2^Z$ is an usco mapping on (Y, τ'), then $(\Phi \circ f) : X \to 2^Y$ defined by, $(\Phi \circ f)(x) := \Phi(f(x))$ for $x \in X$ is an usco on (X, τ). More generally, we have the following.

(v) If $\Phi : X \to 2^Y$ is an usco on (X, τ), (Z, τ'') is a topological space, and $\Psi : Y \to 2^Z$ is an usco on (Y, τ'), then $(\Psi \circ \Phi) : X \to 2^Z$ defined by $(\Psi \circ \Phi)(x) := \Psi(\Phi(x))$ for $x \in X$ is an usco on (X, τ).

(vi) If F is a nonempty finite set and for each $i \in F$, $\Phi_i : X \to 2^{Y_i}$ is an usco on (X, τ) mapping into subsets of a topological space (Y_i, τ_i), then $(\Delta_{i\in F}\Phi_i) : X \to 2^{\prod_{i\in F} Y_i}$ defined by $(\Delta_{i\in F}\Phi_i)(x) := \prod_{i\in F} \Phi_i(x)$ for $x \in X$ is an usco on (X, τ).

(vii) If A is an arbitrary nonempty set and for each $i \in A$, $\Phi_i : X \to 2^{Y_i}$ is an usco on (X, τ) mapping into subsets of a topological space (Y_i, τ_i), then $(\Delta_{i\in A}\Phi_i) : X \to 2^{\prod_{i\in A} Y_i}$ defined by $(\Delta_{i\in A}\Phi_i)(x) := \prod_{i\in A} \Phi_i(x)$ for $x \in X$ is an usco on (X, τ).

(viii) By (iii) and (vi) we have the following. If $(G, \cdot, \tau'')$ is a topological group and $\Phi : X \to 2^G$ and $\Psi : X \to 2^G$ are uscos on (X, τ), then $(\Phi \cdot \Psi) : X \to 2^G$ defined by $(\Phi \cdot \Psi)(x) := \Phi(x) \cdot \Psi(x)$ for $x \in X$ is an usco on (X, τ).

(ix) By Example (i) and item (viii) we have the following. If $(G, \cdot, \tau'')$ is a topological group, $\Phi : X \to 2^G$ is an usco on (X, τ), and $f : X \to G$ is continuous on (X, τ), then $(f \cdot \Phi) : X \to 2^G$ defined by $(f \cdot \Phi)(x) := f(x) \cdot \Phi(x)$ for $x \in X$ is an usco on (X, τ).

(x) If A is an arbitrary nonempty set, (Y, τ') is a Hausdorff space and for each $i \in A$, $\Phi_i : X \to 2^Y$ is an usco on (X, τ), then $(\bigwedge_{i\in A} \Phi_i) : X \to 2^Y$ defined by $(\bigwedge_{i\in A} \Phi_i)(x) := \bigcap_{i\in A} \Phi_i(x)$ for $x \in X$ is an usco on (X, τ), provided that for each $x \in X$ and each finite subset F of A, $\bigcap_{i\in F} \Phi_i(x) \neq \varnothing$.

(xi) If F is a nonempty finite set and for each $i \in F$, $\Phi_i : X_i \to 2^{Y_i}$ is an usco acting from a topological space (X_i, τ_i) into subsets of a topological space (Y_i, τ_i'), then

$$\prod_{i\in F} \Phi_i : \prod_{i\in F} X_i \to 2^{\prod_{i\in F} Y_i}$$

defined by $(\prod_{i\in F} \Phi_i)(x) := \prod_{i\in F} \Phi_i(x(i))$ for $x \in \prod_{i\in F} X_i$ is an usco on $\prod_{i\in F} X_i$.

(xii) If A is an arbitrary nonempty set and for each $i \in A$, $\Phi_i : X_i \to 2^{Y_i}$ is an usco acting from a topological space (X_i, τ_i) into subsets of a topological space (Y_i, τ_i'), then

$$\prod_{i\in A} \Phi_i : \prod_{i\in A} X_i \to 2^{\prod_{i\in A} Y_i}$$

defined by $(\prod_{i\in A} \Phi_i)(x) := \prod_{i\in A} \Phi_i(x(i))$ for $x \in \prod_{i\in A} X_i$ is an usco on $\prod_{i\in A} X_i$.

The next proposition is the set-valued mapping equivalent of the fact that the sum of a compact set and a closed set is again a closed set.

Proposition 1.1.18. *If $(G, \cdot, \tau)$ is a topological group, $\Phi : X \to 2^G$ is an usco on a topological space (X, τ'), and $\Psi : X \to 2^G$ has nonempty images and a closed graph, then the mapping $(\Phi \cdot \Psi) : X \to 2^G$ defined by $(\Phi \cdot \Psi)(x) := \Phi(x) \cdot \Psi(x)$ for $x \in X$ has a closed graph.*

Proof. We will show that $(X \times G) \setminus \mathrm{Gr}(\Phi \cdot \Psi)$ is open in $X \times G$ with the product topology of (X, τ') and (G, τ). Suppose that $(x, y) \notin \mathrm{Gr}(\Phi \cdot \Psi)$, that is, $y \notin (\Phi \cdot \Psi)(x) = \Phi(x) \cdot \Psi(x)$. Then

$$[\Phi(x)]^{-1} \cdot y \cap \Psi(x) = \varnothing.$$

Since Ψ has a closed graph, for each $z \in [\Phi(x)]^{-1} \cdot y$, there exist an open neighbourhood V_z of z and an open neighbourhood U_z of x such that $(U_z \times V_z) \cap \mathrm{Gr}(\Psi) = \varnothing$, that is, $\Psi(U_z) \cap V_z = \varnothing$. Now because $[\Phi(x)]^{-1} \cdot y$ is compact and $[\Phi(x)]^{-1} \cdot y \subseteq \bigcup\{V_z : z \in [\Phi(x)]^{-1} \cdot y\}$, there exists a finite subcover $\{V_{z_j} : 1 \le j \le n\}$ of $\{V_z : z \in [\Phi(x)]^{-1} \cdot y\}$. Let $U_1 := \bigcap_{1\le j\le n} U_{z_j}$ and observe that $\Psi(U_1) \cap \bigcup_{1\le j\le n} V_{z_j} = \varnothing$. Since the mapping $\varphi : G \times G \to G$ defined by

$$\varphi(g, h) := g^{-1} \cdot h \quad \text{for } (g, h) \in G \times G$$

is continuous and $\Phi(x) \times \{y\}$ is a compact subset of $\varphi^{-1}(\bigcup_{1\le j\le n} V_{z_j})$, there exist (see Exercise 1.1.9) an open set W containing $\Phi(x)$ and an open neighbourhood N of y such that $W \times N \subseteq \varphi^{-1}(\bigcup_{1\le j\le n} V_{z_j})$. Therefore $W^{-1} \cdot N = \varphi(W \times N) \subseteq \bigcup_{1\le j\le n} V_{z_j}$. Further, since Φ is upper semicontinuous, there exists an open neighbourhood U_2 of x such that $\Phi(U_2) \subseteq W$. Let $U := U_1 \cap U_2$. Then U is an open neighbourhood of x, and

$$\begin{aligned}\varnothing \subseteq ([\Phi(U)]^{-1} \cdot N) \cap \Psi(U) &\subseteq ([\Phi(U_2)]^{-1} \cdot N) \cap \Psi(U) \\ &\subseteq (W^{-1} \cdot N) \cap \Psi(U) \\ &\subseteq \bigcup_{1\le j\le n} V_{z_j} \cap \Psi(U) \subseteq \bigcup_{1\le j\le n} V_{z_j} \cap \Psi(U_1) = \varnothing.\end{aligned}$$

Hence $N \cap [\Phi(U) \cdot \Psi(U)] = \varnothing$, and so $(U \times N) \cap \mathrm{Gr}(\Phi \cdot \Psi) = \varnothing$. This completes the proof since $(x, y) \in U \times N$. □

Corollary 1.1.19. *Let $(G, \cdot, \tau)$ be a topological group, let $\Phi : X \to 2^G$ be an usco on a topological space (X, τ'), and let N be a nonempty closed subset of (G, τ). If K is a compact subset of (G, τ) and $\{x \in X : [\Phi(x) \cdot N] \cap K \neq \varnothing\}$ is dense in (X, τ'), then $\Psi : X \to 2^G$ defined by $\Psi(x) := [\Phi(x) \cdot N] \cap K$ for $x \in X$ is an usco on (X, τ').*

Proof. Firstly, note that the set-valued mapping $x \mapsto N$ from (X, τ') into subsets of (G, τ) has a closed graph. Hence by Proposition 1.1.18 the mapping $x \mapsto \Phi(x) \cdot N$ from (X, τ') into subsets of (G, τ) has a closed graph. The result now follows from Theorem 1.1.11, bearing in mind the fact that the mapping $x \mapsto K$ from (X, τ') into subsets of (G, τ) is an usco mapping on (X, τ'). □

1.2 The Kakutani–Glicksberg–Fan fixed-point theorem

Fixed-point theorems are among the most powerful and, arguably, the most useful results to emanate from topology. The prince, among all the various fixed-point theorems, is Brouwer's fixed-point theorem. In this section we show that Brouwer's fixed-point theorem can be extended to convex-valued usco mappings defined on compact convex subsets of Hausdorff locally convex topological spaces. Recall that a subset C of a vector space $(V, +, \cdot)$ over the real numbers is called *convex* if $\lambda x + (1 - \lambda)y \in C$ for all $x, y \in C$ and $0 < \lambda < 1$.

We will start by giving a precise definition of a fixed point. If $f : X \to X$ is a function, then we call a point $x_0 \in X$ such that $f(x_0) = x_0$ a *fixed point of* f. In the case where f has at least one fixed point, we say that f *has a fixed point*. We denote by $\mathrm{Fix}(f)$ the set of all fixed points of f in X. We can now state Brouwer's fixed-point theorem.

Theorem 1.2.1 (Brouwer's fixed-point theorem, [49]). *If K is a nonempty compact convex subset of a finite-dimensional Banach space $(X, \|\cdot\|)$ and $f : K \to K$ is a continuous function, then f has a fixed point.*

Of course, it follows that if (K, τ) is any compact topological space homeomorphic to a nonempty compact convex subset of a finite-dimensional Banach space and $f : K \to K$ is any continuous function, then f has a fixed point.

With a view to extending this theorem beyond finite-dimensional spaces, we need to recall some basic facts concerning linear topological spaces.

Let $(V, +, \cdot)$ be a vector space over the field of real numbers, and let τ be a topology on V. Then $(V, +, \cdot, \tau)$ is called a *linear topological space* or a *topological vector space* if vector addition from $V \times V$ into V is continuous, when $V \times V$ is considered with the product topology, and scalar multiplication from $\mathbb{R} \times V$ into V is continuous, again when we consider $\mathbb{R} \times V$ with the product topology and $\mathbb{R}$ with the usual topology.

An important feature of linear topological spaces is that they are always *regular*, that is, if $(X, +, \cdot, \tau)$ is linear topological space, C is a closed subset of X, and $x \in X \setminus C$, then there exist disjoint open sets U and V such that $x \in U$ and $C \subseteq V$. To see this, suppose that $x = x + 0 \in X \setminus C$, which is open. Therefore by the continuity of addition there exist open neighbourhoods U of x and W of 0 such that $U + W \subseteq X \setminus C$, that is, $(U + W) \cap C = \varnothing$. Therefore $U \cap (C + (-W)) = \varnothing$. Let $V := C + (-W) = \bigcup_{c \in C} c - W$. Then V is an open set containing the set C, and $U \cap V = \varnothing$. Thus (X, τ) is a regular topological space.

A linear topological space $(V, +, \cdot, \tau)$ is called a *locally convex space* if for each open set W containing the zero vector $0 \in V$, there exists a convex open subset N of V such that $0 \in N \subseteq W$, that is, 0 has a local base consisting of open convex sets. If $(V, +, \cdot, \tau)$ is a locally convex space, then 0 has a local base consisting of convex symmetric open sets (recall that a set S is called *symmetric* if $S = -S$). To see this, we simply note that if $\mathcal{N}$ is a local base for 0 consisting of open convex subsets, then $\{N \cap -N \in 2^V : N \in \mathcal{N}\}$ is a local base for 0 consisting of open convex symmetric subsets of V.

Every linear topological space admits an important family of sublinear functionals. Given an open convex neighbourhood N of 0, in a linear topological space $(V, +, \cdot, \tau)$, we define the *Minkowski functional for* N, $p_N : V \to [0, \infty)$ by

$$p_N(x) := \inf\{\lambda \in (0, \infty) : x \in \lambda N\} \quad \text{for } x \in V.$$

It is well known that (i) p_N is continuous, (ii) sublinear (i. e., $p_N(tx) = tp_N(x)$ for all $x \in V$ and $0 \le t < \infty$, and $p_N(x + y) \le p_N(x) + p_N(y)$ for all $x, y \in V$) and (iii) $p_N(x) < 1$ if and only if $x \in N$; see any of [75, 80, 160, 280] for the proofs of these facts. Of course, to ensure that we have an ample supply of Minkowski functionals, we should insist that the space $(V, +, \cdot, \tau)$ is locally convex.

Exercise 1.2.2. Let K be a nonempty compact subset of a locally convex space $(X, +, \cdot, \tau)$, and let W be an open set containing K. Show that there exists an open convex neighbourhood N of 0 such that $K + N \subseteq W$.

Exercise 1.2.3. Let $(X, +, \cdot, \tau)$ be a Hausdorff locally convex space. Show that X^2 is also a Hausdorff locally convex space when vector addition is defined by

$$(x_1, y_1) + (x_2, y_2) := (x_1 + x_2, y_1 + y_2) \quad \text{for } x_1, x_2, y_2, y_2 \in X,$$

scalar multiplication is defined by

$$\lambda(x, y) := (\lambda x, \lambda y) \quad \text{for } \lambda \in \mathbb{R} \text{ and } (x, y) \in X^2,$$

and the topology on X^2 is the product topology of (X, τ) and (X, τ).

Proposition 1.2.4. *Suppose that K is a nonempty compact convex subset of a Hausdorff locally convex space $(X, +, \cdot, \tau)$ and $f : K \to K$ is a continuous function. If $Fix(f) = \varnothing$, then there exists a convex open neighbourhood N of 0 such that $f(x) - x \notin N$ for each $x \in K$.*

Proof. Consider X^2 with the locally convex structure defined above. Since $\mathrm{Gr}(f)$ is a closed subset of $K \times K$ that is disjoint from Δ_K (which is compact), there exists an open convex neighbourhood N of 0 such that $(\Delta_K + N \times N) \cap \mathrm{Gr}(f) = \varnothing$. Fix $x \in K$. Now since

$$\bigcup_{k \in K} (k + N) \times (k + N) = \Delta_K + (N \times N),$$

we have that $(x, f(x)) \notin (x + N) \times (x + N)$. Since $x \in (x + N)$, we have that $f(x) \notin x + N$ or, equivalently, $f(x) - x \notin N$. □

By applying this proposition we can reduce the proof of the next theorem to an application of Brouwer's fixed-point theorem, but we first recall a few facts concerning the convex hull of a set. We start with a definition. If A is a subset of a vector space $(V, +, \cdot)$ over the real numbers, then the *convex hull* of A, denoted $\mathrm{co}(A)$, is the smallest convex subset of V containing the set A.

Exercise 1.2.5. Let $(X, +, \cdot, \tau)$ be a Hausdorff linear topological space, and let $\{x_1, \ldots, x_n\}$ be a finite subset of X. Furthermore, let $D := \{(\lambda_1, \lambda_2, \ldots, \lambda_n) \in [0, 1]^n : \sum_{j=1}^n \lambda_j = 1\}$. Show that:
(i) D is compact;
(ii) $\varphi : D \to X$ defined by $\varphi(\lambda_1, \lambda_2, \ldots, \lambda_n) := \sum_{j=1}^n \lambda_j x_j$ is continuous on D;
(iii) $\varphi(D) = \mathrm{co}(\{x_1, x_2, \ldots, x_n\})$;
(iv) $\mathrm{co}(\{x_1, x_2, \ldots, x_n\})$ is homeomorphic to a nonempty compact convex subset of a finite-dimensional Banach space. *Hint:* By Theorem 11 in [91, page 51] every finite-dimensional Hausdorff linear topological space has its topology generated by a norm.

Theorem 1.2.6 (Schauder–Tychonoff fixed-point theorem, [281, 303]). *If K is a nonempty compact convex subset of a Hausdorff locally convex space and $f : K \to K$ is a continuous function, then f has a fixed point.*

Proof. By Proposition 1.2.4 it is sufficient to show that for each open convex neighbourhood N of 0, there exists $x \in K$ such that $f(x) - x \in N$. To this end, let N be an open convex neighbourhood of 0. Let $F := \{k_1, k_2, \ldots, k_m\}$ be a finite subset of K such that $K \subseteq \bigcup_{1\le j\le m} k_j + N$; note that this is possible since K is compact and hence totally bounded. Let $Q = \mathrm{co}(F)$. Then by Exercise 1.2.5 Q is a compact subset of K. Furthermore, Q is homeomorphic to a nonempty compact convex subset of a finite-dimensional Banach space. Therefore by Brouwer's fixed-point theorem every continuous function $h : Q \to Q$ has a fixed point in Q. For each $1 \le j \le m$, let $g_j : X \to [0,1]$ be defined by

$$g_j(x) := \max\{1 - p_N(x - k_j), 0\} \quad \text{for } x \in X,$$

where p_N is the Minkowski functional for N. Then $\mathrm{supp}(g_j) = k_j + N$ for each $1 \le j \le m$, and so $0 < \sum_{j=1}^m g_j(x)$ for all $x \in K$ since $\{k_j + N : 1 \le j \le m\}$ is an open cover of K.

Next, we define $p_j : K \to [0,1]$ by

$$p_j(x) := \frac{g_j(x)}{\sum_{i=1}^m g_i(x)} \quad \text{for } x \in K \text{ and } 1 \le j \le m.$$

Then $\sum_{j=1}^m p_j(x) = 1$ for all $x \in K$, and $\mathrm{supp}(p_j) = \mathrm{supp}(g_j) = k_j + N$ for each $1 \le j \le m$.

Finally, we define $q : K \to Q$ by

$$q(x) := \sum_{j=1}^m p_j(x) k_j \quad \text{for } x \in K.$$

Note that q is well defined and continuous. Furthermore, $x - q(x) \in N$ for all $x \in K$ (i. e., q is approximately the identity mapping). To see this, let $x \in K$, and let

$$J := \{j : 1 \le j \le m \text{ and } p_j(x) \ne 0\} = \{j : 1 \le j \le m \text{ and } x - k_j \in N\}.$$

Now

$$x - q(x) = \sum_{j=1}^m p_j(x)x - \sum_{j=1}^m p_j(x)k_j = \sum_{j=1}^m p_j(x)(x - k_j) = \sum_{j\in J} p_j(x)(x - k_j) \in N$$

since $\sum_{j\in J} p_j(x) = 1$ and N is convex. Next, define $f_N : Q \to Q$ by

$$f_N(x) := q(f(x)) \quad \text{for } x \in Q.$$

By Brouwer's fixed-point theorem (Theorem 1.2.1) f_N has a fixed point $x_N \in Q$. Therefore

$$f(x_N) - x_N = [f(x_N) - f_N(x_N)] + [f_N(x_N) - x_N] = f(x_N) - q(f(x_N)) \in N.$$

This completes the proof. □

Before we can state our main fixed-point theorem for set-valued mappings, we need to make precise what we mean by a fixed point for a set-valued mapping. If $\Phi : X \to 2^X$ is a set-valued mapping, then we call any point $x_0 \in X$ such that $x_0 \in \Phi(x_0)$ a *fixed point of* Φ. In the case where Φ has at least one fixed point we say that Φ *has a fixed point*.

As is often the case with fixed-point theorems, convexity plays a central role. The situation with set-valued mappings is not different. Hence we need to consider the following special class of set-valued mappings. We say that a set-valued mapping $\Phi : X \to 2^Y$ acting from a topological space (X, τ) into subsets of a linear topological space $(Y, +, \cdot, \tau')$ is a *cusco* if (i) it is an usco on (X, τ) and (ii) $\Phi(x)$ is convex for each $x \in X$.

Lemma 1.2.7 (Cellina, [56]). *If K is a nonempty compact convex subset of a Hausdorff locally convex space $(X, +, \cdot, \tau)$ and $\Phi : K \to 2^K$ is a cusco, then for each convex symmetric open neighbourhood N of 0, there exists a continuous function $f : K \to K$ that is an "approximate selection" of Φ, that is, $[(x + N) \times (f(x) + N)] \cap Gr(\Phi) \neq \varnothing$ for all $x \in K$.*

Proof. Let N be a convex symmetric open neighbourhood of 0. For each $x \in K$, let U_x be an open symmetric neighbourhood of 0 such that $U_x \subseteq N$ and

$$\Phi(x + U_x + U_x) \subseteq \Phi(x) + N.$$

Note that this is possible since Φ is τ-upper semicontinuous. Now $\{x + U_x : x \in K\}$ is an open cover of K. Therefore there exists a finite subset F of K such that $K \subseteq \bigcup_{x \in F} x + U_x$.

Let $V := \bigcap_{x \in F} U_x$. Then V is a symmetric open neighbourhood of 0, and so $\{k + V : k \in K\}$ is an open cover of K. Hence there exists a finite subset $\{k_1, k_2, \ldots, k_m\}$ of K such that $K \subseteq \bigcup_{j=1}^m k_j + V$. For each $1 \leq j \leq m$, choose $y_j \in \Phi(k_j)$.

For each $1 \leq j \leq m$, let $g_j : X \to [0, 1]$ be defined by

$$g_j(x) := \max\{1 - p_V(x - k_j), 0\} \quad \text{for } x \in X,$$

where p_V is the Minkowski functional for V. Then $\mathrm{supp}(g_j) = k_j + V$ for each $1 \leq j \leq m$, and so $0 < \sum_{j=1}^m g_j(x)$ for all $x \in K$, since $\{k_j + V : 1 \leq j \leq m\}$ is an open cover of K. Next, we define $p_j : K \to [0, 1]$ by

$$p_j(x) := \frac{g_j(x)}{\sum_{i=1}^m g_i(x)} \quad \text{for } x \in K \text{ and } 1 \leq j \leq m.$$

Then $\sum_{j=1}^m p_j(x) = 1$ for all $x \in K$, and $\mathrm{supp}(p_j) = \mathrm{supp}(g_j) = k_j + V$ for each $1 \leq j \leq m$.

Finally, we define $f : K \to K$ by

$$f(x) := \sum_{j=1}^m p_j(x) y_j \in \mathrm{co}(\Phi(K)) \subseteq K.$$

Note that f is well defined and continuous. Further, $(x + N) \times (f(x) + N) \cap \mathrm{Gr}(\Phi) \neq \varnothing$ for all $x \in K$ (i. e., f is an "approximate selection" of Φ). To see this, let $x \in K$, and let

$$\begin{aligned} J &:= \{j : 1 \le j \le m \text{ and } p_j(x) \neq 0\} \\ &= \{j : 1 \le j \le m \text{ and } x \in k_j + V\} \\ &= \{j : 1 \le j \le m \text{ and } k_j \in x + V\} \quad \text{since } V \text{ is symmetric.} \end{aligned}$$

Choose $x' \in F$ so that $x \in x' + U_{x'}$. We will now show that $f(x) \in \Phi(x') + N$ as follows:

$$\{k_j : j \in J\} \subseteq x + V \subseteq (x' + U_{x'}) + V \subseteq x' + U_{x'} + U_{x'},$$

and so

$$\begin{aligned} f(x) &= \sum_{j=1}^{n} p_j(x)y_j = \sum_{j\in J} p_j(x)y_j \\ &\subseteq \mathrm{co}\{y_j : j \in J\} \subseteq \mathrm{co}\Big\{\bigcup\{\Phi(k_j) : j \in J\}\Big\} = \mathrm{co}\{\Phi(\{k_j : j \in J\})\} \\ &\subseteq \mathrm{co}\{\Phi(x' + U_{x'} + U_{x'})\} \subseteq \mathrm{co}\{\Phi(x') + N\} = \Phi(x') + N. \end{aligned}$$

Therefore $(x, f(x)) \in (x' + U_{x'}) \times (\Phi(x') + N)$. Now since $U_{x'}$ and N are both symmetric, $x' \in x + U_{x'}$ and $\Phi(x') \cap [f(x) + N] \neq \varnothing$. Let $y' \in \Phi(x') \cap [f(x) + N]$; then $(x', y') \in \mathrm{Gr}(\Phi)$ and $(x', y') \in (x + U_{x'}) \times (f(x) + N) \subseteq (x + N) \times (f(x) + N)$. Therefore

$$(x', y') \in \big[(x + N) \times (f(x) + N)\big] \cap \mathrm{Gr}(\Phi),$$

which shows that $[(x + N) \times (f(x) + N)] \cap \mathrm{Gr}(\Phi) \neq \varnothing$. This completes the proof. □

The following proposition is the set-valued mapping version of Proposition 1.2.4.

Proposition 1.2.8. *Let K be a nonempty compact convex subset of a Hausdorff locally convex space, and let $\Phi : K \to 2^K$ be a cusco. If $Fix(\Phi) = \varnothing$, then there exists a convex symmetric open neighbourhood N of 0 such that $[(x+N)\times(x+N)] \cap Gr(\Phi) = \varnothing$ for each $x \in K$.*

Proof. Since $\mathrm{Gr}(\Phi)$ is a closed subset of $K \times K$ that is disjoint from Δ_K (which is compact), there exists an convex symmetric neighbourhood N of 0 such that

$$(\Delta_K + N \times N) \cap \mathrm{Gr}(\Phi) = \varnothing.$$

Now since $\bigcup_{k\in K}(k + N) \times (k + N) = \Delta_K + (N \times N)$, we have the desired result. □

We can now state and prove the Kakutani–Glicksberg–Fan fixed-point theorem.

Theorem 1.2.9 (Kakutani–Glicksberg–Fan fixed-point theorem, [82, 97, 155]). *If K is a nonempty compact convex subset of a Hausdorff locally convex space and $\Phi : K \to 2^K$ is a cusco, then Φ has a fixed point.*

Proof. By Proposition 1.2.8 it is sufficient to show that for each convex symmetric open neighbourhood N of 0, there exists $x \in K$ such that $(x + N) \times (x + N) \cap \mathrm{Gr}(\Phi) \neq \varnothing$. To this end, let N be a convex symmetric open neighbourhood of 0. By Lemma 1.2.7 there exists a continuous function $f : K \to K$ such that $(x+N)\times(f(x)+N)\cap\mathrm{Gr}(\Phi) \neq \varnothing$ for all $x \in K$. However, by the Schauder–Tychonoff fixed-point theorem (Theorem 1.2.6) there exists a point $x_N \in K$ such that $f(x_N) = x_N$. Thus $(x_N + N) \times (x_N + N) \cap \mathrm{Gr}(\Phi) \neq \varnothing$. □

This theorem has had countless applications, both in mathematics and outside it, but perhaps the most famous application of Kakutani's fixed-point theorem (i. e., the finite-dimensional version of the previous theorem) outside of mathematics is due to John Nash. Nash [246] famously used Kakutani's fixed-point theorem to establish the well-known "Nash equilibrium theorem". For further information on this topic, see [189]. We apply the Kakutani–Glicksberg–Fan fixed-point theorem to the geometry of Banach spaces in Chapter 3, and we note here that it is also extensively used within variational analysis [35, p. 201]. However, it would be completely futile for us to try to list all applications of this theorem, as there are simply too many of them.

We will now use the Kakutani–Glicksberg–Fan fixed-point theorem to establish the following result, which will be used in turn to prove a "min-max theorem".

Theorem 1.2.10 ([155, Theorem 2]). *Let K and L be bounded closed convex sets in the Euclidean spaces $\mathbb{R}^m$ and $\mathbb{R}^n$, respectively, and let us consider their Cartesian product $K \times L$ in $\mathbb{R}^{m+n}$. Let U and V be closed subsets of $K \times L$ such that for any $x_0 \in K$, the set U_{x_0} of all $y \in L$ such that $(x_0, y) \in U$ is nonempty and convex and such that for any $y_0 \in L$, the set V_{y_0}, of all $x \in K$ such that $(x, y_0) \in V$ is nonempty and convex. Then $U \cap V \neq \varnothing$.*

Proof. Put $S := K\times L$ and define $\Phi_1 : K \to 2^L$ by $\Phi_1(x) := U_x$ and $\Phi_2 : L \to 2^K$ by $\Phi_2(y) := V_y$. Then $\mathrm{Gr}(\Phi_1) = U$ and $\mathrm{Gr}(\Phi_2) = V$. Therefore by Corollary 1.1.15 both Φ_1 and Φ_2 are cuscos. Hence by Construction (xi) in Section 1.1.3 the mapping $(x, y) \mapsto \Phi_1(x) \times \Phi_2(y)$ is also a cusco. Now the mapping $s : L\times K \to K\times L$ defined by $s(x, y) := (y, x)$ is a linear homeomorphism and, in particular, continuous. Therefore, if $\Phi : S \to 2^S$ is defined by

$$\Phi(x, y) := s(\Phi_1(x) \times \Phi_2(y)) = \Phi_2(y) \times \Phi_1(x) = V_y \times U_x \quad \text{for } (x, y) \in S,$$

then Φ is a cusco on S. Hence by Theorem 1.2.9 there exists a point $z_0 = (x_0, y_0) \in S$ such that $z_0 \in \Phi(z_0)$. In other words, $x_0 \in V_{y_0}$ and $y_0 \in U_{x_0}$. Therefore $(x_0, y_0) \in V$ and $(x_0, y_0) \in U$, that is, $(x_0, y_0) \in V \cap U$. □

The following minimax theorem is one of the fundamental tools used in the early theory of games, as defined and developed by von Neumann [307].

Theorem 1.2.11 (Minimax theorem [155, 308]). *Let K and L be bounded closed convex sets in the Euclidean spaces $\mathbb{R}^m$ and $\mathbb{R}^n$, respectively, and let $f : K \times L \to \mathbb{R}$ be a continuous function. If for each $x_0 \in K$ and for every real number α, the set of all $y \in L$ such* *that $f(x_0, y) \le \alpha$ is convex and if for every $y_0 \in L$ and every real number β, the set of all $x \in K$ such that $\beta \le f(x, y_0)$ is convex, then we have*

$$\max_{x\in K}\min_{y\in L} f(x,y) = \min_{y\in L}\max_{x\in K} f(x,y).$$

Proof. Let U and V be the sets of all $z_0 = (x_0, y_0) \in K \times L$ such that

$$f(x_0,y_0) = \min_{y\in L} f(x_0,y) \quad \text{and} \quad f(x_0,y_0) = \max_{x\in K} f(x,y_0), \quad \text{respectively.}$$

Then we easily see that both U and V satisfy the conditions of Theorem 1.2.10. Hence by Theorem 1.2.10 there exists a point $z_0 = (x_0, y_0) \in K \times L$ such that $z_0 \in U \cap V$ or, equivalently,

$$f(x_0,y_0) = \min_{y\in L} f(x_0,y) = \max_{x\in K} f(x,y_0).$$

Consequently, we have that

$$\min_{y\in L}\max_{x\in K} f(x,y) \le \max_{x\in K} f(x,y_0) = f(x_0,y_0) = \min_{y\in L} f(x_0,y) \le \max_{x\in K}\min_{y\in L} f(x,y).$$

On the other hand, we have the following. For any $y' \in L$ and $x \in K$,

$$\min_{y\in L} f(x,y) \le f(x,y') \quad \text{and so} \quad \max_{x\in K}\min_{y\in L} f(x,y) \le \max_{x\in K} f(x,y').$$

Since $y' \in L$ was arbitrary,

$$\max_{x\in K}\min_{y\in L} f(x,y) \le \min_{y'\in L}\max_{x\in K} f(x,y') = \min_{y\in L}\max_{x\in K} f(x,y).$$

This completes the proof. □

Since this theorem first appeared, it has undergone several improvements; see, for example, [284, 308, 309]. Some of these have been utilised to establish equilibrium points for games with mixed strategies.

1.3 Minimal usco mappings

In this section, we consider a particular class of uscos, whose behaviour resembles that of single-valued quasicontinuous functions.

Let (X, τ) and (Y, τ') be topological spaces, and let $\Phi : X \to 2^Y$ be an usco on (X, τ). Then we say that Φ is a *minimal usco* if the graph of Φ does not contain, as a proper subset, the graph of any other usco defined on (X, τ).

It is clear from this definition that all single-valued uscos are minimal. However, many of the most interesting minimal uscos are not everywhere single-valued. We illustrate this by two simple examples.

Example 1.3.1. Define $\Phi : \mathbb{R} \to 2^{\mathbb{R}}$ by

$$\Phi(x) := \begin{cases} \sin(\frac{1}{x}) & \text{for } x \neq 0, \\ [-1, 1] & \text{for } x = 0. \end{cases}$$

Define $\Psi : \mathbb{R} \to 2^{\mathbb{R}}$ by

$$\Psi(x) := \begin{cases} \sin(\frac{1}{d(x,C)}) & \text{for } x \notin C, \\ [-1, 1] & \text{for } x \in C, \end{cases}$$

where C is any compact nowhere dense subset of $\mathbb{R}$, and $d(x, C) := \inf\{|x - y| : y \in C\}$.

Note that even for minimal uscos on $\mathbb{R}$, the set $\{x \in \mathbb{R} : \Phi(x)$ is not single-valued$\}$ may have positive measure. However, as we will see later (Corollary 1.5.18), this set must be small in terms of Baire category, that is, this set must be of the first Baire category whenever the range space of the minimal usco is metrisable.

Exercise 1.3.2. Let (X, τ) and (Y, τ') be topological spaces, and let $\Phi : X \to 2^Y$ be an usco on (X, τ). Show that if for each $x \in X$, there exists an open neighbourhood U_x of x such that $\Phi|_{U_x}$ is a minimal usco on (U_x, τ''), then Φ is a minimal usco on (X, τ). Here τ'' denotes the relative τ-topology on the set U_x.

The following proposition ensures that there is an ample supply of minimal uscos.

Proposition 1.3.3 ([59]). *Let (X, τ) and (Y, τ') be topological spaces, and let $\Phi : X \to 2^Y$ be an usco on (X, τ). If (Y, τ') is Hausdorff, then there exists a minimal usco mapping $\Psi : X \to 2^Y$ such that $Gr(\Psi) \subseteq Gr(\Phi)$ (i. e., every usco contains a minimal usco).*

Proof. Let $\mathcal{U}$ denote the family of all usco mappings $\Psi : X \to 2^Y$ such that $\mathrm{Gr}(\Psi) \subseteq \mathrm{Gr}(\Phi)$. Obviously, $\mathcal{U} \neq \varnothing$ as $\Phi \in \mathcal{U}$. We may now partially order $\mathcal{U}$ as follows. If Ψ_1 and Ψ_2 are members of $\mathcal{U}$, then we write $\Psi_1 \leq \Psi_2$ if $\Psi_1(x) \subseteq \Psi_2(x)$ for each $x \in X$. Next, we apply Zorn's lemma to show that $(\mathcal{U}, \leq)$ possesses a minimal element. To this end, let $\{\Psi_\gamma : \gamma \in \Gamma\}$ be a totally ordered subset of $\mathcal{U}$, and let $\Phi_M : X \to 2^Y$ be defined by $\Phi_M(x) := \bigcap\{\Psi_\gamma(x) : \gamma \in \Gamma\}$. Since each $\Psi_\gamma(x)$ is nonempty and compact, $\Phi_M(x)$ also is nonempty and compact. Let W be an open subset of Y and consider $U := \{x \in X : \Phi_M(x) \subseteq W\}$. We need to show that U is open in (X, τ). We may, without loss of generality, assume that $U \neq \varnothing$ and consider $x_0 \in U$. By the finite intersection property there exists $\gamma_0 \in \Gamma$ such that $\Psi_{\gamma_0}(x_0) \subseteq W$. Hence there exists an open neighbourhood U_0 of x_0 such that $\Psi_{\gamma_0}(U_0) \subseteq W$, which means that $\Phi_M(U_0) \subseteq W$. Therefore $x_0 \in U_0 \subseteq U$, and so U is open in (X, τ). From this it follows that $\Phi_M \in \mathcal{U}$ (an alternative way of showing this is appealing to Corollary 1.1.14 and using the fact that Φ_M has a closed graph) and $\Phi_M \leq \Psi_\gamma$ for each $\gamma \in \Gamma$. Thus by Zorn's lemma $(\mathcal{U}, \leq)$ possesses a minimal element. It is now easy to see that this element is in fact a minimal usco. □

Proposition 1.3.4 ([157]). *Let (X, τ) and (Y, τ') be topological spaces, and let $\Phi : X \to 2^Y$ be a minimal usco on (X, τ). If $A \subseteq X$ is either (i) open or (ii) dense in (X, τ) and*

(Y,τ') is Hausdorff, then $\Phi|_A$ is a minimal usco on (A,τ''), where τ'' denotes the relative τ-topology on the set A.

Proof. (i) Let A be a nonempty open subset of (X,τ). Clearly, $\Phi|_A$ is an usco on (A,τ''); see Proposition 1.1.5. Now suppose that $\Omega : A \to 2^Y$ is an usco such that $\mathrm{Gr}(\Omega) \subseteq \mathrm{Gr}(\Phi|_A)$. Let $\Psi : X \to 2^Y$ be defined by

$$\Psi(x) := \begin{cases} \Omega(x) & \text{if } x \in A, \\ \Phi(x) & \text{if } x \notin A. \end{cases}$$

Then as in Construction (i) in Section 1.1.3, we see that Ψ is an usco. Moreover, $\mathrm{Gr}(\Psi) \subseteq \mathrm{Gr}(\Phi)$. Therefore by the minimality of Φ we have that $\Psi = \Phi$. Thus, in particular, $\Phi(x) = \Psi(x) = \Omega(x)$ for all $x \in A$, that is, $\Phi|_A = \Omega$. This shows that $\Phi|_A$ is a minimal usco on (A,τ'').

(ii) Let A be a dense subset of (X,τ) and suppose that (Y,τ') is Hausdorff. Clearly, $\Phi|_A$ is an usco on (A,τ''); see Proposition 1.1.5. Now suppose that $\Omega : A \to 2^Y$ is an usco such that $\mathrm{Gr}(\Omega) \subseteq \mathrm{Gr}(\Phi|_A)$. Let $\Psi : X \to 2^Y$ be defined by $\Psi(x) := \mathrm{USC}(\Omega)(x)$. Since the graph of Φ is closed (see Proposition 1.1.8), we have that $\mathrm{Gr}(\Psi) = \overline{\mathrm{Gr}(\Omega)} \subseteq \mathrm{Gr}(\Phi)$. It then follows from Corollary 1.1.14 that Ψ is an usco on (X,τ). Hence by the minimality of Φ we must have that $\Psi = \Phi$. Furthermore, by Exercise 1.1.16 we have that $\Phi(x) = \Psi(x) = \mathrm{USC}(\Omega)(x) = \Omega(x)$ for each $x \in A$, that is, $\Phi|_A = \Omega$. This shows that $\Phi|_A$ is a minimal usco on (A,τ''). □

Now that we have established the existence of minimal usco mappings, it is interesting to characterise when an usco is a minimal usco.

Proposition 1.3.5 ([59, 94]). *Let (X,τ) and (Y,τ') be topological spaces, and let $\Phi : X \to 2^Y$ be an usco on (X,τ). If (Y,τ') is Hausdorff, then Φ is a minimal usco on (X,τ) if and only if for each pair of open subsets U of (X,τ) and W of (Y,τ') with $\Phi(U) \cap W \neq \varnothing$, there exists a nonempty τ-open subset V of U such that $\Phi(V) \subseteq W$.*

Proof. Suppose that Φ is a minimal usco on (X,τ). Let U be an open subset of (X,τ), and let W be an open subset of (Y,τ') such that $\Phi(U) \cap W \neq \varnothing$. We consider two cases.

Case I: If there exists $x \in U$ such that $\Phi(x) \subseteq W$, then the result follows directly from the upper semicontinuity of Φ.

Case II: Suppose that $\Phi(x) \nsubseteq W$ for all $x \in U$. Let $\Psi : X \to 2^X$ be defined by $\Psi(x) := \Phi(x) \cap (X \setminus W)$ if $x \in U$ and by $\Psi(x) := \Phi(x)$ if $x \notin U$. Then, by assumption, Ψ has nonempty compact images. In fact, we claim that Ψ is an usco on (X,τ). To show this, we need only show that Ψ is upper semicontinuous. Let $x_0 \in X$, and let W' be an open set in (Y,τ') containing $\Psi(x_0)$. If $x_0 \notin U$, then $\Psi(x_0) = \Phi(x_0)$, and so $\Phi(x_0) \subseteq W'$. Since Φ is upper semicontinuous at x_0, there exists an open neighbourhood V of x_0 such that $\Phi(V) \subseteq W'$. Therefore $\Psi(V) \subseteq W'$ since $\Psi(x) \subseteq \Phi(x)$ for all $x \in X$. So we are left to consider the case where $x_0 \in U$. Suppose $x_0 \in U$. Then $\Phi(x_0) \subseteq W' \cup W$, since

$\Phi(x_0) \cap (X \setminus W) = \Psi(x_0) \subseteq W'$. Since Φ is upper semicontinuous, there exists an open neighbourhood V of x_0 contained in U such that $\Phi(V) \subseteq W' \cup W$. Therefore

$$\Psi(V) = \Phi(V) \cap (X \setminus W) \subseteq (W' \cup W) \cap (X \setminus W) = W' \cap (X \setminus W) \subseteq W'.$$

This shows that Ψ is a upper semicontinuous and hence an usco on (X, τ). Since Φ is a minimal usco, we must have that $\Phi = \Psi$, but then $\Phi(U) = \Psi(U) \subseteq (X \setminus W)$, which contradicts our original assumption that $\Phi(U) \cap W \neq \varnothing$. Therefore case II does not occur, and so the result follows from case I.

We now consider the converse. Suppose that Φ is an usco on (X, τ) that satisfies the property that for each pair of open subsets U of (X, τ) and W of (Y, τ') with $\Phi(U) \cap W \neq \varnothing$, there exists a nonempty τ-open subset V of U such that $\Phi(V) \subseteq W$. We wish to show that Φ is a minimal usco on (X, τ). To obtain a contradiction, let us suppose that Φ is not a minimal usco. Then there exist an usco Ψ on (X, τ) with $\mathrm{Gr}(\Psi) \subseteq \mathrm{Gr}(\Phi)$ and a point $x_0 \in X$ such that $\Psi(x_0)$ is a proper subset of $\Phi(x_0)$ (i. e., $\Psi(x_0) \neq \Phi(x_0)$). Choose $y_0 \in \Phi(x_0) \setminus \Psi(x_0)$. Since (Y, τ') is Hausdorff and $\Psi(x_0)$ is compact, there exist disjoint open subsets W and W' of (Y, τ') such that $\Psi(x_0) \subseteq W'$ and $y_0 \in W$. By the upper semicontinuity of Ψ there exists an open neighbourhood U of x_0 such that $\Psi(U) \subseteq W'$. On the other hand, $y_0 \in \Phi(U) \cap W$, and so by our assumptions it follows that there exists a nonempty open subset V of U such that $\Phi(V) \subseteq W$. Therefore $\Psi(V) \subseteq \Phi(V) \subseteq W$, and so

$$\varnothing \neq \Psi(V) \subseteq W' \cap W = \varnothing,$$

which is clearly false. Thus our assumption that Φ is not a minimal usco was false. Hence Φ is a minimal usco on (X, τ). □

The next result demonstrates that the minimality of an usco is preserved under a continuous mapping.

Proposition 1.3.6 ([243]). *Let $\Phi : X \to 2^Y$ be a minimal usco acting from a topological space (X, τ) into subsets of a Hausdorff space (Y, τ'), and let $f : Y \to Z$ be a continuous mapping from (Y, τ') into a Hausdorff topological space (Z, τ''). Then the mapping $(f \circ \Phi) : X \to 2^Z$ is a minimal usco on (X, τ).*

Proof. It follows from Construction (iii) in Section 1.1.3, that $(f \circ \Phi) : X \to 2^Z$ is an usco on (X, τ). So it remains to show the minimality of $(f \circ \Phi) : X \to 2^Z$. We will proceed via Proposition 1.3.5. Consider an open set U in (X, τ) and an open set W in (Z, τ'') such that $(f \circ \Phi)(U) \cap W \neq \varnothing$. Let $W' := f^{-1}(W)$. Then W' is an open subset of (Y, τ'), since f is continuous. Moreover, $\Phi(U) \cap W' \neq \varnothing$. Therefore by the minimality of Φ and Proposition 1.3.5 we see that there exists a nonempty τ-open subset V of U such that $\Phi(V) \subseteq W'$. Thus

$$(f \circ \Phi)(V) = f(\Phi(V)) \subseteq f(W') \subseteq W,$$

as required. □

The following theorem, although not surprising, has wide utility.

Theorem 1.3.7. *Let (X,τ) and (Y,τ') be topological spaces, and let $\Phi : X \to 2^Y$ be an usco on (X,τ). If (Y,τ') is completely regular, then Φ is a minimal usco on (X,τ) if and only if for each $f \in C(Y)$, $(f \circ \Phi) : X \to 2^{\mathbb{R}}$ is a minimal usco on (X,τ).*

Proof. Suppose that Φ is a minimal usco on (X,τ). Then by Proposition 1.3.6 we have that for each $f \in C(Y)$, $(f \circ \Phi) : X \to 2^{\mathbb{R}}$ is a minimal usco on (X,τ). For the converse statement, consider an usco $\Phi : X \to 2^Y$ such that for each $f \in C(Y)$, $(f \circ \Phi) : X \to 2^{\mathbb{R}}$ is a minimal usco on (X,τ). To obtain a contradiction, let us also assume that Φ is not a minimal usco. Then there exist an usco $\Psi : X \to 2^Y$ such that $\mathrm{Gr}(\Psi) \subseteq \mathrm{Gr}(\Phi)$ and an element $x_0 \in X$ such that $\Psi(x_0)$ is a proper subset of $\Phi(x_0)$, that is, $\Psi(x_0) \neq \Phi(x_0)$. Let $y_0 \in \Phi(x_0) \setminus \Psi(x_0)$. Then choose $f \in C(Y)$ such that $f(y_0) = 1$ and $f(y) = 0$ for all $y \in \Psi(x_0)$. Note that such f exists, since (Y,τ') is completely regular. Now by Construction (iii) in Section 1.1.3, both $(f \circ \Psi) : X \to 2^{\mathbb{R}}$ and $(f \circ \Phi) : X \to 2^{\mathbb{R}}$ are uscos on (X,τ) and clearly $\mathrm{Gr}(f \circ \Psi) \subseteq \mathrm{Gr}(f \circ \Phi)$. However, $(f \circ \Psi)(x_0) \neq (f \circ \Phi)(x_0)$ since $1 \in (f \circ \Phi)(x_0)$ while $(f \circ \Psi)(x_0) = \{0\}$. This contradicts our assumption that $(f \circ \Phi)$ is a minimal usco on (X,τ). Thus Φ is a minimal usco on (X,τ). □

Proposition 1.3.8. *Let $\Phi : Y \to 2^Z$ be a minimal usco acting from a topological space (Y,τ') into subsets of a Hausdorff space (Z,τ''), and let $f : X \to Y$ be a continuous open mapping from a topological space (X,τ) into (Y,τ'). Then the mapping $(\Phi \circ f) : X \to 2^Z$ is a minimal usco on (X,τ).*

Proof. It follows from Construction (iv) in Section 1.1.3, that $(\Phi \circ f) : X \to 2^Z$ is an usco on (X,τ). So it remains to show the minimality of $(\Phi \circ f) : X \to 2^Z$. We will proceed via Proposition 1.3.5. Consider an open set U in (X,τ) and an open set W in (Z,τ'') such that $(\Phi \circ f)(U) \cap W \neq \varnothing$. Let $U' := f(U)$. Then U' is an open subset of (Y,τ'), since f is an open mapping. Moreover, $\Phi(U') \cap W \neq \varnothing$. Therefore by the minimality of Φ and Proposition 1.3.5 we see that there exists a nonempty τ'-open subset V' of U' such that $\Phi(V') \subseteq W$. Let $V := f^{-1}(V') \cap U$. Then V is τ-open, since f is continuous, and nonempty, since V' is a nonempty subset of $U' = f(U)$. Furthermore, $(\Phi \circ f)(V) = \Phi(f(V)) \subseteq \Phi(V') \subseteq W$, as required. □

An important aspect of the study of minimal uscos is the investigation of their interactions with their selections. Indeed, the minimality of an usco mapping can be characterised in terms of its selections.

Suppose that $\Phi : X \to 2^Y$ is a set-valued mapping acting from a set X into subsets of a set Y and suppose also that D is a subset of X. Then we call any function $\sigma : D \to Y$ such that $\sigma(x) \in \Phi(x)$ for all $x \in D$ a *selection of Φ on D*. If (X,τ) is a topological space and the domain of the function σ is dense in (X,τ), then we say that σ is a *densely defined selection*. Frequently, with abuse of notation, we will identify a function $f : D \to Y$ defined on a subset D of X with the set-valued mapping defined on X that maps

x to $\{f(x)\}$ if $x \in D$ and to $\varnothing$ if $x \notin D$. We will denote this set-valued mapping also by f (i. e., the same name as the original function $f : D \to Y$).

Suppose that (X, τ) and (Y, τ') are topological spaces and $f : X \to Y$. If $x_0 \in X$, then we say that f is *τ'-quasicontinuous at x_0* if for every $W \in \tau'$ that contains $f(x_0)$ and every τ-open neighbourhood U of x_0, there exists a nonempty τ-open subset V of U $f(V) \subseteq W$. If f is τ'-quasicontinuous at every point of X, then we say that *f is τ'-quasicontinuous on (X, τ)*. When there is no ambiguity concerning the topology τ', we will simply say that f is quasicontinuous on (X, τ).

Exercise 1.3.9. Let (X, τ) and (Y, τ') be topological spaces, and let $f : X \to Y$. Show that $f : X \to Y$ is quasicontinuous on (X, τ) if and only if for each open subset W of (Y, τ') and each open subset U of (X, τ) such that $f(U) \cap W \neq \varnothing$, there exists a nonempty τ-open subset V of U such that $f(V) \subseteq W$.

Theorem 1.3.10 ([36, 37]). *Let $\Phi : X \to 2^Y$ be an usco mapping acting from a topological space (X, τ) into subsets of a regular Hausdorff topological space (Y, τ'). Then the following properties are equivalent:*
(i) *Φ is a minimal usco on (X, τ);*
(ii) *for every densely defined selection σ of Φ, $USC(\sigma) = \Phi$;*
(iii) *there exists a densely defined selection σ of Φ such that $USC(\sigma|_{D'}) = \Phi$ for every dense subset D' of $Dom(\sigma)$;*
(iv) *there exists a densely defined quasicontinuous selection σ of Φ such that $USC(\sigma) = \Phi$.*

Proof. (i)⇒(ii). Suppose that Φ is a minimal usco on (X, τ). Let $\sigma : D \to Y$ be a densely defined selection of Φ. By Exercise 1.1.16 and Proposition 1.1.8 we have that $\mathrm{Gr}(\mathrm{USC}(\sigma)) = \overline{\mathrm{Gr}(\sigma)} \subseteq \mathrm{Gr}(\Phi)$, and by Corollary 1.1.14 we have that $\mathrm{USC}(\sigma)$ is an usco on (X, τ). Furthermore, since Φ is a minimal usco on (X, τ), we have that $\mathrm{USC}(\sigma) = \Phi$.

(ii)⇒(iii). This implication is obvious.

(iii)⇒(iv). Let $\sigma : D \to Y$ be a densely defined selection of Φ such that $\mathrm{USC}(\sigma|_{D'}) = \Phi$ for every dense subset D' of D. We claim that σ is quasicontinuous on D. We will verify this indirectly. To obtain a contradiction, suppose that σ is not quasicontinuous on D. Then there exist $x_0 \in D$, an open neighbourhood U of x_0 and an open neighbourhood W of $\sigma(x_0)$ such that for every nonempty open subset V of U, $\sigma(V \cap D) \not\subseteq W$. This means that $D^* := \{x \in U \cap D : \sigma(x) \notin W\}$ is dense in $U \cap D$. Let $D' := (D \setminus U) \cup D^*$. Then D' is dense in D. Furthermore, $\overline{\sigma|_{D'}(U \cap D')} = \overline{\sigma|_{D'}(D^*)} \subseteq Y \setminus W$, and so

$$\sigma(x_0) \in \Phi(x_0) = \mathrm{USC}(\sigma|_{D'})(x_0) \subseteq \overline{\sigma|_{D'}(U \cap D')} \subseteq Y \setminus W.$$

However, this contradicts the fact that $\sigma(x_0) \in W$. Hence σ must be quasicontinuous on D. Finally, by assumption, $\Phi = \mathrm{USC}(\sigma)$.

(iv)⇒(i). Suppose that $\sigma : D \to Y$ is a densely defined quasicontinuous selection of Φ such that $\Phi = \mathrm{USC}(\sigma)$. We will proceed via Proposition 1.3.5. To this end, suppose that U is a nonempty open subset of (X, τ) and W is a nonempty open subset of (Y, τ') such that $\Phi(U) \cap W \neq \varnothing$. Since $\mathrm{Gr}(\Phi) = \mathrm{Gr}(\mathrm{USC}(\sigma)) = \overline{\mathrm{Gr}(\sigma)}$ and $(U \times W) \cap \mathrm{Gr}(\Phi) \neq \varnothing$, there exists $x \in D \cap U$ such that $(x, \sigma(x)) \in \mathrm{Gr}(\sigma) \cap (U \times W)$. Since (Y, τ') is regular, we obtain an open neighbourhood W' of $\sigma(x)$ such that $\overline{W'} \subseteq W$. Since σ is quasicontinuous on D, there exists a nonempty open subset V of U such that $\sigma(V \cap D) \subseteq W'$. Now by the definition of $\mathrm{USC}(\sigma)$ we have that $\Phi(V) = \mathrm{USC}(\sigma)(V) \subseteq \overline{\sigma(V \cap D)} \subseteq \overline{W'} \subseteq W$. □

Corollary 1.3.11. *Let $\Phi : X \to 2^Y$ be an usco mapping acting from a topological space (X, τ) into subsets of a regular Hausdorff topological space (Y, τ'). If $\sigma : D \to Y$ is a densely defined continuous selection of Φ (i. e., a continuous function that is also a selection of Φ) and $\mathrm{USC}(\sigma) = \Phi$, then Φ is a minimal usco. In particular, if (Y, τ') is a compact Hausdorff space and $f : D \to Y$ is a continuous function defined on a dense subset D of (X, τ), then $\mathrm{USC}(f)$ is a minimal usco (i. e., every densely defined continuous function that maps into a compact Hausdorff space, has an extension to a minimal usco defined on the whole space).*

Corollary 1.3.12. *Let $\Phi : X \to 2^Y$ be a minimal usco mapping acting from a topological space (X, τ) into subsets of a Hausdorff topological space (Y, τ'). If $\sigma : D \to Y$ is a densely defined selection of Φ and σ is continuous at $x_0 \in D$, then $\Phi(x_0) = \{\sigma(x_0)\}$.*

Proof. Clearly, it is sufficient to show that $\Phi(x_0) \subseteq \{\sigma(x_0)\}$. So let us suppose, to obtain a contradiction, that this is not the case. Then there exists $y_0 \in \Phi(x_0) \setminus \{\sigma(x_0)\}$. Since (Y, τ') is Hausdorff there exist disjoint τ'-open neighbourhoods W of $\sigma(x_0)$ and W' of y_0. Now, as σ is τ'-continuous at x_0, there exists a τ-open neighbourhood U of x_0 such that $\sigma(U \cap D) \subseteq W$. On the other hand, $\Phi(U) \cap W' \neq \varnothing$, and so by Proposition 1.3.5 there exists a nonempty τ-open subset V of U such that $\Phi(V) \subseteq W'$. In particular, $\sigma(V \cap D) \subseteq \Phi(V \cap D) \subseteq W'$. Thus

$$\varnothing \neq \sigma(V \cap D) \subseteq W \cap W' = \varnothing,$$

which is impossible. Therefore $\Phi(x_0) = \{\sigma(x_0)\}$. □

Theorem 1.3.13. *Let (X, τ) and (Y, τ') be topological spaces, and let $\Phi : X \to 2^Y$ be an usco on (X, τ). If (Y, τ') is Hausdorff, $\Omega : X \to 2^Y$ is a minimal usco and $\{x \in X : \Phi(x) \cap \Omega(x) \neq \varnothing\}$ is dense in (X, τ), then $\Omega(x) \subseteq \Phi(x)$ for all $x \in X$.*

Proof. By Proposition 1.1.8 Ω has a closed graph. Let $\Psi : X \to 2^Y$ be defined by $\Psi(x) := \Phi(x) \cap \Omega(x)$ for $x \in X$. Then,by Theorem 1.1.11 Ψ is an usco on (X, τ), and, moreover, $\mathrm{Gr}(\Psi) \subseteq \mathrm{Gr}(\Omega)$. Therefore $\Psi = \Omega$ by the minimality of Ω. Thus $\Omega(x) = \Psi(x) = \Phi(x) \cap \Omega(x)$ for all $x \in X$, which implies that $\Omega(x) \subseteq \Phi(x)$ for all $x \in X$. □

Corollary 1.3.14. *Let (X, τ) and (Y, τ') be topological spaces, and let $\Phi : X \to 2^Y$ be an usco on (X, τ). If (Y, τ') is Hausdorff, $f : X \to Y$ is a continuous function, and $\{x \in X :$*

$f(x) \in \Phi(x)\}$ is dense in (X, τ), then $f(x) \in \Phi(x)$ for all $x \in X$ (i. e., f is a continuous selection of Φ).

Proof. Let $\Omega : X \to 2^Y$ be defined by $\Omega(x) := \{f(x)\}$ for all $x \in X$. By Example (i) in Section 1.1.2, we see that Ω is a minimal usco on (X, τ). Then the result follows from the previous theorem. □

1.4 Selection theorems

One of the largest and most studied areas of research within set-valued analysis is the study of selections. Therefore it would be inappropriate for us to completely ignore this important research area. So, in the next section of this chapter, we will briefly touch upon some of the most important selections.

The first selection theorem we will consider is "Michael's Selection Theorem". This theorem is undoubtedly the most cited selection theorem in mathematics.

1.4.1 Michael's selection theorem

To state and prove Michael's selection theorem, we must first recall the definition of a paracompact topological space and then derive some of their most basic properties.

A *cover* of a set X is a collection of subsets of X whose union contains X. In symbols, if $\mathcal{U} := \{U_\alpha : \alpha \in A\}$ is an indexed family of subsets of X, then $\mathcal{U}$ is a cover of X if $X \subseteq \bigcup_{\alpha \in A} U_\alpha$. A cover of a topological space is an *open cover* if all its members are open sets. A *refinement* of a cover of a topological space (X, τ) is a new cover of the same space such that every set in the new cover is a subset of some set in the old cover. In symbols, the cover $\mathcal{V} := \{V_\beta : \beta \in B\}$ is a refinement of the cover $\mathcal{U} := \{U_\alpha : \alpha \in A\}$ if and only if for any $V_\beta \in \mathcal{V}$, there exists some $U_\alpha \in \mathcal{U}$ such that $V_\beta \subseteq U_\alpha$.

A collection of subsets of a space (X, τ) is *locally finite* if every point of the space has a neighbourhood that intersects only finitely many subsets in the cover. In symbols, $\mathcal{U} := \{U_\alpha : \alpha \in A\}$ is locally finite if and only if for any $x \in X$, there exists a neighbourhood V of x such that the set $\{\alpha \in A : U_\alpha \cap V \neq \varnothing\}$ is finite. A topological space (X, τ) is now said to be *paracompact* if every open cover of (X, τ) has a locally finite open refinement [71].

Clearly, all compact spaces are paracompact, but we also have the following important class of paracompact spaces. In the proof of the next theorem, we will employ the following standard notation. If (M, d) is a metric space, $x \in M$ and $0 < r$, then $B(x, r) := \{y \in M : d(x, y) < r\}$.

Theorem 1.4.1 ([279, 294]). *Every metric space (M, d) is paracompact.*

Proof. Assume that (M, d) is a metric space and $\{C_\alpha : \alpha \in A\}$ is an open cover of M. We may and do assume that the set A is well ordered. From this it follows that $\{C_\alpha \setminus \bigcup_{\beta < \alpha} C_\beta : \alpha \in A\}$ is a disjoint cover of M.

We will inductively, upon $n \in \mathbb{N}$, define some collections of subsets $\{A_{\alpha,n} : (\alpha, n) \in A \times \mathbb{N}\}$ and $\{D_{\alpha,n} : (\alpha, n) \in A \times \mathbb{N}\}$ of M.

Case $n = 1$. Let $\alpha \in A$. Then define

$$A_{\alpha,1} := \left\{x \in C_\alpha \setminus \bigcup_{\beta<\alpha} C_\beta : B(x, 3/2) \subseteq C_\alpha\right\} \quad \text{and} \quad D_{\alpha,1} := \bigcup_{x\in A_{\alpha,1}} B(x, 1/2).$$

Case $n = 2$. Let $\alpha \in A$. Then define

$$A_{\alpha,2} := \left\{x \in C_\alpha \setminus \bigcup_{\beta<\alpha} C_\beta : B(x, 3/2^2) \subseteq C_\alpha\right\} \setminus \bigcup_{\beta\in A} D_{\beta,1} \quad \text{and} \quad D_{\alpha,2} := \bigcup_{x\in A_{\alpha,2}} B(x, 1/2^2).$$

In general, for $1 < n \in \mathbb{N}$, we define, for each $\alpha \in A$,

$$A_{\alpha,n} := \left\{x \in C_\alpha \setminus \bigcup_{\beta<\alpha} C_\beta : B(x, 3/2^n) \subseteq C_\alpha\right\} \setminus \bigcup\{D_{\beta,j} : \beta \in A, 1 \le j < n\}$$

and

$$D_{\alpha,n} := \bigcup_{x\in A_{\alpha,n}} B(x, 1/2^n).$$

Note that for every $(\alpha, n) \in A \times \mathbb{N}$, $D_{\alpha,n}$ is open, and $A_{\alpha,n} \subseteq D_{\alpha,n} \subseteq C_\alpha$. Moreover, if $x \in M$, then $x \in C_\alpha \setminus \bigcup_{\beta<\alpha} C_\beta$ for some $\alpha \in A$, in fact $\alpha = \min_{\beta\in A}(x \in C_\beta)$. Since C_α is an open subset of M, there exists $n \in \mathbb{N}$ such that $B(x, 3/2^n) \subseteq C_\alpha$. Hence either $x \in \bigcup\{D_{\beta,j} : \beta \in A, 1 \le j < n\}$ or $x \in A_{\alpha,n} \subseteq D_{\alpha,n}$. Thus in either case,

$$x \in \bigcup\{D_{\beta,j} : \beta \in A, 1 \le j \le n\} \subseteq \bigcup\{D_{\beta,n} : (\beta, n) \in A \times \mathbb{N}\}.$$

This shows that $\{D_{\beta,n} : (\beta, n) \in A \times \mathbb{N}\}$ is an open refinement of $\{C_\alpha : \alpha \in A\}$.

Let us now observe that for each $n \in \mathbb{N}$, $\{D_{\alpha,n} : \alpha \in A\}$ is discrete in M, in the sense that if $\alpha \neq \beta$, $p \in D_{\alpha,n}$ and $q \in D_{\beta,n}$ then $1/2^n \le d(p, q)$. To see this, consider $\alpha, \beta \in A$ with $\alpha \neq \beta$ such that $p \in D_{\alpha,n}$ and $q \in D_{\beta,n}$. Note that by possibly relabelling we may assume that $\alpha < \beta$. Now by the definitions of the sets $D_{\alpha,n}$ and $D_{\beta,n}$ there exist points $x \in A_{\alpha,n}$ and $y \in A_{\beta,n}$ such that $p \in B(x, 1/2^n)$ and $q \in B(y, 1/2^n)$. Since $y \in A_{\beta,n} \subseteq C_\beta \setminus \bigcup_{\gamma<\beta} C_\gamma$, $y \notin C_\alpha$. Thus $3/2^n \le d(x, y)$, since $B(x, 3/2^n) \subseteq C_\alpha$. We now use the triangle inequality to estimate $d(p, q)$:

$$3/2^n \le d(x, y) \le d(x, p) + d(p, q) + d(q, y) < 1/2^n + d(p, q) + 1/2^n = d(p, q) + 2/2^n.$$

Therefore $1/2^n \le d(p, q)$, and we are done. It is important that this implies that for any $x_0 \in M$, $B(x_0, 1/2^{n+1})$ intersects at most one element of the set $\{D_{\alpha,n} : \alpha \in A\}$.

We now show that $\{D_{\alpha,n} : (\alpha, n) \in A \times \mathbb{N}\}$ is locally finite. To this end, let $x \in M$. Then $x \in D_{\alpha,n}$ for some $(\alpha, n) \in A \times \mathbb{N}$. Since $D_{\alpha,n}$ is an open subset of M, there exists

$m \in \mathbb{N}$ such that $B(x, 1/2^m) \subseteq D_{\alpha,n}$. Observe that

$$\begin{aligned}&\left|\{(\beta,j) \in A \times \mathbb{N} : B(x,1/2^{n+m}) \cap D_{\beta,j} \neq \varnothing : \text{ for some } 1 \leq j < n+m\}\right| \\ &\leq \left|\{(\beta,j) \in A \times \mathbb{N} : B(x,1/2^{j+1}) \cap D_{\beta,j} \neq \varnothing : \text{ for some } 1 \leq j < n+m\}\right| \\ &= \sum_{1 \leq j < m+n} \left|\{\beta \in A : B(x,1/2^{j+1}) \cap D_{\beta,j}\}\right| \leq m+n-1.\end{aligned}$$

On the other hand, if $(\beta, j) \in A \times \mathbb{N}$ and $n + m \leq j$, then $B(x, 1/2^{n+m}) \cap D_{\beta,j} = \varnothing$. Indeed, suppose, to obtain a contradiction, that for some $(\beta, j) \in A \times \mathbb{N}$ with $n + m \leq j$, there exists $y \in B(x, 1/2^{n+m}) \cap D_{\beta,j}$. Then by the definition $D_{\beta,j}$ there exists $z \in A_{\beta,j} \subseteq M \setminus D_{\alpha,n}$ (since $n < j$) such that $y \in B(z, 1/2^j)$. Since $B(x, 1/2^m) \subseteq D_{\alpha,n}$, this means that $1/2^m \leq d(x, z)$. Thus by the triangle inequality we have that

$$1/2^m \leq d(x,z) \leq d(x,y) + d(y,z) < 1/2^{n+m} + 1/2^j \leq 1/2^{m+1} + 1/2^{m+1} = 1/2^m,$$

which is impossible. Therefore $B(x, 1/2^{n+m}) \cap D_{\beta,j} = \varnothing$. This shows that $B(x, 1/2^{n+m})$ intersects at most $n + m - 1$ elements of the cover $\{D_{\alpha,n} : (\alpha, n) \in A \times \mathbb{N}\}$. □

As paracompactness is a generalisation of compactness, it is not surprising that many of the properties of compact spaces carry over to paracompact spaces.

Proposition 1.4.2 ([71]). *Every Hausdorff paracompact space (X, τ) is normal.*

Proof. We prove this result in two parts.

Part I. (We show that (X, τ) is regular.) Let A be a closed subset of (X, τ), and let $x \in X \setminus A$. Since (X, τ) is Hausdorff, for each $a \in A$, there exist disjoint open sets U_a and V_a such that $x \in U_a$ and $a \in V_a$. Now $\mathcal{U} := \{X \setminus A\} \cup \{V_a : a \in A\}$ is an open cover of X. Therefore $\mathcal{U}$ has a locally finite open refinement $\mathcal{V}$. Let $\mathcal{V}^* := \{V \in \mathcal{V} : V \cap A \neq \varnothing\}$ and, for each $V \in \mathcal{V}^*$, choose $a_V \in A$ such that $V \subseteq V_{a_V}$. Note that $V \cap U_{a_V} = \varnothing$ for all $V \in \mathcal{V}^*$. Now $\mathcal{V}^*$ is locally finite (and an open cover of A), so there exists an open neighbourhood W of x such that $\mathcal{P} := \{V \in \mathcal{V}^* : V \cap W \neq \varnothing\}$ has only finitely many elements. Let $U := W \cap \bigcap\{U_{a_V} : V \in \mathcal{P}\}$, and let $V := \bigcup\{V' : V' \in \mathcal{V}^*\}$. Then U and V are open, $x \in U$, $A \subseteq V$ and $U \cap V = \varnothing$. To see that $U \cap V = \varnothing$, note that $V = \bigcup\{V' : V' \in \mathcal{P}\} \cup \bigcup\{V' : V' \in \mathcal{V}^* \setminus \mathcal{P}\}$. This shows that (X, τ) is regular.

Part II. (We show that (X, τ) is normal.) Suppose that A and B are disjoint closed subsets of (X, τ). Since (X, τ) is regular, for each $a \in A$, there exist disjoint open sets U_a and V_a such that $a \in U_a$ and $B \subseteq V_a$. Now $\mathcal{U} := \{X \setminus A\} \cup \{U_a : a \in A\}$ is an open cover of X. Therefore $\mathcal{U}$ has a locally finite open refinement $\mathcal{V}$. Let $\mathcal{V}^* := \{U \in \mathcal{V} : U \cap A \neq \varnothing\}$ and for each $U \in \mathcal{V}^*$ choose $a_U \in A$ such that $U \subseteq U_{a_U}$. Note that $U \cap V_{a_U} = \varnothing$ for all $U \in \mathcal{V}^*$. Now $\mathcal{V}^*$ is a locally finite open cover of A. Let $U_A := \bigcup\{U \in \mathcal{V}^*\}$. Then U_A is an open neighbourhood of A. For every $b \in B$, there exists an open neighbourhood N_b of b such that $N_b \cap U_A = \varnothing$. To see this, first note that there exists an open neighbourhood

W_b of b such that $\mathcal{P} := \{U \in \mathcal{V}^* : U \cap W_b \neq \varnothing\}$ has only finitely many members. Let $N_b := W_b \cap \bigcap\{V_{a_U} : U \in \mathcal{P}\}$. Then N_b is an open neighbourhood of b, and $N_b \cap U_A = \varnothing$. To see that $N_b \cap U_A = \varnothing$, note that $U_A = \bigcup\{U : U \in \mathcal{P}\} \cup \bigcup\{U : U \in \mathcal{V}^* \setminus \mathcal{P}\}$. Let $V_B := \bigcup_{b \in B} N_b$. Then V_B is an open neighbourhood of B, and $V_B \cap U_A = \varnothing$. This shows that (X, τ) is normal. □

Exercise 1.4.3. Let (X, τ) be a topological space, and let $\{A_\gamma : \gamma \in \Gamma\}$ be a locally finite collection of subsets of X. Show that for any subset $\varnothing \neq \Gamma_0 \subseteq \Gamma$,

$$\overline{\bigcup\{A_\gamma : \gamma \in \Gamma_0\}} = \bigcup\{\overline{A_\gamma} : \gamma \in \Gamma_0\}.$$

Proposition 1.4.4. *Let $\{V_\alpha : \alpha \in A\}$ be an open covering of a Hausdorff paracompact space (X, τ). Then there exists and open refinement $\{W_\alpha : \alpha \in A\}$ of $\{V_\alpha : \alpha \in A\}$ such that $\overline{W_\alpha} \subseteq V_\alpha$ for every $\alpha \in A$.*

Proof. Let $\mathcal{U} := \{U \in \tau : \overline{U} \subseteq V_\alpha$ for some $\alpha \in A\}$. By Proposition 1.4.2 X is regular, and so $\mathcal{U}$ is an open refinement of $\{V_\alpha : \alpha \in A\}$. Since (X, τ) is paracompact, $\mathcal{U}$ has a locally finite open refinement, which we call $\mathcal{U}^*$. Notice that $\mathcal{U}^* \subseteq \mathcal{U}$, that is, for each $U \in \mathcal{U}^*$, $\overline{U} \subseteq V_\alpha$ for some $\alpha \in A$. We now define our desired open refinement as follows. For each $\alpha \in A$, let $W_\alpha := \bigcup\{U \in \mathcal{U}^* : \overline{U} \subseteq V_\alpha\}$. Then clearly the elements of $\{W_\alpha : \alpha \in A\}$ are open, and by Exercise 1.4.3 $\overline{W_\alpha} = \bigcup\{\overline{U} \in 2^X : U \in \mathcal{U}^*$ and $\overline{U} \subseteq V_\alpha\} \subseteq V_\alpha$ for each $\alpha \in A$. So it remains to show that $\{W_\alpha : \alpha \in A\}$ is a cover of X. Let $x \in X$. Then $x \in U$ for some U in $\mathcal{U}^*$. Hence, by the definition of $\mathcal{U}^*$, $x \in U \subseteq \overline{U} \subseteq V_\alpha$ for some $\alpha \in A$. Therefore $x \in U \subseteq W_\alpha$. □

Proposition 1.4.5 ([209]). *If $\{V_\alpha : \alpha \in A\}$ is a locally finite open covering of a Hausdorff paracompact space (X, τ), then there exists a family $\{p_\alpha : \alpha \in A\}$ of continuous functions from (X, τ) into $[0, 1]$ with the following properties:*

(a) *each p_α vanishes outside V_α;*

(b) *$\sum_{\alpha \in A} p_\alpha(x) = 1$ for every $x \in X$.*

Proof. By Proposition 1.4.4 there exists an open covering $\{W_\alpha : \alpha \in A\}$ of (X, τ) such that $\overline{W_\alpha} \subseteq V_\alpha$ for every $\alpha \in A$. By the normality of (X, τ) (see Proposition 1.4.2) we can find continuous functions $q_\alpha : X \to [0, 1]$ such that $q_\alpha(x) = 1$ if $x \in \overline{W_\alpha}$ and $q_\alpha(x) = 0$ if $x \in X \setminus V_\alpha$. Now define $p_\alpha : X \to [0, 1]$ by

$$p_\alpha(x) := \frac{q_\alpha(x)}{\sum_{\beta \in A} q_\beta(x)} \quad \text{for } x \in X.$$

The family $\{p_\alpha : \alpha \in A\}$ satisfies all the requirements. □

Remarks 1.4.6. In many situations that occur in analysis the spaces under consideration are metric spaces. In this case, Proposition 1.4.5 has a very simple proof. For each

$\alpha \in A$, we let $q_\alpha(x) := d(x, X \setminus V_\alpha)$ for each $x \in X$. Then we define, as before,

$$p_\alpha(x) := \frac{q_\alpha(x)}{\sum_{\beta \in A} q_\beta(x)} \quad \text{for } x \in X,$$

and we are done.

We next give the required notion of continuity for Michael's selection theorem.

Suppose that (X,τ) and (Y,τ') are topological spaces and $\Phi : X \to 2^Y$. If $x_0 \in X$, then we say that Φ is *τ'-lower semicontinuous at x_0* if for every $W \in \tau'$ such that $\Phi(x_0) \cap W \neq \varnothing$, there exists a τ-neighbourhood U of x_0 such that $\Phi(x) \cap W \neq \varnothing$ for all $x \in U$. If Φ is both τ'-upper and τ'-lower semicontinuous at a point x_0, then we simply say that Φ is *τ'-continuous* at x_0. If Φ is τ'-lower semicontinuous (τ'-continuous) at every point of X, then we say that Φ *is τ'-lower semicontinuous on* (X,τ) (*τ'-continuous on* (X,τ)). When there is no ambiguity concerning the topology τ', we will simply say that Φ is lower semicontinuous (continuous) on (X,τ).

In the proof of the next lemma, we will employ the following standard notation. If (M,d) is a metric space, A is a subset of M, and $0 < r$, then $B(A,r) := \bigcup_{a \in A} B(a,r)$.

Lemma 1.4.7 ([207]). *If (X,τ) is Hausdorff and paracompact, $(Y,\|\cdot\|)$ is a normed linear space, $\Phi : X \to 2^Y$ is lower semicontinuous from (X,τ) into nonempty convex subsets of (Y,τ'), and $0 < r$, then there exists a continuous function $f : X \to Y$ such that $B(f(x),r) \cap \Phi(x) \neq \varnothing$ for every $x \in X$.*

Proof. For every $y \in Y$, let $U_y = \{x \in X : y \in B(\Phi(x),r)\}$. Then $U_y := \{x \in X : B(y,r) \cap \Phi(x) \neq \varnothing\}$, and it follows from the definition of lower semicontinuity that every U_y is open in (X,τ). Hence $\{U_y : y \in Y\}$ is an open cover of (X,τ) and therefore has an open locally finite refinement $\{V_\alpha : \alpha \in A\}$. Note that by removing any redundant elements on the indexing set A we may assume that $V_\alpha = V_\beta$ if and only if $\alpha = \beta$. Now pick the family $\{p_\alpha : \alpha \in A\}$ of continuous functions from X into $[0,1]$ as in Proposition 1.4.5. Also pick, for each $\alpha \in A$, $y(\alpha) \in Y$ such that $V_\alpha \subseteq U_{y(\alpha)}$. The desired function $f : X \to Y$ can now be defined by

$$f(x) := \sum_{\alpha \in A} p_\alpha(x) y(\alpha) \quad \text{for each } x \in X.$$

To see that this works, observe first that each $x \in X$ has a neighbourhood U intersecting only finitely many V_α, and on this U, f is the sum of finitely many continuous functions. Hence every x has a neighbourhood on which f is continuous, and therefore f is continuous on (X,τ). Moreover, for each $x \in X$, $f(x)$ is a convex combination of finitely many $y(\alpha)$, all of which lie in the convex set $B(\Phi(x),r) = \Phi(x) + rB_Y$, and hence $f(x) \in B(\Phi(x),r)$. This completes the proof. □

To extend this "approximate selection" theorem into a fully fledged selection theorem, we will need to deduce some further elementary results that will enable us to repeatedly apply Lemma 1.4.7 to a succession of refinements of the original set-valued mapping Φ.

Exercise 1.4.8. Let $\Phi : X \to 2^Y$ be a lower semicontinuous set-valued mapping with nonempty images acting from a topological space (X, τ) into subsets of a linear topological space $(Y, +, \cdot, \tau')$, and let $f : X \to Y$ be a continuous function. Show that the set-valued mapping Ψ defined by

$$\Psi(x) := \Phi(x) + f(x) \quad \text{for } x \in X$$

is also a lower semicontinuous set-valued mapping on (X, τ). *Hint:* It may help to first consider the particular case where $\Psi(x) = \Phi(x) + y_0$ for some fixed $y_0 \in Y$.

Proposition 1.4.9. *Let $\Phi : X \to 2^Y$ be a lower semicontinuous set-valued mapping acting from a topological space (X, τ) into subsets of a normed linear topological space $(Y, \|\cdot\|)$, $\varepsilon > 0$, and let $f : X \to Y$ be a continuous function. If $B(f(x), \varepsilon) \cap \Phi(x) \neq \varnothing$ for all $x \in X$, then $\Psi : X \to 2^Y$ defined by*

$$\Psi(x) := \Phi(x) \cap B(f(x), \varepsilon) \quad \textit{for } x \in X$$

is also a lower semicontinuous set-valued mapping on (X, τ).

Proof. Let us start with some general observations. If $F : A \to 2^B$ is a lower semicontinuous mapping from a topological space (A, τ') into subsets of a topological space (B, τ'') and if V is an open subset of (B, τ''), then the mapping $F_V : A \to 2^B$ defined by $F_V(x) := F(x) \cap V$ for $x \in A$ is also lower semicontinuous on (A, τ'). If S and T are subsets of a vector space V and $x \in V$, then $(S + x) \cap (T + x) = (S \cap T) + x$. With these observations out of the way, let us begin the proof. Let $F : X \to 2^Y$ be defined by $F(x) := \Phi(x) - f(x)$ for $x \in X$, and let $V := B(0, \varepsilon)$. By Exercise 1.4.8 F is lower semicontinuous. Furthermore, from above we see that F_V is lower semicontinuous on (X, τ). We can again apply Exercise 1.4.8 to deduce that $F_V + f$ is lower semicontinuous on (X, τ). However, if $x \in X$, then

$$\begin{aligned} F_V(x) + f(x) &= F(x) \cap B(0, \varepsilon) + f(x) \\ &= ([\Phi(x) - f(x)] \cap B(0, \varepsilon)) + f(x) \\ &= ([\Phi(x) - f(x)] + f(x)) \cap (B(0, \varepsilon) + f(x)) \\ &= \Phi(x) \cap B(f(x), \varepsilon) = \Psi(x). \end{aligned}$$

Thus $\Psi = F_V + f$, which completes the proof. □

Exercise 1.4.10. Let $\Phi : X \to 2^Y$ be a lower semicontinuous set-valued mapping acting between topological spaces (X, τ) and (Y, τ'). If $\Phi^* : X \to 2^Y$ is defined by $\Phi^*(x) := \overline{\Phi(x)}$ for $x \in X$, then show that Φ^* is also lower semicontinuous on (X, τ).

Theorem 1.4.11 (Michael's selection theorem, [207]). *If (X,τ) is a Hausdorff paracompact space, then every lower semicontinuous set-valued mapping $\Phi : X \to 2^Y$ acting from X into the nonempty closed and convex subsets of a Banach space $(Y, \|\cdot\|)$ admits a continuous selection.*

Proof. By recursively applying Lemma 1.4.7 we generate a sequence $((\Phi_n, f_n) : n \in \mathbb{N})$ of ordered pairs such that for every $n \in \mathbb{N}$:
(a_n) $f_n : X \to Y$ is a continuous function;
(b_n) $\Phi_n(x) := \overline{\Phi_{n-1}(x) \cap B(f_n(x), 1/n)}$ for all $x \in X$;
(c_n) $\Phi_n : X \to 2^Y$ is lower semicontinuous with nonempty closed convex images.

Base Step. Let $\Phi_0 := \Phi$.

Step 1. By Lemma 1.4.7 there exists a continuous function $f_1 : X \to Y$ such that $\Phi_0(x) \cap B(f_1(x), 1) \neq \varnothing$ for all $x \in X$. Thus by Proposition 1.4.9 and Exercise 1.4.10 the mapping $\Phi_1 : X \to 2^Y$ defined by

$$\Phi_1(x) := \overline{\Phi_0(x) \cap B(f_1(x), 1)} \quad \text{for } x \in X$$

is a lower semicontinuous mapping with nonempty closed and convex images.

Assuming that we have constructed the ordered pairs (Φ_k, f_k) in the sequence satisfying the properties (a_k), (b_k), and (c_k) up to, and including, the nth step, we proceed to construct the next step.

Step $n+1$. By Lemma 1.4.7 there exists a continuous function $f_{n+1} : X \to Y$ such that $\Phi_n(x) \cap B(f_{n+1}(x), 1/(n+1)) \neq \varnothing$ for all $x \in X$. Thus by Proposition 1.4.9 and Exercise 1.4.10 the mapping $\Phi_{n+1} : X \to 2^Y$ defined by

$$\Phi_{n+1}(x) := \overline{\Phi_n(x) \cap B(f_{n+1}(x), 1/(n+1))} \quad \text{for } x \in X$$

is a lower semicontinuous mapping with nonempty closed and convex images.

This completes the induction. Let $\sigma : X \to Y$ be defined by $\sigma(x) \in \bigcap_{n \in \mathbb{N}} \Phi_n(x)$, for $x \in X$. This definition is well-defined, since for each $x \in X$, $(\Phi_n(x) : n \in \mathbb{N})$ is a decreasing sequence of nonempty closed subsets of Y with $\lim_{n\to\infty} \operatorname{diam}[\Phi_n(x)] = 0$. Therefore, by the Cantor intersection property, $\bigcap_{n \in \mathbb{N}} \Phi_n(x)$ is a singleton set. Further, if x is any element of X and n is any element of $\mathbb{N}$, then

$$\sigma(x) \in \Phi_n(x) \subseteq \overline{\Phi_{n-1}(x) \cap B(f_n(x), 1/n)} \subseteq B[f_n(x), 1/n],$$

and so $\|\sigma(x) - f_n(x)\| \leq 1/n$. Hence σ is continuous as the uniform limit of the sequence of continuous functions $(f_n : n \in \mathbb{N})$. □

The following exercise provides an important source of lower semicontinuous set-valued mappings.

Exercise 1.4.12. Let (X,τ) and (Y,τ') be topological spaces, and let $f : X \to Y$ be an open mapping (i. e., maps open sets in X to open sets in Y). Then $f^{-1} : Y \to 2^X$ defined by $f^{-1}(y) := \{x \in X : f(x) = y\}$ for $y \in Y$ is a lower semicontinuous mapping on (Y,τ').

To fully exploit Exercise 1.4.12, we need to recall a fundamental result from the theory of Banach spaces, the open mapping theorem, which states that if $(X, \|\cdot\|)$ and $(Y, |\!|\!|\cdot|\!|\!|)$ are Banach spaces and $T : X \to Y$ is a continuous linear surjection, then T is an open mapping (i. e., maps open sets in X to open sets in Y); see [280, Theorem 2.11], [80] or [75, Chapter 2].

Corollary 1.4.13 (Bartle–Graves theorem [17]). *Let $(X, \|\cdot\|)$ and $(Y, |\!|\!|\cdot|\!|\!|)$ be Banach spaces, and let $T : X \to Y$ be a continuous linear mapping onto Y. Then there exists a continuous function $r : Y \to X$ such that $T(r(y)) = y$ for all $y \in Y$.*

Proof. By the open mapping theorem $T : X \to Y$ is an open mapping. Hence by Exercise 1.4.12 the mapping $T^{-1} : Y \to 2^X$ defined by $T^{-1}(y) := \{x \in X : T(x) = y\}$ for $y \in Y$ is lower semicontinuous, and because T is surjective, T^{-1} has nonempty images. Further, since T is continuous and linear, T^{-1} has closed and convex images. Thus by Michael's selection theorem (Theorem 1.4.11) there exists a continuous function $r : Y \to X$ such that $r(y) \in T^{-1}(y)$ for all $y \in Y$, that is, $T(r(y)) = y$ for all $y \in Y$. □

1.4.2 Jayne–Rogers' selection theorem

We start this subsection with a simple example.

Example 1.4.14. Let $\Phi : \mathbb{R} \to 2^{\mathbb{R}}$ be defined by

$$\Phi(x) := \begin{cases} \{-1\} & \text{if } x < 0, \\ [-1, 1] & \text{if } x = 0, \\ \{1\} & \text{if } 0 < x. \end{cases}$$

Then Φ is a cusco mapping. However, Φ does not admit any continuous selections. Thus there is no hope of obtaining a result like Michael's selection theorem for cusco mappings.

Example 1.4.14 shows that in general usco mappings do not always admit continuous selections. However, in some important situations occurring in functional analysis, usco mappings do have selections that are "almost continuous" in the following sense.

Let (X, τ) be a topological space, and let (Y, μ) be a uniform space. Following [149, 204], we will say that a mapping $f : X \to Y$ is (τ, μ)*-fragmented* if for every nonempty subset A of X and every $\varepsilon \in \mu$, there exists a τ-open subset U of X such that $U \cap A \neq \varnothing$ and $f(U \cap A)$ is ε-small in (Y, μ). When the topology τ and the uniformity μ are understood (from the context), then we simply say that the function f is *fragmented*. In the particular case where the range space (Y, d) is a metric space, this definition simplifies to the following: a mapping $f : X \to Y$ is fragmented if for every nonempty subset A of X and every $\varepsilon > 0$, there exists a τ-open subset U of X such that

$U \cap A \neq \varnothing$ and $\operatorname{diam}[f(U \cap A)] < \varepsilon$. Continuing further in this direction, we can define a fragmentable topological space. If (X, τ) is a topological space, then we say that (X, τ) is *fragmented* or *fragmentable* if there exists a metric d on X such that the identity mapping $i : (X, \tau) \to (X, d)$ is a fragmented mapping. Clearly, every metric space (X, d) is fragmentable, but, importantly, there are many fragmentable spaces that are not metrisable.

The origin of this definition most likely stems from the definition of a scattered space, where the identity mapping is "fragmented" by the discrete metric, or perhaps from [300], or even perhaps from functional analysis; see [271].

We next give a closely related technical notion required for the proof of Theorem 1.4.15.

Let (X, τ) be a topological space. Then we call $\mathcal{P} \subseteq 2^X \setminus \{\varnothing\}$ a *partial exhaustive partition of* X [223] if:

(i) $\bigcup_{P \in \mathcal{P}} P \in \tau$;
(ii) the members of $\mathcal{P}$ are pairwise disjoint;
(iii) for every nonempty subset A of $\bigcup_{P \in \mathcal{P}} P$, there exists $P \in \mathcal{P}$ such that $A \cap P$ is a nonempty relatively open subset of A.

If $\bigcup_{P \in \mathcal{P}} P = X$, then we simply call $\mathcal{P}$ an *exhaustive partition of* (X, τ).

Theorem 1.4.15. *Let (X, τ) be a regular topological space, let (M, d) be a complete metric space, and let $\Phi : X \to 2^M$ be a set-valued mapping with nonempty images that satisfies property $(*)$:*
for each $0 < \varepsilon$, each nonempty closed subset C of X and each closed ball B of M, with perhaps infinite radius, such that $\Phi(t) \cap B \neq \varnothing$ for all $t \in C$, there exist an open subset U of X and an element $x \in B$ such that $U \cap C \neq \varnothing$ and $\Phi(t) \cap B[x, \varepsilon] \neq \varnothing$ for all $t \in C \cap U$. Then there exists a fragmented function $\sigma : X \to M$, such that $\sigma(x) \in \overline{\Phi(x)}$ for all $x \in X$.

Proof. Let $y_0 \in M$ and $r_0 := \infty$, and let $(r_n : n \in \mathbb{N})$ be a sequence of positive numbers such that $\sum_{n=1}^{\infty} r_n < \infty$. For reasons of notational consistency, we set $B[y_0, r_0] := M$. To prove our theorem, we inductively construct a sequence of families of ordered pairs $\mathcal{F}^n := \{(U_\alpha^n, y_\alpha^n) : \alpha \in \Lambda^n\}$ such that $\{U_\alpha^n : \alpha \in \Lambda^n\}$ is an exhaustive partition of (X, τ) and $y_\alpha^n \in M$ for all $\alpha \in \Lambda^n$.

Base step. Consider $\Lambda^0 := \{\varnothing\}$, $U_\varnothing^0 = X$ and $y_\varnothing^0 := y_0$ and define $\mathcal{F}^0 := \{(U_\alpha^0, y_\alpha^0) : \alpha \in \Lambda^0\}$.

For each $n \in \mathbb{N}$, we require the family $\mathcal{F}^n$ to have the following properties:

(a_n) $\{U_\alpha^n : \alpha \in \Lambda^n\}$ is an exhaustive partition of (X, τ);
(b_n) for each $\alpha \in \Lambda^n$, there exists $\beta \in \Lambda^{n-1}$ such that $U_\alpha^n \subseteq U_\beta^{n-1}$ and $y_\alpha^n \in B[y_\beta^{n-1}, r_{n-1}]$;
(c_n) for each $\alpha \in \Lambda^n$, $\Phi(x) \cap B[y_\alpha^n, r_n] \neq \varnothing$ for all $x \in U_\alpha^n$.

Step 1. Consider $\mathcal{F}^1 := \{(U_\alpha^1, y_\alpha^1) : \alpha \in \Lambda^1\}$, a maximal, with respect to set inclusion, family of ordered pairs such that $\{U_\alpha^1 : \alpha \in \Lambda^1\}$ is a partial exhaustive partition of (X, τ) and properties (b_1) and (c_1) are satisfied. By Zorn's lemma such a maximal family exists. We will show that $\{U^1_\alpha : \alpha \in \Lambda^1\}$ is in fact an exhaustive partition of (X,τ). If not, then $C := X \setminus \bigcup_{\alpha\in\Lambda^1} U^1_\alpha$ is a nonempty closed subset of (X,τ). Then by property $(*)$ there exist an open set U and an element $y \in B[y_0, r_0]$ such that $U \cap C \neq \varnothing$ and $\Phi(x) \cap B[y, r_1] \neq \varnothing$ for all $x \in C \cap U$. Let $U^1_{\Lambda^1} := C \cap U$ and $y^1_{\Lambda^1} := y$. Then $\mathcal{F} := \{(U^1_\alpha, y^1_\alpha) : \alpha \in \Lambda^1 \cup \{\Lambda^1\}\}$ satisfies (b_1) and (c_1), and $\{U^1_\alpha : \alpha \in \Lambda^1 \cup \{\Lambda^1\}\}$ is a partial exhaustive partition of (X,τ). This contradicts the maximality of $\mathcal{F}^1$. Hence we may conclude that $\mathcal{F}^1$ satisfies the property (a_1) as well.

Assuming that we have constructed the families $\mathcal{F}^k$ in the sequence satisfying properties (a_k), (b_k) and (c_k) up to and including the nth step, we proceed to construct the next step.

Step $n+1$. Consider $\mathcal{F}^{n+1} := \{(U^{n+1}_\alpha, y^{n+1}_\alpha) : \alpha \in \Lambda^{n+1}\}$, a maximal, with respect to set inclusion, family of ordered pairs such that $\{U^{n+1}_\alpha : \alpha \in \Lambda^{n+1}\}$ is a partial exhaustive partition of (X,τ) and properties (b_{n+1}) and (c_{n+1}) are satisfied. By Zorn's lemma such a maximal family exists. We will show that $\{U^{n+1}_\alpha : \alpha \in \Lambda^{n+1}\}$ is in fact an exhaustive partition of (X,τ). If not, then $D := X \setminus \bigcup_{\alpha\in\Lambda^{n+1}} U^{n+1}_\alpha$ is a nonempty closed subset of (X,τ). Since $\{U^n_\alpha : \alpha \in \Lambda^n\}$ is an exhaustive partition of (X,τ), there exists $\beta \in \Lambda^n$ such that $D \cap U^n_\beta$ is a nonempty relatively open subset of D, and since $(U^n_\beta, y^n_\beta) \in \mathcal{F}^n$, $\Phi(x) \cap B[y^n_\beta, r_n] \neq \varnothing$ for all $x \in U^n_\beta$. It now follows from the regularity[1] of (X,τ) that there exists an open subset V of (X,τ) such that $\varnothing \neq D \cap V$ and $D \cap \overline{V} \subseteq U^n_\beta$. Therefore

$$\varnothing \neq D \cap V \subseteq \overline{D \cap V} \subseteq D \cap \overline{V} \subseteq U^n_\beta.$$

By property $(*)$ there exist an open set U and an element $y \in B[y^n_\beta, r_n]$ such that $\varnothing \neq U \cap \overline{D \cap V}$ and $\Phi(x) \cap B[y, r_{n+1}] \neq \varnothing$ for all $x \in U \cap \overline{D \cap V}$. Let $U^{n+1}_{\Lambda^{n+1}} := U \cap V \cap D \neq \varnothing$ and $y^{n+1}_{\Lambda^{n+1}} := y$. Then $\mathcal{F} := \{(U^{n+1}_\alpha, y^{n+1}_\alpha) : \alpha \in \Lambda^{n+1} \cup \{\Lambda^{n+1}\}\}$ satisfies (b_{n+1}) and (c_{n+1}), and $\{U^{n+1}_\alpha : \alpha \in \Lambda^{n+1} \cup \{\Lambda^{n+1}\}\}$ is a partial exhaustive partition of (X,τ). This contradicts the maximality of $\mathcal{F}^{n+1}$. Hence we may conclude that $\mathcal{F}^{n+1}$ satisfies property (a_{n+1}) as well. This completes the induction.

Next, we use the families $\{\mathcal{F}^n : n \in \mathbb{N}\}$ to construct our desired selection $\sigma : X \to M$.

For any $x \in X$ and $n \in \mathbb{N}$, there exists a unique $\alpha_n(x) \in \Lambda^n$ such that $x \in U^n_{\alpha_n(x)}$. Let x be an arbitrary element of X. We claim that $(y^n_{\alpha_n(x)} : n \in \mathbb{N})$ is a Cauchy sequence in (M,d). To justify this claim, consider an arbitrary $\varepsilon > 0$. Since $\sum_{n=1}^\infty r_n < \infty$, there exists $N \in \mathbb{N}$ such that $\sum_{n=N}^\infty r_n < \varepsilon$. Let $N < n < m$. Then

$$d(y^m_{\alpha_m(x)}, y^n_{\alpha_n(x)}) \leq \sum_{k=n+1}^{m} d(y^k_{\alpha_k(x)}, y^{k-1}_{\alpha_{k-1}(x)}) \leq \sum_{k=N+1}^{m} d(y^k_{\alpha_k(x)}, y^{k-1}_{\alpha_{k-1}(x)}) \leq \sum_{k=N}^{\infty} r_k < \varepsilon.$$

Thus $(y^n_{\alpha_n(x)} : n \in \mathbb{N})$ is indeed a Cauchy sequence in (M,d). We now define $\sigma : X \to M$ as follows. For each $x \in X$, let $\sigma(x) := \lim_{n\to\infty} y^n_{\alpha_n(x)}$. From above it follows that σ is well-defined, and by property (c_n) for all $n \in \mathbb{N}$ it follows that $\sigma(x) \in \overline{\Phi(x)}$. So it remains to

1 This is the only place where we use the regularity of X.

show that σ is a fragmented mapping. To this end, let $\varepsilon > 0$, and let C be a nonempty closed subset of (X,τ). Choose $n_0 \in \mathbb{N}$ such that $\sum_{k=n_0}^{\infty} r_k < \varepsilon/2$ and choose $\alpha_0 \in \Lambda^{n_0}$ so that $C \cap U_{\alpha_0}^{n_0}$ is nonempty and relatively open in C. Note that this is possible since $\{U_\alpha^{n_0} : \alpha \in \Lambda^{n_0}\}$ is an exhaustive partition of (X,τ). We claim that $\operatorname{diam}[\sigma(C \cap U_{\alpha_0}^{n_0})] \le \varepsilon$. So suppose that $x, z \in C \cap U_{\alpha_0}^{n_0}$. Note that $\alpha_{n_0}(x) = \alpha_0 = \alpha_{n_0}(z)$. Let $n_0 < m$, then

$$\begin{aligned} d(y^m_{\alpha_m(x)}, y^m_{\alpha_m(z)}) &\le d(y^m_{\alpha_m(x)}, y^{n_0}_{\alpha_0}) + d(y^{n_0}_{\alpha_0}, y^m_{\alpha_m(z)}) \\ &\le d(y^m_{\alpha_m(x)}, y^{n_0}_{\alpha_{n_0}(x)}) + d(y^{n_0}_{\alpha_{n_0}(z)}, y^m_{\alpha_m(z)}) \\ &\le \sum_{k=n_0+1}^{m} d(y^k_{\alpha_k(x)}, y^{k-1}_{\alpha_{k-1}(x)}) + \sum_{k=n_0+1}^{m} d(y^k_{\alpha_k(z)}, y^{k-1}_{\alpha_{k-1}(z)}) \\ &\le \sum_{k=n_0+1}^{m} r_{k-1} + \sum_{k=n_0+1}^{m} r_{k-1} \\ &= \sum_{k=n_0}^{m} r_k + \sum_{k=n_0}^{m} r_k < \varepsilon. \end{aligned}$$

If we take the limit as m tends to ∞, then

$$d(\sigma(x), \sigma(z)) = d\Big(\lim_{m\to\infty} y^m_{\alpha_m(x)}, \lim_{m\to\infty} y^m_{\alpha_m(z)}\Big) = \lim_{m\to\infty} d(y^m_{\alpha_m(x)}, y^m_{\alpha_m(z)}) \le \varepsilon.$$

Since $x, z \in C \cap U_{\alpha_0}^{n_0}$ were arbitrary, $\operatorname{diam}[\sigma(C \cap U_{\alpha_0}^{n_0})] \le \varepsilon$. This completes the proof. □

This theorem has a sister theorem, which is very similar, except that we relax the condition that (X,τ) is regular and strengthen property $(*)$ so that we no longer require the subsets C of X to be closed.

Theorem 1.4.16. *Let (X,τ) be a topological space, let (M,d) be a complete metric space, and let $\Phi : X \to 2^M$ be a set-valued mapping with nonempty images that satisfies property $(**)$:*
for each $0 < \varepsilon$, each nonempty subset C of X and each closed ball B of M, with perhaps infinite radius, such that $\Phi(t) \cap B \neq \varnothing$ for all $t \in C$, there exist an open subset U of (X,τ) and an element $x \in B$ such that $U \cap C \neq \varnothing$ and $\Phi(t) \cap B[x,\varepsilon] \neq \varnothing$ for all $t \in C \cap U$.
Then there exists a fragmented function $\sigma : X \to M$ such that $\sigma(x) \in \overline{\Phi(x)}$ for all $x \in X$.

Proof. As the proof of this theorem is essentially identical to that of the last theorem, we do not give the details here, but we encourage the interested reader to provide the details by themselves. □

We can now present our most fundamental selection theorem concerning usco mappings.

One of the key ingredients in the proof is the fact that if U is a nonempty open subset of a topological space (X,τ) and $\{A_j : 1 \le j \le n\}$ is a closed cover of U, with respect to the relative topology on U, then for some $1 \le j_0 \le n$, $\operatorname{int}(A_{j_0}) \neq \varnothing$ (the interior being with respect to the relative topology on U).

Theorem 1.4.17 (Usco selection theorem). *Let (X, τ) be a topological space, let (M, d) be a complete metric space, and let $\Phi : X \to 2^M$ be an usco. Then Φ admits a fragmented selection.*

Proof. To prove this theorem, we will appeal to Theorem 1.4.16. Suppose $\varepsilon > 0$, C is a nonempty subset of X, and B is a closed ball of M, with perhaps infinite radius, such that $\Phi(x) \cap B \neq \varnothing$ for all $x \in C$. Define $\Psi : C \to 2^B$ by $\Psi(x) := \Phi(x) \cap B$ for $x \in C$. Then, by Proposition 1.1.5 and Corollary 1.1.13, Ψ is an usco on (C, τ'), where τ' denotes the relative τ-topology on C. Let $x_0 \in C$, and let $F := \{y_1, y_2, \ldots, y_n\}$ be a finite subset of $\Psi(x_0)$ such that $\Psi(x_0) \subseteq \bigcup_{1 \le j \le n} B(y_j, \varepsilon)$. Since Ψ is upper semicontinuous, there exists an open subset V of (X, τ) such that $x_0 \in C \cap V$ and $\Psi(V \cap C) \subseteq \bigcup_{1 \le j \le n} B(y_j, \varepsilon)$. For each $1 \le j \le n$, let

$$A_j := \{x \in C \cap V : \Psi(x) \cap B[y_j, \varepsilon] \neq \varnothing\}.$$

Then $C \cap V = \bigcup_{1 \le j \le n} A_j$. It now follows by a simple induction argument that for some $1 \le j_0 \le n$, $\text{int}(A_{j_0}) \neq \varnothing$ (the interior being with respect to the relative topology on $C \cap V$). Therefore there exists an open set U' in (X, τ) such that $\varnothing \neq U' \cap (V \cap C) = (U' \cap V) \cap C \subseteq A_{j_0}$. Thus

$$\varnothing \neq \Psi(x) \cap B[y_{j_0}, \varepsilon] \subseteq \Phi(x) \cap B[y_{j_0}, \varepsilon]$$

for all $x \in U \cap C$, where $U := U' \cap V$. This completes the proof. □

This theorem may appear less than impressive until one realises that fragmented mappings are closely related to Baire one functions. So in the next subsection, we will show that under appropriate conditions, fragmented mappings are Baire one.

A function of the *first Baire class* (briefly said to be *Baire one*) is the pointwise limit of a sequence of continuous functions.

Before we dive into the main theorems, we start with some preliminary definitions and facts. If (X, τ) is a topological space and (M, d) is a metric space, then we denote by $\mathcal{B}^1(X, M)$ the set of all Baire one functions from X into M. It is not hard to see that if $(Y, \|\cdot\|)$ is a normed linear space and $f, g \in \mathcal{B}^1(X, Y)$, then so is $f + g$. Similarly, if $\lambda \in \mathbb{R}$ and $f \in \mathcal{B}^1(X, Y)$, then $\lambda f \in \mathcal{B}^1(X, Y)$. It is also easy to see that if $f \in \mathcal{B}^1(X) := \mathcal{B}^1(X, \mathbb{R})$ and $g \in \mathcal{B}^1(X, Y)$, then $fg \in \mathcal{B}^1(X, Y)$ (i. e., $\mathcal{B}^1(X, Y)$ is closed under multiplication by real-valued Baire one functions).

Proposition 1.4.18. *Let (X, τ) be a topological space, and let $(Y, \|\cdot\|)$ be a normed linear space. If $f \in \mathcal{B}^1(X, Y)$ is bounded, then there exists a bounded sequence $(f_n : n \in \mathbb{N})$ of continuous functions $f_n : X \to Y$ such that for each $x \in X$, $f(x) = \lim_{n \to \infty} f_n(x)$ and $\|f_n\|_\infty \le 2\|f\|_\infty$ for all $n \in \mathbb{N}$, where $\|f_n\|_\infty := \sup\{\|f_n(x)\| : x \in X\}$ for all $n \in \mathbb{N}$, and $\|f\|_\infty := \sup\{\|f(x)\| : x \in X\}$.*

Proof. If $\|f\|_\infty = 0$, then f is the zero function, that is, $f(x) = 0$ for all $x \in X$, and the result is clear. So let us consider the case $0 < \|f\|_\infty$. Let $g : Y \to Y$ be defined by

$$g(y) := \begin{cases} y & \text{if } \|y\| \le 2\|f\|_\infty, \\ \frac{2\|f\|_\infty y}{\|y\|} & \text{if } 2\|f\|_\infty < \|y\| \end{cases} \qquad \text{for all } y \in Y.$$

Then g is continuous, and $\|g(y)\| \le 2\|f\|_\infty$ for all $y \in Y$. Since $f \in \mathcal{B}^1(X,Y)$, there exists a sequence $(g_n : n \in \mathbb{N})$ of continuous functions $g_n : X \to Y$ such that for each $x \in X$, $f(x) = \lim_{n\to\infty} g_n(x)$. Now, for each $n \in \mathbb{N}$, let $f_n : X \to Y$ be defined by $f_n(x) := (g \circ g_n)(x)$ for $x \in X$. Then $(f_n : n \in \mathbb{N})$ has the desired properties. □

Proposition 1.4.19. *Let (X,τ) be a topological space, and let $(Y, \|\cdot\|)$ be a Banach space. If $\{f_n : n \in \mathbb{N}\} \subseteq \mathcal{B}^1(X,Y)$, $g \in Y^X$ and $\lim_{n\to\infty} \|f_n - g\|_\infty = 0$, then $g \in \mathcal{B}^1(X,Y)$.*

Proof. By passing to a subsequence and relabelling if necessary, we may assume that $\|f_n - g\|_\infty < 1/2^{n+1}$ for each $n \in \mathbb{N}$. Furthermore, note that for all $2 \le n \in \mathbb{N}$ and $x \in X$, $f_n(x) = \sum_{k=1}^{n-1}[f_{k+1}(x) - f_k(x)] + f_1(x)$, and so $g(x) = \sum_{k=1}^{\infty}[f_{k+1}(x) - f_k(x)] + f_1(x)$ for all $x \in X$. Thus, to show that $g \in \mathcal{B}^1(X,Y)$, it is sufficient to show that $h := \sum_{n=1}^{\infty}[f_{n+1} - f_n] \in \mathcal{B}^1(X,Y)$. For each $n \in \mathbb{N}$, let $g_n : X \to Y$ be defined by $g_n(x) := f_{n+1}(x) - f_n(x)$ for $x \in X$. Then $g_n \in \mathcal{B}^1(X,Y)$ and $\|g_n\|_\infty \le 1/2^n$ (by the triangle inequality) for all $n \in \mathbb{N}$.

Now by Proposition 1.4.18 for each $n \in \mathbb{N}$, there exists a sequence $(g_n^k : k \in \mathbb{N})$ of continuous functions $g_n^k : X \to Y$ such that $\|g_n^k\|_\infty \le 2\|g_n\|_\infty$ for all $k \in \mathbb{N}$ and $\lim_{k\to\infty} g_n^k(x) = g_n(x)$ for all $x \in X$. We will now use these functions to construct a sequence $(c^k : k \in \mathbb{N})$ of continuous functions $c^k : X \to Y$ such that $\lim_{k\to\infty} c^k(x) = h(x)$ for all $x \in X$.

For each $k \in \mathbb{N}$, let $c^k : X \to Y$ be defined by $c^k(x) := \sum_{n=1}^{\infty} g_n^k(x)$ for $x \in X$. Then each c^k is well-defined since $\sum_{n=1}^{\infty} \|g_n^k\|_\infty \le 2\sum_{n=1}^{\infty} \|g_n\|_\infty \le 2$. Furthermore, by Weierstrass' M-test each function c^k is continuous on (X,τ). We claim that $h(x) = \lim_{k\to\infty} c^k(x)$ for each $x \in X$. To verify this statement, let us fix arbitrary $x \in X$ and $\varepsilon > 0$. Then choose $N_0 \in \mathbb{N}$ such that $\sum_{n=N_0+1}^{\infty} \|g_n\|_\infty < \varepsilon/4$ and let $k \in \mathbb{N}$. Then

$$\begin{aligned}
\|h(x) - c^k(x)\| &= \left\| \sum_{n=1}^{\infty} g_n(x) - \sum_{n=1}^{\infty} g_n^k(x) \right\| \\
&\le \left\| \sum_{n=N_0+1}^{\infty} g_n(x) + \sum_{n=1}^{N_0} [g_n(x) - g_n^k(x)] - \sum_{n=N_0+1}^{\infty} g_n^k(x) \right\| \\
&\le \sum_{n=N_0+1}^{\infty} \|g_n(x)\| + \sum_{n=1}^{N_0} \|g_n(x) - g_n^k(x)\| + \sum_{n=N_0+1}^{\infty} \|g_n^k(x)\| \\
&< \varepsilon/4 + \sum_{n=1}^{N_0} \|g_n(x) - g_n^k(x)\| + 2 \cdot (\varepsilon/4) \\
&= (3/4) \cdot \varepsilon + \sum_{n=1}^{N_0} \|g_n(x) - g_n^k(x)\|.
\end{aligned}$$

For each $1 \le n \le N_0$, choose $M_n \in \mathbb{N}$ such that $\|g_n(x) - g_n^k(x)\| < \varepsilon/(4N_0)$ for all $M_n \le k$. Let $N := \max\{M_1, M_2, \ldots, M_{N_0}\}$. Then for any $N \le k$, $\|h(x) - c^k(x)\| < \varepsilon$. □

Let us also observe that if (M, d) is a metric space and U is an open subset of (M, d), then $\chi_U \in \mathcal{B}^1(M)$. In fact, if we define $f_n : M \to \mathbb{R}$ by

$$f_n(x) := \min\{1, \max\{0, nd(x, M \setminus U) - 1\}\} \quad \text{for } x \in M,$$

then χ_U is the pointwise limit of the sequence $(f_n : n \in \mathbb{N})$, each of which vanishes on $\{x \in M : d(x, M \setminus U) \le 1/n\}$, and $0 \le f_n \le \chi_U$ for all $n \in \mathbb{N}$.

From this it follows that if C is a closed subset of (M, d), then $\chi_C \in \mathcal{B}^1(M)$ as $\chi_C = 1 - \chi_{M\setminus C}$.

Furthermore, if $\chi_A \in \mathcal{B}^1(M)$ and $\chi_B \in \mathcal{B}^1(M)$, then $\chi_{A\cap B} \in \mathcal{B}^1(M)$, as $\chi_{A\cap B} = \chi_A \cdot \chi_B$.

To expedite the presentation of the next two results, we will introduce some new notation. If $f : M \to X$ is a function acting from a topological space (M, τ) into a normed linear space $(X, \|\cdot\|)$ and $\varepsilon > 0$, then

$$\mathcal{U}_\varepsilon := \{U \in \tau : \exists f_U \in \mathcal{B}^1(M, X) \text{ with } \|f_U(x) - f(x)\| \le \varepsilon \text{ for all } x \in U\}.$$

It is immediate that $\mathcal{U}_\varepsilon \ne \varnothing$, since if $f_\varnothing : M \to X$ is the zero function, that is, $f_\varnothing(x) := 0$ for all $x \in M$, then $\varnothing \in \mathcal{U}_\varepsilon$, as vacuously, $\|f_\varnothing(x) - f(x)\| \le \varepsilon$ for all $x \in \varnothing$.

The following result has many fathers; see, for example, [106, 150, 181, 287, 293].

Lemma 1.4.20. *Let (M, d) be a metric space, let $(X, \|\cdot\|)$ be a normed linear space, and let $f : M \to X$ be a function. Then for each $\varepsilon > 0$, there exists a Baire one function $f_{U_\varepsilon} : M \to X$ such that $\|f_{U_\varepsilon}(x) - f(x)\| \le \varepsilon$ for all $x \in U_\varepsilon := \bigcup\{U : U \in \mathcal{U}_\varepsilon\}$, that is, $U_\varepsilon \in \mathcal{U}_\varepsilon$.*

Proof. Let $\varepsilon > 0$. Since the metric space U_ε is paracompact (Theorem 1.4.1), there exists a locally finite open refinement $\{V_\gamma : \gamma \in \Gamma\}$ of $\mathcal{U}_\varepsilon$. Note that in fact $\{V_\gamma : \gamma \in \Gamma\} \subseteq \mathcal{U}_\varepsilon$, that is, for each $\gamma \in \Gamma$, there exists $f_{V_\gamma} \in \mathcal{B}^1(M, X)$ such that $\|f_{V_\gamma}(x) - f(x)\| \le \varepsilon$ for all $x \in V_\gamma$.

We may and do assume that Γ is a well-ordered set. From this it follows that $\{V_\gamma \setminus \bigcup_{\beta<\gamma} V_\beta : \gamma \in \Gamma\}$ is a disjoint cover of U_ε. If we define $f_{U_\varepsilon} : M \to X$ by

$$f_{U_\varepsilon}(x) := \sum_{\gamma\in\Gamma} \chi_{V_\gamma\setminus\bigcup_{\beta<\gamma} V_\beta}(x) f_{V_\gamma}(x) \quad \text{for } x \in M,$$

then f_{U_ε} is well-defined (i. e., the pointwise sum makes sense) since $\{V_\gamma \setminus \bigcup_{\beta<\gamma} V_\beta : \gamma \in \Gamma\}$ is a disjoint cover of U_ε. Furthermore, $\|f_{U_\varepsilon}(x) - f(x)\| \le \varepsilon$ for all $x \in U_\varepsilon$. So to show that $U_\varepsilon \in \mathcal{U}_\varepsilon$, it is sufficient to show that $f_{U_\varepsilon} \in \mathcal{B}^1(M, X)$. This is the place where we use the fact that $\{V_\gamma : \gamma \in \Gamma\}$ is locally finite.

For each $\gamma \in \Gamma$, choose a sequence of continuous functions $r_{\gamma,n} : M \to \mathbb{R}$ such that each $r_{\gamma,n}$ is supported on V_γ and $(r_{\gamma,n} : n \in \mathbb{N})$ converges pointwise to $\chi_{V_\gamma\setminus\bigcup_{\beta<\gamma} V_\beta}$. Next, choose continuous functions $f_{\gamma,n} : M \to X$ so that $(f_{\gamma,n} : n \in \mathbb{N})$ converges pointwise to f_{V_γ}. Finally, choose continuous functions $s_n : M \to \mathbb{R}$ such that $(s_n :$

$n \in \mathbb{N}$) converges pointwise to χ_{U_ε} and $s_n(x) = 0$ if $d(x, M \setminus U_\varepsilon) \le 1/n$. Now, for each $n \in \mathbb{N}$, define

$$g_n(x) := s_n(x) \cdot \sum_{\gamma \in \Gamma} r_{\gamma,n}(x) f_{\gamma,n}(x).$$

Since $\{V_\gamma : \gamma \in \Gamma\}$ is locally finite on U_ε, each g_n is well-defined and continuous at each point of U_ε, and since $s_n(x) = 0$ on a neighbourhood of $M \setminus U_\varepsilon$, g_n is well-defined and continuous at each point of $M \setminus U_\varepsilon$. Hence g_n is well-defined and continuous on (M, d). It is now routine to check that for each $x \in M$,

$$f_{U_\varepsilon}(x) = \sum_{\gamma \in \Gamma} \chi_{V_\gamma \setminus \bigcup_{\beta<\gamma} V_\beta}(x) f_{V_\gamma}(x) = \chi_{U_\varepsilon}(x) \cdot \sum_{\gamma \in \Gamma} \chi_{V_\gamma \setminus \bigcup_{\beta<\gamma} V_\beta}(x) f_{V_\gamma}(x) = \lim_{n \to \infty} g_n(x).$$

This shows that f_{U_ε} is indeed Baire one. □

Note the important role played by the functions $\{s_n : n \in \mathbb{N}\}$ in the previous lemma, because although $\{V_\gamma : \gamma \in \Gamma\}$ is locally finite on U_ε, it is not necessarily locally finite on M or even on $\overline{U_\varepsilon}$.

Theorem 1.4.21 ([13]). *Let (M, d) be a metric space, let $(X, \|\cdot\|)$ be a Banach space, and let $f : M \to X$ be a fragmented mapping. Then $f \in \mathcal{B}^1(M, X)$.*

Proof. It follows from Proposition 1.4.19 that we need only show that for every $n \in \mathbb{N}$, there exists a function $f_n \in \mathcal{B}^1(M, X)$ such that $\|f_n(x) - f(x)\| \le 1/n$ for all $x \in M$. To this end, let $n \in \mathbb{N}$. We claim that $M = U_{1/n} \in \mathcal{U}_{1/n}$ (by Lemma 1.4.20), that is, there exists a Baire one function $f_n : M \to X$ such that $\|f_n(x) - f(x)\| \le 1/n$ for all $x \in M$. Indeed, if $M \setminus U_{1/n} \ne \varnothing$, then $C := M \setminus U_{1/n}$ is a nonempty closed subset of (M, d). Since f is fragmented, there exists an open subset U of (M, d) such that $\varnothing \ne C \cap U$ and $\operatorname{diam}[f(C \cap U)] < 1/n$. Let $x_0 \in C \cap U$ and set $y_0 := f(x_0)$. Then $U_{1/n} \cup (C \cap U) = U_{1/n} \cup U \in \mathcal{U}_{1/n}$, since $f_{U_{1/n}} \cdot \chi_{U_{1/n}} + y_0 \cdot \chi_{C \cap U}$ is Baire one and $\|[f_{U_{1/n}}(x) \cdot \chi_{U_{1/n}}(x) + y_0 \cdot \chi_{C \cap U}(x)] - f(x)\| \le 1/n$ for all $x \in U_{1/n} \cup U$. However, this contradicts the definition of $U_{1/n}$. Thus $M \in \mathcal{U}_{1/n}$. □

It is interesting to see that in the context of complete metric domain spaces, the notions of fragmentability and Baire one coincide. To demonstrate this, we need to recall a few more classical results.

Theorem 1.4.22 (Baire–Osgood theorem [13, 255]). *Let (X, τ) be a topological space, and let (M, d) be a metric space. If $f \in \mathcal{B}^1(X; M)$, then there exists a countable family $\mathcal{F}$ of closed sets, each without interior, such that f is continuous at each point of $X \setminus \bigcup_{F \in \mathcal{F}} F$.*

Proof. Let $(f_n : n \in \mathbb{N})$ be any sequence of continuous functions that map from (X, τ) into (M, d) and have the property that $f(x) = \lim_{n \to \infty} f_n(x)$ for all $x \in X$. For $(m, n) \in \mathbb{N}^2$, we define

$$D_{(m,n)} := \{x \in X : d(f_k(x), f_m(x)) \le 1/n \text{ for all } m \le k\}$$
$$= \bigcap_{m \le k} \{x \in X : d(f_k(x), f_m(x)) \le 1/n\}.$$

Clearly, each set $D_{(m,n)}$ is closed. For each $(m,n) \in \mathbb{N}^2$, let

$$F_{(m,n)} := \mathrm{Bd}(D_{(m,n)}) = D_{(m,n)} \setminus \mathrm{int}(D_{(m,n)}).$$

Then each $F_{(m,n)}$ is closed and has no interior. We claim that f is continuous at each point of $X \setminus \bigcup_{(m,n)\in\mathbb{N}^2} F_{(m,n)}$. To this end, let $x \in X \setminus \bigcup_{(m,n)\in\mathbb{N}^2} F_{(m,n)}$, and let $\varepsilon > 0$. We will construct an open neighbourhood U of x such that $d(f(y), f(x)) < \varepsilon$ for all $y \in U$. Firstly, we choose $n \in \mathbb{N}$ so that $1/n < \varepsilon/3$. Since $(f_k(x) : k \in \mathbb{N})$ is a Cauchy sequence, there exists $m \in \mathbb{N}$ such that $d(f_k(x), f_m(x)) \le 1/n$ for all $m < k$. Therefore $x \in D_{(m,n)}$. On the other hand, $x \notin F_{(m,n)} = \mathrm{Bd}(D_{(m,n)})$. Therefore $x \in \mathrm{int}(D_{(m,n)})$. Now f_m is continuous at x, so there exists an open neighbourhood V of x such that $d(f_m(y), f_m(x)) < \varepsilon/3$ for all $y \in V$. Let $U := V \cap \mathrm{int}(D_{(m,n)})$, $y \in U$ and $m < k$. Then

$$d(f_k(y), f_k(x)) \le d(f_k(y), f_m(y)) + d(f_m(y), f_m(x)) + d(f_m(x), f_k(x))$$
$$< 1/n + \varepsilon/3 + 1/n.$$

Taking the limit as k tends to ∞, we get $d(f(y), f(x)) \le 2/n + \varepsilon/3 < \varepsilon$. □

In order for this theorem to have some teeth, we need to know that the space (X, τ) cannot be covered by a countable family of closed subsets with no interior. Otherwise, the conclusion could be vacuous. Indeed, on $\mathbb{Q}$, with the usual topology inherited from $\mathbb{R}$, there exists a Baire one function with no points of continuity (we leave this as an exercise for the interested reader).

However, the following theorem, due to R. Baire, comes to our rescue.

Theorem 1.4.23 ([12]). *Let (M, d) be a complete metric space, and let $\mathcal{F}$ be a countable family of closed sets, each without interior. Then $M \setminus \bigcup_{F\in\mathcal{F}} F$ is dense in (M, d).*

Proof. Let W be a nonempty open subset of M and suppose $\mathcal{F} := \{F_n : n \in \mathbb{N}\}$, where each F_n is closed with empty interior. We will inductively define a nested sequence $(B[x_n, r_n] : n \in \mathbb{N})$ of closed balls in W such that $B[x_n, r_n] \cap F_n = \varnothing$ and $0 < r_n < 1/n$ for all $n \in \mathbb{N}$.

Case $n = 1$. Since $\mathrm{int}(F_1) = \varnothing$, $W \not\subseteq F_1$. Therefore there exists a point $x_1 \in W \setminus F_1$. Furthermore, since F_1 is a closed subset of M, there exists $0 < r_1 < 1$ such that $B[x_1, r_1] \subseteq W \setminus F_1$.

Next, suppose that we have defined

$$B[x_n, r_n] \subseteq B[x_{n-1}, r_{n-1}] \subseteq \cdots \subseteq B[x_1, r_1] \subseteq W$$

such that $B[x_k, r_k] \cap F_k = \varnothing$ and $0 < r_k < 1/k$ for all $1 \le k \le n$.

Case $n+1$. Since $\operatorname{int}(F_{n+1}) = \varnothing$, $B(x_n, r_n) \nsubseteq F_{n+1}$. Therefore there exists a point $x_{n+1} \in B(x_n, r_n) \setminus F_{n+1}$. Furthermore, since F_{n+1} is a closed subset of M, there exists $0 < r_{n+1} < 1/(n+1)$ such that $B[x_{n+1}, r_{n+1}] \subseteq B(x_n, r_n) \setminus F_{n+1}$.

This completes the induction. Now, as (M, d) is a complete metric space, by the Cantor intersection property we have that $\bigcap_{n\in\mathbb{N}} B[x_n, r_n]$ is a singleton. Let $\{x_\infty\} := \bigcap_{n\in\mathbb{N}} B[x_n, r_n]$. Note that $x_\infty \in B[x_1, r_1] \subseteq W$, that is, $x_\infty \in W$. More generally, note that if $n \in \mathbb{N}$, then $x_\infty \in B[x_n, r_n] \subseteq M \setminus F_n$, that is, $x_\infty \notin F_n$. Therefore $x_\infty \in W \setminus \bigcup_{n\in\mathbb{N}} F_n$. □

In terms of sets of the first Baire category, the previous theorem says that in every nonempty complete metric space (M, d), the complement of every first Baire category set is dense in (M, d). Recall that (i) a subset F of a topological space (X, τ) is said to be *nowhere dense* in (X, τ) if $\operatorname{int}(\overline{F}) = \varnothing$ and (ii) a subset F of a topological space (X, τ) is a *first category* subset of (X, τ) or is of the *first Baire category in* (X, τ) if there exists a countable family $\{F_n : n \in \mathbb{N}\}$ of nowhere dense subsets of (X, τ) such that $F \subseteq \bigcup_{n\in\mathbb{N}} F_n$.

By applying de Morgan's laws to the previous result we obtain the following corollary.

Corollary 1.4.24 (Baire category theorem). *Let (M, d) be a complete metric space, and let $\mathcal{F}$ be a countable family of dense open sets. Then $\bigcap_{O\in\mathcal{F}} O$ is dense in (M, d).*

Note that by combining the Baire–Osgood theorem with Theorem 1.4.15 and Theorem 1.4.21 we obtain the following well-known Baire theorem.

Theorem 1.4.25 (Baire's great theorem [13]). *Ley (M, d) be a complete metric space, let $(X, \|\cdot\|)$ be a Banach space, and let $f : M \to X$. Then $f \in \mathcal{B}^1(M, X)$ if and only if for every nonempty closed subset C of (M, d), there exists a point $x \in C$, where $f|_C$ is continuous.*

Exercise 1.4.26. Let (T, τ) be a topological space, let (M, d) be a metric space, and let $f \in \mathcal{B}^1(T, M)$. Show that the range of f,

$$R(f) := \{f(x) \in M : x \in T\},$$

is separable if T is either (i) separable or (ii) compact.
Hint: if f is the pointwise limit of a sequence of functions $(f_n : n \in \mathbb{N})$ from T into M, then $R(f) \subseteq \overline{\bigcup_{n\in\mathbb{N}} R(f_n)}$.

By combining the previous exercise with the definition of a scattered space we can establish some of the limitations of Theorem 1.4.21. Recall that a topological space (T, τ) is *scattered* if every nonempty subset A of T has an isolated point, that is, there exists an open subset U of T such that $A \cap U$ is a singleton. It is not hard to see that if $(T, \leq)$ is a well-ordered set, then T, with the order topology, is scattered. Indeed, if A is a nonempty subset of T, then the "least" element in A is an isolated point.

Exercise 1.4.27. Show that in Theorem 1.4.21 we cannot change the hypothesis on the domain space T to being compact, as opposed to being completely metrizable.

Hint: Let ω_1 denote the first uncountable ordinal number. Then $T := [0, \omega_1]$, with the order topology, is a compact scattered Hausdorff space. Let $(Y, \|\cdot\|)$ denote the dual space of the Banach space $(C(T), \|\cdot\|_\infty)$, and let $f : T \to Y$ be defined by $f(x) := \delta_x$, where $\delta_x : C(T) \to \mathbb{R}$ is defined by $\delta_x(f) := f(x)$ for every $f \in C(T)$. Then f is not Baire one (see Exercise 1.4.26, part (ii)) but $f|_C$ has a point of continuity for every nonempty subset C of T.

We end the section by giving some applications of Theorem 1.4.15. In this regard, the following particular case of Theorem 1.4.15 is very useful.

Theorem 1.4.28 (Jayne–Rogers' selection theorem [150]). *Let (M, d) be a nonempty complete metric space, and let $(X, \|\cdot\|)$ be a Banach space. Suppose also that τ is a Hausdorff topology on X, weaker or equal to the norm topology on X, such that $B[x, r]$ is τ-closed for each $x \in X$ and $r > 0$. If (X, τ) has the property that every minimal τ-usco $\Phi : A \to 2^X$ from a nonempty complete metric space (A, ρ) into subsets of X is single-valued and norm upper semicontinuous at some point of A, then every τ-usco $\Psi : M \to 2^X$ has a Baire one selection.*

Proof. To prove this theorem, we will appeal to Theorem 1.4.15. Suppose $\varepsilon > 0$, C is a nonempty closed subset of (M, d), and B is a closed ball of X, with perhaps infinite radius, such that $\Phi(x) \cap B \neq \varnothing$ for all $x \in C$. Define $\Psi : C \to 2^B$ by $\Psi(x) := \Phi(x) \cap B$ for all $x \in C$. Then, by Proposition 1.1.5 and Corollary 1.1.13, Ψ is a τ-usco on (C, τ'), where τ' denotes the relative d-topology on the set C. Furthermore, by Proposition 1.3.3 there exists a minimal τ-usco $\Psi_M : C \to 2^B$ such that $\mathrm{Gr}(\Psi_M) \subseteq \mathrm{Gr}(\Psi)$. Now by assumption there exists a point $x_0 \in C$ such that Ψ_M is single-valued and norm upper semicontinuous at x_0. If $\{y_0\} := \Psi_M(x_0)$, then $y_0 \in B$. Moreover, since Ψ_M is norm upper semicontinuous at x_0, there exists an open subset U of (M, d) such that $x_0 \in C \cap U$ and $\Psi_M(C \cap U) \subseteq B[y_0, \varepsilon]$. Hence $\Phi(x) \cap B[y_0, \varepsilon] \neq \varnothing$ for all $x \in C \cap U$. □

Corollary 1.4.29. *Let $(X, \|\cdot\|)$ be a Banach space whose dual space $(X^*, \|\cdot\|)$ is separable. Then every weak* usco $\Phi : M \to 2^{X^*}$ from a complete metric space (M, d) into the subsets of X^* admits a Baire one selection.*

Proof. Let us start with the preliminary observation that for any $x^* \in X^*$ and $r > 0$, $B[x^*, r] = x^* + rB_{X^*}$ is weak* closed (see the Banach-Alaoglu theorem, [1]). Then let us consider an arbitrary minimal weak* usco $\Phi : A \to 2^{X^*}$ acting from a nonempty complete metric space (A, ρ) into subsets of X^*. We will show that Φ is single-valued and norm upper semicontinuous at the points of a dense subset of (A, ρ). As our first step in this direction, let us fix an $\varepsilon > 0$ and consider the set

$$O_\varepsilon := \bigcup \{U \in 2^A : U \text{ is open and } \mathrm{diam}[\Phi(U)] \leq 2\varepsilon\}.$$

Clearly, O_ε is open. We will show that O_ε is dense in (A, ρ). To this end, let W be a nonempty open subset of A, and let $\{B[x_n^*, \varepsilon] : n \in \mathbb{N}\}$ be a cover of X^*. For each $n \in \mathbb{N}$, let $F_n := \{x \in A : \Phi(x) \cap B[x_n^*, \varepsilon] \neq \varnothing\}$. Then each set F_n is a closed subset of

(A,ρ). Further, for each $n \in \mathbb{N}$, we let $D_n := \mathrm{Bd}(F_n) = F_n \setminus \mathrm{int}(F_n)$. Then each D_n is closed and has no interior. By Theorem 1.4.23 $A \setminus \bigcup_{n\in\mathbb{N}} D_n$ is dense in (A,ρ). In particular, there exists a point $x_0 \in W \cap (A \setminus \bigcup_{n\in\mathbb{N}} D_n)$. Now, because $\Phi(x_0) \neq \varnothing$, there exists $n_0 \in \mathbb{N}$ such that $\Phi(x_0) \cap B[x^*_{n_0}, \varepsilon] \neq \varnothing$. Therefore $x_0 \in F_{n_0} \setminus D_{n_0} = \mathrm{int}(F_{n_0})$. Since $\Phi(x) \cap B[x^*_{n_0}, \varepsilon] \neq \varnothing$ for all $x \in \mathrm{int}(F_{n_0})$, it follows from Proposition 1.3.5 that $\Phi(\mathrm{int}(F_{n_0})) \subseteq B[x^*_{n_0}, \varepsilon]$. Thus

$$\varnothing \neq \mathrm{int}(F_{n_0}) \cap W \subseteq O_\varepsilon \cap W,$$

and so O_ε is dense in (A,ρ). It now only remains to observe that Φ is single-valued and norm upper semicontinuous at each point of $\bigcap_{n\in\mathbb{N}} O_{1/n}$, which is dense in (A,ρ) by Corollary 1.4.24. □

As our last application of Theorems 1.4.15 and 1.4.21, we will show that every weakly upper semicontinuous set-valued mapping, with nonempty closed images, that maps from a complete metric space (M,d) into a Banach space $(X, \|\cdot\|)$ admits a Baire one selection.

Lemma 1.4.30. *Let $\Phi : T \to 2^X$ be a weak upper semicontinuous set-valued mapping, with nonempty images, from a topological space (T,τ) into subsets of a normed linear space $(X, \|\cdot\|)$. Suppose also that $\varepsilon > 0$, U is a nonempty open subset of (T,τ), and K is a compact subset of $(X, \|\cdot\|)$. If $\Phi(t) \cap [K + (\varepsilon/2)B_X] \neq \varnothing$ for all $t \in U$, then there exist a point $x_0 \in K$ and a nonempty open subset V of U such that $\Phi(t) \cap B[x_0, \varepsilon] \neq \varnothing$ for all $t \in V$.*

Proof. Since K is compact, there exists a finite subset $F := \{k_1, k_2, \dots, k_n\}$ of K such that $K \subseteq \bigcup_{1\le j\le n} B(k_j, \varepsilon/2)$. Therefore

$$K + (\varepsilon/2)B_X = \bigcup_{k\in K} B[k, \varepsilon/2] \subseteq \bigcup_{1\le j\le n} B[k_j, \varepsilon].$$

For each $1 \le j \le n$, let

$$A_j := \{x \in U : \Phi(x) \cap B[k_j, \varepsilon] \neq \varnothing\}.$$

Then $U = \bigcup_{1\le j\le n} A_j$. It now follows by a simple induction argument that for some $1 \le j_0 \le n$, $\mathrm{int}(A_{j_0}) \neq \varnothing$ (the interior being with respect to the relative topology on U). Therefore there exists an open set V' in (T,τ) such that $\varnothing \neq V' \cap U \subseteq A_{j_0}$. Thus $\varnothing \neq \Phi(t) \cap B[x_0, \varepsilon]$ for all $t \in V$, where $x_0 := k_{j_0}$ and $V := V' \cap U$. □

Theorem 1.4.31. *Let (M,d) be a nonempty complete metric space, and let $(X, \|\cdot\|)$ be a Banach space. If $\Phi : M \to 2^X$ is a weak upper semicontinuous set-valued mapping, with nonempty closed images, then Φ admits a Baire one selection.*

Proof. To prove this theorem, we will appeal to Theorem 1.4.15 and then to Theorem 1.4.21. Suppose $\varepsilon > 0$, C is a nonempty closed subset of (M,d), and B is a closed ball of X, with perhaps infinite radius, such that $\Phi(t) \cap B \neq \varnothing$ for all $t \in C$. We wish to show that there exist an open subset U of (M,d) and a point $x_0 \in B$ such that $\varnothing \neq C \cap U$ and $\Phi(t) \cap B[x_0, \varepsilon] \neq \varnothing$ for all $t \in C \cap U$. So let us suppose, to obtain a contradiction, that for every $x \in B$, $\{t \in C : \Phi(t) \cap B[x, \varepsilon] \neq \varnothing\}$ has no interior, relative to (C, τ'), where τ' denotes the relative d-topology on the set C. Therefore, by Lemma 1.4.30, for every compact subset K of B, $\{t \in C : \Phi(t) \cap [K + (\varepsilon/2)B_X] \neq \varnothing\}$ has no interior in (C, τ'). Hence $\{t \in C : \Phi(t) \cap [K + (\varepsilon/2)B_X] = \varnothing\}$ is a dense subset of (C, τ').

We will use this fact to inductively define sequences $(t_n : n \in \mathbb{N})$ in C, $(x_n : n \in \mathbb{N})$ in B and $(r_n : n \in \mathbb{N})$ in $(0, \infty)$ such that the following conditions are satisfied for every $n \in \mathbb{N}$:

(a_n) $x_n \in \Phi(t_{n-1})$;

(b_n) $\Phi(B[t_n, r_n] \cap C) \cap [\mathrm{co}\{x_1, x_2, \dots, x_n\} + (\varepsilon/2)B_X] = \varnothing$;

(c_n) $B[t_n, r_n] \subseteq B(t_{n-1}, r_{n-1})$ and $0 < r_n < 1/(n+1)$.

Base step. Let t_0 be any element of C, and let $r_0 := 1$.

Step 1. Let $x_1 \in \Phi(t_0) \cap B$. Then by assumption there exists $t_1 \in B(t_0, 1) \cap C$ such that $\Phi(t_1) \cap B[x_1, \varepsilon/2] = \varnothing$. By the weak upper semicontinuity of Φ there exists $0 < r_1 < 1/2$ such that $B[t_1, r_1] \subseteq B(t_0, r_0)$ and $\Phi(B[t_1, r_1] \cap C) \cap B[x_1, \varepsilon/2] = \varnothing$.

Assuming that we have constructed the t_k, x_k and r_k in the sequence satisfying the properties (a_k), (b_k) and (c_k) up to and including the nth step, we proceed to construct the next step.

Step $n+1$. Let $x_{n+1} \in \Phi(t_n) \cap B$. Then by assumption there exists $t_{n+1} \in B(t_n, r_n) \cap C$ such that $\Phi(t_{n+1}) \cap [\mathrm{co}\{x_1, x_2, \dots, x_{n+1}\} + (\varepsilon/2)B_X] = \varnothing$, since $\mathrm{co}\{x_1, x_2, \dots, x_{n+1}\}$ is compact. Furthermore, using the fact that in a Hausdorff linear topological space the sum of a closed set and a compact set is closed, we have that $\mathrm{co}\{x_1, x_2, \dots, x_{n+1}\} + (\varepsilon/2)B_X$ is closed in the weak topology on X, as $\mathrm{co}\{x_1, x_2, \dots, x_{n+1}\}$ is norm compact and hence compact in the weak topology on X and $(\varepsilon/2)B_X$ is closed in the weak topology on X, since Mazur's theorem says that all norm closed and convex subsets of a normed linear space are weakly closed. Therefore by the weak upper semicontinuity of Φ there exists $0 < r_{n+1} < 1/(n+2)$ such that $B[t_{n+1}, r_{n+1}] \subseteq B(t_n, r_n)$ and $\Phi(B[t_{n+1}, r_{n+1}] \cap C) \cap [\mathrm{co}\{x_1, x_2, \dots, x_{n+1}\} + (\varepsilon/2)B_X] = \varnothing$. This completes the induction.

Let $\{t_\infty\} := \bigcap_{n \in \mathbb{N}} B[t_n, r_n] \cap C$, and let $n \in \mathbb{N}$. Then

$$\Phi(t_\infty) \cap [\mathrm{co}\{x_1, x_2, \dots, x_n\} + (\varepsilon/2)B_X] \subseteq \Phi(B[t_n, r_n] \cap C) \cap [\mathrm{co}\{x_1, x_2, \dots, x_n\} + (\varepsilon/2)B_X] = \varnothing,$$

that is, $[\Phi(t_\infty) + (\varepsilon/2)B_X] \cap \mathrm{co}\{x_1, x_2, \dots, x_n\} = \varnothing$. Hence

$$[\Phi(t_\infty) + (\varepsilon/2)B_X] \cap \mathrm{co}\{x_k : k \in \mathbb{N}\} = [\Phi(t_\infty) + (\varepsilon/2)B_X] \cap \bigcup_{n \in \mathbb{N}} \mathrm{co}\{x_1, x_2, \dots, x_n\} = \varnothing.$$

Thus $\Phi(t_\infty) \cap \overline{\mathrm{co}}\{x_k : k \in \mathbb{N}\} = \varnothing$. Since Φ is weak upper semicontinuous and $\overline{\mathrm{co}}\{x_k : k \in \mathbb{N}\}$ is closed with respect to the weak topology on X (by Mazur's theorem), there exists an open neighbourhood U of t_∞ such that $\Phi(U) \cap \overline{\mathrm{co}}\{x_k : k \in \mathbb{N}\} = \varnothing$. However, for n large enough, $t_n \in U$, and so

$$x_{n+1} \in \Phi(t_n) \cap \{x_k : k \in \mathbb{N}\} \subseteq \Phi(U) \cap \overline{\mathrm{co}}\{x_k : k \in \mathbb{N}\} = \varnothing,$$

which is impossible. Therefore there exists $x_0 \in B$ such that $\{t \in C : \Phi(t) \cap B[x_0, \varepsilon] \neq \varnothing\}$ has a nonempty interior relative to (C, τ'). This completes the proof. □

There are several refinements of this result in the literature; see [106, 149, 151, 217, 287, 293]. For a more thorough discussion of selection theorems, see [151, 207, 208, 210, 265].

1.5 Metric-valued mappings

In this section, we solely focus on mappings whose range space is a metric space.

1.5.1 Metric-valued usco mappings

In many applications of usco mappings the range space under consideration is not an abstract topological space, but rather a well-behaved metric space (e. g. a normed linear space). In fact, in many applications, it is in fact a complete metric space (e. g. a Banach space). Therefore, in this section, we will investigate the usconess (if such a word exists) of a set-valued mapping in this setting. Interestingly, in this case, usconess of a set-valued mapping can be characterised in terms of Kuratowski's index of noncompactness.

For a *bounded set* E in a metric space (X, d) (i. e. there exists a point $x \in X$ and $r > 0$ such that $E \subseteq B[x; r]$), the *Kuratowski index of non-compactness* is:

$$\alpha(E) := \inf\{0 < r : E \text{ is covered by a finite family of sets of diameter less than } r\}.$$

If E is not a bounded subset of (X, d), then we define $\alpha(E) := \infty$.

This index of non-compactness was first introduced by Casimir Kuratowski in 1930, (see [176]) and has subsequently been used in many areas of functional analysis, including fixed-point theory [299], the study of the Drop property, [277, 278], the differentiability theory of convex functions on Banach spaces, [94] and more recently, in the renorming theory in Banach spaces, [88, 302].

We shall begin our investigation of Kuratowski's index of non-compactness by recalling some of its basic properties..

Proposition 1.5.1. *For a metric space (X, d), Kuratowski's index of non-compactness α, satisfies the following properties:*

(i) $0 \le \alpha(E)$ *for any bounded subset E of (X, d);*
(ii) $\alpha(E) \le diam(E)$ *for any bounded subset E of (X, d);*
(iii) $\alpha(E) \le \alpha(F)$ *if $E \subseteq F$ and F is bounded in (X, d);*
(iv) $\alpha(E) = \alpha(\overline{E})$ *for any bounded subset E of (X, d);*
(v) $\alpha(\bigcup_{j=1}^{n} E_j) = max\{\alpha(E_j) : j \in \{1, 2, \ldots, n\}\}$ *for any bounded subsets E_j of (X, d);*

(vi) $\alpha(E \cap F) \leq min\{\alpha(E), \alpha(F)\}$ *for any bounded subsets E and F of* (X, d);
(vii) $\alpha(E) = 0$ *if, and only if, E is a totally bounded subset of* (X, d).

We omit the proof of the properties (i) to (vii) as they are straightforward. However, we observe that if (X, d) is a complete metric space, then property (vii) gives us the following result: (vii') $\alpha(E) = 0$ if, and only if, $\overline{E}$ is a compact subset of (X, d).

The following important theorem shows that Kuratowski's index of non-compactness satisfies a generalised Cantor's intersection property, [176, p. 303].

Theorem 1.5.2. *Let* $(F_n : n \in \mathbb{N})$ *be a decreasing sequence of nonempty closed subsets of a complete metric space* (X, d). *If* $lim_{n\to\infty}\alpha(F_n) = 0$, *then* $\bigcap_{n\in\mathbb{N}} F_n$ *is nonempty and compact.*

Proof. Consider a sequence $(x_n : n \in \mathbb{N})$ in X such that $x_n \in F_n$ for each $n \in \mathbb{N}$, then

$$\begin{aligned}\alpha(\{x_n : n \in \mathbb{N}\}) &= \max\{\alpha(\{x_1, x_2, \ldots, x_{k-1}\}), \alpha(\{x_n : k \leq n\})\}\\ &= \alpha(\{x_n : k \leq n\}) \quad \text{since } \alpha(\{x_1, x_2, \ldots, x_{k-1}\}) = 0\\ &\leq \alpha(F_k) \quad \text{for each } k \in \mathbb{N}, \text{ since } \{x_n : k \leq n\} \subseteq F_k.\end{aligned}$$

So $\alpha(\{x_n : n \in \mathbb{N}\}) = 0$ and thus, by Property (vii'), $\overline{\{x_n : n \in \mathbb{N}\}}$ is compact. Therefore, the sequence $(x_n : n \in \mathbb{N})$ contains a convergent subsequence $(x_{n_k} : k \in \mathbb{N})$. Denote the limit of $(x_{n_k} : k \in \mathbb{N})$ by x_∞. Clearly $x_\infty \in F_n$ for each $n \in \mathbb{N}$, therefore $x_\infty \in \bigcap_{n\in\mathbb{N}} F_n$ and so $\bigcap_{n\in\mathbb{N}} F_n$ is nonempty. Now, $\bigcap_{n\in\mathbb{N}} F_n \subseteq F_k$ for each $k \in \mathbb{N}$, so $\alpha(\bigcap_{n\in\mathbb{N}} F_n) = 0$, and therefore, we have that $\overline{\bigcap_{n\in\mathbb{N}} F_n} = \bigcap_{n\in\mathbb{N}} F_n$ is compact. □

Corollary 1.5.3. *Let* $(\mathcal{P}, \subseteq)$ *be a downwardly directed family (i. e., for each* $C_1, C_2 \in \mathcal{P}$ *there exists a* $C_3 \in \mathcal{P}$ *such that* $C_3 \subseteq C_1 \cap C_2$*) of nonempty closed subsets of a complete metric space* (X, d). *If* $inf\{\alpha(C) : C \in \mathcal{P}\} = 0$, *then* $\bigcap\{C : C \in \mathcal{P}\}$ *is nonempty and compact.*

Proof. Choose a sequence $(C_n : n \in \mathbb{N})$ in $\mathcal{P}$ such that $\lim_{n\to\infty} \alpha(C_n) = 0$ and $C_{n+1} \subseteq C_n$ for all $n \in \mathbb{N}$. We see then that $C_\infty := \bigcap_{n\in\mathbb{N}} C_n$ is nonempty and compact. Now, for a fixed $C \in \mathcal{P}$, the sequence $(C \cap C_n : n \in \mathbb{N})$ is a decreasing sequence of nonempty closed subsets such that $\lim_{n\to\infty} \alpha(C \cap C_n) = 0$. Therefore, $C \cap C_\infty = \bigcap_{n\in\mathbb{N}}(C \cap C_n) \neq \varnothing$. Consider $\{K_j : 1 \leq j \leq m\}$, a finite subfamily of $\mathcal{P}$ and choose $K \in \mathcal{P}$ such that $K \subseteq K_j$ for each $1 \leq j \leq m$, then

$$\varnothing \neq K \cap C_\infty \subseteq \Big(\bigcap_{1\leq j\leq m} K_j\Big) \cap C_\infty = \bigcap_{1\leq j\leq m} (K_j \cap C_\infty).$$

So, by the finite intersection property, (since C_∞ is compact)

$$\varnothing \neq \bigcap\{C \cap C_\infty : C \in \mathcal{P}\} \subseteq \bigcap\{C : C \in \mathcal{P}\}.$$

Clearly, $\bigcap\{C : C \in \mathcal{P}\}$ is compact as $\bigcap\{C : C \in \mathcal{P}\}$ is a closed subset of C_∞. □

Using Kuratowski's index of non-compactness we can introduce a new continuity property for set-valued mappings. Suppose that (X,τ) is a topological space, (M,d) is a metric space and $\Phi : X \to 2^M$. If $x_0 \in X$, then we say that Φ is *α upper semicontinuous at x_0* if, for every $0 < \varepsilon$ there exists a neighbourhood U of x_0 such that $\alpha(\Phi(U)) < \varepsilon$. If Φ is α upper semicontinuous at every point of X, then we say that Φ is *α upper semicontinuous on (X,τ)*.

The next result reveals the connection between a set-valued mapping being an usco at a point x_0 and being α upper semicontinuous at a point x_0.

Theorem 1.5.4. *Suppose that (X,τ) is a topological space, (M,d) is a complete metric space and $\Phi : X \to 2^M$ is densely defined. Then Φ is an usco at x_0 if, and only if, Φ is α upper semicontinuous at x_0 and $\Phi(x_0) = \bigcap\{\overline{\Phi(U)} : U$ is an open neighbourhood of $x_0\}$.*

Proof. Suppose that Φ is an usco at x_0. Since metric spaces are Hausdorff, it follows from Exercise 1.1.16 that $\Phi(x_0) = \bigcap\{\overline{\Phi(U)} : U$ is an open neighbourhood of $x_0\}$. Furthermore, if $0 < \varepsilon$, then since $\Phi(x_0)$ is compact, there exists a finite set F in M such that $\Phi(x_0) \subseteq \bigcup_{x\in F} B(x,\varepsilon/3)$. Since Φ is upper semicontinuous at x_0, there exists an open neighbourhood U of x_0 such that $\Phi(U) \subseteq \bigcup_{x\in F} B(x,\varepsilon/3)$. Therefore, $\alpha(\Phi(U)) \le \alpha(\bigcup_{x\in F} B(x,\varepsilon/3)) < \varepsilon$. Hence, Φ is α upper semicontinuous at x_0.

Conversely, suppose that $\Phi(x_0) = \bigcap\{\overline{\Phi(U)} : U$ is an open neighbourhood of $x_0\}$ and Φ α upper semicontinuous at x_0. Firstly, let us note that if

$$\mathcal{P} := \{\overline{\Phi(U)} : U \text{ is an open neighbourhood of } x_0\}$$

then $(\mathcal{P},\subseteq)$ is a downwardly directed family of nonempty closed subsets of (M,d) and $\inf\{\alpha(C) : C \in \mathcal{P}\} = 0$. Thus, by Corollary 1.5.3,

$$\Phi(x_0) = \bigcap\{\overline{\Phi(U)} : U \text{ is an open neighbourhood of } x_0\} = \bigcap\{C : C \in \mathcal{P}\}$$

is nonempty and compact. We now need to show that Φ is upper semicontinuous at x_0. To this end, let W be an open subset of M containing $\Phi(x_0)$. Then either, for some open neighbourhood U of x_0, $\overline{\Phi(U)} \subseteq W$ (in which case we are done), or else

$$\mathcal{P}^* := \{\overline{\Phi(U)} \setminus W : U \text{ is an open neighbourhood of } x_0\}$$

is a downwardly directed family of nonempty closed subsets of (M,d) which also has the property that $\inf\{\alpha(C) : C \in \mathcal{P}^*\} = 0$. In the latter case, we obtain a contradiction since by Corollary 1.5.3,

$$\varnothing \neq \bigcap\{C : C \in \mathcal{P}^*\} = \bigcap\{C : C \in \mathcal{P}\} \setminus W = \Phi(x_0) \setminus W = \varnothing.$$

Therefore, it must be the case that for some open neighbourhood U of x_0, $\overline{\Phi(U)} \subseteq W$. □

Corollary 1.5.5. *Suppose that (X,τ) is a topological space, (M,d) is a complete metric space and $\Phi : X \to 2^M$ is a densely defined mapping with a closed graph. Then Φ is an usco at x_0 if, and only if, Φ is α upper semicontinuous at x_0. Consequently,*

$$\{x \in X : \Phi(x) \text{ is an usco at } x\} = \bigcap_{n\in\mathbb{N}} \Big(\bigcup\{U \in \tau : \alpha(\Phi(U)) < 1/n\}\Big)$$

is a G_δ subset of (X,τ).

Proof. To see that this last equation is true we simply observe that Φ is α upper semicontinuous at a point $x \in X$ if, and only if, $x \in \bigcap_{n\in\mathbb{N}}(\bigcup\{U \in \tau : \alpha(\Phi(U)) < 1/n\})$. □

Consider two set-valued mappings $\Phi : A \to 2^X$ and $\Psi : A \to 2^X$ acting from a set A into a set X. We will say that Φ is *embedded in* Ψ if, for each $a \in A$, $\Phi(a) \subseteq \Psi(a)$.

Theorem 1.5.6. *Suppose that (X,τ) is a topological space, (M,d) is a complete metric space and $\Psi : X \to 2^M$ is densely defined. Then Ψ can be embedded in an usco if, and only if, Ψ is α upper semicontinuous on (X,τ).*

Proof. Suppose that $\Psi : X \to 2^M$ is densely defined and embedded in an usco $\Phi : X \to 2^M$. Let $x_0 \in X$ and $0 < \varepsilon$. By Theorem 1.5.4, Φ is α upper semicontinuous at $x_0 \in X$. Therefore, there exists an open neighbourhood U of x_0 such that $\alpha(\Phi(U)) < \varepsilon$. Hence,

$$\alpha(\Psi(U)) \le \alpha(\Phi(U)) < \varepsilon \quad \text{since } \Psi(U) \subseteq \Phi(U).$$

This shows that Ψ is α upper semicontinuous on (X,τ). Conversely, suppose that $\Psi : X \to 2^M$ is densely defined and α upper semicontinuous on (X,τ). Let $\Phi := \mathrm{USC}(\Psi)$ then, by Exercise 1.1.16, $\mathrm{Gr}(\Psi) \subseteq \overline{\mathrm{Gr}(\Psi)} = \mathrm{Gr}(\Phi)$. Therefore, Ψ is embedded in Φ. We will show that Φ is an usco on (X,τ), by applying Theorem 1.5.4, but first we will show that for any nonempty open subset V of (X,τ), $\overline{\Phi(V)} = \overline{\Psi(V)}$. To this end, let V be a nonempty open subset of (X,τ). Since Ψ is embedded in Φ and $\overline{\Psi(V)}$ is closed it is sufficient to show that $\Phi(V) \subseteq \overline{\Psi(V)}$. Let $y \in \Phi(V)$. Then there exists an $x \in V$ such that $y \in \Phi(x)$. Therefore,

$$y \in \Phi(x) = \mathrm{USC}(\Psi)(x) = \bigcap\{\overline{\Psi(U)} : U \text{ is an open neighbourhood of } x\} \subseteq \overline{\Psi(V)}$$

since V is an open neighbourhood of x. Thus, $\overline{\Phi(V)} = \overline{\Psi(V)}$. Let $x_0 \in X$, then

$$\begin{aligned}\Phi(x_0) &\subseteq \bigcap\{\overline{\Phi(U)} : U \text{ is an open neighbourhood of } x_0\}\\ &= \bigcap\{\overline{\Psi(U)} : U \text{ is an open neighbourhood of } x_0\}\\ &= \mathrm{USC}(\Psi)(x_0) = \Phi(x_0)\end{aligned}$$

and so $\Phi(x_0) = \bigcap\{\overline{\Phi(U)} : U \text{ is an open neighbourhood of } x_0\}$. Let $0 < \varepsilon$, then since Ψ is α upper semicontinuous at $x_0 \in X$, there exists an open neighbourhood U of x_0 such that $\alpha(\Psi(U)) < \varepsilon$. Thus,

$$\alpha(\Phi(U)) = \alpha(\overline{\Phi(U)}) = \alpha(\overline{\Psi(U)}) = \alpha(\Psi(U)) < \varepsilon.$$

This shows that Φ is α upper semicontinuous at x_0. It now follows from Theorem 1.5.4 that Φ is an usco on (X, τ). This completes the proof. □

This characterisation of usconess has been exploited in functional analysis to answer some problems on the drop property, the weak* drop property, near uniform convexity and the differentiability of continuous convex functions defined on Banach spaces, see [94, 214, 215, 277, 278].

1.5.2 Fort's Theorem

Throughout this subsection we shall denote by $C(\Phi) := \{x \in X : \Phi \text{ is continuous at } x\}$. We shall be particularly interested in set-valued mappings that are defined on Baire spaces. So let us recall that a topological space (X, τ) is called a *Baire space* if, for every countable family $\{O_n : n \in \mathbb{N}\}$ of dense open subsets, $\bigcap_{n\in\mathbb{N}} O_n$ is dense in (X, τ). Importantly, it follows from the Baire Category Theorem (see Corollary 1.4.24) that all nonempty complete metric spaces are Baire spaces.

Exercise 1.5.7. Suppose that (X, τ) and (Y, τ') are topological spaces and $\Phi : X \to 2^Y$. Show that $\Phi : X \to 2^Y$ is lower semicontinuous on (X, τ) if, and only if, for each open subset W of (Y, τ'), $\{x \in X : \Phi(x) \cap W \neq \varnothing\}$ is an open subset of (X, τ).

Exercise 1.5.8. Suppose that (X, τ) and (Y, τ') are topological spaces and $\Phi : X \to 2^Y$. Show that $\Phi : X \to 2^Y$ is lower semicontinuous on (X, τ) if, and only if, for each closed subset C of (Y, τ'), $\{x \in X : \Phi(x) \subseteq C\}$ is a closed subset of (X, τ).

Exercise 1.5.9. Suppose that (X, τ) and (Y, τ') are topological spaces and $\Phi : X \to 2^Y$. Show that if $\Phi : X \to 2^Y$ is lower semicontinuous on (X, τ), then $\mathrm{Dom}(\Phi)$ is an open subset of (X, τ).

Example 1.5.10. Let $X := Y := \mathbb{R}$, let τ be the Sorgenfrey topology on $\mathbb{R}$ (i. e., each point x of $\mathbb{R}$ has a τ-neighbourhood base comprising of sets of the form: $(a, x]$ where $a < x$) and let $\rho : Y \times Y \to \{0, 1\}$ be the discrete metric on Y (i. e., $\rho(x, y) = 1$ if, and only if, $x \neq y$). Let us also consider the following set-valued mappings. $\Phi_1 : (X, \tau) \to 2^Y$ defined by $\Phi_1(x) := (-\infty, x]$ and $\Phi_2 : (X, \tau) \to 2^Y$ defined by $\Phi_2(x) = [x, \infty)$. Then (X, τ) is a Baire space, Φ_1 is upper semicontinuous, but $C(\Phi_1) = \varnothing$ and Φ_2 is lower semicontinuous, but $C(\Phi_2) = \varnothing$.

Although the notions of upper semicontinuity and lower semicontinuity are in general completely distinct there is an interesting relationship between these notions when the domain space of the mapping under consideration is a Baire space, the range space is metric and the mapping has nonempty compact images.

Let (X, τ) be a topological space and let (Y, d) be a metric space. We shall say that $\Phi : X \to 2^Y$ is *quasi-totally-bounded* if for every $0 < \varepsilon$ and every nonempty open subset U of (X, τ) there exists a nonempty τ-open subset V of U and a finite subset F of Y such that $\Phi(V) \subseteq B(F,\varepsilon)$. Here we are employing the notation: $B(A,\varepsilon) := \bigcup_{a\in A} B(a,\varepsilon)$ and $B[A,\varepsilon] := \bigcup_{a\in A} B[a,\varepsilon]$, where A is any nonempty subset of the metric space (Y,d).

A subset C of a topological space (X,τ) is called *comeager* or *residual* in (X,τ) if the complement $X \setminus C$ is of the first Baire category in (X,τ), or, equivalently, if C contains the intersection of a countable family of dense open subsets of (X,τ).

Proposition 1.5.11. *Let (X,τ) be a topological space, (Y,d) be a metric space and $\Phi : X \to 2^Y$ be a quasi-totally-bounded mapping. If Φ is upper semicontinuous (lower semicontinuous) on (X,τ) and possesses nonempty compact images, then there exists a residual subset R of (X,τ) such that Φ is continuous (i. e., both upper and lower semicontinuous) at each point of R.*

Proof. Fix $0 < \varepsilon$ and define $O_\varepsilon := \bigcup\{U \in \tau : \Phi(U) \subseteq B(\Phi(t),2\varepsilon) \text{ for all } t \in U\}$. Clearly, O_ε is open. We will also show that O_ε is dense in (X,τ). To this end, let W be a nonempty open subset of (X,τ). Let

$$m := \min\{|T| : T \text{ is any finite subset of } Y \text{ with the property that exists a nonempty}$$
$$\tau\text{-open subset } V \text{ of } W \text{ such that } \Phi(V) \subseteq B(T,\varepsilon)\}.$$

Since Φ is quasi-totally-bounded this number is well-defined. Next, let us choose $T \subseteq Y$ and a nonempty τ-open subset V of W such that $m = |T|$ and $\Phi(V) \subseteq B(T,\varepsilon)$. It follows from the definition of m that for each $t \in T$, $\{x \in V : \Phi(x) \cap B(t,\varepsilon) \neq \varnothing\}$ is dense in (V,τ'), where τ' denotes the relative τ-topology on the set V.

(i) If Φ is upper semicontinuous, then $\Phi(x) \cap B[t,\varepsilon] \neq \varnothing$ for all $(x,t) \in V \times T$ and so

$$\Phi(V) \subseteq B(T,\varepsilon) \subseteq B(B[\Phi(x),\varepsilon],\varepsilon) \subseteq B(\Phi(x),2\varepsilon) \quad \text{for all } x \in V.$$

Therefore, $\varnothing \neq V \subseteq O_\varepsilon \cap W$.

(ii) On the other hand, if Φ is lower semicontinuous, then there exists a nonempty τ-open subset V' of V such that $\Phi(x) \cap B(t,\varepsilon) \neq \varnothing$ for all $(x,t) \in V' \times T$ and so

$$\Phi(V') \subseteq \Phi(V) \subseteq B(T,\varepsilon) \subseteq B(B(\Phi(x),\varepsilon),\varepsilon) \subseteq B(\Phi(x),2\varepsilon) \quad \text{for all } x \in V'.$$

Therefore, $\varnothing \neq V' \subseteq O_\varepsilon \cap W$.

Hence, in both cases, O_ε is dense in (X,τ). It now only remains to see that Φ is continuous at each point of $R := \bigcap_{n\in\mathbb{N}} O_{\frac{1}{n}}$. □

Theorem 1.5.12 (Fort's Theorem, [85]). *Let (X,τ) be a Baire space, (Y,d) be a metric space and $\Phi : X \to 2^Y$ be an upper semicontinuous (lower semicontinuous) mapping. If Φ possesses nonempty compact images, then there exists a residual subset R of (X,τ) such that Φ is continuous (i. e., both upper and lower semicontinuous) at each point of R.*

Proof. In the case when Φ is upper semicontinuous on (X,τ) it is easy to show (and hence left as an exercise for the reader) that Φ is quasi-totally-bounded on (X,τ).

Hence, by Proposition 1.5.11, Φ is continuous on a residual subset of (X, τ). The case in which Φ is lower semicontinuous on (X, τ) is a little more delicate. So suppose that Φ is lower semicontinuous on (X, τ). Furthermore, suppose in order to obtain a contradiction, that Φ is not quasi-totally-bounded. Then there exists a nonempty open subset U of (X, τ) and an $0 < \varepsilon$ such that, for every nonempty τ-open subset V of U and every finite subset F of Y, $\Phi(V) \nsubseteq B(F, \varepsilon)$. For each $n \in \mathbb{N}$, let

$$F_n := \{x \in U : \max\{|T| : T \in S_\varepsilon(Y) \text{ and } T \subseteq B(\Phi(x), \varepsilon/3)\} \leq n\}.$$

Recall that $S_\varepsilon(Y) := \{A \in 2^Y : \text{ for every distinct } x, y \in A, \varepsilon \leq d(x, y)\}$. From the compactness of the images of Φ, it follows that for each $x \in X$,

$$\max\{|T| : T \in S_\varepsilon(Y) \text{ and } T \subseteq B(\Phi(x), \varepsilon/3)\}$$

is a finite number. By the lower semicontinuity of Φ, it follows that each set F_n is closed in U. Furthermore, $U := \bigcup_{n \in \mathbb{N}} F_n$. Hence, there exists a $k \in \mathbb{N}$ and a nonempty τ-open subset V_0 such that $V_0 \subseteq F_k$. We shall now inductively define a sequence $(x_n : n \in \mathbb{N})$ in X, a sequence $(y_n : n \in \mathbb{N})$ in Y and a sequence $(V_n : n \in \mathbb{N})$ of nonempty open subsets of (X, τ) such that:
(i) $V_n \subseteq V_{n-1}$ for all $n \in \mathbb{N}$;
(ii) $x_n \in V_{n-1}$ and $y_n \in \Phi(x_n)$ for all $n \in \mathbb{N}$;
(iii) $\{y_1, y_2, \ldots, y_n\} \in S_\varepsilon(Y)$ and $\{y_1, y_2, \ldots, y_n\} \subseteq B(\Phi(x), \varepsilon/3)$ for all $x \in V_n$.

Base Step. Let $x_1 \in V_0$, let $y_1 \in \Phi(x_1)$ and choose an open neighbourhood V_1 of x_1, contained in V_0, such that $\{y_1\} \subseteq B(\Phi(x), \varepsilon/3)$ for all $x \in V_1$. Note that this is possible since Φ is lower semicontinuous.

Now suppose that $m \in \mathbb{N}$ and $\{x_1, x_2, \ldots, x_m\}$, $\{y_1, y_2, \ldots, y_m\}$ and $\{V_1, V_2, \ldots, V_m\}$ have been defined so that:
(i) $V_j \subseteq V_{j-1}$ for all $1 \leq j \leq m$;
(ii) $x_j \in V_{j-1}$ and $y_j \in \Phi(x_j)$ for all $1 \leq j \leq m$;
(iii) $\{y_1, y_2, \ldots, y_m\} \in S_\varepsilon(Y)$ and $\{y_1, y_2, \ldots, y_m\} \subseteq B(\Phi(x), \varepsilon/3)$ for all $x \in V_m$.

Inductive Step. By assumption, $\Phi(V_m) \nsubseteq B(\{y_1, y_2, \ldots, y_m\}, \varepsilon)$. Therefore there exists $x_{m+1} \in V_m$ and $y_{m+1} \in \Phi(x_{m+1})$ such that $y_{m+1} \notin B(\{y_1, y_2, \ldots, y_m\}, \varepsilon)$. Then $\{y_1, y_2, \ldots, y_m, y_{m+1}\} \in S_\varepsilon(Y)$. By the lower semicontinuity of Φ there exists an open neighbourhood V_{m+1} of x_{m+1}, contained in V_m such that $\{y_1, y_2, \ldots, y_{m+1}\} \subseteq B(\Phi(x), \varepsilon/3)$ for all $x \in V_{m+1}$. This complete the induction. Now, when $n = k + 1$ we have a contradiction, since $V_n \subseteq V_0 \subseteq F_k$, $\{y_1, y_2, \ldots, y_n\} \in S_\varepsilon(Y)$ and $\{y_1, y_2, \ldots, y_n\} \subseteq B(\Phi(x), \varepsilon/3)$ for all $x \in V_n$. Hence, Φ must be quasi-totally-bounded on (X, τ). The result now follows from Proposition 1.5.11. □

Remarks 1.5.13. Let (Y, d) be a metric space and let $K(Y)$ denote the set of all nonempty compact subsets of (Y, d). For any nonempty subset A of Y we can always define the *distance function* $d(\cdot, A) : Y \to [0, \infty)$ by, $d(x, A) := \inf\{d(x, a) : a \in A\}$ for all $x \in Y$. Then we can define the *Hausdorff metric* D on $K(Y)$ by,

$$D(A, B) := \max\left\{\sup_{x\in A} d(x, B), \sup_{x\in B} d(x, A)\right\} \quad \text{for all } A, B \in K(Y).$$

Now, for any mapping $\Phi : (X, \tau) \to K(Y)$, emanating from a topological space (X, τ) we have that $C(\Phi) = \bigcap_{n\in\mathbb{N}} O_{1/n}$ where, for each $0 < \varepsilon$,

$$O_\varepsilon := \bigcup\{U \in \tau : D - \operatorname{diam}\{\Phi(t) : t \in U\} < \varepsilon\}.$$

Thus, we see that for any set-valued mapping $\Phi : (X, \tau) \to K(Y)$, $C(\Phi)$ is always a G_δ subset of (X, τ).

Exercise 1.5.14. Let (X, τ) be a topological space. Show that (X, τ) is a Baire space if, and only if, $\bigcap_{n\in\mathbb{N}} O_n$ is dense in (X, τ), for each decreasing sequence (i. e., $O_{n+1} \subseteq O_n$ for all $n \in \mathbb{N}$) of dense open subsets $(O_n : n \in \mathbb{N})$ of (X, τ), with $O_1 = X$. Hint: the family of all dense open subsets on a topological space (X, τ) is closed under finite intersections.

It is known that Fort's Theorem provides a characterisation for the class of Baire spaces in the following sense, (see [221, 313]).

Theorem 1.5.15. *Let (X, τ) be a topological space and (Y, d) be a metric space with at least one non-isolated point. Then (X, τ) is a Baire space if, and only if, for every upper semicontinuous mapping $\Phi : X \to K(Y)$, $C(\Phi)$ is dense in (X, τ).*

Proof. If (X, τ) is a Baire space, then $C(\Phi)$ is dense in (X, τ) by Fort's Theorem. So let us consider the converse. Let $(O_n : n \in \mathbb{N})$ be a decreasing sequence of dense open subsets of (X, τ), with $O_1 = X$. We need to show that $\bigcap_{n\in\mathbb{N}} O_n$ is dense in (X, τ). Let $y_\infty \in Y$ be a non-isolated point of (Y, d) and let $(y_n : n \in \mathbb{N})$ be a sequence of distinct points of $Y \setminus \{y_\infty\}$ that converge to y_∞. For each $n \in \mathbb{N}$, let $Y_n := \{y_k : n \le k\} \cup \{y_\infty\}$.

Define $\Phi : X \to K(Y)$ by,

$$\Phi(x) := \begin{cases} \{y_\infty\} & \text{if } x \in \bigcap_{n\in\mathbb{N}} O_n \\ Y_n & \text{if } x \in O_n \setminus O_{n+1}. \end{cases}$$

Then Φ is upper semicontinuous on (X, τ) and $C(\Phi) = \bigcap_{n\in\mathbb{N}} O_n$. Therefore, if $C(\Phi)$ is dense in (X, τ), then $\bigcap_{n\in\mathbb{N}} O_n$ is dense in (X, τ); which implies that (X, τ) is a Baire space. □

Theorem 1.5.16. *Let (X, τ) be a topological space and (Y, d) be a metric space with at least one non-isolated point. Then (X, τ) is a Baire space if, and only if, for every lower semicontinuous mapping $\Phi : X \to K(Y)$, $C(\Phi)$ is dense in (X, τ).*

Proof. If (X, τ) is a Baire space, then $C(\Phi)$ is dense in (X, τ) by Fort's Theorem. So let us consider the converse. Let $(O_n : n \in \mathbb{N})$ be a decreasing sequence of dense open subsets of (X,τ), with $O_1 = X$. We need to show that $\bigcap_{n\in\mathbb{N}} O_n$ is dense in (X,τ). Let $y_\infty \in Y$ be a non-isolated point of (Y,d) and let $(y_n : n \in \mathbb{N})$ be a sequence of distinct points of $Y \setminus \{y_\infty\}$ that converge to y_∞. For each $n \in \mathbb{N}$, let $Y_n := \{y_k : 1 \le k \le n\}$ and let $Y_\infty := \{y_k : k \in \mathbb{N}\} \cup \{y_\infty\}$.

Define $\Phi : X \to K(Y)$ by,

$$\Phi(x) := \begin{cases} Y_\infty & \text{if } x \in \bigcap_{n\in\mathbb{N}} O_n \\ Y_n & \text{if } x \in O_n \setminus O_{n+1}. \end{cases}$$

Then Φ is lower semicontinuous on (X,τ) and $C(\Phi) = \bigcap_{n\in\mathbb{N}} O_n$. Therefore, if $C(\Phi)$ is dense in (X,τ), then $\bigcap_{n\in\mathbb{N}} O_n$ is dense in (X,τ); which implies that (X,τ) is a Baire space. □

In order to see an interesting application of Fort's Theorem to minimal uscos we will need the following general fact concerning minimal uscos and lower semicontinuity.

Proposition 1.5.17. *Suppose that (X,τ) and (Y,τ') are topological spaces and $\Phi : X \to 2^Y$ is a minimal usco on (X,τ). If (Y,τ') is a Hausdorff space and Φ is lower semicontinuous at $x_0 \in X$, then $\Phi(x_0)$ is a singleton.*

Proof. Suppose that (X,τ) and (Y,τ') are topological spaces and $\Phi : X \to 2^Y$ is a minimal usco on (X,τ). Suppose also that Φ is lower semicontinuous at $x_0 \in X$. We wish to show that $\Phi(x_0)$ is a singleton. Next, let us assume, in order to obtain a contradiction, that $\Phi(x_0)$ is not a singleton. Then there exist distinct points $y_1, y_2 \in \Phi(x_0)$. Since (Y,τ') is Hausdorff there exist disjoint open neighbourhoods W_1 of y_1 and W_2 of y_2. Now, because Φ is lower semicontinuous at x_0, there exists an open neighbourhood U of x_0 such that $\Phi(x) \cap W_1 \neq \varnothing$ and $\Phi(x) \cap W_2 \neq \varnothing$ for all $x \in U$. On the other hand, by Proposition 1.3.5, there exists a nonempty τ-open subset V of U such that $\Phi(V) \subseteq W_1$. However,

$$\varnothing \neq \Phi(x) \cap W_2 \subseteq W_1 \cap W_2 = \varnothing \quad \text{for all } x \in V \neq \varnothing;$$

which is impossible. Therefore, $\Phi(x_0)$ must be a singleton. □

Corollary 1.5.18. *Let $\Phi : X \to 2^Y$ be a minimal usco acting from a Baire space (X,τ) into subsets of a metric space (Y,d). Then Φ is single-valued at the points of a dense G_δ subset of (X,τ).*

Proof. This follows directly from Theorem 1.5.12 and Proposition 1.5.17. □

It is interesting to see that by Example 1.3.1, we cannot say that Φ is single-valued *almost everywhere*, (i. e., everywhere except for a set of measure zero), even in the case when $X = \mathbb{R}$, endowed with the usual topology and $Y = \mathbb{R}$, also endowed with the usual topology.

1.6 Exercises and commentary

It is difficult to say when the notion of "upper semicontinuity" first appeared. In [73] it is claimed "Historically, minimal usco mappings seem to have appeared first in complex analysis (in the second half of the 19th century) in the form of a bounded holomorphic function and its *cluster sets* see, [64]." What we do know is that in [177] C. Kuratowski considered what we now call usconess, in the special case of set-valued mappings whose range is a compact metric space. For the notion of upper semi-continuity there seem to have been many names: G. Choquet [58] called them *strongly upper semi-continuous*, Strother [297] *weakly continuous*, Ponomarev [263] *continuous*, Michael [206] and others used the term *upper semicontinuous*. In some other places, for example in economics, they use the name *upper hemi-continuous*. In the setting when the range space is a metric space (M, d) there is the related notion of *metric upper semicontinuous* or *Hausdorff upper semicontinuous*. That is, a set-valued mapping $\Phi : X \to 2^M$ from a topological space (X, τ) into nonempty subsets of a metric space (M, d) is *metrically upper semicontinuous at a point* $x_0 \in X$ (or *Hausdorff upper semicontinuous at a point* $x_0 \in X$) if for each $0 < \varepsilon$ there exists an open neighbourhood U of x_0 such that $\Phi(U) \subseteq \bigcup\{B(m, \varepsilon) : m \in \Phi(x_0)\}$. Finally, if the range space is a linear topological space $(V, +, \cdot, \tau')$ then we call a set-valued mapping $\Phi : X \to 2^V$ *Hausdorff upper semicontinuous at a point* $x_0 \in X$ if for each open neighbourhood W of $0 \in V$, there exists an open neighbourhood U of x_0 such that $\Phi(U) \subseteq \Phi(x_0) + W$. Of course, if $\Phi(x_0)$ is nonempty and compact then both metric upper semicontinuous and Hausdorff upper semicontinuity (both meanings) coincide with the upper semicontinuity at x_0. If $\Phi(x_0)$ is not compact, then these notions may be weaker.

As with the origins of the concept/name of upper semicontinuity it is not easy to trace the exact origins of many of the basic facts concerning upper semicontinuous and usco mappings. So, for the most part, we have either, not given a reference, or else, we have given a reference that is easy to access. In this direction, two extremely useful monographs that contain a considerable volume of information on the basic theory of set-valued mappings are [24, 28].

In Subsection 1.1.2 we presented a selection of examples of usco mappings. This list is in no way purporting to be an exhaustive list, but merely a collection of examples that are, or have been, of interest to the authors.

Subsection 1.1.3 presents some of the basic constructions of usco mappings. In most cases the proofs are straightforward and are left as exercises for the reader.

After having established the basic properties, examples and constructions we present some classical fixed-point theorem for set-valued mappings. We acknowledge that this section is not self-contained. Indeed, all the results in Section 1.2 rely upon Brouwer's Fixed-Point Theorem, which does not currently appear to have a proof that is both short and simple, and elementary. Having said that, there are many proofs of Brouwer's Fixed-Point Theorem in the literature that are in some way, depending upon your background, more accessible than the standard proof that relies upon algebraic topology, [194]. For example, if you have a background in combinatorics, then the proof using Sperner's Lemma is perhaps more accessible, [286, 298], whereas if you have a background in analysis, then there are several proofs that rely upon the change of variable formula for integration [180, 275].

From Brouwer's Fixed-Point Theorem we prove Schauder's Fixed-Point Theorem and then the Kakutani–Glicksberg–Fan Fixed-Point Theorem. The finite dimensional version of the Kakutani–Glicksberg–Fan Fixed-Point Theorem is called the "Kakutani Fixed-Point Theorem". We then use the Kakutani Fixed-Point Theorem to prove the "Minimax Theorem" which has applications to the theory of zero-sum games, [155].

The mathematician John Nash used the Kakutani Fixed-Point Theorem to prove a major result in game theory, [247]. Stated informally, the theorem implies the existence of a Nash equilibrium in every finite game with mixed strategies for any number of players. This work later earned Nash a Nobel Prize in Economics.

In general equilibrium theory in economics, Kakutani's theorem has been used to prove the existence of a set of prices which simultaneously equate supply with demand in all markets of an economy, [288]. The existence of such prices had been an open question in economics dating back to, at least Walras. The first proof of this result was constructed by Lionel McKenzie in [202]. However, another proof may be found in [8].

Kakutani's Fixed-Point Theorem can also be used for proving the existence of cake allocations that are "fair". This result is known as Weller's Theorem, [311].

Fair cake-cutting has been studied since the 1940s. There is a heterogeneous divisible resource, such as a cake or a land-estate. There are n partners, each of whom has a personal value-density function over the cake. The value of a piece to a partner is the integral of his value-density over that piece (this means that the value is a nonatomic measure over the cake). The envy-free cake-cutting problem is to partition the cake to n disjoint pieces, one piece per agent, such that for each agent, the value of his piece is greater than, or equal to, the values of all the other pieces (so no agent envies another agent's share).

Following on from the section on fixed-point theorems we considered "minimal uscos". In this section we provided the fundamental results behind the theory of minimal uscos.

Then, in Section 1.4, we considered several selection theorems. We note however, that we presented only a very small cross section of this vast area of research and some very important results were not even mentioned. For example: the Kuratowski-Ryll Nardzewski selection theorem [178], which gives the existence of measurable selectors for certain mappings that map into separable metric spaces, (for an up-to-date description of this area of research see [54]); results on the existence of quasicontinuous selections, [53, 190]; results on the existence of densely defined selections [63, 95, 172, 232], etc.

Finally, we considered uscos mappings in the setting of metric spaces. Here, the most well-known result is Fort's Theorem [85]. Since this result first appeared it has undergone several generalisations, see for example, [170, 221, 313].

Exercises

1. (Sequential characterisation of uscos) Let $\Psi : X \to 2^Y$ be a densely defined set-valued mapping acting from a first countable space (X, τ_X) into subsets of a topological space (Y, τ_Y), and let $x_0 \in X$. Suppose that every countably compact subset of (Y, τ_Y) is compact. Prove that the following are equivalent:
   (i) Ψ is a τ_Y-usco at $x_0 \in X$;
   (ii) every sequence $(y_n : n \in \mathbb{N})$ in Y with $y_n \in \Psi(x_n)$ for all $n \in \mathbb{N}$, for some sequence $(x_n : n \in \mathbb{N})$ in X, converging to x_0, has a τ_Y-cluster point in $\Psi(x_0)$.

   If one wishes to provide a convergence-type characterisation of usconess in a more general setting, then one is obliged to use nets.
2. Prove the following statement. Suppose that (X, τ) and (Y, τ') are topological spaces and $\Phi : X \to 2^Y$. If (Y, τ') is compact and Φ has a closed graph, then Ω is upper semicontinuous on (X, τ). *Hint:* First use Corollary 1.1.15 to show that if $A := \overline{\mathrm{Dom}(\Phi)}$, then $\Phi|_A : A \to 2^Y$ is an usco on A, with the relative τ-topology. Then note that Φ is trivially upper semicontinuous at each point of $X \setminus A$ and then finally note that Φ is in fact upper semicontinuous at each point of A.
3. Verify that all the examples in Section 1.1.2 are indeed uscos.
4. Provide a proof for all the constructions of uscos given is Section 1.1.3
5. Suppose that (X, τ) and (Y, τ') are topological spaces and $\Phi : X \to 2^Y$. Suppose also that Φ is upper semicontinuous on (X, τ) and has connected images (i. e., $\Phi(x)$ is a connected subset of (Y, τ') for each $x \in X$). Show that if (X, τ) is connected, then $\Phi(X)$ is a connected subset of (Y, τ').
6. Suppose that (X, τ) and (Y, τ') are topological spaces and $\Phi : X \to 2^Y$ is an usco on (X, τ).
   (a) Show that $\tilde{\Phi} : X \to 2^{X \times Y}$ defined by, $\tilde{\Phi}(x) := \{x\} \times \Phi(x)$ for all $x \in X$, is an usco on (X, τ).
   (b) Show moreover that if (X, τ) is compact, then $\mathrm{Gr}(\Phi) = \tilde{\Phi}(X)$ is a compact subset of $X \times Y$, endowed with the product topology. *Hint:* Consider Proposition 1.1.7.
   (c) Finally, deduce that if (Y, τ') is also (nonempty) compact, then $X \times Y$ is compact by considering the usco mapping $\Phi : X \to 2^Y$ defined by, $\Phi(x) := Y$ for all $x \in X$.
7. Suppose that (X, τ) and (Y, τ') are topological spaces and $\Phi : X \to 2^Y$. Show that if Φ has connected images, then $\tilde{\Phi} : X \to 2^{X \times Y}$ defined by, $\tilde{\Phi}(x) := \{x\} \times \Phi(x)$ has connected images. Hence show that if $\Phi : X \to 2^Y$ is an usco with connected images and (X, τ) is connected, then $\mathrm{Gr}(\Phi) = \tilde{\Phi}(X)$ is a connected subset of $X \times Y$.

Hint: use your answers to Questions 5 and 6. Finally, show that

$$\{0\} \times [-1,1] \cup \{(x, \sin(1/x)) : x \in \mathbb{R} \setminus \{0\}\}$$

is a connected subset of $\mathbb{R}^2$ by considering the mapping $\Phi : \mathbb{R} \to 2^{\mathbb{R}}$ be defined by,

$$\Phi(x) := \begin{cases} \sin(1/x) & \text{if } x \neq 0 \\ [-1,1] & \text{if } x = 0. \end{cases}$$

8. Suppose that (X,τ) and (Y,τ') are topological spaces and $\Phi : X \to 2^Y$ is an usco on (X,τ). Show that if K is a Lindelöf subset of (X,τ), then $\Phi(K)$ is a Lindelöf subset of (Y,τ').
9. Suppose that (X,τ) and (Y,τ') are topological spaces and $\Phi : X \to 2^Y$ is an usco on (X,τ). Show that if K is a closed subset of (Y,τ') and $C := \{x \in X : \Phi(x) \cap K \neq \varnothing\}$ is nonempty, then $\Phi_C : C \to 2^K$ defined by, $\Phi_C(x) := \Phi(x) \cap K$ for all $x \in C$ is an usco on (C,τ''). Here τ'' denotes the relative τ-topology on the set C. *Hint:* see Section 1.1.2 Example (ii) and Theorem 1.1.11.
10. Suppose that (X,τ) and (Y,τ') are topological spaces and $f : X \to Y$ is a continuous function. Show that f is a perfect mapping onto (Y,τ') if, and only if, $f^{-1} : Y \to 2^X$ defined by, $f^{-1}(y) := \{x \in X : f(x) = y\}$ for all $y \in Y$, is an usco on (Y,τ').
11. Suppose that (X,τ) and (Y,τ') are topological spaces and $\Phi : X \to 2^Y$ is an usco on (X,τ). Show that $f : \mathrm{Gr}(\Phi) \to X$ defined by, $f(x,y) := x$ for all $(x,y) \in \mathrm{Gr}(\Phi)$, is a perfect mapping onto (X,τ). *Hint:* use the fact that $f^{-1}(x) = \{x\} \times \Phi(x)$ for all $x \in X$, Exercise 6 part(a) and Exercise 10.
12. Prove the Kakutani–Glicksberg–Fan Fixed-Point Theorem from Kakutani's Fixed-Point Theorem which says "If K is a nonempty compact convex subset of a finite dimensional normed linear space and $\Phi : K \to 2^K$ is a cusco, then Φ has a fixed-point". *Hint:* For each convex symmetric closed neighbourhood N of 0, let K_N be a finite dimensional compact convex subset of K such that $K \subseteq K_N + N$ and consider the mapping $\Phi_N : K_N \to 2^{K_N}$ defined by, $\Phi_N(k) := (\Phi(k) + N) \cap K_N$. Then apply Corollary 1.1.19, to deduce that Φ_N is a cusco.
13. Show that the mappings Φ and Ψ in Example 1.3.1 are indeed minimal uscos.
14. Suppose that (X,τ) and (Y,τ') are topological spaces and $\Phi : X \to 2^Y$ is a minimal usco on (X,τ). Show that if (X,τ) is a separable space, then $\Phi(X)$ is a separable subspace of (Y,τ').
15. Suppose that (X,τ) and (Y,τ') are topological spaces and $\Phi : X \to 2^Y$ is a minimal usco on (X,τ). Show that if (X,τ) satisfies the countable chain condition, then so does $\Phi(X)$, with the relative topology inherited from (Y,τ'). Recall that a topological space (X,τ) is said to satisfy the *countable chain condition* if it does not possess an uncountable family of nonempty disjoint open subsets.

16. Suppose that (X,τ) and (Y,τ') are Hausdorff topological spaces and $f : X \to Y$ is a perfect mapping. Show that if $\Phi : A \to 2^Y$ is a minimal usco defined on a topological space (A,τ''), then there exists a minimal usco $\Psi : A \to 2^X$ such that $\Phi = f \circ \Psi$.
17. Suppose that (X,d) and (Y,ρ) are metric spaces and $\Phi : X \to 2^Y$. We will say that Φ is *uniformly upper semicontinuous on* (X,d) if for every $0 < \varepsilon$ there exists a $0 < \delta$ such that, for every $x,y \in X$, if $d(x,y) < \delta$, then $\Phi(x) \subseteq B(\Phi(y),\varepsilon)$. Show that every uniformly upper semicontinuous, minimal usco acting between metric spaces (X,d) and (Y,ρ) is everywhere single-valued on X.
18. Suppose that (X,τ) is a first countable topological space, $(Y,\|\cdot\|)$ is a Banach space and $\Psi : X \to 2^{Y^*}$ is densely defined. Show that Ψ can be embedded in a weak* usco if, and only if, Ψ is *locally bounded on* X (i.e., for each $x \in X$, there exists a neighbourhood U of x and an $0 < r$ such that $\Psi(U) \subseteq rB_{Y^*}$). *Hint:* Recall that a weak* compact subset of the dual of a Banach space is bounded.
19. Suppose that (X,τ) and (Y,τ') are topological spaces and $\Phi : X \to 2^Y$ is an usco on (X,τ). Suppose also that $f : Y \to \mathbb{R}$ is a lower semicontinuous function.
    (a) Show that $\varphi : X \to \mathbb{R}$ defined by, $\varphi(x) := \min\{f(y) : y \in \Phi(x)\}$ for each $x \in X$, is well-defined and lower semi-continuous on (X,τ).
    (b) Show that if φ is continuous at a point $x \in X$ and Φ is a minimal usco, then $\Phi(x) \subseteq f^{-1}(\varphi(x))$.
    (c) Show that if (X,τ) is a Baire space and Φ is a minimal usco, then $f \circ \Phi$ is single-valued at the points of a dense G_δ subset of (X,τ). *Hint:* Recall that a lower semicontinuous function from a Baire space into $\mathbb{R}$ is continuous at the points of a dense G_δ subset of its domain, see Lemma 3.3.28.
20. Let (X,τ) be a compact Hausdorff space and let $(C(X),\|\cdot\|_\infty)$ be the Banach space of all real-valued continuous functions defined on (X,τ), endowed with the supremum norm. Use Michael's Selection Theorem to show that if Y is a nonempty closed subset of (X,τ), then there exists a norm-to-norm continuous function $s : C(Y) \to C(X)$ such that $s(f)|_Y = f$ for every $f \in C(Y)$.
21. Let $\{O_\alpha : \alpha \in A\}$ be a point-finite open cover of a Baire space (X,τ). Show that there exists a dense open subset of (X,τ) on which $\{O_\alpha : \alpha \in A\}$ is locally-finite. Recall that a cover $\{O_\alpha : \alpha \in A\}$ of a set X is *point-finite* if for each $x \in X$, $\{\alpha \in A : x \in O_\alpha\}$ is a finite set and a cover $\{O_\alpha : \alpha \in A\}$ of a topological space (X,τ) is *locally-finite* if for each $x \in X$, there exists an open neighbourhood U of x such that $\{\alpha \in A : O_\alpha \cap U \neq \varnothing\}$ is a finite set. *Hint:* Consider the set A endowed with the discrete metric and apply Fort's Theorem to the mapping $\Phi : X \to 2^A$ defined by, $\Phi(x) := \{\alpha \in A : x \in O_\alpha\}$ for all $x \in X$.

2 Quasicontinuity

2.1 Introduction

As far as we know, the first mention of the condition of quasicontinuity can be found in the paper of Baire [12] in the study of continuity points of separately continuous functions from $\mathbb{R}^2$ into $\mathbb{R}$.

A function $f : \mathbb{R}^2 \to \mathbb{R}$ is said to be *separately continuous* on $\mathbb{R}^2$ if for all $x_0 \in X$ and $y_0 \in Y$, the functions $y \mapsto f(x_0, y)$ and $x \mapsto f(x, y_0)$ are both continuous on $\mathbb{R}$.

In 1899, Baire [12] proved the following result. Let $f : \mathbb{R}^2 \to \mathbb{R}$ be a separately continuous function. Then f has the following property:

(QC) For every point $(x_0, y_0) \in \mathbb{R}^2$, for every disc K centred at (x_0, y_0) and for every $\varepsilon > 0$, there is a disc $K_1 \subseteq K$ such that $|f(x, y) - f(x_0, y_0)| < \varepsilon$ for every $(x, y) \in K_1$.

Baire [12, p. 95] also indicated that condition (QC) was suggested to him by Vito Volterra. Condition (QC) is the property of quasicontinuity of $f : \mathbb{R}^2 \to \mathbb{R}$ at (x_0, y_0).

However, the notion of quasicontinuity was not formally defined until 1932, when Kempisty introduced this notion for real-valued functions of real variables (see [161]). Kempisty [161] mentioned that the notion of quasicontinuity has its origin in the paper of Hahn [101], which seems to indicate that Kempisty was unaware of the earlier paper of Baire [12]. Therefore the official birth date of the notion of quasicontinuity seems to be 1899.

2.2 Quasicontinuous functions

We start this section by recalling the definition, given earlier on page 23, of a quasicontinuous mapping.

Let (X, τ) and (Y, τ') be topological spaces, and let $f : X \to Y$. If $x_0 \in X$, then f is *quasicontinuous at* $x_0 \in X$ if for every open neighbourhood U of x_0 in (X, τ) and for every open neighbourhood W of $f(x_0)$ in (Y, τ'), there is a nonempty τ-open set $V \subseteq U$ such that $f(x) \in W$ for every $x \in V$. The function $f : X \to Y$ is *quasicontinuous on* (X, τ) if it is quasicontinuous at every $x \in X$ [187, 188].

The following equivalent formulation of quasicontinuity (for metric spaces) was introduced by Bledsoe [30], who used the notion of "neighbourly function".

Let (X, d) and (Y, ρ) be metric spaces, and let $f : X \to Y$. If $x_0 \in X$, then the function f is *neighbourly* at x_0 if for every $\varepsilon > 0$, there is a nonempty d-open subset V of X such that $d(x, x_0) + \rho(f(x), f(x_0)) < \varepsilon$ for every $x \in V$ [30].

Marcus [187] proved that the notions of neighbourly and quasicontinuous are equivalent.

Another equivalent formulation of quasicontinuity can be found in the paper of Levine [185].

https://doi.org/10.1515/9783110750188-002

Levine introduced the notion of a semi-continuous function using the notion of a semi-open set. We say that a subset A of a topological space (X, τ) is *semi-open* or *quasi-open* if $A \subseteq \overline{\operatorname{int}(A)}$, [249].

Let (X, τ) and (Y, τ') be topological spaces, and let $f : X \to Y$. The function f is *semi-continuous* if $f^{-1}(V)$ is semi-open in (X, τ) for every open set V of (Y, τ') [185].

A collection $\mathcal{G}$ of nonempty open subsets of a space (X, τ) is called a *π-base*,[201] (also known as a *pseudo-base*) for (X, τ), provided that every nonempty open subset of (X, τ) contains some element of $\mathcal{G}$. There is a characterisation of quasicontinuity in terms of π-bases.

Proposition 2.2.1. *Let (X, τ) and (Y, τ') be topological spaces, and let $f : X \to Y$. Then f is quasicontinuous if and only if there exists a π-base τ^* for (X, τ) such that whenever $U \in \tau$ and $V \in \tau'$ with $f(U) \cap V \neq \varnothing$, there exists $T \in \tau^*$ with $T \subseteq U$ such that $f(T) \subseteq V$.*

Proof. Suppose first that $f : X \to Y$ is quasicontinuous. Put

$$\tau^* = \{\operatorname{int}(f^{-1}(V) \cap U) : f^{-1}(V) \cap U \neq \varnothing, V \in \tau', U \in \tau\}.$$

To prove that τ^* is a π-base for (X, τ), let $U \in \tau$. Take $x \in U$. Let $V \in \tau'$ be such that $f(x) \in V$. The quasicontinuity of f at x implies that there is a nonempty open set H in (X, τ) such that $H \subseteq U$ and $f(H) \subseteq V$. Then $H \subseteq f^{-1}(V) \cap U$. Thus $T = \operatorname{int}(f^{-1}(V) \cap U) \neq \varnothing$ and $T \subseteq U$. Thus τ^* is a π-base for (X, τ). It is easy to verify that τ^* satisfies the condition of the proposition.

Suppose now that $f : X \to Y$ is a function such that there is a π-base τ^* for (X, τ) such that whenever U is open in (X, τ) and V is open in (Y, τ') with $f(U) \cap V \neq \varnothing$, there exists $T \in \tau^*$ with $T \subseteq U$ such that $f(T) \subseteq V$. We show that f is quasicontinuous. Let $x \in X$. Let U be an open neighbourhood of x in (X, τ), and let V be an open neighbourhood of $f(x)$ in (Y, τ'). By the assumption there is $T \in \tau^*$ such that $T \subseteq U$ and $f(T) \subseteq V$. Since T is open in (X, τ), the quasicontinuity of f at x is proved. □

Let (X, τ) and (Y, τ') be topological spaces. It is easy to see that if a function $f : X \to Y$ is quasicontinuous and $U \in \tau$, then the restriction $f|_U$ of the function f to U is quasicontinuous (see Exercise 7). Of course, this is not true for closed subsets of (X, τ).

We have the following interesting theorem.

Theorem 2.2.2 ([188]). *Let (X, τ) and (Y, τ') be topological spaces, and let $f : X \to Y$ be a function whose restriction to each closed subset of (X, τ) is quasicontinuous. Then f is continuous on (X, τ).*

Proof. We will prove the contrapositive statement. Suppose f is not continuous on (X, τ). Then there is $x \in X$ such that f is not continuous at x. Therefore there exists a neighbourhood W of $f(x)$ in (Y, τ') such that for every neighbourhood U of x in (X, τ), there is $x_U \in U$ such that $f(x_U) \notin W$. Put

$$E = \{x_U : U \text{ is a neighbourhood of } x \text{ in } (X, \tau) \text{ and } f(x_U) \notin W\}.$$

We claim that $f|_{\overline{E}}$ is not quasicontinuous at x. To this end, let U be any neighbourhood of x in $(\overline{E}, \tau'')$, where τ'' denotes the relative τ-topology on $\overline{E}$, and let V be a nonempty τ''-open subset of U. Then there exists $z \in V \cap E$, since E is dense in $(\overline{E}, \tau'')$. However, $f(z) \notin W$, which shows that $f|_{\overline{E}}$ is not quasicontinuous at x. □

Quasicontinuous functions were used by Kempisty [161] to extend some results of Hahn and Baire to real-valued functions of several real variables that are continuous in each variable separately. Martin [188] showed that some of the results of Kempisty hold for more general spaces.

Let X, Y, Z be sets, and let $f : X \times Y \to Z$. For $x \in X$, f_x denotes the function from Y to Z defined by

$$f_x(y) := f(x,y) \quad \text{for } y \in Y.$$

Similarly, for $y \in Y$, f^y denotes the function from X to Z defined by

$$f^y(x) := f(x,y) \quad \text{for } x \in X.$$

If (X,τ), (Y,τ') and (Z,τ'') are topological spaces, then we say that f is *separately continuous at* $(x_0, y_0) \in X \times Y$ if $f_{x_0} : (Y,\tau') \to (Z,\tau'')$ is continuous at y_0 and $f^{y_0} : (X,\tau) \to (Z,\tau'')$ is continuous at x_0. If f_x is continuous on (Y,τ') and f^y is continuous on (X,τ) for every $(x,y) \in X \times Y$, then we say that f is *separately continuous on* $X \times Y$.

Exercise 2.2.3. Let (X,τ) be a topological space. We say that a subset F of (X,τ) is *nowhere dense* in X if $\operatorname{int}(\overline{F}) = \varnothing$. Further, we say that a subset F of X is a *first category* subset of (X,τ) (or is of the *first Baire category* in (X,τ)) if there exists a countable family $\{F_n : n \in \mathbb{N}\}$ of nowhere dense subsets of (X,τ) such that $F \subseteq \bigcup_{n\in\mathbb{N}} F_n$. A subset S of (X,τ) is a *second category* subset of (X,τ) (or of the *second Baire category* in (X,τ)) if S is not of the first Baire category in (X,τ). Show that if (X,τ) is a topological space, then:
(i) $\varnothing$ is a nowhere dense subset of (X,τ);
(ii) if $F' \subseteq F$ and F is of the first Baire category in (X,τ), then so is F';
(iii) if $\{F_n : n \in \mathbb{N}\}$ are first category sets, then so is $\bigcup_{n\in\mathbb{N}} F_n$;
(iv) (X,τ) is a Baire space if and only if every nonempty open subset U of (X,τ) is a second category subset of (X,τ).

Theorem 2.2.4 ([188]). *Let (X,τ) be a Baire space, let (Y,τ') be second countable, and let (Z,ρ) be a metric space. If $f : X \times Y \to Z$ is a function such that f_x and f^y are quasicontinuous for all $x \in X$ and $y \in Y$, then f is a quasicontinuous function from $(X \times Y, \tau \times \tau')$ to (Z,ρ).*

Proof. Suppose there is a point $(p,q) \in X \times Y$ such that f is not quasicontinuous at (p,q). Then there are $\varepsilon > 0$ and open sets U and V in (X,τ) and (Y,τ'), respectively, with $p \in U$ and $q \in V$ such that for every nonempty open set $G \subseteq U \times V$ in $(X \times Y, \tau \times \tau')$, there is a point $(x, y) \in G$ such that

$$\varepsilon \leq \rho(f(x,y), f(p,q)).$$

Since f_q is quasicontinuous at p, there is a nonempty open set $W \subseteq U$ in (X, τ) such that for all $x \in W$,

$$\rho(f(x,q), f(p,q)) < \varepsilon/3.$$

Let $\mathcal{V}$ be a countable base for (Y, τ), and let $\{V_n : n \in \mathbb{N}\}$ be those elements from $\mathcal{V}$ that are contained in V. For each positive integer n, let

$$A_n := \{x \in W : \rho(f(x,y), f(x,q)) < \varepsilon/3 \text{ for all } y \in V_n\}.$$

Let $x \in W$. Since f_x is quasicontinuous at q, there is a nonempty open set $E \subseteq V$ in (Y, τ') such that

$$\rho(f(x,y), f(x,q)) < \varepsilon/3 \quad \text{for all } y \in E.$$

But there is $k \in \mathbb{N}$ such that $V_k \subseteq E$. Thus $x \in A_k$, and

$$W = \bigcup_{n \in \mathbb{N}} A_n.$$

Let W' be any open subset of W in (X, τ), and let $n \in \mathbb{N}$. Then $W' \times V_n \subseteq U \times V$, and there is a point $(x', y') \in W' \times V_n$ such that

$$\varepsilon \leq \rho(f(x', y'), f(p,q)).$$

Since $f^{y'}$ is quasicontinuous at x', there is a nonempty open set $W'' \subseteq W'$ in (X, τ) such that for all $x \in W''$,

$$\rho(f(x, y'), f(x', y')) < \varepsilon/3.$$

Let $x \in W''$. Then by applying the triangle inequality twice we have that

$$\rho(f(x', y'), f(p,q)) \leq \rho(f(x', y'), f(x, y')) + \rho(f(x, y'), f(x,q)) + \rho(f(x,q), f(p,q)).$$

Therefore

$$\rho(f(x', y'), f(p,q)) - \rho(f(x', y'), f(x, y')) - \rho(f(x,q), f(p,q)) \leq \rho(f(x, y'), f(x,q)).$$

However,

$$\varepsilon/3 < \rho(f(x', y'), f(p,q)) - \rho(f(x', y'), f(x, y')) - \rho(f(x,q), f(p,q)),$$

and so $\varepsilon/3 < \rho(f(x, y'), f(x,q))$. Since $y' \in V_n$, $x \notin A_n$, and we have $W'' \cap A_n = \emptyset$. Therefore A_n is nowhere dense, and W is of the first category. A contradiction. □

Neubrunn [248] generalised Theorem 2.2.4 by relaxing the hypothesis on Z from being metrizable to being regular. Bouziad and Troallic [47] further generalized Theorem 2.2.4 by relaxing the assumption that (Y, τ') is second countable to the assumption that every $y \in Y$ has a neighbourhood in (Y, τ') with a countable pseudo-base.

Corollary 2.2.5. *Let (X_i, τ_i), $i = 1, 2, \ldots, n$, be second countable Baire spaces, let (Z, ρ) be a metric space, and let $f : X_1 \times X_2 \times \cdots \times X_n \to (Z, \rho)$ be a function such that f_x^i is quasicontinuous for all $x \in X_1 \times \cdots \times X_{i-1} \times X_{i+1} \cdots \times X_n$, $i = 1, 2, \ldots, n$, where $f_x^i : X_i \to Z$ is defined by*

$$f_x^i(y) := f(x_1, \ldots, x_{i-1}, y, x_{i+1}, \ldots, x_n) \quad \text{for } y \in X_i$$

and $x := (x_1, \ldots, x_{i-1}, x_{i+1}, \ldots, x_n) \in X_1 \times \cdots \times X_{i-1} \times X_{i+1} \cdots \times X_n$. Then f is a quasicontinuous function from $(X_1 \times X_2 \times \cdots \times X_n, \tau_1 \times \tau_2 \times \cdots \times \tau_n)$ to (Z, ρ).

The converse of Theorem 2.2.4 is not true as the following example shows.

Example 2.2.6. Let $X := Y := (0, 1)$, and let e be the usual Euclidean topology on $(0, 1)$. Let $Z := \{0, 1\}$, and let d_0 be the 0–1-metric on Z. Let $f : X \times Y \to Z$ be the function defined by

$$f(x, y) := \begin{cases} 1 & \text{if } 0 < x < 1/2,\ 0 < y < 1, \\ 0 & \text{if } 1/2 < x < 1,\ 0 < y < 1, \\ 1 & \text{if } x = 1/2,\ y \text{ rational in } (0, 1), \\ 0 & \text{if } x = 1/2,\ y \text{ irrational in } (0, 1). \end{cases}$$

Then f is a quasicontinuous function from $(X \times Y, e \times e)$ to (Z, d_0), but $f_{1/2}$ defined by $f_{1/2}(y) = f(1/2, y)$ for $y \in Y$ is not a quasicontinuous function from (Y, e) to (Z, d_0).

For functions defined in product spaces, there is a stronger notion of quasicontinuity than the normal definition of quasicontinuity.

Let (X, τ), (Y, τ') and (Z, τ'') be topological spaces. We say that a function $f : X \times Y \to Z$ is *strongly quasicontinuous, with respect to the second variable, at* $(x, y) \in X \times Y$ if for each neighbourhood W of $f(x, y)$ in (Z, τ'') and each open neighbourhood U of x in (X, τ), there exist a nonempty τ-open subset $U' \subseteq U$ and a neighbourhood V' of y in (Y, τ') such that $f(U' \times V') \subseteq W$ [222]. If f is strongly quasicontinuous with respect to second variable at each point of its domain, then f is called *strongly quasicontinuous with respect to the second variable on $X \times Y$*.

A corresponding definition may be formulated for a function to be strongly quasicontinuous with respect to the first variable.

The metric-valued variant of the above definition was introduced in 1932 by Kempisty [161]. Kempisty called such functions *symmetrically quasicontinuous*.

Let (X, τ) and (Y, τ') be topological spaces, and let (Z, ρ) be a metric one. A function $f : X \times Y \to Z$ is called *symmetrically quasicontinuous at* $(p, q) \in X \times Y$ *with respect to* y if for every $\varepsilon > 0$ and every neighbourhood $U \times V$ of (p, q) in $(X \times Y, \tau \times \tau')$, there are a neighbourhood V' of q in (Y, τ') contained in V and a nonempty open set $U' \subseteq U$ in (X, τ) such that $\rho(f(x,y), f(p,q)) < \varepsilon$ for all $(x,y) \in U' \times V'$.

If f is symmetrically quasicontinuous with respect to y at each point of its domain, then f is called *symmetrically quasicontinuous with respect to y on $X \times Y$.*

This notion has been rediscovered on many occasions. For example, in 1980, Piotrowski [259, p. 114] introduced the phrase *quasicontinuous with respect to the variable x* in a general topological setting. Then, in 1992, the phrase *strongly quasicontinuous* was used in [104, p. 140]. Finally, in [222, p. 2043] the phrase strongly quasicontinuous with respect to the second variable at (x, y) was used.

Theorem 2.2.7 ([188, 205, 259, 260]). *Let (X, τ), (Y, τ') and (Z, τ'') be topological spaces, and let $f : X \times Y \to Z$ be a separately continuous mapping. If*
(i) *(X, τ) is a Baire space,*
(ii) *$y_0 \in Y$ has a countable local base, and*
(iii) *(Z, τ'') is regular,*

then f is strongly quasicontinuous with respect to the second variable at each point of $X \times \{y_0\}$.

Proof. Let $x_0 \in X$, let W be an open neighbourhood of $f(x_0, y_0)$, and let U be an open neighbourhood of x_0. Let $(V_n : n \in \mathbb{N})$ be a countable local base for the topology at y_0. Since $x \mapsto f(x, y_0)$ is continuous and $f(x_0, y_0) \in W$, we may assume, by possibly making U smaller, that $f(U \times \{y_0\}) \subseteq W$. For each $n \in \mathbb{N}$, let $F_n := \{x \in U : f(\{x\} \times V_n) \subseteq W\}$. Then $U = \bigcup_{n \in \mathbb{N}} F_n$. Now because (X, τ) is a Baire space, there exists $k \in \mathbb{N}$ such that $\operatorname{int}(\overline{F_k}) \cap U \neq \varnothing$. We claim that if $U' := \operatorname{int}(\overline{F_k}) \cap U$ and $V' := V_k$, then $f(U' \times V') \subseteq \overline{W}$. To see this, consider $(x, y) \in U' \times V'$ such that $f(x, y) \notin \overline{W}$. Since f is separately continuous, there exists an open neighbourhood N of x, contained in U', such that $f(N \times \{y\}) \subseteq Z \setminus \overline{W}$. However, this is impossible since $N \cap F_k \neq \varnothing$ and

$$f((N \cap F_k) \times \{y\}) \subseteq f((N \cap F_k) \times V_k) \subseteq f(F_k \times V_k) \subseteq W.$$

Therefore $f(U' \times V') \subseteq \overline{W}$. Finally, since (Z, τ'') is regular, it is sufficient to show that f is strongly quasicontinuous with respect to the second variable at (x_0, y_0). □

Example 2.2.8. Let $f : \mathbb{R} \times \mathbb{R} \to \mathbb{R}$ be defined by

$$f(x,y) := \begin{cases} \sin(1/(x^2 + y^2)) & \text{if } x^2 + y^2 \neq 0, \\ 0 & \text{if } x^2 + y^2 = 0. \end{cases}$$

This function is strongly quasicontinuous with respect to the first and second variables, but the function $f^0 : \mathbb{R} \to \mathbb{R}$ defined by $f^0(x) := f(x, 0)$ for all $x \in \mathbb{R}$, is not continuous. Thus the converse of Theorem 2.2.7 is not true.

Theorem 2.2.9 ([188]). *Let (X, τ) be a Baire space, let (Y, τ') be first countable, and let (Z, ρ) be a metric space. If $f : X \times Y \to Z$ is a function such that f_x is continuous for all*

$x \in X$ and f^y is quasicontinuous for all $y \in Y$, then f is strongly quasicontinuous with respect to the second variable.

Proof. This may be proved by a slight modification of the proof of Theorem 2.2.7. □

Corollary 2.2.10. *Let (X_i, τ_i), $i = 1, 2, \ldots, n$, be first countable Baire spaces, and let (Z, ρ) be a metric space. If $f : X_1 \times X_2 \times \cdots \times X_n \to (Z, \rho)$ is a function that is continuous in each variable separately, then f is a quasicontinuous function from $(X_1 \times X_2 \times \cdots \times X_n, \tau_1 \times \tau_2 \times \cdots \times \tau_n)$ to (Z, ρ).*

We call a topological space (X, τ) *quasi-regular* if for every $U \in \tau$ with $U \neq \emptyset$, there is $V \in \tau$, $V \neq \emptyset$, such that $\overline{V} \subseteq U$.

Theorem 2.2.11 ([188]). *Let (X, τ) be a countably compact quasi-regular space, let (Y, τ') be a topological space, and let (Z, ρ) be a metric space. If $f : X \times Y \to Z$ is a strongly quasicontinuous function with respect to the second variable, then the points of continuity of f are dense in $X \times \{p\}$ for all $p \in Y$.*

Proof. Let $x \in X$ and $p \in Y$, and let $U \times V$ be any neighbourhood of (x, p) in $(X \times Y, \tau \times \tau')$. Since f is strongly quasicontinuous with respect to the second variable, there are a neighbourhood V^1 of p contained in V and a nonempty open set $U^1 \subseteq U$ in (X, τ) such that for all (x', y') and $(x'', y'') \in U^1 \times V^1$, $\rho(f(x', y'), f(x'', y'')) < 1$. Let W^1 be a nonempty open set such that $\overline{W^1} \subseteq U^1$. Then $W^1 \times V^1$ is a neighbourhood of (x_1, p), where $x_1 \in W^1$, and since f is strongly quasicontinuous with respect to the second variable at (x_1, p), there are a neighbourhood V^2 of p contained in V^1 and a nonempty open set $U^2 \subseteq W^1$ such that for all (x', y') and $(x'', y'') \in U^2 \times V^2$, $\rho(f(x', y'), f(x'', y'')) < 1/2$. Let W^2 be a nonempty open set such that $\overline{W^2} \subseteq U^2$. Proceed by induction and get a neighbourhood $U^n \times V^n$ of (x_n, p), $x_n \in U_n$, such that for all (x', y') and $(x'', y'') \in U^n \times V^n$, $\rho(f(x', y'), f(x'', y'')) < 1/n$ and a nonempty open set W^n such that $\overline{W^n} \subseteq U^n$. Since (X, τ) is countably compact, there is a point $x^* \in \bigcap_{n \in \mathbb{N}} \overline{W^n}$. Then $(x^*, p) \in \bigcap_{n \in \mathbb{N}} (\overline{W^n} \times V^n) \subseteq \bigcap_{n \in \mathbb{N}} (U^n \times V^n) \subseteq U \times V$. Thus $(x^*, p) \in (U \times V) \cap (X \times \{p\})$, and (x^*, p) is a point of continuity of f. □

Piotrowski [261] and Mirmostafaee [213] generalised Theorem 2.2.11 by relaxing the hypothesis on (X, τ) from being countably compact quasi-regular to being Baire space and by relaxing the hypothesis on (Z, ρ) from being metrizable to being a generalised metric space.

Theorem 2.2.12. *Let (X, τ) be a countably compact quasi-regular space, let (Y, τ') be first countable, and let (Z, ρ) be a metric space. If $f : X \times Y \to Z$ is a function such that f_x is continuous for all $x \in X$ and f^y is quasicontinuous for all $y \in Y$, then the points of continuity of f are dense in $X \times \{p\}$ for all $p \in Y$.*

Proof. Since every quasi-regular countably compact space is a Baire space, the proof immediately follows from Theorems 2.2.9 and 2.2.11. □

Theorem 2.2.13. *Let (X_i, τ_i), $i = 1, 2, \ldots, n$, be first countable and countably compact quasi-regular spaces, and let (Z, ρ) be a metric space. If $f : X_1 \times X_2 \times \cdots \times X_n \to (Z, \rho)$ is a function that is continuous in each variable separately, then the points of continuity of f are dense in $X_1 \times \cdots \times X_{i-1} \times \{p\} \times X_{i+1} \times \cdots \times X_n$ for all $p \in X_i$, $i = 1, 2, \ldots, n$.*

If we wish to escape the realm of first countable spaces, then we are obliged to work harder.

We call a topological space (X, τ) *countably Čech-complete* if there exists a sequence $(\mathscr{A}_n : n \in \mathbb{N})$ of open covers of X such that every decreasing sequence $(F_n : n \in \mathbb{N})$ of nonempty closed subsets of X has $\bigcap_{n \in \mathbb{N}} F_n \neq \varnothing$, provided that each F_n is $\mathscr{A}_n$-small, that is, for each $n \in \mathbb{N}$, there exists $A_n \in \mathscr{A}_n$ such that $F_n \subseteq A_n$. Further, we say that a point $x \in X$ is a *q-point* if there exists a sequence $(U_n : n \in \mathbb{N})$ of neighbourhoods of x such that every sequence $(x_n : n \in \mathbb{N})$ in X with $x_n \in U_n$ for all $n \in \mathbb{N}$ has a cluster point in X. If every point of X is a q-point, then we call (X, τ) a *q-space*.

Exercise 2.2.14. Show that every regular countably Čech-complete topological space (X, τ) is a Baire space.

Theorem 2.2.15 ([45, Theorem 1]). *Let (X, τ), (Y, τ') and (Z, τ'') be topological spaces, and let $f : X \times Y \to Z$ be a separately continuous mapping. If*
(i) *(X, τ) is a regular countably Čech-complete space,*
(ii) *$y_0 \in Y$ is a q-point, and*
(iii) *(Z, τ'') is regular,*

then f is strongly quasicontinuous with respect to the second variable at each point of $X \times \{y_0\}$.

Proof. The proof given here follows that of [165, Lemma 1]. Firstly, because (X, τ) is countably Čech-complete space, there exists a sequence $\{\mathcal{F}_n : n \in \mathbb{N}\}$ of open covers of X such that every sequence of nonempty closed subsets $(F_n : n \in \mathbb{N})$ of X with $F_{n+1} \subseteq F_n$ for all $n \in \mathbb{N}$ has a nonempty intersection, provided that each set F_n is $\mathcal{F}_n$-small.

Suppose, to obtain a contradiction, that f is not strongly quasicontinuous with respect to the second variable at some point (x_0, y_0) of $X \times \{y_0\}$. Then by the regularity of (Z, τ'') there exist open neighbourhoods W of $f(x_0, y_0)$ and U of x_0 such that $f(U' \times V') \nsubseteq \overline{W}$ for each nonempty open subset U' of U and each open neighbourhood V' of y_0. Again, by the regularity of (Z, τ'') there exists an open neighbourhood W' of $f(x_0, y_0)$ such that $\overline{W'} \subseteq W$. Note that by possibly making U smaller we may assume that $f(U \times \{y_0\}) \subseteq W'$.

To prove the theorem, we will inductively define sequences $(x_n : n \in \mathbb{N})$ in X, $(U_n : n \in \mathbb{N})$ of nonempty open subsets of (X, τ), $(y_n : n \in \mathbb{N})$ in Y, and $(V_n : n \in \mathbb{N})$ of open neighbourhoods of y_0, but before that, we will denote by $(O_n : n \in \mathbb{N})$ any sequence of open neighbourhoods of y_0 such that each sequence $(y_n : n \in \mathbb{N})$ in Y

with $y_n \in O_n$ for all $n \in \mathbb{N}$ has a cluster point in (Y, τ'), and for notional consistency, we will denote U by U_0 and Y by V_0.

Step 1. Let $V_1 := \{y \in V_0 \cap O_1 : f(x_0, y) \in W'\}$. Then $y_0 \in V_1$ as $f(x_0, y_0) \in W'$. Now by our hypotheses there exists $(x_1, y_1) \in U_0 \times V_1$ such that $f(x_1, y_1) \notin \overline{W}$. Next, we choose an open neighbourhood U_1 of x_1 such that $\overline{U_1}$ is $\mathcal{F}_1$-small and $U_1 \subseteq \{x \in U_0 : f(x, y_1) \notin \overline{W}\}$.

Now suppose that $(x_j, y_j) \in U \times Y$, V_j and U_j have been defined for each $1 \le j \le n$ so that:

(i) $y_0 \in V_j := \{y \in V_{j-1} \cap O_j : f(x_{j-1}, y) \in W'\}$;
(ii) $(x_j, y_j) \in U_{j-1} \times V_j$ and $f(x_j, y_j) \notin \overline{W}$;
(iii) U_j is an open set, $\overline{U_j}$ is $\mathcal{F}_j$-small, and $x_j \in U_j \subseteq \{x \in U_{j-1} : f(x, y_j) \notin \overline{W}\}$.

Step $n+1$. Let $V_{n+1} := \{y \in V_n \cap O_{n+1} : f(x_n, y) \in W'\}$. Note that since $f(U \times \{y_0\}) \subseteq W'$ and $x_n \in U_{n-1} \subseteq U$, $y_0 \in V_{n+1}$. Now by our hypotheses there exists $(x_{n+1}, y_{n+1}) \in U_n \times V_{n+1}$ such that $f(x_{n+1}, y_{n+1}) \notin \overline{W}$. Next, we choose an open neighbourhood U_{n+1} of x_{n+1} such that $\overline{U_{n+1}}$ is $\mathcal{F}_{n+1}$-small and $U_{n+1} \subseteq \{x \in U_0 : f(x, y_{n+1}) \notin \overline{W}\}$.

This completes the induction. Since $y_n \in O_n$ for all $n \in \mathbb{N}$, $(y_n : n \in \mathbb{N})$ has a cluster point $y_\infty \in Y$. Similarly, since $\overline{\{x_k : n < k\}} \subseteq \overline{U_n}$ is $\mathcal{F}_n$-small for all $n \in \mathbb{N}$, $(x_n : n \in \mathbb{N})$ has a cluster point $x_\infty \in X$. Then for each fixed $n \in \mathbb{N}$,

$$f(x_n, y_k) \in f(\{x_n\} \times V_k) \subseteq f(\{x_n\} \times V_{n+1}) \subseteq W' \quad \text{for all } n < k, \text{ since } y_k \in V_k \subseteq V_{n+1}.$$

Therefore $f(x_n, y_\infty) \in \overline{W'}$ for each $n \in \mathbb{N}$, and so $f(x_\infty, y_\infty) \in \overline{W'} \subseteq W$. On the other hand, if we again fix $n \in \mathbb{N}$, then

$$f(x_{k+1}, y_n) \in f(U_k \times \{y_n\}) \subseteq f(U_n \times \{y_n\}) \subseteq Z \setminus \overline{W} \quad \text{for all } n \le k, \text{ since } x_{k+1} \in U_k \subseteq U_n.$$

Therefore $f(x_\infty, y_n) \in Z \setminus W$ for each $n \in \mathbb{N}$, and so $f(x_\infty, y_\infty) \in Z \setminus W$. However, this contradicts our earlier conclusion that $f(x_\infty, y_\infty) \in W$. Hence f must be strongly quasicontinuous with respect to the second variable at each point of $X \times \{y_0\}$. □

There is a rich literature concerning quasicontinuous functions and multifunctions. As a reference, we would like to mention (at least) the survey paper of Neubrunn [249].

Exercises

1. Let (X, τ) and (Y, τ') be topological spaces, and let $f : X \to Y$. Show that f is quasicontinuous if, and only if, f is semi-continuous.
2. Let (X, d) and (Y, ρ) be metric spaces and let $f : X \to Y$. Show that if $x_0 \in X$, then f is neighbourly at x_0 if and only if f is quasicontinuous at x_0.
3. Let (X, τ) and (Y, τ') be topological spaces, and let $f : X \to Y$. Show that if $x_0 \in X$, then f is quasicontinuous at x_0 if and only if for every $U \in \tau'$, $f(x_0) \in U$, there exists $V \in \tau$ such that $x_0 \in \overline{V}$ and $f(V) \subseteq U$.

4. Let (X, τ) and (Y, τ') be topological spaces. A function $f : X \to Y$ is said to be *somewhat continuous* [90] if for every $U \in \tau'$ such that $f^{-1}(U) \neq \emptyset$, there exists $V \in \tau$ such that $V \neq \emptyset$ and $V \subseteq f^{-1}(U)$. Let $g : X \to Y$ be a function. Show that g is quasicontinuous if and only if there is a base $\mathcal{B} \subseteq \tau$ such that $g|_B$ is somewhat continuous for every $B \in \mathcal{B}$ (see [248]).
5. Let $(X, \cdot, \tau)$ and $(Y, \cdot, \tau')$ be topological groups. Show that if $f : (X, \cdot, \tau) \to (Y, \cdot, \tau')$ is somewhat continuous homomorphism, then f is continuous [90].
6. Give an example of somewhat continuous function that is not quasicontinuous.
7. Let (X, τ) and (Y, τ') be topological spaces, and let $f : X \to Y$ be a quasicontinuous function. Show that if $U \subseteq X$ is either open or dense, then $f|_U$ is also quasicontinuous.
8. Show that the claim in Exercise 7 does not hold for closed sets.
9. Let (X, τ) and (Y, τ') be topological spaces, and let $f : X \to Y$ be a function. Show that f is quasicontinuous if and only if for every dense set $D \subseteq X$, $f(D \cap G)$ is dense in $f(G)$ for every $G \in \tau$.
10. Let (X, τ) be a separable topological space, and let (Y, τ') be a topological space. Let $f : X \to Y$ be a somewhat continuous function onto (Y, τ'). Show that (Y, τ') is separable [86].
11. Prove Theorem 2.2.9.

2.3 The set of points of continuity of a quasicontinuous function

The problem of evaluation of the sets of discontinuity points of quasicontinuous functions is classical in real analysis and traces its history back to Volterra, Baire and Kempisty.

Let (X, τ) and (Y, τ') be topological spaces, and let $f : X \to Y$ be a function. We denote

$$C(f) := \{x \in X : f \text{ is continuous at } x\} \quad \text{and} \quad D(f) := X \setminus C(f).$$

The following definition will be useful in the study of quasicontinuous functions.

Let (X, τ) be a topological space, let (Y, d) be a metric space, and let $f : X \to Y$. The function f is *cliquish at* $x_0 \in X$ if for every $\varepsilon > 0$ and every open neighbourhood U of x_0 in (X, τ), there exists a nonempty τ-open subset V of U such that for all $x_1, x_2 \in V$, $d(f(x_1), f(x_2)) < \varepsilon$. The function $f : X \to Y$ is *cliquish on* (X, τ) if it is cliquish at every $x \in X$ [300].

If a function f is continuous at a point x_0, then it is quasicontinuous at x_0. If a function f is quasicontinuous at a point x_0 and maps into a metric space, then it is cliquish at x_0.

Exercise 2.3.1. Find a function f that is cliquish but not quasicontinuous at a point x_0. Also, find a function f that is quasicontinuous but not continuous at a point x_0.

For a function $f : X \to Y$ from a topological space (X, τ) into a metric space (Y, d), we denote by $A(f) := \{x \in X : f \text{ is cliquish at } x\}$.

Proposition 2.3.2 ([251]). *Let (X, τ) be a topological space, let (Y, d) be a metric space, and let $f : X \to Y$ be a function. Then $A(f)$ is a closed set.*

Proof. Let $x_0 \in \overline{A(f)}$. Let $\varepsilon > 0$, and let $U(x_0)$ be an open neighbourhood of x_0. Then there exists a point $x \in A(f) \cap U(x_0)$. Since f is cliquish at x, there exists a nonempty τ-open subset V of $U(x_0)$ such that $d(f(x_1), f(x_2)) < \varepsilon$ for every $x_1, x_2 \in V$. From this it follows that f is cliquish at x_0, and hence $x_0 \in A(f)$. □

Let (X, τ) be a topological space, and let (Y, d) be a metric space. For each $x \in X$, we denote by $\mathcal{U}(x)$ a base of open neighbourhoods of x in (X, τ). Let $A \subseteq X$, and let $f : A \to Y$ be a function. Now suppose that $x_0 \in \overline{A}$ and $U \in \mathcal{U}(x_0)$. Then the *oscillation* w_f *of* f *at* x_0, is defined as $w_f(x_0) := \inf_{U \in \mathcal{U}(x_0)} d_U$, where $d_U := \operatorname{diam} f(A \cap U)$.

The following results show that under some assumptions the set of discontinuity points of a quasicontinuous function is a small set from the point of view of the Baire category.

Theorem 2.3.3 ([251]). *Let (X, τ) be a topological space, and let (Y, d) be a metric space. If $f : X \to Y$, then the set $A(f) \setminus C(f)$ is an F_σ-set of the first Baire category in (X, τ).*

Proof. Let $f : X \to Y$ be a function. Then $D(f) = \bigcup_{k=1}^{\infty} \{x \in X : 1/k \le w_f(x)\}$ and

$$A(f) \setminus C(f) = A(f) \cap \bigcup_{k=1}^{\infty} \left\{x \in X : \frac{1}{k} \le w_f(x)\right\} = \bigcup_{k=1}^{\infty} A(f) \cap \left\{x \in X : \frac{1}{k} \le w_f(x)\right\}.$$

So it is sufficient to prove that $M_k = A(f) \cap \{x \in X : \frac{1}{k} \le w_f(x)\}$ is nowhere dense set for every $k \in \mathbb{N}$. Let $x \in X$, let U be an open neighbourhood of x in (X, τ), and let $k \in \mathbb{N}$. If $x \notin A(f)$, then $x \in (X \setminus A(f)) \cap U$, and $(X \setminus A(f)) \cap U$ is an open set in (X, τ), which has an empty intersection with M_k. If $x \in A(f)$, then there exists a nonempty open set $V \subseteq U$ such that for all $x_1, x_2 \in V$, $d(f(x_1), f(x_2)) < \frac{1}{2k}$. From this it follows that $w_f(z) \le \frac{1}{2k} < \frac{1}{k}$ for every point $z \in V$. Hence $V \cap M_k = \emptyset$. □

Corollary 2.3.4. *Let (X, τ) be a topological space, and let (Y, d) be a metric space. If $f : X \to Y$ is quasicontinuous, then the set $D(f)$ of all discontinuity points of f is a first Baire category subset of (X, τ).*

Note that Corollary 2.3.4 was proved by Bledsoe in 1952 even for the pointwise limit of a sequence of quasicontinuous functions, as we will see later in Theorem 2.5.1.

Corollary 2.3.5 ([30]). *Let (X, τ) be a Baire space, and let (Y, d) be a metric space. If $f : X \to Y$ is a quasicontinuous function, then the set $C(f)$ of all points of continuity of f is a dense G_δ-subset of (X, τ)*

Levine [185] proved the following result.

Theorem 2.3.6. *Let (X, τ) be a topological space, and let (Y, τ') be a topological space with countable base. If $f : X \to Y$ is a quasicontinuous function, then the set $D(f)$ of all points of discontinuity of f is a set of the first Baire category in (X, τ).*

Proof. Let $(V_n : n \in \mathbb{N})$ be a countable base of (Y, τ). It is easy to verify that

$$D(f) = \bigcup_{n \in \mathbb{N}} f^{-1}(V_n) \setminus \operatorname{int}(f^{-1}(V_n)).$$

The quasicontinuity of f implies that $f^{-1}(V_n) \subseteq \overline{\operatorname{int}(f^{-1}(V_n))}$ for every $n \in \mathbb{N}$. Thus $f^{-1}(V_n) \setminus \operatorname{int}(f^{-1}(V_n))$ is a nowhere dense set for every $n \in \mathbb{N}$. It follows that $D(f)$ is the set of the first Baire category. □

Corollary 2.3.7. *Let (X, τ) be a Baire space, and let (Y, τ') be a topological space with countable base. If $f : X \to Y$ is a quasicontinuous function, then the set $C(f)$ of all points of continuity of f is a residual subset of (X, τ), that is, $X \setminus C(f)$ is a first Baire category subset of (X, τ).*

The following example shows that the assumptions on (Y, τ') in Corollaries 2.3.5 and 2.3.7 are essential.

Example 2.3.8 ([262]). Let $X := \mathbb{R}$ with the usual Euclidean topology, and let $Y := \mathbb{R}$ with the Sorgenfrey topology. Let $f : X \to Y$ be the identity function. Then f is quasicontinuous, but the set $C(f) = \varnothing$.

In the literature, we can find another very important notion of generalised continuity, the so-called almost continuity. The notion of an almost continuous mapping from a topological space (X, τ) to a topological space (Y, τ') was introduced in 1966 by Husain [145] and studied by many authors (see [51, 110, 112, 253]). Almost continuous mappings are often also called in the literature *nearly continuous mappings*. It is interesting to observe that quasicontinuity and almost continuity in some spaces form a decomposition of continuity.

Let (X, τ) and (Y, τ') be topological spaces. The function $f : X \to Y$ is *almost continuous at* $x \in X$ if for each $V \in \tau'$ containing $f(x)$, the closure of $f^{-1}(V)$ is a neighbourhood of x in (X, τ) [145]. If f is almost continuous at each point x of X, then it is called *almost continuous on* (X, τ).

If f is a function from a topological space (X, τ) to a topological space (Y, τ'), then denote

$$Q(f) := \{x \in X : f \text{ is quasicontinuous at } x\}.$$

Theorem 2.3.9 ([110]). *Let (X, τ) and (Y, τ') be topological spaces and suppose that (Y, τ') is regular. If $f : X \to Y$ is a function, then f is continuous if and only if f is almost continuous, and the set $Q(f)$ is dense in (X, τ).*

Proof. Clearly, if f is continuous on (X, τ), then f is both almost continuous and quasicontinuous on (X, τ). So we will consider the converse. To that end, suppose that f is almost continuous and the set $Q(f)$ is dense in (X, τ). Then, to obtain a contradiction, let us suppose that there is $x_0 \in X$ such that f is not continuous at x_0. Then there exists $V \in \tau'$ such that $f(x_0) \in V$ but

$$f^{-1}(V) \text{ is not a neighbourhood of } x_0. \qquad \text{(i)}$$

Since (Y, τ') is a regular space, there exist $V_1, V_2 \in \tau'$ such that

$$f(x_0) \in V_1 \subseteq \overline{V_1} \subseteq V_2 \subseteq \overline{V_2} \subseteq V. \qquad \text{(ii)}$$

The almost continuity of f at x_0 implies that there is $U \in \tau$ such that $x_0 \in U$ and

$$U \subseteq \overline{f^{-1}(V_1)}. \qquad \text{(iii)}$$

We claim that if $s \in Q(f) \cap U$, then $f(s) \in \overline{V_1}$. (iv)

Suppose (iv) does not hold. Then there is $s \in Q(f) \cap U$ such that $f(s) \notin \overline{V_1}$. The quasicontinuity of f at s implies that there exists a nonempty open set $W \subseteq U$ such that

$$W \subseteq f^{-1}(Y \setminus \overline{V_1}) \subseteq f^{-1}(Y \setminus V_1). \qquad \text{(v)}$$

Since (v) is in contradiction with (iii), (iv) holds. By (i) there exists $x \in U$ such that $f(x) \notin V$. By (ii), $Y \setminus V \subseteq Y \setminus \overline{V_2}$. The almost continuity of f at x implies that there exists an open set G in X such that

$$x \in G,\ G \subseteq U \quad \text{and} \quad G \subseteq \overline{f^{-1}(Y \setminus V_2)}. \qquad \text{(vi)}$$

Let $z \in G \cap Q(f)$. By (iv), $f(z) \in \overline{V_1}$. Thus $f(z) \in V_2$. Since f is quasicontinuous at z, there exists $H \in \tau$ such that $H \neq \emptyset$, $H \subseteq G$ and $f(H) \subseteq V_2$, which contradicts (vi). □

Corollary 2.3.10. *Let (X, τ) and (Y, τ') be topological spaces and suppose that (Y, τ') is regular. If $f : X \to Y$ is a function, then f is continuous on (X, τ) if and only if f is both almost continuous and quasicontinuous on (X, τ).*

The following example shows that the condition of regularity of (Y, τ') in Theorem 2.3.9 and Corollary 2.3.10 is essential.

Example 2.3.11. Let $X := [0, 1]$ with the usual topology. Put $Y := [0, 1]$, where the topology of Y consists of the usual topology and, moreover, of all the sets

$$G \setminus \{1, 1/2, \ldots, 1/n, \ldots\},$$

where G is open in the usual topology. It is easy to verify that the space Y is not regular. The identity mapping $I : X \to Y$ is quasicontinuous and almost continuous, but I is not continuous at 0, since the set $V := [0, 1] \setminus \{1, 1/2, \ldots, 1/n, \ldots\}$ is not a neighbourhood of 0 in the usual topology.

Piotrowski [262] asked the following question: Let (X, τ) be a Baire space. What are "large spaces" (Y, τ') such that every quasicontinuous function $f : X \to Y$ has the nonempty set $C(f)$?

Of course, by "large spaces"' we mean spaces that are neither metrizable nor possess a countable base, since as we saw before, for such a space (Y, τ') and every quasicontinuous function f from a Baire space (X, τ) into (Y, τ'), the set $C(f)$ is a residual subset of (X, τ), that is, $X \setminus C(f)$ is a first category subset of (X, τ).

In [136], there is a partial solution of Piotrowski's question for p-spaces with a G_δ-diagonal.

A topological space (X, τ) is said to have a G_δ*-diagonal* [100] if the diagonal $\triangle := \{(x, x) : x \in X\}$ of the space (X, τ) is a G_δ-set in $(X \times X, \tau \times \tau)$.

A completely regular space (X, τ) is a p-space [100] if there exists a sequence $(\mathcal{U}_n : n \in \mathbb{N})$ of families of open subsets of βX such that:

(i) each $\mathcal{U}_n$ covers X;

(ii) for each $x \in X$, $\bigcap_n st(x, \mathcal{U}_n) \subseteq X$. (By $st(x, \mathcal{U}_n)$ we mean the set $\bigcup\{U \in \mathcal{U}_n, x \in U\}$.)

The p-spaces were introduced by Arhangelskiĭ [5] in 1963. Every Čech-complete space is a p-space, and every Moore space is a p-space.

Proposition 2.3.12 ([136]). *Let (X, τ) be a Baire space, and let (Y, τ') be a p-space with a G_δ-diagonal. If $f : X \to Y$ is a quasicontinuous function, then $C(f)$ is a dense G_δ-set in (X, τ).*

Honouring the contribution of Zbigniew Piotrowski to the study of quasicontinuous functions, Taras Banakh in [15] suggested the following definition. A topological space (Y, τ') is called a *Piotrowski space* if every quasicontinuous function $f : X \to Y$ from a Baire space (X, τ) has a point of continuity.

In fact, if (Y, τ') is a regular Piotrowski space, then for every Baire space (X, τ) and every quasicontinuous function $f : X \to Y$, the set $C(f)$ of all points of continuity of f is comeager in (X, τ). To prove this fact, we need the following useful lemma.

Lemma 2.3.13 ([15]). *Let (X, τ) be a topological space, and let (Y, τ') be a regular topological space. If $f : X \to Y$ is a quasicontinuous function, then for every dense subset Z of (X, τ), we have $C(f) \cap Z = C(f|_Z)$.*

Proof. Given a continuity point $z \in Z$ of the restriction $f|_Z$, we should prove that z remains a continuity point of the function f. Given any neighbourhood O of the point $f(z)$ in (Y, τ'), we should find a neighbourhood U of z in (X, τ) such that $f(U) \subseteq O$. By the regularity of (Y, τ') there is an open neighbourhood W of $f(z)$ such that

$$\overline{W} \subseteq O.$$

By the continuity of $f|_Z$ at z there exists a neighbourhood V of z in (X, τ) such that

$$f(V \cap Z) \subseteq W.$$

We claim that $f(V) \subseteq \overline{W} \subseteq O$. To derive a contradiction, assume that $f(v) \notin \overline{W}$ for some point $v \in V$. By the quasicontinuity of f there exists $V' \in \tau$, $V' \neq \varnothing$, $V' \subseteq V$, such that

$$f(V') \subseteq Y \setminus \overline{W},$$

a contradiction. □

Proposition 2.3.14. *Let (X, τ) be a Baire space, and let (Y, τ') be a regular Piotrowski space. If $f : X \to Y$ is a quasicontinuous function, then the set $C(f)$ of all the points of continuity of f is comeager in (X, τ).*

Proof. Suppose that $C(f)$ is not a comeager set in (X, τ). Then $X \setminus C(f)$ is of the second Baire category in (X, τ). By [159] there is $V \in \tau$, $V \neq \varnothing$, such that

$$(X \setminus C(f)) \cap G_x \quad \text{is second category}$$

for every neighbourhood G_x of every point $x \in \overline{V}$. Then $G = V \cap (X \setminus C(f))$ is a dense Baire subspace of a Baire space V with the subspace topology. Since (Y, τ') is a Piotrowski space, $C(f|_G) \neq \varnothing$. By Lemma 2.3.13, $C(f|_G) = C(f) \cap G$, a contradiction. □

It turns out that Piotrowski spaces are tightly connected with fragmentable and Stegall spaces, which are well-known spaces in general topology and its applications to functional analysis [164].

A topological space (Y, τ') is called a *Stegall space* [81] if for every Baire topological space (X, τ) and every minimal usco mapping $F : X \to 2^Y$, F is single-valued at some point of (X, τ).

Proposition 2.3.15 ([15]). *Every Hausdorff Piotrowski space (Y, τ') is a Stegall space.*

Proof. Let (X, τ) be a Baire topological space, and let $F : X \to 2^Y$ be a minimal usco mapping. By Proposition 1.3.5 any selection $f : X \to Y$ of the set-valued mapping F is quasicontinuous. Since the space (Y, τ') is Piotrowski, the set $C(f) \neq \varnothing$. By Corollary 1.3.12, $F(x) = \{f(x)\}$ for every $x \in C(f)$. Thus F is single-valued at every $x \in C(f)$. Since $C(f) \neq \varnothing$, (Y, τ') is Stegall. □

Proposition 2.3.16 ([268]). *Let (X, τ) be a topological space, and let (Y, τ') be a fragmentable topological space. Let $F : X \to 2^Y$ be a minimal usco. Then F is single-valued at the points of a dense G_δ-subset on (X, τ). Thus (Y, τ') is a Stegall space.*

Proof. Let ρ be a metric that fragments (Y, τ'), let (X, τ) be a Baire space, and let $F : X \to 2^Y$ be a minimal usco. For every $\varepsilon > 0$ denote by $\mathcal{U}_\varepsilon$ the family of all open subsets U of (X, τ) such that ρ-diam$[F(U)] < \varepsilon$. We claim that $\bigcup \mathcal{U}_\varepsilon$ is dense in (X, τ). Indeed, take $W \in \tau$, $W \neq \varnothing$. By the fragmentability of (Y, τ') the set $F(W)$ contains a nonempty relatively open subset V of $F(W)$ with ρ-diam$(V) < \varepsilon$. By Proposition 1.3.5 there exists $U \in \tau$, $U \neq \varnothing$, $U \subseteq W$, such that $F(U) \subseteq V$, and hence $U \in \mathcal{U}_\varepsilon$. So $\bigcup \mathcal{U}_\varepsilon$ is dense in (X, τ), and $G := \bigcap_{n \in \mathbb{N}} \mathcal{U}_{1/n}$ is a dense G_δ-set in (X, τ). It is clear that for every $x \in G$, ρ-diam$[F(x)] = 0$, which means that $F(x)$ is a singleton. □

Proposition 2.3.17. *Let (X,τ) be a Baire space, and let (Y,τ') be a Hausdorff Stegall space. If $F : X \to 2^Y$ is a minimal usco, then $\{x \in X : F(x)$ is a singleton$\}$ is comeager in (X,τ).*

Proof. Suppose that the set $L := \{x \in X : F(x)$ is a singleton$\}$ is not comeager in (X,τ). Then $X \setminus L$ is a set of the second Baire category in (X,τ). By [159] there is $V \in \tau$, $V \neq \varnothing$, such that $(X \setminus L) \cap G_x$ is a second category subset of (X,τ) for every neighbourhood G_x of every point $x \in \overline{V}$. Then $G := V \cap (X \setminus L)$ is a dense Baire subspace of the Baire space V with the subspace topology. Since (Y,τ') is a Hausdorff space and G is dense in V, we have by Proposition 1.3.4, that $F|_G$ is a minimal usco. Since (Y,τ') is a Stegall space, there is a point $x \in G$ such that $F|_G(x)$ is a singleton. Since $F|_G(x) = F(x)$, $F(x)$ is a singleton, which contradicts the fact that $x \notin L$. □

In [157] consistent examples of nonfragmentable compact Hausdorff spaces belonging to the class of Stegall spaces are given.

Kalenda [156] introduced the class of weakly Stegall spaces. A topological space (Y,τ') belongs to the class of *weakly Stegall spaces* if for every nonempty complete metric space (X,d) and every minimal usco mapping $F : X \to 2^Y$, F is single-valued at some point of (X,d).

In [233], it is shown that under some additional set-theoretic assumptions that are equiconsistent with the existence of a measurable cardinal, there is a Banach space $(X,\|\cdot\|)$ whose dual space equipped with the weak* topology is in the class of weakly Stegall spaces but not in the class of Stegall spaces.

The proof of the next simple result shows some techniques associated with quasicontinuity of mappings and fragmentability of spaces.

Theorem 2.3.18 ([164]). *Let (X,τ) be a Baire space, and let $f : X \to Y$ be a quasicontinuous mapping from (X,τ) into a topological space (Y,τ') fragmented by some metric ρ. Then there exists a dense G_δ subset C in (X,τ) at the points of which $f : X \to (Y,\rho)$ is continuous. In particular, if the topology generated by the metric ρ on Y contains the topology τ', then $f : X \to Y$ is continuous at every point of the set C.*

Proof. Consider, for every $n = 1,2,\ldots$, the set

$$V_n := \bigcup\{V : V \in \tau \text{ and } \rho - \operatorname{diam}[f(V)] \leq 1/n\}.$$

Then $V_n \in \tau$. It is also dense in (X,τ). Indeed, suppose $W \in \tau$, $W \neq \varnothing$. Consider the set $A := f(W) \subseteq Y$. By the fragmentability of (Y,τ') there is a relatively open subset $B := A \cap U = f(W) \cap U$, where U is open in (Y,τ'), such that ρ-diam$[B] \leq 1/n$. The quasicontinuity of f implies that there is a nonempty τ-open $V \subseteq W$ with $f(V) \subseteq U \cap f(W) = B$. This shows that $\varnothing \neq V \subseteq V_n \cap W$. Hence V_n is dense in (X,τ). Obviously, at each point of $C := \bigcap_{n\geq 1} V_n$, the mapping f is ρ-continuous. □

Note that, according to a result of Ribarska [268, 269], if the space (Y,τ) is compact, Hausdorff and fragmentable, then it is also fragmentable by some metric that majorizes the topology of Y, that is, the metric topology generated by the new fragmenting metric contains the topology of the compact space (Y,τ). Therefore we have the following result for compact spaces (Y,τ).

Corollary 2.3.19. *Let (X,τ) be a Baire space. If $f : X \to Y$ is a quasicontinuous mapping from (X,τ) into a fragmentable compact Hausdorff space (Y,τ'), then there exists a dense G_δ subset C in (X,τ) at the points of which $f : X \to Y$ is continuous.*

To formulate further results, we need to recast the fragmentability of (Y,τ) in terms of a topological *fragmenting game* $G(Y)$ in the space (Y,τ), [167, 168]. This game involves two players Σ and Ω. The players select, one after the other, nonempty subsets of (Y,τ). Ω starts the game by selecting the whole space Y. Σ answers by choosing any nonempty subset A_1 of Y, and Ω goes on by taking a nonempty subset $B_1 \subseteq A_1$ that is relatively open in A_1. After that, on the nth stage of the development of the game, Σ takes any nonempty subset A_n of the last move B_{n-1} of Ω, and the latter answers by taking again a nonempty relatively open subset B_n of the set A_n just chosen by Σ. Acting this way, the players produce a sequence of nonempty sets

$$A_1 \supseteq B_1 \supseteq A_2 \supseteq \cdots \supseteq A_n \supseteq B_n \supseteq \cdots,$$

which is called a *play* and will be denoted by $p := (A_i, B_i)_{i\geq 1}$ (in this notation, there is no need to include the space Y, which is the first and obligatory move of Ω). The player Ω is said to have *won* this play if the set $\bigcap_{n\geq 1} A_n$ contains at most one point. Otherwise, the player Σ is said to have *won* the play.

A *partial play* is a finite sequence that consists of the first several moves

$$A_1 \supseteq B_1 \supseteq A_2 \supseteq \cdots \supseteq A_n \quad \text{or} \quad A_1 \supseteq B_1 \supseteq A_2 \supseteq \cdots \supseteq B_n$$

of a play. A *strategy* ω for the player Ω is a mapping that assigns to each partial play $A_1 \supseteq B_1 \supseteq A_2 \supseteq \cdots \supseteq A_n$ a nonempty relatively open set B_n such that $A_1 \supseteq B_1 \supseteq A_2 \supseteq \cdots \supseteq A_n \supseteq B_n$ is again a partial play. A *strategy* σ for Σ is defined in a similar way. Sometimes we will denote the first choice A_1 under a strategy σ by $\Sigma(Y)$. A *σ-play* (*ω-play*) is a play in which Σ (Ω) selects his/her moves according to σ (ω). The strategy ω (σ) is said to be a *winning strategy* if every ω-play (σ-play) is won by Ω (Σ). The game $G(Y)$ or the space Y is called *Ω-favourable* (*Σ-favourable*) if there is a winning strategy for the player Ω (Σ). The game $G(Y)$ (or the space (Y,τ)) is called *Σ-unfavourable* if there does not exist a winning strategy for the player Σ. Examples show (see [164, Example 1]) that there are compact spaces (Y,τ) that are unfavourable for both players.

It was proven in [168] that the fragmentability of a given topological space (Y,τ) is equivalent to the existence of a winning strategy for the player Ω in the game $G(Y)$, that is, (Y,τ) is fragmentable if and only if the game $G(Y)$ is Ω-favourable. By a change of the rule for winning a play in the game $G(Y)$ (but keeping intact the rules for the moves of the players) we can express in a similar way the existence of a fragmenting metric that majorizes τ, the topology of the space Y. We denote by $G^*(Y)$ the game in which the plays are the same as in $G(Y)$ but for which the rule for winning a play is the following one. The player Ω is said to have *won* the play $p := (A_i, B_i)_{i\geq 1}$ in the game $G^*(Y)$ if the set $\bigcap_{n\geq 1} A_n$ is either empty or consists of exactly one point y such that for every open neighbourhood U of y, there is a positive integer n with $A_n \subseteq U$. Otherwise, the player Σ is said to have *won* the play $(A_i, B_i)_{i\geq 1}$. As shown in [167], a topological space (Y, τ) is fragmentable by a metric that majorizes τ if and only if the player Ω has a winning strategy in the game $G^*(Y)$.

The next result shows what we can expect from spaces (Y, τ) in which the other player Σ does not have a winning strategy in $G^*(Y)$. As already mentioned, the absence of a winning strategy for Σ does not necessarily imply that Ω has a winning strategy in $G^*(Y)$, that is, the condition that the game $G^*(Y)$ is Σ-unfavourable (or the space (Y, τ) is Σ-unfavourable) is weaker than the condition that (Y, τ) is fragmentable by a metric majorizing its topology. Correspondingly, the conclusion is also weaker. The set of continuity points $C(f)$ is not necessarily residual in (X, τ). It is however of the second Baire category in every nonempty open subset of (X, τ), that is, for every nonempty τ-open subset $V \subseteq X$, the set $C(f) \cap V$ is not of the first Baire category (equivalently, the set $C(f) \cap V$ cannot be covered by a countable union of subsets whose closures in (X, τ) have no interior points).

Theorem 2.3.20 ([164]). *For a T_1 regular topological space (Y, τ'), the following are equivalent:*
(i) *$G^*(Y)$ is Σ-unfavourable,*
(ii) *every quasicontinuous mapping $f : X \to Y$ from the complete metric space (X, d) into (Y, τ') is continuous at least one point of X,*
(iii) *every quasicontinuous mapping $f : X \to Y$ from the complete metric space (X, d) into (Y, τ') is continuous at the points of some subset of the second Baire category in every nonempty open subset of (X, d),*
(iv) *every quasicontinuous mapping $f : X \to Y$ from an α-favourable space (X, τ) into (Y, τ') is continuous at the points of some subset of the second Baire category in every nonempty open subset of (X, τ).*

Before we recall the concept of α-favourability, we need to first review the well-known Banach–Mazur game. Let (X, τ) be a topological space. The *Banach–Mazur game* BM(X) is played by two players α and β, who alternately select nonempty open subsets of (X, τ). The player α starts the game by selecting $W_0 := X$. Then the player β answers by taking some $V_0 \in \tau, V_0 \neq \emptyset$. On the nth move, $n \geq 1$, the player α takes $W_n \in \tau, W_n \neq \emptyset$, with $W_n \subseteq V_{n-1}$, and β answers by taking $V_n \in \tau, V_n \neq \emptyset$, with $V_n \subseteq W_n$. Using this way of selection, the players get a sequence $(W_n, V_n)_{n=0}^{\infty}$, which is called a *play*. The player β is said to have *won* this play if $\bigcap_{n\geq 1} W_n = \emptyset$; otherwise, this play is *won* by α. A *partial play* is a finite sequence that consists of the first several consecutive moves in the game. A strategy ψ for the player α is a mapping that assigns to each partial play $(V_0, W_1, V_1, W_2, V_2, \dots, W_{n-1}, V_{n-1})$ a nonempty open subset W_n of

V_{n-1}. A *ψ-play* is a play in which α selects his/her moves according to ψ. A strategy ψ is said to be a *winning strategy* if every ψ-play is won by α. A topological space (X,τ) is called *α-favourable* if there exists a winning strategy for α in BM(X).

We now remind the reader that a completely regular topological space (X,τ) is called *Čech-complete* if it is a G_δ subset of its Stone–Čech compactification; see page 129 for the details.

A completely regular topological space (X,τ) is said to be *almost Čech-complete* if it contains a dense Čech-complete subset. It is known that complete metric spaces are Čech-complete ([55] or [77, p. 190]) and that every almost Čech-complete space is α-favourable.

Next, we prove Theorem 2.3.20.

Proof. We show that (i) ⇒ (iv) and (ii) ⇒ (i). The implications (iv) ⇒ (iii) ⇒ (ii) are obvious. (i) ⇒ (iv) Suppose Y is Σ-unfavourable for $G^*(Y)$ and $f : X \to Y$ is a quasicontinuous mapping from the α-favourable space (X,τ). Let H be a first category subset of (X,τ). Then there is some winning strategy ψ for the player α in BM(X) that avoids the set H, that is, $\bigcap_{i\geq 0} W_i \neq \varnothing$ and $H \cap (\bigcap_{i\geq 0} W_i) = \varnothing$ whenever $(V_i, W_i)_{i\geq 0}$ is a ψ-play. Take an open set $V_0 \neq \varnothing$, $V_0 \subseteq X$. We will show that f is continuous at some point of $V_0 \setminus H$. To do this we first construct a strategy σ for the player Σ in $G^*(Y)$ and then use the fact that Σ does not win some σ-play. Put the first move of β in BM(X) to be V_0 and let $W_1 := \psi(V_0)$ be the answer of α. Assign $A_1 := f(W_1)$ to be the first move in the strategy Σ. Suppose that the answer of Ω in $G^*(Y)$ is B_1, a nonempty relatively open subset of A_1. The quasicontinuity of f implies that there exists a nonempty open subset V_1 of W_1 such that $f(V_1) \subseteq B_1$. Suppose the set V_1 is the next move of the player β in the game BM(X). The player α, of course, uses the strategy ψ to answer this move and selects the set $W_2 := \psi(V_0, W_1, V_1)$. Then we define the second move of Σ in $G^*(Y)$ to be $A_2 := \sigma(A_1, B_1) = f(W_2)$. Proceeding like this, we inductively construct the strategy σ. Together with each σ-play $(A_i, B_i)_{i\geq 1}$ in $G^*(Y)$, we also construct a ψ-play in BM(X) with $A_n := f(W_n)$ and $W_n := \psi(V_0, W_1, V_1, \dots W_{n-1}, V_{n-1})$ for all $n \in N$. As ψ is a winning strategy for α, we have $\bigcap_{i\geq 1} W_i \neq \varnothing$. Therefore

$$\varnothing \neq f\Big(\bigcap_{i\geq 1} W_i\Big) \subseteq \bigcap_{i\geq 1} f(W_i) = \bigcap_{i\geq 1} A_i.$$

Since (Y,τ') is Σ-unfavourable, there is some σ-play $(A_i, B_i)_{i\geq 1}$ that is won by Ω, and hence the nonempty set $\bigcap_{i\geq 1} A_i$ has just one point y, and, for every $U \in \tau$ with $y \in U$, there is some n with $A_n = f(W_n) \subseteq U$. All this means that $f(x) = y$ for every $x \in \bigcap_{i\geq 1} W_i \subseteq V_0 \setminus H$ and that f is continuous at such x.

(ii) ⇒ (i) Let σ be an arbitrary strategy for the player Σ in $G^*(Y)$. We will show that it is not a winning strategy. Consider the space P of all σ-plays $p := (A_i, B_i)_{i\geq 1}$ endowed with the Baire metric d, that is, if $p := (A_i, B_i)_{i\geq 1} \in P$ and $p' := (A_i', B_i')_{i\geq 1} \in P$, then $d(p,p') := 0$ if $p = p'$, and otherwise $d(p,p') := 1/n$, where $n := \min\{k \in \mathbb{N} : B_k \neq B_k'\}$.

Note that all the plays in P start with the same set $A_1 := \sigma(Y)$, the first choice of the strategy σ. Also, if $A_i = A_i'$ and $B_i = B_i'$ for all $i \le n$, then

$$A_{n+1} = \sigma(A_1, B_1, \ldots, A_n, B_n) = \sigma(A_1', B_1', \ldots, A_n', B_n') = A_{n+1}'.$$

In other words, if $p \neq p'$, then there is some n such that $B_n \neq B_n'$ and $A_i = A_i'$ for $i \le n$ and $B_i = B_i'$ for $i < n$. It is easy to verify that (P, d) is a complete metric space.

Consider the set-valued mapping $F : P \to 2^Y$ defined by $F((A_i, B_i)_{i\ge1}) := \bigcap_{i\ge1} A_i$. If for some σ-play p, we have $F(p) = \varnothing$, then the play p is won by Ω, and there is nothing to prove. Therefore, without loss of generality, we may assume that F is nonempty valued at every point of P. Let $f : P \to Y$ be an arbitrary selection of F. We will show that f is quasicontinuous. Then by (ii) f will turn out to be continuous at some point $p_0 \in P$. Finally, we will show (see Proposition 2.3.24) that the play p_0 is won by Ω. This will show that Σ is not a winning strategy and will complete the proof. □

Let (X, τ) and (Y, τ') be topological spaces. A set-valued mapping $G : X \to 2^Y$ is said to be *minimal* at $x \in X$ if for every $U \in \tau'$ with $G(x) \cap U \neq \varnothing$, there exists $V \in \tau$ such that $x \in \overline{V}$ and $G(V) = \bigcup\{G(v) : v \in V\} \subseteq U$. A mapping $G : X \to 2^Y$ is said to be *minimal on* (X, τ) if it is minimal at each point of (X, τ).

Proposition 2.3.21. *Let (X, τ) and (Y, τ') be topological spaces. A set-valued mapping $G : X \to 2^Y$ is minimal on (X, τ) if for each pair of open subsets U of (X, τ) and W of (Y, τ') such that $G(U) \cap W \neq \varnothing$, there exists a nonempty τ-open subset $V \subseteq U$ such that $G(V) \subseteq W$.*

Corollary 2.3.22. *Let (X, τ) and (Y, τ') be topological spaces. If $G : X \to 2^Y$ is a minimal mapping and $g : X \to Y$ is any selection of G, that is, $g(x) \in G(x)$ for all $x \in X$, then g is quasicontinuous on (X, τ).*

Lemma 2.3.23. *The set-valued mapping $F : P \to 2^Y$ defined in the proof of Theorem 2.3.20 is minimal on (P, d).*

Proof. We will prove the following claim: Let the play $p_0 := (A_i, B_i)_{i\ge1}$ be an element of the space P and $U \in \tau'$ with $U \cap A_n \neq \varnothing$ for every $n \in \mathbb{N}$. Then there is an open subset V in X such that $p_0 \in \overline{V}$ and $F(V) \subseteq U$. Let $p_0 := (A_i, B_i)_{i\ge1}$, and let U be an open set in (Y, τ') with the above property. Given a positive integer n, consider the nonempty set $B_n' := A_n \cap U$ (which is nonempty and relatively open in A_n, and is a possible move of the player Ω). Denote $A_{n+1}' := \sigma(A_1, \ldots, B_n')$, which is the answer of the player Σ by means of the strategy σ. Let $p' \in P$ be some play in $G^*(Y)$ that starts with the partial play $(A_1, \ldots, A_{n+1}')$. Clearly, $d(p_0, p') \le 1/n$. Moreover, the closed ball $B[p_0, 1/n] := \{p \in P : d(p_0, p) \le 1/n\}$ contains the ball $B[p', 1/(n+1)]$, and for every play p'' in the latter ball, we have $F(p'') \subseteq B_n' \subseteq U$. Put V_n to be the interior of $B[p', 1/(n+1)]$. Thus, for every integer $n \in \mathbb{N}$, we have found a nonempty open subset $V_n \subseteq B[p_0, 1/n]$ such that $F(V_n) \subseteq U$. The set $V := \bigcup_{n\ge1} V_n$ satisfies the requirements. □

Proposition 2.3.24. *Let f be an arbitrary selection of the minimal mapping $F : P \to 2^Y$ defined in the proof of Theorem 2.3.20. If f is continuous at some point $p_0 \in P$, then the play $p_0 := (A_i, B_i)_{i\geq 1}$ is won by the player Ω in the game $G^*(Y)$, and the mapping $F : P \to 2^Y$ is single-valued and upper semi-continuous at p_0.*

Proof. Let W be an open subset of (Y, τ') with $f(p_0) \in W$. Since f is continuous at $p_0 := (A_i, B_i)_{i\geq 1}$, there exists some open V', $p_0 \in V'$, with $f(V') \subseteq W$. We will show that there is an integer $n \in \mathbb{N}$ for which $A_n \subseteq \overline{W}$. Suppose this is not the case. Then the open set $U := Y \setminus \overline{W}$ intersects all sets A_n with $n \in \mathbb{N}$. By Lemma 2.3.23 there is an open set $V \subseteq P$ such that $p_0 \in \overline{V}$ and $f(V) \subseteq F(V) \subseteq U$. Then there is a point $p' \in V \cap V'$. For p', we have the contradiction: $f(p') \in U \cap W = \varnothing$. This shows that for some $n \in \mathbb{N}$, $A_n \subseteq \overline{W}$. In other words, $F(B[p_0, 1/n)] \subseteq \overline{W}$. Since W was an arbitrary open neighbourhood of $f(p_0)$, we derive that $F(p_0) = \{f(p_0)\}$, that F is upper semi-continuous at p_0 and that the play p_0 is won by the player Ω in the game $G^*(Y)$. This completes the proof of Theorem 2.3.20. □

There are also some interesting results concerning the norm continuity of quasi-continuous mappings into the space $C(Y)$ of continuous real-valued functions on a compact space Y equipped with the topology of pointwise convergence [133, 166].

Exercises

1. Show that the composition of two almost continuous functions need not be almost continuous.
2. Show that the Dirichlet function is not cliquish at any point.
3. Prove that every α-favourable space is a Baire space.
4. Prove that every Čech-complete space is α-favourable.
5. Prove Proposition 2.3.21.

2.4 Quasicontinuity and measurability

It seems from the previous results that quasicontinuous functions are closely related to continuous functions. Thus we could expect that quasicontinuous functions have some known properties of continuous functions, for example, the measurability. However, it is known that there is a quasicontinuous function $g : [0, 1] \to \mathbb{R}$ that is not Lebesgue measurable [66, 187]. Marcus [188] used in his example a fat Cantor set. A *fat Cantor set* is a closed nowhere dense subset of $\mathbb{R}$ of positive Lebesgue measure. Fat Cantor sets are also called *Smith–Volterra–Cantor sets*. We offer a very simple example of a quasicontinuous non-Lebesgue-measurable function using a fat Cantor set.

Example 2.4.1. Let D be a fat Cantor set (i. e., a nowhere dense closed set of positive Lebesgue measure) in $[0, 1]$. Let N be a non-Lebesgue-measurable subset of D (here we use the fact the every Borel subset of $\mathbb{R}$ of positive Lebesgue measure contains a non-Lebesgue measurable set). Denote $d(x, D) := \inf\{|x - d| : d \in D\}$ for $x \in \mathbb{R}$. Let $f : [0,1] \to [-1,1]$ be defined by

$$f(x) := \begin{cases} \sin(1/d(x,D)) & \text{if } x \notin D, \\ 1 & \text{if } x \in N, \\ 0 & \text{if } x \in D \setminus N. \end{cases}$$

It is easy to verify that f is quasicontinuous and that the set $f^{-1}(\{1\})$ is non-Lebesgue measurable. Thus f is a non-Lebesgue-measurable quasicontinuous function.

Next, we show that if (X, τ) is either an uncountable Polish space or a locally pathwise connected perfectly normal topological space with at least one non-isolated point, then there is a quasicontinuous non-Borel-measurable function from X to $[0,1]$.

Most of the results in this part are contained in [108].

We call a subset E of a pseudometric space (X, d) *ε-discrete* if $d(x, y) > \varepsilon$ whenever $x, y \in E$ and $x \neq y$; see also ε-nets and ε-separated sets.

First, we introduce the following lemma, which will be useful in the next section.

Lemma 2.4.2 ([32]). *Let (X, d) be a pseudometric space. Let F be a closed nowhere dense subset of (X, d). Then there is a quasicontinuous function $f : X \to [0,1]$ such that f is continuous at each point $x \notin F$, $f(z) = 0$ for every $z \in F$ and in every neighbourhood V of $z \in F$, there are $x, y \in V \setminus F$ with $f(x) = 0$ and $f(y) = 1$.*

Proof. Let F be a closed nowhere dense subset of (X, d). We will construct sets $(S : n \in \mathbb{N})_n$ in the following way. Let $S_1 := \varnothing$. Assume that we have constructed S_i for $i < n$. Denote

$$T_n := \left\{x \in X \setminus \left(F \cup \bigcup_{i<n} S_i\right) : d(x, F) < \frac{1}{n}\right\} \text{ and } \mathfrak{P}_n := \{P \subseteq T_n : P \text{ is } 1/n - \text{discrete}\}.$$

According to the Zorn lemma, there is a maximal element S_n of $\mathfrak{P}_n$. Denote $A := \bigcup_{n=1}^{\infty} S_{2n}$ and $B := \bigcup_{n=1}^{\infty} S_{2n+1}$. We will show that $\overline{A} = A \cup F$.

(i) Let $x \in F$. Let U be a neighbourhood of x. Then there is an even number n such that $B(x, \frac{2}{n}) \subseteq U$ ($B(u, \varepsilon)$ is the open ball of radius $\varepsilon > 0$ about u). If $B(x, \frac{2}{n}) \cap S_n = \varnothing$, then for $i < n$, we have $d(u, v) = 0$ for each $u, v \in S_i \cap B(x, \frac{1}{2n})$. Therefore there is $y \in B(x, \frac{1}{2n}) \cap T_n$. However, then $d(y, z) > \frac{1}{n}$ for each $z \in S_n$, and hence S_n is not maximal in $\mathfrak{P}_n$, a contradiction. Therefore $B(x, \frac{2}{n}) \cap S_n \neq \varnothing$, and hence $x \in \overline{A}$.

(ii) Now let $x \in \overline{A} \setminus F$. Then $d(x, F) > \frac{1}{n}$ for some $n \in \mathbb{N}$. Since $x \in \overline{A}$, there is a sequence $(z_k : k \in \mathbb{N})$ in $B(x, \frac{1}{2n})$ converging to x. However, $z_k \notin S_i$ for $i \geqq n$, and hence we may assume that $z_k \in S_j$ for some even $j < n$ and each $k \in \mathbb{N}$. Then $d(z_s, z_t) = 0$ for all $s, t \in \mathbb{N}$, and hence $d(z_t, x) = 0$ for each $t \in \mathbb{N}$. Since S_j is a maximal element in $\mathfrak{P}_j$, we have $x \in S_j \subseteq A$.

Therefore we have shown that $\overline{A} = A \cup F$. Similarly, we can prove that $\overline{B} = B \cup F$. Since the sets $(S_i : i \in \mathbb{N})$ are mutually disjoint, $A \cap B = \emptyset$. Thus $\overline{A} \cap \overline{B} = F$. Therefore A and B are disjoint closed sets in $X \setminus F$, and hence there is a continuous function $g : X \setminus F \to [0,1]$ such that

$$g(x) := \begin{cases} 0 & \text{if } x \in A, \\ 1 & \text{if } x \in B. \end{cases}$$

Next, we define $f : X \to [0,1]$ as $f(x) := g(x)$ for $x \notin F$ and $f(x) := 0$ for $x \in F$. Then f has the desired properties. □

Proposition 2.4.3 ([108]). *Let (X, τ) be a separable metrizable space in which there is a closed nowhere dense subset F of (X, τ) with cardinality c. Then there is a quasicontinuous non-Borel-measurable function $g : X \to [0,1]$.*

Proof. Let $f : X \to [0,1]$ be a quasicontinuous function from Lemma 2.4.2. Since the cardinality $|\mathcal{B}|$ of Borel sets in a separable metrizable space is c, there must exist a non-Borel subset N of F. Define the function $g : X \to [0,1]$ as follows: $g(x) := f(x)$ for every $x \notin F$, $g(x) := 1$ for $x \in N$ and $g(x) := 0$ for $x \in F \setminus N$. It is easy to verify that g is a quasicontinuous function and the set $g^{-1}(\{1\})$ is non-Borel. Thus g is a non-Borel-measurable quasicontinuous function. □

A topological space (X, τ) is *completely metrizable* if it admits a compatible metric d such that (X, d) is complete. A separable completely metrizable space is called a *Polish space* [158].

Corollary 2.4.4 ([108]). *Let (X, τ) be an uncountable Polish space. Then there is a quasicontinuous non-Borel-measurable function $g : X \to [0,1]$.*

Proof. (M. Sleziak), [158, Theorem 6.2], shows that every uncountable Polish space contains a closed subspace homeomorphic to $2^{\mathbb{N}}$. Put

$$F := \{(x_0, x_0, x_1, x_1, x_2, x_2, \ldots, x_n, x_n, \ldots) : (x_0, x_1, x_2, \ldots, x_n, \ldots) \in 2^{\mathbb{N}}\}.$$

It is easy to verify that F is a closed nowhere dense set in $2^{\mathbb{N}}$ with cardinality c. Thus a homeomorphic image of F is also a closed nowhere dense set in (X, τ). Now we can use Proposition 2.4.3. □

Lemma 2.4.5 ([32]). *Let (X, τ) be a locally connected perfectly normal topological space. Let F be a closed nowhere dense subset of (X, τ). Then there is a quasicontinuous function $f : X \to [0,1]$ such that f is continuous at each point $x \notin F$, $f(z) = 0$ for every $z \in F$ and in every neighbourhood V of $z \in F$, there are $x, y \in V \setminus F$ with $f(x) = 0$ and $f(y) = 1$.*

Proof. Let F be a closed nowhere dense set in (X, τ). Then there is a continuous function $h : X \to [0,1]$ such that $h^{-1}(0) = F$. Define $g : X \setminus F \to [0,1]$ as $g(x) = |\sin(1/h(x))|$

for all $x \in X \setminus F$. Then evidently g is continuous. Let $x \in F$, and let U be a neighbourhood of x. We can assume that U is open and connected. There is $\beta > 0$ such that $[0, \beta) \subseteq h(U)$. This yields that there are points $u, v \in U \setminus F$ such that $g(u) = 0$ and $g(v) = 1$. Next, define $f : X \to [0,1]$ by $f(x) := g(x)$ for $x \notin F$ and $f(x) := 0$ for $x \in F$. Then f has the desired properties. □

Proposition 2.4.6 ([108]). *Let (X, τ) be a locally pathwise connected perfectly normal topological space with at least one non-isolated point. Then there is a quasicontinuous non-Borel-measurable function $g : X \to [0,1]$.*

Proof. Let $x \in X$ be a non-isolated point. There exists a neighbourhood V of x such that for any $y \in V \setminus \{x\}$, there exists a homeomorphic embedding $h : [0,1] \to X$ satisfying $h(0) = x$ and $h(1) = y$. Fix $y \in V \setminus \{x\}$ and a homeomorphic embedding $h : [0,1] \to X$ with $h(0) = x$ and $h(1) = y$. Let C be the Cantor set in $[0,1]$. Then $h(C)$ is a closed nowhere dense set in X with cardinality c. Since $h([0,1])$ has a countable base, there is a non-Borel subset L of $h(C)$. By Lemma 2.4.5 there is a quasicontinuous function $f : X \to [0,1]$ such that f is continuous at each point $x \notin h(C)$, $f(z) = 0$ for every $z \in h(C)$, and in every neighbourhood V of $z \in h(C)$, there are $x, y \in V \setminus h(C)$ with $f(x) = 0$ and $f(y) = 1$. Define the function $g : X \to [0,1]$ as follows: $g(x) := f(x)$ for $x \notin h(C)$, $g(x) := 0$ for $x \in L$ and $g(x) := 1$ for $x \in h(C) \setminus L$.

It is easy to verify that g is a quasicontinuous function and the set $g^{-1}(\{0\})$ is a non-Borel subset. Thus g is a non-Borel measurable function. □

We say that a topological space (X, τ) has the property *CP* (*QP*) [33] if for every nonempty nowhere dense closed set F in (X, τ), there is a continuous (quasicontinuous) function $g : (X \setminus F) \to [0,1]$ such that $w_g(x) = 1$ for every $x \in F$. Thus every pseudometrizable topological space, and every perfectly normal locally connected topological space has the property CP, [32].

Evidently, every space with the property CP has the property QP. It was proven in [33] that the Niemytzki plane is a space with the property QP but not the CP property, whereas $\beta\omega$, the Čech–Stone compactification of the space ω, is a space without the QP property.

It was proved in [66] that if (X, d) and (Y, ρ) are compact metric spaces and $g : X \to Y$ is a quasicontinuous (and possibly even non-Borel-measurable) function, then there is a quasicontinuous Borel-measurable function $h : X \to Y$ such that $\overline{\mathrm{Gr}(h)} = \overline{\mathrm{Gr}(g)}$ [66].

We have an even more general result, which generalises the result from [66]. To prove it, we will use the following result proved by Matejdes [191].

Proposition 2.4.7 ([191]). *Let (X, τ) and (Y, τ') be topological spaces and suppose that (Y, τ') is regular. Let $f : X \to Y$ be a quasicontinuous function. If $g : X \to Y$ is a function such that $Gr(g) \subseteq \overline{Gr(f)}$, then g is quasicontinuous, and $\overline{Gr(g)} = \overline{Gr(f)}$.*

Proof. Let $g : X \to Y$ be a function such that $\mathrm{Gr}(g) \subseteq \overline{\mathrm{Gr}(f)}$. To prove the quasicontinuity of g, let $x \in X$, let U be an open neighbourhood of x, and V be an open neighbourhood of $g(x)$. Let G be an open neighbourhood of $g(x)$ such that $g(x) \in G \subseteq \overline{G} \subseteq V$. There is $z \in U$ such that $f(z) \in G$. The quasicontinuity of f at z implies that there is $H \in \tau, H \neq \emptyset, H \subseteq U$, such that $f(H) \subseteq G$, and thus $g(H) \subseteq V$. To prove that $\mathrm{Gr}(f) \subseteq \overline{\mathrm{Gr}(g)}$, let $(x, f(x)) \in \mathrm{Gr}(f)$, let U be an open neighbourhood of x, and V be an open neighbourhood of $f(x)$. Again, let $G \in \tau'$ be such that $f(x) \in G \subseteq \overline{G} \subseteq V$. The quasicontinuity of f at x implies that there is $H \in \tau, H \neq \emptyset, H \subseteq U$, such that $f(H) \subseteq G$. Thus $g(H) \subseteq V$, that is, $\mathrm{Gr}(f) \subseteq \overline{\mathrm{Gr}(g)}$. □

Theorem 2.4.8 ([108]). *Let (X,τ) be a topological space, and let (Y,τ') be a completely metrizable σ-compact space. If $f : X \to Y$ is a quasicontinuous function, then there is a quasicontinuous Borel-measurable function $g : X \to Y$ such that $\overline{\mathrm{Gr}(f)} = \overline{\mathrm{Gr}(g)}$.*

Proof. Define the set-valued mapping $F : X \to 2^Y$ by $F(x) := \mathrm{USC}(f)(x)$ for every $x \in X$. By Exercise 1.1.16, $\mathrm{Gr}(F) = \overline{\mathrm{Gr}(f)}$. Since (Y,τ') is a σ-compact metrizable space, F is Borel measurable. (Let $V \in \tau'$. It is easy to verify that there is a sequence $\{K_n : n \in \mathbb{N}\}$ of compact sets in Y such that $V = \bigcup\{K_n : n \in \mathbb{N}\}$. Then $F^{-1}(V) = \bigcup\{F^{-1}(K_n) : n \in \mathbb{N}\}$, and $F^{-1}(K_n)$ is a closed set in (X,τ) for every $n \in \mathbb{N}$.) By [24, Theorem 6.6.7] F has a Borel-measurable selection g, that is, $\mathrm{Gr}(g) \subseteq \mathrm{Gr}(F) = \overline{\mathrm{Gr}(f)}$. By Proposition 2.4.7, g is quasicontinuous and $\overline{\mathrm{Gr}(g)} = \overline{\mathrm{Gr}(f)}$. □

Corollary 2.4.9 ([66]). *Let (X,d) and (Y,ρ) be compact metric spaces. If $f : X \to Y$ is a quasicontinuous function, then there is a quasicontinuous Borel-measurable function $g : X \to Y$ such that $\overline{\mathrm{Gr}(f)} = \overline{\mathrm{Gr}(g)}$.*

We need to recall some definitions before we state our next results. In a topological space (X,τ) the *Borel sets of additive class* 0 are just the open sets, and the *Borel sets of multiplicative class* 0 are just the closed sets. For each ordinal α with $0 < \alpha < \omega_1$ (here ω_1 denotes the first uncountable ordinal), the *Borel sets of additive class α* are just the countable unions of Borel sets of multiplicative class less than α, and the *Borel sets of multiplicative class α* are just the countable intersections of Borel sets of additive class less than α. The Borel sets of *ambiguous class α* are those that are simultaneously of additive class α and multiplicative class α. In particular, the sets of *additive class* 1 are the F_σ-sets, and the sets of *multiplicative class* 1 are the G_δ-sets.

Let (X,τ) be a topological space, and let (Y,d) be a metric space. By a *Borel-measurable function of class α* $(0 \le \alpha < \omega_1)$ [105] we mean a function $f : X \to Y$ such that $f^{-1}(G)$ is a Borel set of additive class α in (X,τ) whenever G is an open set in (Y,d). In particular, if $f : X \to Y$ is a *Borel-measurable function of the first class*, then $f^{-1}(G)$ is an F_σ-subset of (X,τ) for each open subset G in (Y,d).

Theorem 2.4.10 ([108]). *Let (X,τ) be a Baire space, and let (Y,d) be a separable metric space. If $f : X \to Y$ is a Borel-measurable function of the first class, then f is cliquish.*

Proof. Let $x \in X$ and $\varepsilon > 0$, and let O be an open neighbourhood of x. Let $\{y_i : i \in \mathbb{N}\}$ be a countable dense set in (Y, d). Therefore $Y = \bigcup_{i\in\mathbb{N}} B(y_i, \varepsilon/2)$. Then $O = (\bigcup_{i\in\mathbb{N}} f^{-1}(B(y_i, \varepsilon/2))) \cap O$. Since f is a Borel-measurable function of the first class, for each $i \in \mathbb{N}$, we have $f^{-1}(B(y_i, \varepsilon/2)) = \bigcup_{n\in\mathbb{N}} F_n^i$, where each F_n^i is a closed set in (X, τ) for every $n \in \mathbb{N}$. Thus $O = (\bigcup_{i\in\mathbb{N}} \bigcup_{n\in\mathbb{N}} F_n^i) \cap O$.

The Baireness of (X, τ) implies that there are $i, n \in \mathbb{N}$ such that $F_n^i \cap O$ is not nowhere dense. Therefore there exists $V \in \tau$ such that $\emptyset \neq V \subseteq \overline{F_n^i \cap O}$. Thus $V \cap O \neq \emptyset$, and $V \cap O \subseteq F_n^i \subseteq f^{-1}(B(y_i, \varepsilon/2))$, i. e., $f(V \cap O) \subseteq B(y_i, \varepsilon/2)$. Then, by the triangle inequality,

$$d(f(z), f(v)) \leq d(f(z), y_i) + d(y_i, f(v)) < \varepsilon/2 + \varepsilon/2 = \varepsilon \quad \text{for every } z, v \in V \cap O,$$

and thus f is cliquish at $x \in X$. □

Theorem 2.4.11 ([108]). *Let (X, τ) be a topological space in which every open set is an F_σ set. The following are equivalent:*
(i) *(X, τ) is a Baire space;*
(ii) *Every Borel-measurable function of the first class from (X, τ) to $\mathbb{R}$ is cliquish.*

Proof. It is sufficient to prove that (ii)⇒(i). To do this, we will prove the contrapositive statement. Suppose that (X, τ) is not a Baire space. Then there is a nonempty open set U in (X, τ) of the first Baire category in (X, τ). Thus there is a sequence $(L_n : n \in \mathbb{N})$ of nowhere dense sets such that $U = \bigcup_{n\in\mathbb{N}} L_n$. Furthermore, since every open subset of (X, τ) is an F_σ set, there is a sequence $(F_k : k \in \mathbb{N})$ of closed subsets of (X, τ) such that $U = \bigcup_{k\in\mathbb{N}} F_k$. Thus $U = \bigcup_{k,n\in\mathbb{N}} F_k \cap \overline{L_n}$. Hence, without loss of generality, we can suppose that there is a sequence $(H_n : n \in \mathbb{N})$ of closed nowhere dense sets such that

$$U = \bigcup_{n\in\mathbb{N}} H_n \quad \text{and} \quad H_n \subseteq H_{n+1} \quad \text{for all } n \in \mathbb{N}.$$

We will define the function $f : X \to \mathbb{R}$ as follows: $f(x) := 0$ for $x \in X \setminus U$, $f(x) := 1$ for $x \in H_1$ and $f(x) := n$ for $x \in H_n \setminus H_{n-1}$. For every $n \geq 2$, the set $H_n \setminus H_{n-1}$ is an F_σ set. Thus f is a Borel-measurable function of the first class. However, it is easy to verify that f is not cliquish on (X, τ). □

By Corollary 2.4.4, for every uncountable Polish space, there is a quasicontinuous function $f : X \to \mathbb{R}$ that is not Borel measurable and thus not Borel measurable of the first class.

Exercises

1. Let (X, τ) and (Y, τ') be topological spaces. Let $F : X \to 2^Y$ be a minimal set-valued mapping such that $F(x) \neq \emptyset$ for every $x \in X$. Prove that every selection $f : X \to Y$ of F is quasicontinuous.

2. Let (X, τ) and (Y, τ') be topological spaces, and let (Y, τ') be a regular space. Then the following are equivalent [191]:
   (a) $f : X \to Y$ is quasicontinuous,
   (b) $\mathrm{USC}(f) : X \to 2^Y$ is minimal,
   (c) any selection g of $\mathrm{USC}(f)$ is quasicontinuous, and $\mathrm{Gr}(f) \subseteq \overline{\mathrm{Gr}(g)}$,
   (d) there is a quasicontinuous selection g of $\mathrm{USC}(f)$ for which $\mathrm{Gr}(f) \subseteq \overline{\mathrm{Gr}(g)}$.
3. Let (X, τ) be a topological space, and let (Y, d) be a metric space. Show that if $f : X \to Y$ and $Q(f)$ is dense in (X, τ), then f is cliquish.
4. Let (X, τ) and (Y, τ') be topological spaces. Show that if $f : X \to Y$ is quasicontinuous and D is a dense set in (X, τ), then $\overline{\mathrm{Gr}(f)} = \overline{\mathrm{Gr}(f|_D)}$.
5. Define $\Phi : \mathbb{R} \to 2^{\mathbb{R}}$ by

$$\Phi(x) := \begin{cases} \sin(1/d(x, C)) & \text{if } x \notin C, \\ [-1, 1] & \text{if } x \in C, \end{cases}$$

   where C is any compact fat nowhere dense subset of $\mathbb{R}$. Find a non-Lebesgue-measurable selection of Φ.
6. Provide another proof of Corollary 2.4.4. *Hint:* Theorem 6.2 and Exercise 3.12 in [158] show that every uncountable Polish space contains a G_δ subspace homeomorphic to the space $\mathbb{R}\backslash\mathbb{Q}$ of irrational numbers equipped with the usual topology. Let D be a closed nowhere dense subset of $\mathbb{R} \setminus \mathbb{Q}$ with cardinality c. Let F be the closure in X of a homeomorphic image of D. Then F is a closed nowhere dense set in X with cardinality c. Finally, apply Proposition 2.4.3.
7. Let (X, τ) and (Y, τ') be topological spaces, and let $f : X \to Y$. Show that if $H \in \tau$, then

$$\mathrm{USC}(f)(H) \subseteq \overline{f(H)}.$$

8. Let (X, τ) and (Y, τ') be topological spaces. A set-valued mapping $F : X \to 2^Y$ is *lower (resp., upper) quasicontinuous at a point* $x \in X$ [249] if for every $V \in \tau'$ with $F(x) \cap V \neq \emptyset$ (resp., $F(x) \subseteq V$) and every neighbourhood U of x, there is $G \in \tau, G \neq \emptyset, G \subseteq U$, such that

$$F(s) \cap V \neq \emptyset \text{ (resp., } F(s) \subseteq V) \quad \text{for every } s \in G.$$

   A set-valued mapping $F : X \to 2^Y$ is *lower (resp., upper) quasicontinuous* if it is lower (resp., upper) quasicontinuous at each point of X.
   Show that if $f : X \to Y$ is quasicontinuous, then $\mathrm{USC}(f) : X \to 2^Y$ is lower quasicontinuous [127].
9. A function $f : X \to \mathbb{R}$ is said to be *upper (lower) quasicontinuous at* $x \in X$ if for every $\varepsilon > 0$ and for every neighbourhood U of x, there is a nonempty open set $G \subseteq U$ such that $f(v) < f(x) + \varepsilon$ ($f(v) > f(x) - \varepsilon$) for every $v \in G$. A function $f : X \to \mathbb{R}$ is *upper (lower) quasicontinuous* if it is upper (lower) quasicontinuous at every $x \in X$ [79].

Let (Y, ρ) be a metric space. For $y \in Y$ and $\Phi : X \to 2^Y$, we define the function $g_y : X \to \mathbb{R}$ by $g_y(x) = \rho(y, \Phi(x))$ for all $x \in X$.
Prove the following properties [143]: If $\Phi : X \to 2^Y$ is a set-valued mapping, then
(a) Φ is lower semi-continuous if and only if g_y is upper semi-continuous for every $y \in Y$.
(b) If Φ is upper semi-continuous, then g_y is lower semi-continuous for each $y \in Y$.
Prove the following properties [128]. If $\Phi : X \to 2^Y$ is a set-valued mapping, then
(a) Φ is lower quasicontinuous if and only if g_y is upper quasicontinuous for every $y \in Y$.
(b) If Φ is upper quasicontinuous, then g_y is lower quasicontinuous for each $y \in Y$.

2.5 Limits of quasicontinuous functions

Of course, it is very easy to verify that the uniform limit of a net of quasicontinuous functions is quasicontinuous and that the pointwise limit of a sequence of even continuous functions need not be quasicontinuous. However, it is known that the pointwise limit of an equicontinuous net of functions is continuous. Of course, equicontinuity is too strong; it is not necessary to guarantee the continuity of the pointwise limit of a net of continuous functions.

In [26], necessary and sufficient conditions for continuity of the pointwise limit of a net of continuous functions are given. Bledsoe [30] proved the following theorem concerning the pointwise limit of a sequence of quasicontinuous mappings with values in a metric space.

Theorem 2.5.1. *Let (X, τ) be a topological space, and let (Y, d) be a metric space. If $(f_n : n \in \mathbb{N})$ is a sequence of quasicontinuous functions defined on (X, τ) with values in (Y, d) that is pointwise convergent to $f : X \to Y$, then the set $D(f)$ of all points of discontinuity of f is of the first category in (X, τ).*

Proof. We have $D(f) = \bigcup_{k=1}^{\infty} \{x \in X : \frac{1}{k} \le w_f(x)\}$. The desired conclusion is a consequence of the following statement. Let $n \in \mathbb{N}$ and $\varepsilon > 0$. Then the set

$$A_n^\varepsilon := \{x \in X : \varepsilon \le w_f(x) \text{ and } d(f(x), f_m(x)) \le \varepsilon/16 \text{ for all } n \le m\}$$

is nowhere dense in (X, τ). So let us suppose, to obtain a contradiction, that A_n^ε is not nowhere dense. Then there exists $U \in \tau, U \neq \emptyset$, such that A_n^ε is dense in U. Let $x_1 \in A_n^\varepsilon \cap U$. The quasicontinuity of f_n at x_1 implies that there is a nonempty open subset $U_1 \subseteq U$ such that $d(f_n(x_1), f_n(z)) \le \varepsilon/16$ for $z \in U_1$.

Let $x \in U_1$. There is $m \in \mathbb{N}, n \le m$, such that $d(f_m(x), f(x)) \le \varepsilon/16$. The quasicontinuity of f_m at x implies that there is a nonempty open set $U_2 \subseteq U_1$ such that

$d(f_m(x), f_m(z)) \leq \varepsilon/16$ for $z \in U_2$. Let $x_2 \in A_n^\varepsilon \cap U_2$. From above and the fact that $x_2 \in A_n^\varepsilon$ and $x_1 \subset A_n^\varepsilon$ we have

$$d(f(x), f(x_1)) \leq d(f(x), f_m(x)) + d(f_m(x), f_m(x_2)) + d(f_m(x_2), f(x_2)) \\ + d(f(x_2), f_n(x_2)) + d(f_n(x_2), f_n(x_1)) + d(f_n(x_1), f(x_1)) \leq 3\varepsilon/8.$$

Thus $d(f(x), f(x_1)) \leq 3\varepsilon/8$ for $x \in U_1$. Accordingly, $d(f(x), f(y)) \leq 3\varepsilon/4$ for $x \in U_1$ and $y \in U_1$. Hence $w_f(x) \leq 3\varepsilon/4$ for $x \in U_1$. Thus $A_n^\varepsilon \cap U_1 = \emptyset$, which contradicts the density of A_n^ε in U. Therefore A_n^ε is nowhere dense in (X, τ). □

The same result was rediscovered later by Giles and Bartlett [92].

Corollary 2.5.2. *Let (X, τ) be a Baire space, and let (Y, d) be a metric space. If $(f_n : n \in \mathbb{N})$ is a sequence of quasicontinuous functions defined on (X, τ) with values in (Y, d) that is pointwise convergent to $f : X \to Y$, then $C(f)$ is a dense G_δ-set in (X, τ), and f is a cliquish function.*

Conversely, if (X, d) is a metric space and (Y, ρ) is a separable metric space, then each function $f : X \to Y$ with $C(f)$ dense in (X, d) is the pointwise limit of a sequence of quasicontinuous functions from (X, d) to (Y, ρ) (see [270], for $X := \mathbb{R}$ [31, 99]).

The following definition and results are from [119].

Let (X, τ) be a topological space, let (Y, d) be a metric space, and let $(f_n : n \in \mathbb{N})$ be a sequence of functions defined on (X, τ) with values in (Y, d). We say that the sequence $(f_n : n \in \mathbb{N})$ is *equiquasicontinuous at $x \in X$* if for every $\varepsilon > 0$ and every open neighbourhood U of x, there are $n_0 \in \mathbb{N}$ and $W \in \tau, W \neq \emptyset$, $W \subseteq U$, such that $d(f_n(z), f_n(x)) < \varepsilon$ for every $z \in W$ and every $n_0 \leq n$. We say that $(f_n : n \in \mathbb{N})$ is *equiquasicontinuous on* (X, τ) if it is equiquasicontinuous at every $x \in X$.

Remarks 2.5.3. Notice that in a natural way we can also define equiquasicontinuity for a net of functions.

Of course, every equicontinuous sequence is equiquasicontinuous, and there are easy examples of equiquasicontinuous sequences that are not equicontinuous.

Proposition 2.5.4. *Let (X, τ) be a topological space, let (Y, d) be a metric space, and let $(f_n : n \in \mathbb{N})$ be a sequence of functions defined on a topological space (X, τ) with values in (Y, d) that is pointwise convergent to a function $f : X \to Y$. If $(f_n : n \in \mathbb{N})$ is equiquasicontinuous at $x \in X$, then f is quasicontinuous at x.*

Proof. Let $\varepsilon > 0$, and let $U \in \tau$ with $x \in U$. There are $n_0 \in \mathbb{N}$ and $W \in \tau, W \neq \emptyset$, $W \subseteq U$, such that $d(f_n(x), f_n(z)) < \varepsilon/3$ for every $n_0 \leq n$ and every $z \in W$. Let $w \in W$. The pointwise convergence of $(f_n : n \in \mathbb{N})$ to f implies that there is $n_0 \leq n_1$ such that

$$d(f_n(w), f(w)) < \varepsilon/3 \quad \text{and} \quad d(f_n(x), f(x)) < \varepsilon/3 \quad \text{for every } n_1 \leq n.$$

Then we have:

$$d(f(x),f(w)) \le d(f(x),f_{n_1}(x)) + d(f_{n_1}(x),f_{n_1}(w)) + d(f_{n_1}(w),f(w)) < \varepsilon/3 + \varepsilon/3 + \varepsilon/3.$$

This completes the proof. □

The following theorem was first proved in [119] but later simplified in [33].

Theorem 2.5.5. *Let (X,τ) be a Baire space, and let (Y,d) be a metric space and suppose that $(f_n : n \in \mathbb{N})$ is a sequence of quasicontinuous functions defined on (X,τ) with values in (Y,d) that is pointwise convergent to a function $f : X \to Y$. Then the following are equivalent:*
(i) *f is quasicontinuous;*
(ii) *$(f_n : n \in \mathbb{N})$ is equiquasicontinuous.*

Proof. (ii)⇒(i) is established by Proposition 2.5.4. Now we prove (i)⇒(ii). Let $x_0 \in X$, let U be a neighbourhood of x_0, and $\varepsilon > 0$. The quasicontinuity of f at x_0 implies the existence of a nonempty open subset $U_1 \subseteq U$ such that $d(f(x),f(x_0)) < \varepsilon/4$ for every $x \in U_1$. For every $n \in \mathbb{N}$, put

$$A_n^\varepsilon := \{x \in U_1 : \text{ for every } n \le m : d(f_m(x),f(x)) < \varepsilon/4\}.$$

The pointwise convergence of $(f_k : k \in \mathbb{N})$ to f implies that $U_1 = \bigcup_n A_n^\varepsilon$. Since (X,τ) is a Baire space, there is $n_0 \in \mathbb{N}$ such that $A_{n_0}^\varepsilon$ is not nowhere dense in U_1. There is a nonempty open set $V \subseteq U_1$ such that $A_{n_0}^\varepsilon$ is dense in V. There is $n_0 < n_1$ such that $d(f_n(x_0),f(x_0)) < \varepsilon/4$ for every $n_1 \le n$. We claim that for every $n_1 \le n$ and for every $x \in V$, we have $d(f_n(x),f_n(x_0)) < \varepsilon$. Let $n_1 \le n$ and $x \in V$. The quasicontinuity of f_n at x implies the existence of a nonempty open subset $G \subseteq V$ such that $d(f_n(t),f_n(x)) < \varepsilon/4$ for every $t \in G$. Since $A_{n_0}^\varepsilon$ is dense in V, there exists a $z \in A_{n_0}^\varepsilon \cap G$. Thus

$$d(f_n(x),f_n(x_0)) \le d(f_n(x),f_n(z)) + d(f_n(z),f(z)) + d(f(z),f(x_0)) + d(f(x_0),f_n(x_0)) < \varepsilon.$$

This completes the proof. □

A variant of the above theorem was proved by Bouziad and Troallic [47, Theorem 4.5].

Theorem 2.5.6. *Let (X,d) be a metric space. Then the following are equivalent:*
(i) *X is Baire;*
(ii) *If $(f_n : n \in \mathbb{N})$ is a sequence of real-valued quasicontinuous functions defined on (X,d) that is pointwise convergent to a quasicontinuous function $f : X \to \mathbb{R}$, then $(f_n : n \in \mathbb{N})$ is equiquasicontinuous.*

Proof. Only (ii)⇒(i) needs some explanation. Suppose (X,d) is not a Baire space. Let G be a nonempty open set in (X,d) of the first Baire category. Let $(K_n : n \in \mathbb{N})$ be a sequence of nowhere dense subsets of G such that $G = \bigcup\{K_n : n \in \mathbb{N}\}$. For every $n \in \mathbb{N}$, put

$$L_n := \bigcup\{\overline{K_i} : i \leq n\} \cup (\overline{G} \setminus G).$$

Then $\overline{G} = \bigcup\{L_n : n \in \mathbb{N}\}$. Of course, every L_n is a closed nowhere dense set in $\overline{G}$. Let $n \in \mathbb{N}$. We will apply Lemma 2.4.2 on $\overline{G}$ and L_n. There is a quasicontinuous function $f_n^* : \overline{G} \to [0,1]$ such that $f_n^*(z) = 0$ for every $z \in L_n$, every $z \in L_n$ and every neighbourhood V of $z \in L_n$, there is $y \in V$ with $f_n^*(y) = 1$. Let $f_n : X \to [0,1]$ be a function defined as follows: $f_n(x) := f_n^*(x)$ for $x \in \overline{G}$ and $f_n(x) := 0$ otherwise. Of course, f_n is also quasicontinuous. It is easy to verify that the sequence $(f_n : n \in \mathbb{N})$ pointwise converges to the function that is identically equal to 0. However, the sequence $(f_n : n \in \mathbb{N})$ is not equiquasicontinuous. Let $x \in G$ be arbitrary. We will show that $(f_n : n \in \mathbb{N})$ is not equiquasicontinuous at x. In fact, we claim that with $\varepsilon := 1/2$, for every $n \in \mathbb{N}$ and for every nonempty open set $V \subseteq G$, there are $n < k$ and $z_k \in V$ with

$$\varepsilon < 1 = |f_k(z_k) - f_k(x)|.$$

Let $n \in \mathbb{N}$, and let $V \subseteq G$ be a nonempty open set. Let $z \in V$ be arbitrary. There is $n < k$ with $x, z \in L_k$. Thus $f_k(x) = 0$ and also $f_k(z) = 0$. There is $z_k \in V$ with $f_k(z_k) = 1$. □

A notion which will be helpful later is that of a locally countable π-base (see [256] and also [314]). A π-base is *a locally countable π-base* if each its element contains only countably many members of the π-base.

Theorem 2.5.7. *Let (X, τ) be a quasi-regular T_1 topological space that has a locally countable π-base. Then the following are equivalent:*
(i) *(X, τ) is Baire;*
(ii) *If $(f_n : n \in \mathbb{N})$ is a sequence of real-valued quasicontinuous functions defined on (X, τ) that is pointwise convergent to a quasicontinuous function $f : X \to \mathbb{R}$, then $(f_n : n \in \mathbb{N})$ is equiquasicontinuous.*

Proof. Only (ii)⇒(i) needs some explanation. Suppose that (X, τ) is not a Baire space. Let $U \in \tau, U \neq \varnothing$, be of the first Baire category. Let τ be a locally countable π-base of (X, τ). Let $G \in \tau$ be a nonempty subset of U. Then, of course, also G is of the first Baire category.

Let $(V_n : n \in \mathbb{N})$ be a countable π-base of G, and let $(K_n : n \in \mathbb{N})$ be a sequence of nowhere dense subsets of G such that $G = \bigcup\{K_n : n \in \mathbb{N}\}$. For every $n \in \mathbb{N}$, we define the function $f_n : X \to \mathbb{R}$ as follows: $\bigcup\{K_i : i \leq n\}$ is nowhere dense in G (so also in (X, τ)); that is, $V_n \setminus \bigcup\{\overline{K_i} : i \leq n\} \neq \varnothing$. Moreover, the quasi-regularity of (X, τ) implies that there is an open set H_n such that

$$H_n \subseteq \overline{H_n} \subseteq V_n \setminus \bigcup\{\overline{K_i} : i \leq n\}.$$

Define the function f_n as follows: $f_n(x) := 1$ for $x \in \overline{H_n}$ and $f_n(x) := 0$ otherwise. It is easy to verify that each f_n is quasicontinuous and that the sequence $(f_n : n \in \mathbb{N})$ pointwise converges to the function f that is identically equal to 0.

We will show that the sequence $(f_n : n \in \mathbb{N})$ is not equiquasicontinuous. Let $x \in G$ be arbitrary. We will show that $(f_n : n \in \mathbb{N})$ is not equiquasicontinuous at x. In fact, we claim that with $\varepsilon := 1/2$, for every $n \in \mathbb{N}$ and every nonempty open set $V \subseteq G$, there are $n < k$ and $z_k \in V$ such that

$$|f_k(z_k) - f_k(x)| > \varepsilon.$$

Let $n \in \mathbb{N}$, and let $V \subseteq G$ be a nonempty open set. There is $m \in \mathbb{N}$ such that $V_m \subseteq V$. Since G is of the first Baire category it has no isolated points; thus every open subset of G contains infinitely many points and infinitely many elements of the sequence $(V_n : n \in \mathbb{N})$. Let $n_0 \in \mathbb{N}$ be such that $x \in K_{n_0}$. There is $k \in \mathbb{N}$ with $\max\{n, n_0\} < k$ and $V_k \subseteq V_m$. Let $x_k \in V_k$ be such that $f_k(x_k) = 1$. Then we have

$$\varepsilon < 1 = |f_k(x_k) - f_k(x)|.$$

Thus $(f_n : n \in \mathbb{N})$ is not equiquasicontinuous at x, a contradiction with (ii). □

There are interesting connections between quasicontinuity and transfinite convergence.

Let ω_1 be the first uncountable ordinal number. A transfinite sequence $(a_\xi : \xi < \omega_1)$ of elements of a topological space (Y, τ') is said to be *convergent to* $a \in Y$ if for any neighbourhood V of a, there is $0 \leq \xi_0 < \omega_1$ such that $a_\xi \in V$ for every $\xi_0 < \xi$.

A transfinite sequence $(f_\xi : \xi < \omega_1)$ of functions defined on a set X and taking values in a topological space (Y, τ') is said to be *pointwise convergent to* $f : X \to Y$ if $(f_\xi(x) : \xi < \omega_1)$ converges to $f(x)$ for every $x \in X$ [282].

The following is a useful lemma.

Lemma 2.5.8 ([282] for real-valued functions [173]). *Let $(f_\xi : \xi < \omega_1)$ be a transfinite sequence of functions defined on a set X and taking values in a first countable T_1 topological space (Y, τ') that pointwise converges to $f : X \to Y$. Let $S \subseteq X$ be a countable set. Then there is $\xi_0 < \omega_1$ such that $f_\xi(x) = f(x)$ for every $\xi_0 < \xi < \omega_1$ and every $x \in S$.*

The following theorem was proved by Neubrunnová [252].

Theorem 2.5.9. *Let (X, τ) be a locally separable first countable topological space. Let (Y, τ') be a first countable T_1 topological space. If $(f_\xi : \xi < \omega_1)$ is a transfinite sequence of quasicontinuous functions $f_\xi : X \to Y$ that pointwise converges to a function $f : X \to Y$. Then f is quasicontinuous.*

Proof. Suppose that f is not quasicontinuous at x_0. Then there exist a neighbourhood V of $f(x_0)$ and a neighbourhood U of x_0 such that for every nonempty open set $G \subseteq U$, there is $x \in G$ such that $f(x) \notin V$. The neighbourhood U may be supposed to be separable. Let

$$M := \{x : x \in U \text{ and } f(x) \notin V\}.$$

Let D be a countable dense set in U. For $s \in D$, let $(B_n^s \subseteq U : n \in \omega)$ be a countable basis of neighbourhoods of s. There exist $x_n^s \in B_n^s$ such that $x_n^s \in M$. The set

$$T := \{x_n^s : s \in D \text{ and } n \in \omega\}$$

is a countable dense set in U. Using Lemma 2.5.8, we obtain $\xi_0 < \omega_1$ such that

$$f_\xi(x) = f(x) \quad \text{for every } x \in T \cup \{x_0\} \text{ and every } \xi_0 < \xi < \omega_1.$$

Using the quasicontinuity of f_{ξ_0} at x_0, we have a nonempty open set $G \subseteq U$ such that

$$f_{\xi_0}(x) \in V \quad \text{for every } x \in G.$$

However, $G \cap T \neq \emptyset$, so we have $x \in G \cap T \subseteq M$ such that $f(x) = f_{\xi_0}(x) \in V$. This is a contradiction. □

Theorem 2.5.10 ([252]). *A metric space (X, d) is locally separable if and only if for any subspace $Y \subseteq X$, we have the following: If $(f_\xi : \xi < \omega_1)$ is a transfinite sequence of quasicontinuous functions $f_\xi : Y \to \mathbb{R}$ that pointwise converge to a function $f : Y \to \mathbb{R}$, then f is quasicontinuous on Y.*

Proof. The sufficiency follows from Theorem 2.5.9. We give a sketch of proof of necessity (see [252]). To do this, we consider the contrapositive statement. Suppose that (X, d) is not locally separable. Then there exists an isolated set $M \subseteq X$ such that M has a condensation point y [250], that is, $U \cap M$ is uncountable for any neighbourhood U of y. Denote

$$I_n := \{x \in M : \ d(x, y) < 1/n\}.$$

Now we construct a transfinite sequence $(x_\xi : \ \xi < \omega_1)$ in the following way. Choose $x_1 \in M$ arbitrarily. Then suppose that $(x_\eta : \eta < \xi)$ is constructed for some $1 < \xi < \omega_1$. If ξ is not a limit number, then $\xi = \xi_0 + n$, where ξ_0 is a limit number, and n an integer. Then we choose $x_\xi \in I_n$, $x_\xi \neq x_\eta$, for $\eta < \xi$. Put $Z := \{x_\xi : \ \xi < \omega_1\} \cup \{y\}$. Then Z is not a locally separable subspace of (X, d). For each $\xi < \omega_1$, we define $f_\xi : Z \to \mathbb{R}$ by

$$f_\xi(x) := \begin{cases} 0 & \text{if } x = x_\eta \text{ and } \eta < \xi, \\ 1 & \text{if } x = x_\eta \text{ and } \eta \geq \xi, \\ 1 & \text{if } x = y. \end{cases}$$

We omit the details included in [252] showing that f_ξ are quasicontinuous for $\xi < \omega_1$ and that the limit function

$$f(x) := \begin{cases} 0 & \text{if } x \in Z \setminus \{y\}, \\ 1 & \text{if } x = y \end{cases}$$

is not quasicontinuous on Z. □

A sequence of mappings $(f_n : n \in \mathbb{N})$ from a topological space (X, τ) into a metric space (Y, d) converging pointwise to a mapping f is called *quasi-uniformly convergent* if for any $\varepsilon > 0$ and $N \in \mathbb{N}$, there exist a countable open cover $(O_n : n \in \mathbb{N})$ of (X, τ) and a sequence $(n_k : k \in \mathbb{N})$ of natural numbers greater than N such that $d(f(x), f_{n_k}(x)) < \varepsilon$ for every $x \in O_k$.

For sequences of continuous functions, quasi-uniform convergence is a necessary and sufficient condition for the limit function to be continuous (the Arzelá–Aleksandrov theorem [2]).

The quasi-uniform convergence of quasicontinuous functions was discussed in [72]. It was proven in [72] that quasi-uniform convergence does not preserve quasicontinuity. For the convenience of the reader, we present an example substantiating this fact.

Example 2.5.11 ([72]). For every $n \in \mathbb{N}$, let f_n be the functions from $\mathbb{R}$ to $\mathbb{R}$ defined by $f_n := \chi_{[0,1/n]}((-1)^n x)$, where χ_E denotes the characteristic function of a set E. Then $(f_n : n \in \mathbb{N})$ quasi-uniformly converges to $\chi_{\{0\}}$, but $\chi_{\{0\}}$ is not quasicontinuous.

Exercises
1. Show that uniform convergence preserves quasicontinuity.
2. Find examples of equiquasicontinuous sequences that are not equicontinuous.
3. Prove Lemma 2.5.8.
4. Verify Example 2.5.11.

2.6 Applications of quasicontinuity

The notion of quasicontinuity has recently been instrumental in the proof that some semitopological groups are in fact topological groups [7, 45, 46, 165, 220, 222, 225, 262], in proofs of some generalisations of Michael's selection theorem [92, 232], in the characterisations of minimal usco and cusco mappings [36, 37, 118, 122, 192, 193], and in the study of dynamical systems [66]. Quasicontinuity also appears in the study of CHART groups [225] that arise in the study of topological dynamics. In particular, CHART groups appear in the study of distal flows on compact spaces [224, 229, 242, 244].

Our first application of the notion of quasicontinuity is in the realm of semitopological groups. Since semitopological groups are defined in terms of separately continuous functions, we will take this opportunity to restate the definition of a separately continuous function.

Let X, Y, Z be sets, and let $f : X \times Y \to Z$. For $x \in X$, by f_x we denote the function from Y to Z defined by

$$f_x(y) := f(x,y) \quad \text{for } y \in Y.$$

Similarly, for $y \in Y$, by f^y we denote the function from X to Z defined by

$$f^y(x) := f(x,y) \quad \text{for } x \in X.$$

If (X,τ), (Y,τ') and (Z,τ'') are topological spaces, then we say that f is *separately continuous at* $(x_0,y_0) \in X \times Y$ if $f_{x_0} : (Y,\tau') \to (Z,\tau'')$ is continuous at y_0 and $f^{y_0} : (X,\tau) \to (Z,\tau'')$ is continuous at x_0. If f_x is continuous on (Y,τ') and f^y is continuous on (X,τ) for every $(x,y) \in X \times Y$, then we say that f is *separately continuous on* $X \times Y$.

Recall that a triple $(G,\cdot,\tau)$ is called a *semitopological group* if $(G,\cdot)$ is a group, (G,τ) is a topological space, and the multiplication operation $(g,h) \mapsto g \cdot h$, is separately continuous on $G \times G$.

If the multiplication operation is continuous on $(G \times G, \tau \times \tau)$, then we call $(G,\cdot,\tau)$ a *paratopological group*, and we say that the multiplication operation is *jointly continuous on* $G \times G$. We note that there exist paratopological groups that are not topological groups.

Example 2.6.1. $(\mathbb{R},+,\tau_S)$, where τ_S (the Sorgenfrey topology) is the topology on $\mathbb{R}$ generated by the sets $\{[a,b) : a,b \in \mathbb{R} \text{ and } a < b\}$. Note that in this example, $(\mathbb{R},+,\tau_S)$ is a paratopological group (i. e. $(x,y) \mapsto x+y$ is jointly continuous), $(\mathbb{R},\tau_S)$ is a Baire space, but the inversion is not continuous, that is, $x \mapsto (-x)$ is not continuous.

Theorem 2.6.2 ([165, Lemma 4]). *Let $(G,\cdot,\tau)$ be a paratopological group. If the inversion is quasicontinuous at e, then $(G,\cdot,\tau)$ is a topological group.*

Proof. Since $(G,\cdot,\tau)$ is a paratopological group, it suffices to show that the inversion is continuous on (G,τ). In fact, because $(G,\cdot,\tau)$ is a semitopological group, it suffices to show that the inversion is continuous at $e \in G$. To this end, let W be any neighbourhood of e. Since G is a paratopological group, there exists a neighbourhood U of e such that $U \cdot U \subseteq W$. Now, since the inversion is quasicontinuous at e, there is a nonempty open subset V of U such that $V^{-1} \subseteq U$. Hence $V \cdot V^{-1}$ is an open neighbourhood of e, and

$$(V \cdot V^{-1})^{-1} = V \cdot V^{-1} \subseteq U \cdot U \subseteq W.$$

This completes the proof. □

Exercise 2.6.3. Let $(G,\cdot,\tau)$ be a semitopological group. Show that the multiplication operation is quasicontinuous on $G \times G$ if and only if the multiplication operation is strongly quasicontinuous with respect to the second variable at $(e, e) \in G \times G$, and this holds if and only if the multiplication operation is quasicontinuous at $(e, e) \in G \times G$, which in turn holds if and only if the multiplication operation is strongly quasicontinuous on $G \times G$ with respect to the second variable.

Next, we will see the impact that quasicontinuity has upon the study of the continuity of group homomorphisms. We recall, of course, that a function $f : H \to G$ acting between groups $(H, *)$ and $(G, \cdot)$ is a *homomorphism* if $f(g * h) = f(g) \cdot f(h)$ for all $g, h \in H$.

Proposition 2.6.4. *Suppose that $f : H \to G$ is a homomorphism acting between a semitopological group $(H, *, \tau)$ and a topological group $(G, \cdot, \tau')$. Then f is continuous on H if and only if f is quasicontinuous at e_H, the identity element of H.*

Proof. Clearly, if f is continuous on H, then it is quasicontinuous at e_H. So suppose that f is quasicontinuous at e_H. Let us now exploit the easily verifiable fact that it is sufficient to show that f is continuous at e_H. (This is true since f is a group homomorphism.) To this end, let W be an open neighbourhood of $f(e_H) = e_G$, the identity element of G. (This is true since f is a group homomorphism.) Since $(G, \cdot, \tau')$ is a topological group, there exists an open neighbourhood U of e_G such that $U^{-1} \cdot U \subseteq W$. Since f is quasicontinuous at e_H, there exists a nonempty open subset V of H such that $f(V) \subseteq U$. Now $e_H \in V^{-1} * V$, $V^{-1} * V$ is open and $f(V^{-1} * V) = f(V)^{-1} \cdot f(V) \subseteq U^{-1} \cdot U \subseteq W$. This completes the proof. □

Proposition 2.6.5. *Let $(H, *, \tau)$ be a semitopological group, let (X, τ') be a topological space, and let $(G, \cdot, \tau'')$ be a topological group. Suppose also that $f : H \times X \to G$, $x_0 \in X$, and for each $x \in X$, the mapping $h \mapsto f(h, x)$ is a group homomorphism from $(H, *)$ into $(G, \cdot)$. Then f is strongly quasicontinuous with respect to the second variable at (e_H, x_0) if and only if f is continuous at (e_H, x_0).*

Proof. Clearly, if f is continuous at (e_H, x_0), then f is strongly quasicontinuous with respect to the second variable at (e_H, x_0). So we only consider the converse.

Let W be an open neighbourhood of $f(e_H, x_0) = e_G$. (This is true since $h \mapsto f(h, x_0)$ is a group homomorphism.) Since $(G, \cdot, \tau'')$ is a topological group, there exists an open neighbourhood W_0 of e_G such that $e_G \in W_0^{-1} \cdot W_0 \subseteq W$. Now, since f is strongly quasicontinuous with respect to the second variable at (e_H, x_0), there exist a nonempty open subset U of H and a neighbourhood V of x_0 such that $f(U \times V) \subseteq W_0$. We claim that $f(U^{-1} * U \times V) \subseteq W_0^{-1} \cdot W_0 \subseteq W$. Indeed, suppose that $h_1, h_2 \in U$ and $x \in V$. Then

$$f(h_1^{-1} * h_2, x) = f(h_1^{-1}, x) \cdot f(h_2, x) = f(h_1, x)^{-1} \cdot f(h_2, x) \in W_0^{-1} \cdot W_0 \subseteq W.$$

Since $U^{-1} * U$ is a neighbourhood of e_H, this shows that f is continuous at (e_H, x_0). □

Combining Proposition 2.6.5 with Theorem 2.2.7 (Theorem 2.2.15), we can obtain a rather pleasing result concerning joint continuity.

Theorem 2.6.6. *Suppose that $(H,*,\tau)$ is a semitopological group, (X,τ') is a topological space, $(G,\cdot,\tau'')$ is a topological group, and $f : H\times X \to G$ is a separately continuous function such that for each $x\in X$, the mapping $h\mapsto f(h,x)$ is a group homomorphism from $(H,*)$ into $(G,\cdot)$. Suppose also that (H,τ) is a Baire space (a regular countably Čech-complete space) and (X,τ') is a first countable space (a q-space). Then f is continuous on $H\times X$.*

Proof. First, recall that the topology on any topological group is regular. By Theorem 2.2.7 (Theorem 2.2.15) we see that f is strongly quasicontinuous with respect to the second variable at each point of $H\times X$ and, in particular, at each point of $\{e_H\}\times X$. Therefore, by Proposition 2.6.5, f is continuous at each point of $\{e_H\}\times X$.

Let (h,x) be an arbitrary element of $H\times X$, and let W be an open neighbourhood of $f(h,x)$. Note that since $(G,\cdot,\tau'')$ is a topological group, there exist an open neighbourhood W_1 of $f(h,x)$ and an open neighbourhood W_2 of e_G such that $W_1\cdot W_2\subseteq W$.

Firstly, since f is separately continuous, there exists an open neighbourhood V_1 of x such that $f(\{h\}\times V_1)\subseteq W_1$. Secondly, since f is continuous at (e_H,x), there exist an open neighbourhood U_2 of e_H and V_2 of x such that $f(U_2\times V_2)\subseteq W_2$. Let $U' := h*U_2$ and $V' := V_1\cap V_2$. We claim that $f(U'\times V')\subseteq W$. To prove this, consider $g\in U'$ and $y\in V'$. Then $g=h*u$ for some $u\in U$, and

$$f(g,y)=f(h*u,y)=f(h,y)\cdot f(u,y)\in W_1\cdot f(u,y)\in W_1\cdot W_2\subseteq W.$$

Since $h\in U'$ and $x\in V'$, we have that f is continuous at (h,x). □

Our first application of Theorem 2.6.6 involves group endomorphisms.

Let X and Y be arbitrary sets, and let $A\subseteq X$ and $B\subseteq Y$. We denote by $F(A;B)$ for the set of all functions from X to Y that map A into B, that is,

$$F(A;B) := \{f\in Y^X : f(A)\subseteq B\}.$$

If (X,τ) and (Y,τ') are topological spaces, then the *compact-open* (*pointwise*) topology on Y^X is the topology generated by the sets

$$\{F(A;B) : A\in\mathscr{A} \text{ and } B\in\mathscr{G}\},$$

where $\mathscr{A}$ denotes the class of all compact subsets (finite subsets) of (X,τ), and $\mathscr{G}$ denotes the class of all open subsets of (Y,τ').

For a topological group $(G,\cdot,\tau)$, we denote by $End_p(G)$ the space of all continuous *endomorphisms* on G, that is, homomorphisms from G into G, endowed with the topology of pointwise convergence on G.

Corollary 2.6.7 ([52]). *Let $(G,\cdot,\tau)$ be a countably Čech-complete topological group. If Σ is a q-subspace of $End_p(G)$ (i. e. each point in Σ is a q-point), then the mapping $\pi : End_p(G)\times\Sigma\to End_p(G)$ defined by $\pi(m,m') := m'\circ m$ is continuous. In particular, $\pi|_{\Sigma\times\Sigma}$ is continuous on $\Sigma\times\Sigma$.*

Proof. Consider the mapping $f : G\times\Sigma \to G$ defined by $f(g, m) := m(g)$ for $(g, m) \in G\times\Sigma$. Then f satisfies the hypotheses of Theorem 2.6.6 and so is jointly continuous on $G \times \Sigma$. Let $(m, m') \in End_p(G) \times \Sigma$. We will show that π is continuous at (m, m'). To this end, let $g \in G$, and let W be an open neighbourhood of $\pi(m, m')(g)$. Then

$$\pi(m, m')(g) = (m' \circ m)(g) = m'(m(g)) = f(m(g), m') \in W.$$

Since f is continuous at $(m(g), m') \in G \times \Sigma$, there exist open neighbourhoods U' of $m(g)$ in (G, τ) and V of m' in Σ such that $f(U' \times V) \subseteq W$. Furthermore, there exists an open neighbourhood U of m in $End_p(G)$ such that $u(g) \in U'$ for all $u \in U$. Therefore $\pi(U \times V)(g) \subseteq f(U' \times V) \subseteq W$. Hence π is continuous on $End_p(G) \times \Sigma$. □

Let $(G, \cdot, \tau)$ and $(H, *, \tau')$ be topological groups. We denote by $Hom_p(H; G)$ the space of all continuous homomorphisms from H into G endowed with the topology of pointwise convergence on H.

Corollary 2.6.8 ([52]). *Let $(G, \cdot, \tau)$ and $(H, *, \tau')$ be topological groups, and let Σ be a subset of $Hom_p(H; G)$. If (H, τ') is countably Čech-complete and Σ is a q-subspace of $Hom_p(H; G)$, then on Σ the pointwise and compact-open topologies coincide.*

Proof. Since the compact-open topology is always finer than (or equal to) the pointwise topology, it is sufficient to show that for each compact subset $K \subseteq H$ and open set $W \subseteq G$, $F(K; W) \cap \Sigma$ is open in the pointwise topology on Σ. To this end, let K be a nonempty compact subset of H, let W be a nonempty open subset of G, and let $m_0 \in F(K; W) \cap \Sigma$, that is, $m_0 \in \Sigma$ and $m_0(K) \subseteq W$. From Theorem 2.6.6 it follows that the mapping $f : H \times \Sigma \to G$ defined by $f(h, m) := m(h)$ is continuous on $H \times \Sigma$, and so, by a simple compactness argument it follows that there exist an open set U in H and a pointwise open neighbourhood V of m_0 in Σ such that $K \subseteq U$ and $f(U \times V) \subseteq W$. In particular, this means that $m_0 \in V \subseteq F(K; W) \cap \Sigma$, and so $F(K; W) \cap \Sigma$ is open in the pointwise topology on Σ. □

Next, we present a proof of Glicksberg's theorem for locally compact Hausdorff Abelian groups. To state and prove Glicksberg's theorem [96, Theorem 1.2], we first need to recall some facts concerning topological groups.

We denote $\mathbb{T} := \{z \in \mathbb{C} : |z| = 1\}$ and consider this set endowed with the binary operation of complex number multiplication. Then $(\mathbb{T}, \cdot, \tau)$ is a compact Hausdorff Abelian topological group, where the topology τ is the relative topology on $\mathbb{T}$ inherited from $\mathbb{C}$.

If $(G, \cdot, \tau)$ is a topological group, then $Hom(G, \mathbb{T})$ is the set of all *characters* on G, which is sometimes denoted as $\chi(G)$. We note that $\chi(G)$ naturally becomes an Abelian group under pointwise multiplication, denoted by $*$. In fact, if τ' denotes the compact-open topology on $\chi(G)$, then $(\chi(G), *, \tau')$ is a locally compact Hausdorff Abelian topological group [107].

Theorem 2.6.9 (Pontryagin duality theorem [238, 264]). *If $(G, \cdot, \tau)$ is a locally compact Hausdorff Abelian group, then the natural homomorphism $\omega : G \to \chi(\chi(G))$, mapping $g \in G$ to the character $\omega_g : \chi(G) \to \mathbb{T}$, defined by*

$$\omega_g(\alpha) := \alpha(g) \quad \text{for } \alpha \in \chi(G)$$

is an isomorphism of topological groups.

The final ingredient that we require for the proof of Glicksberg's theorem is the following definition. If $(G, \cdot, \tau)$ is a locally compact Hausdorff Abelian group, then the weakest topology of G that makes the function $w : G \to Hom_p(\chi(G), \mathbb{T})$ continuous is called the *Bohr topology on G*.

Theorem 2.6.10 (Glicksberg's theorem [96, Theorem 1.2]). *If $(G, \cdot, \tau)$ is a locally compact Hausdorff Abelian group, then every subset of G that is compact in the Bohr topology on G is τ-compact.*

Proof. Let A be a subset of G that is compact in the Bohr topology on G. Then $\omega(A)$ is a compact subset of $Hom_p(\chi(G), \mathbb{T})$. In particular, every point of $\omega(A)$ is a q-point with respect to the relative pointwise topology on $\omega(A)$. Hence, by Corollary 2.6.8, $\omega(A)$ is compact in $\chi(\chi(G))$. Since ω is a homeomorphism, A is compact with respect to τ. □

Of course, several other similar theorems could be stated using Corollary 2.6.8. For example, every subset of a locally compact Hausdorff Abelian topological group that is countably compact in the Bohr topology is countably τ-compact, and so on.

Our next application of Theorem 2.6.6 involves Banach spaces and differentiability.

We say that a mapping $f : X \to Y$ acting between Banach spaces $(X, \|\cdot\|)$ and $(Y, |\!|\!|\cdot|\!|\!|)$ is *almost C^1* on $(X, \|\cdot\|)$ if for each $y \in X$, the mapping $x \mapsto f'(x; y)$ defined by

$$f'(x; y) := \text{weak-}\lim_{t \to 0} \frac{f(x + ty) - f(x)}{t},$$

where the limit is with respect to the weak topology on $(X, \|\cdot\|)$, is norm-to-norm continuous on $(X, \|\cdot\|)$.

Corollary 2.6.11. *Let $f : X \to Y$ be a continuous mapping acting between Banach spaces $(X, \|\cdot\|)$ and $(Y, |\!|\!|\cdot|\!|\!|)$. If f is almost C^1 on $(X, \|\cdot\|)$, then the mapping $(x, y) \mapsto f'(x; y)$ is jointly norm continuous on $X \times X$.*

Proof. Fix $x_0 \in X$. We will first show that the mapping $y \mapsto f'(x_0; y)$ is linear. To do this, it is sufficient to show that for each $y^* \in Y^*$, the mapping $y \mapsto y^*(f'(x_0; y))$ is linear on X. Fix $y^* \in Y^*$ and let $g : X \to \mathbb{R}$ be defined by $g(x) := (y^* \circ f)(x)$. Then g is continuous and almost C^1 with $g'(x; y) = y^*(f'(x; y))$ for each $y \in X$. It now follows, as in the finite-dimensional case (see [4], p. 261), that the mapping $y \mapsto g'(x_0; y)$ is linear on X (Note that $g'(x_0, y)$ is linear on X if it is linear on every two-dimensional subspace of X.) Next, let $\{t_n : n \in \mathbb{N}\}$ be any sequence of positive numbers converging to 0 and define $f_n : X \to Y$ by

$$f_n(y) := \frac{f(x_0 + t_n y) - f(x_0)}{t_n} \quad \text{for } y \in X.$$

Then each f_n is continuous, and weak-$\lim_{n\to\infty} f_n(y) = f'(x_0; y)$ for each $y \in X$. Therefore by the uniform boundedness theorem [75, page 66] the sequence of functions $(f_n : n \in \mathbb{N})$ is pointwise bounded. For each $n \in \mathbb{N}$, let $F_n := \{x \in X : \|f_k(x)\| \leq n \text{ for all } k \in \mathbb{N}\}$. Then each set F_n is closed, and $X = \bigcup_{n\in\mathbb{N}} F_n$. Therefore by the Baire category theorem there exists $n_0 \in \mathbb{N}$ such that $U := \operatorname{int}(F_{n_0}) \neq \emptyset$. Thus $y \mapsto f'(x_0; y)$ is bounded on U. However, since $y \mapsto f'(x_0; y)$ is linear, it is bounded and thus continuous on X. The result then follows from Theorem 2.6.6. □

A further application of our earlier results on quasicontinuity to the study of differentiability is given next.

Let $f : \mathbb{R}^n \to \mathbb{R}$ $(f(x) := f(x_1, \ldots, x_n))$. A well-known theorem of mathematical analysis states that the continuity of (finite) partial derivatives of f is sufficient for the differentiability of f. On the other hand, there are well-known examples showings that the existence of finite partial derivatives does not imply differentiability. In this connection an interesting result has been achieved by S. Marcus. A simple proof of the mentioned result, which we give for the convenience of the reader, is a simple application of two our theorems on quasicontinuity.

Theorem 2.6.12 ([187]). *Let $n \in \mathbb{N}$. If $f : \mathbb{R}^n \to \mathbb{R}$ possesses finite partial derivatives $\frac{\partial f}{\partial x_i}$ $(i = 1, 2, \ldots, n)$ on $\mathbb{R}^n$, then the set of those points where f is not differentiable is of the first Baire category in $(\mathbb{R}^n, \|\cdot\|_2)$, where $\|\cdot\|_2$ denotes the Euclidean norm on $\mathbb{R}^n$.*

Proof. For $i = 1, 2, \ldots, n$, we have $\frac{\partial f}{\partial x_i} = \lim_{m\to\infty} g_{m,i}$, where

$$g_{m,i}(x_1, \ldots, x_n) := \frac{f(x_1, \ldots, x_i + \frac{1}{m}, \ldots, x_n) - f(x_1, \ldots, x_n)}{1/m} \quad \text{for } (x_1, \ldots, x_n) \in \mathbb{R}^n.$$

The functions $g_{m,i}$ are separately continuous, and hence by Theorem 2.2.4 they are quasicontinuous on $(\mathbb{R}^n, \|\cdot\|_2)$. So $\frac{\partial f}{\partial x_i}$ is the limit of a sequence of quasicontinuous functions. By Theorem 2.5.1 the set of discontinuity points of each partial derivative $\frac{\partial f}{\partial x_i}$ is of the first Baire category in $(\mathbb{R}^n, \|\cdot\|_2)$. Thus, with the exception of a set of first Baire category, all the partial derivatives are continuous, and hence f is differentiable with the exception of a set of first Baire category. □

Note that a theorem more general than Theorem 2.6.12 was obtained in [310] by a different method. Namely, the existence of partial derivatives on a dense G_δ subset of $\mathbb{R}^n$ implies the differentiability on a dense G_δ subset of $\mathbb{R}^n$.

In the subsequent part of this section, we examine the impact of quasicontinuity on the study of group actions.

Let $(G, \cdot)$ be a group, and let X be a set. Then any mapping $\pi : G \times X \to X$ such that

(i) $\pi(e,x) = x$ for all $x \in X$, where e denotes the identity element of G, and
(ii) $\pi(gh,x) = \pi(g,\pi(h,x))$ for all $g,h \in G$ and $x \in X$

is called a *group action on* X.

The following example is the canonical example of a group action.

Example 2.6.13. Let $(G,\cdot,\tau)$ be a semitopological group and define $\pi : G \times G \to G$ by

$$\pi(h,g) := h \cdot g \quad \text{for all } (h,g) \in G \times G.$$

Then π is a group action on the set G. In fact, π is a separately continuous group action on the space (G,τ).

As the next proposition shows, group actions possess desirable continuity properties.

Proposition 2.6.14. *Let $(G,\cdot,\tau)$ be a semitopological group, and let (X,τ') be a topological space. If $x_0 \in X$ and $\pi : G\times X \to X$ is a separately continuous group action on (X,τ), then π is continuous on $G \times \{x_0\}$ if and only if there exists a point $g_0 \in G$ such that π is continuous at (g_0,x_0).*

Proof. Suppose that π is continuous at $(g_0,x_0) \in G \times X$. Let g be an arbitrary element of G, and let W be an open neigbourhood of $\pi(g,x_0)$. Then $\pi(\{g_0g^{-1}\} \times W)$ is an open neighbourhood of $\pi(g_0,x_0)$. To see this, note that:
(i) for each $h \in G$, the mapping $x \mapsto \pi(h,x)$ is a homeomorphism on (X,τ') with inverse $x \mapsto \pi(h^{-1},x)$;
(ii) $\pi(g_0,x_0) = \pi(g_0g^{-1},\pi(g,x_0)) \in \pi(\{g_0g^{-1}\} \times W)$.

Since π is continuous at (g_0,x_0), there exist open neighbourhoods U of g_0 and V of x_0 such that $\pi(U \times V) \subseteq \pi(\{g_0g^{-1}\} \times W)$. Now $g \in g(g_0)^{-1}U$, and

$$\begin{aligned}\pi(g(g_0)^{-1}U \times V) &= \pi(\{g(g_0)^{-1}\} \times \pi(U \times V))\\ &\subseteq \pi(\{g(g_0)^{-1}\} \times \pi(\{g_0g^{-1}\} \times W))\\ &= \pi(\{e_G\} \times W) = W.\end{aligned}$$

This shows that π is continuous at (g,x_0) and hence continuous on $G \times \{x_0\}$. The converse is obvious. □

We may exploit Proposition 2.6.14 to obtain the following elegant theorem.

Theorem 2.6.15. *Let $(G,\cdot,\tau)$ be a Baire semitopological group, and let (X,τ') be a metrizable topological space. Then every separately continuous group action $\pi : G\times X \to X$ is continuous on $G \times X$.*

Proof. Let $\pi : G \times X \to X$ be a separately continuous group action on X, and let d be a metric on X that generates the topology τ'. Let (g,x) be any element of $G \times X$. By

Proposition 2.6.14, to show that π is continuous at (g, x), we need only show that there exists a point $g^* \in G$ such that π is continuous at (g^*, x). To this end, fix $n \in \mathbb{N}$ and consider the open set

$$O_n := \bigcup\{U \in \tau : \exists \text{ a neighbourhood } V \text{ of } x \text{ with } d\text{-diam}[\pi(U \times V)] < 1/n\}.$$

We will show that O_n is dense in G. To achieve this, consider a nonempty open subset U_0 of G. Let $g_0 \in U_0$. Since, by Theorem 2.2.7, π is strongly quasicontinuous with respect to the second variable at (g_0, x), there exist a nonempty open subset U of U_0 and a neighbourhood V of x such that $\pi(U \times V) \subseteq B(\pi(g_0, x), 1/(3n))$. Therefore $d\text{-diam}[\pi(U \times V)] < 1/n$, and so $\varnothing \neq U \subseteq O_n \cap U_0$. It now only remains to observe that π is continuous at each point of $(\bigcap_{n\in\mathbb{N}} O_n) \times \{x\}$, which is nonempty, since (G, τ) is a Baire space. □

As before, to escape beyond the realm of first countable spaces, we need to work harder.

Proposition 2.6.16. *Let $(G, \cdot, \tau)$ be a semitopological group, and let (X, τ') be a regular topological space. Suppose also that $\pi : G \times X \to X$ is a separately continuous group action on X, $x_0 \in X$, and $e_G \in G$ is a q-point. Then π is continuous on $G \times \{x_0\}$ if:*
(i) *π is strongly quasicontinuous with respect to the second variable at (e_G, x_0) and*
(ii) *there exist sequences of open neighbourhoods $(U_n : n \in \mathbb{N})$ of e_G and $(V_n : n \in \mathbb{N})$ of x_0 such that every sequence $(z_n : n \in \mathbb{N})$ in X with $z_n \in \pi(U_n \times V_n)$ for all $n \in \mathbb{N}$ has a cluster point in X.*

Proof. Let us first recall that by Proposition 2.6.14 it is sufficient to show that π is continuous at (e_G, x_0).

To this end, suppose that π is strongly quasicontinuous with respect to the second variable at (e_G, x_0) and there exist sequences of open neighbourhoods $(U_n : n \in \mathbb{N})$ of e_G and $(V_n : n \in \mathbb{N})$ of x_0 such that every sequence $(z_n : n \in \mathbb{N})$ in X with $z_n \in \pi(U_n \times V_n)$ for all $n \in \mathbb{N}$ has a cluster point in (X, τ'). Furthermore, since e_G is a q-point, there exists a sequence of open neighbourhoods $(O_n : n \in \mathbb{N})$ of e_G such that every sequence $(g_n : n \in \mathbb{N})$ in G with $g_n \in O_n$ for all $n \in \mathbb{N}$ has a cluster point in (G, τ).

To obtain a contradiction, we will assume that π is not continuous at (e_G, x_0). Then, since (X, τ') is regular, there exists an open neighbourhood W of $\pi(e_G, x_0) = x_0$ such that for every pair of open neighbourhoods U' of e_G and V' of x_0, $\pi(U' \times V') \not\subseteq \overline{W}$. Again, by the regularity of (X, τ') there exists an open neighbourhood V of x_0 such that $x_0 \in V \subseteq \overline{V} \subseteq W$. Let

$$V^* := \{g \in G : (g, x_0) \in \operatorname{int}(\pi^{-1}(V))\}.$$

Note that here the interior is with respect to the product topology on $G \times X$. Then, by the strong quasicontinuous of π with respect to the second variable, $e_G \in \overline{V^*}$. We will now inductively define sequences $(z_n : n \in \mathbb{N})$ in X and $(g_n : n \in \mathbb{N})$ in G and decreasing open neighbourhoods $(G_n : n \in \mathbb{N})$ and $(U_n' : n \in \mathbb{N})$ of e_G and $(V_n' : n \in \mathbb{N})$ of x_0.

Step 1. Choose $g_1 \in V^* \cap O_1$ and open neighbourhoods U_1' of e_G and V_1' of x_0 such that $U_1' \subseteq U_1$, $V_1' \subseteq V_1$, and $\pi(g_1 U_1' \times V_1') \subseteq V$. Then choose $z_1 \in \pi(U_1' \times V_1') \setminus \overline{W}$ and an open neighbourhood G_1 of e_G so that $\pi(G_1 \times \{z_1\}) \subseteq X \setminus \overline{W}$. For purely notational reasons, we denote $G_0 := G$, $U_0' := G$, and $V_0' := X$.

Now suppose that $g_j \in G$, $z_j \in X$, G_j, U_j', and V_j' have been defined for each $1 \le j \le n$, so that:

(i) $g_j \in V^* \cap (O_j \cap G_{j-1})$ and $\pi(g_j U_j' \times V_j') \subseteq V$;
(ii) $U_j' \subseteq U_{j-1}' \cap U_j$, $V_j' \subseteq V_{j-1}' \cap V_j$ and $G_j \subseteq G_{j-1}$.
(iii) $z_j \in \pi(U_j' \times V_j') \setminus \overline{W}$ and $\pi(G_j \times \{z_j\}) \subseteq X \setminus \overline{W}$.

Step $n+1$. Choose $g_{n+1} \in V^* \cap (O_{n+1} \cap G_n)$ and open neighbourhoods U_{n+1}' of e_G and V_{n+1}' of x_0 so that $U_{n+1}' \subseteq U_n' \cap U_{n+1}$, $V_{n+1}' \subseteq V_n' \cap V_{n+1}$, and $\pi(g_{n+1} U_{n+1}' \times V_{n+1}') \subseteq V$. Then choose $z_{n+1} \in \pi(U_{n+1}' \times V_{n+1}') \setminus \overline{W}$ and an open neighbourhood G_{n+1} of e_G so that $G_{n+1} \subseteq G_n$ and $\pi(G_{n+1} \times \{z_{n+1}\}) \subseteq X \setminus \overline{W}$.

This completes the induction.

Now since $z_j \in \pi(U_j' \times V_j') \subseteq \pi(U_j \times V_j)$ and $g_j \in O_j$ for all $j \in \mathbb{N}$, both sequences $(g_n : n \in \mathbb{N})$ and $(z_n : n \in \mathbb{N})$ have cluster points. Let z_∞ be any cluster point of $(z_n : n \in \mathbb{N})$, and let g_∞ be any cluster point of $(g_n : n \in \mathbb{N})$. Then, for each fixed $n \in \mathbb{N}$,

$$\pi(g_n, z_k) \subseteq \pi(\{g_n\} \times \pi(U_n' \times V_n')) = \pi(g_n U_n' \times V_n') \subseteq V \quad \text{for all } n \le k$$

since $U_k' \times V_k' \subseteq U_n' \times V_n'$. Therefore, $\pi(g_n, z_\infty) \in \overline{V}$ for each $n \in \mathbb{N}$ and so $\pi(g_\infty, z_\infty) \in \overline{V} \subseteq W$. On the other hand, if we again fix $n \in \mathbb{N}$, then

$$\pi(g_{k+1}, z_n) \in \pi(G_k \times \{z_n\}) \subseteq \pi(G_n \times \{z_n\}) \subseteq X \setminus \overline{W} \quad \text{for all } n \le k$$

since $g_{k+1} \in G_k \subseteq G_n$. Therefore, $\pi(g_\infty, z_n) \in X \setminus W$ for each $n \in \mathbb{N}$, and so $\pi(g_\infty, z_\infty) \in X \setminus W$. However, this contradicts our earlier conclusion that $\pi(g_\infty, z_\infty) \in W$. Hence π must be continuous at (e_G, x_0). □

We call a topological space (X, τ) *pointwise countably complete* if it is regular, and there exists a sequence $(\mathscr{A}_n : n \in \mathbb{N})$ of open covers of X such that every decreasing sequence $(F_n : n \in \mathbb{N})$ of nonempty subsets of X has $\bigcap_{n\in\mathbb{N}} \overline{F_n} \neq \varnothing$, provided that

(i) for each $n \in \mathbb{N}$, there exists $A_n \in \mathscr{A}_n$ such that $F_n \subseteq A_n$ and
(ii) $\bigcap_{n\in\mathbb{N}} A_n \neq \varnothing$.

Clearly, all metric spaces and regular countably Čech-complete spaces are pointwise countably complete. In the other direction, all pointwise countably complete spaces are q-spaces.

Theorem 2.6.17. *Let $(G, \cdot, \tau)$ be a regular countably Čech-complete semitopological group, and let (X, τ') be a pointwise countably complete topological space. Then every separately continuous group action $\pi : G \times X \to X$ is continuous on $G \times X$.*

Proof. Let x_0 be an arbitrary element of X. By Theorem 2.2.15, π is strongly quasicontinuous at $G \times \{x_0\}$, as X is a regular q-space. So by Proposition 2.6.16 we need only construct sequences of open neighhbourhoods $(U_n : n \in \mathbb{N})$ of e_G and $(V_n : n \in \mathbb{N})$ of x_0 such that every sequence $(z_n : n \in \mathbb{N})$ in X with $z_n \in \pi(U_n \times V_n)$ for all $n \in \mathbb{N}$ has a cluster point in (X, τ').

Let $(\mathscr{A}_n : n \in \mathbb{N})$ be a sequence of open covers of X such that every decreasing sequence $(F_n : n \in \mathbb{N})$ of nonempty subsets of X has $\bigcap_{n\in\mathbb{N}} \overline{F_n} \neq \varnothing$, provided that
(i) for each $n \in \mathbb{N}$, there exists $A_n \in \mathscr{A}_n$ such that $F_n \subseteq A_n$ and
(ii) $\bigcap_{n\in\mathbb{N}} A_n \neq \varnothing$.

Of course, such a sequence of open covers is guaranteed by the fact that (X, τ') is pointwise countably complete. Fix $n \in \mathbb{N}$, and consider the open set

$$O_n := \bigcup\{U \in \tau : \exists \text{ a neighbourhood } V \text{ of } x_0 \text{ with } \pi(U \times V) \subseteq A_n \text{ for some } A_n \in \mathscr{A}_n\}.$$

We will show that O_n is dense in (G, τ). To this end, let U_0 be a nonempty open subset of G, and let $g_0 \in U_0$. Since $\mathscr{A}_n$ is a cover of X, there exists $A_n \in \mathscr{A}_n$ such that $\pi(g_0, x_0) \in A_n$. Since π is strongly quasicontinuous with respect to the second variable at (g_0, x_0), there exist a nonempty open subset U of U_0 and an open neighbourhood V of x_0 such that $\pi(U \times V) \subseteq A_n$. Then $\varnothing \neq U \subseteq O_n \cap U_0$. Hence $\bigcap_{n\in\mathbb{N}} O_n$ is dense in (G, τ). Let $g^* \in \bigcap_{n\in\mathbb{N}} O_n$. Then, for each $n \in \mathbb{N}$, we may inductively define open neighbourhoods $(U'_n : n \in \mathbb{N})$ of g^* and $(V'_n : n \in \mathbb{N})$ of x_0 such that for all $n \in \mathbb{N}$:
(i) $g^* \in U'_{n+1} \subseteq U'_n$,
(ii) $x_0 \in V'_{n+1} \subseteq V'_n$, and
(iii) $\pi(U'_n \times V'_n) \subseteq A_n$ for some $A_n \in \mathscr{A}_n$.

Note that any sequence $(z_n : n \in \mathbb{N})$ in X with $z_n \in \pi(U'_n \times V'_n)$ for all $n \in \mathbb{N}$ has a cluster point in (X, τ').

For each $n \in \mathbb{N}$, let $U_n := (g^*)^{-1}U'_n$ and $V_n := V'_n$. Then $e_G \in U_n \in \tau$ and $x_0 \in V_n \in \tau'$ for all $n \in \mathbb{N}$, and any sequence $(z_n : n \in \mathbb{N})$ in X with $z_n \in \pi(U_n \times V_n)$ for all $n \in \mathbb{N}$ has a cluster point in (X, τ'). This completes the proof. □

As we will see further, quasicontinuity plays a significant role in the study of semitopological groups.

Next, we give the canonical construction of a semitopological group.

Example 2.6.18. Let (X, τ) be a nonempty topological space, and let G be a nonempty subset of X^X. If $(G, \circ)$ is a group (where "$\circ$" denotes the binary relation of function composition) and τ_p denotes the topology on X^X of pointwise convergence on X, then $(G, \circ, \tau_p)$ is a semitopological group, provided that the elements of G are continuous functions.

Not surprisingly, not all the semitopological groups described in Example 2.6.18 are topological groups.

Example 2.6.19. Let G denote the set of all homeomorphisms on $(\mathbb{R}, \tau_S)$. From Example 2.6.18 we see that $(G, \circ, \tau_p)$ is a semitopological group. However, $(G, \circ, \tau_p)$ is not a paratopological group.

To see this, define $g_n : \mathbb{R} \to \mathbb{R}$ by, $g_n(x) := [1 + 1/(n+1)]x$, $a_n := 1 + 1/(2n)$ and

$$f_n(x) := \begin{cases} x & \text{if } x \notin [a_n, a_n + 1/(2n)) \cup [n, n + 1/(2n)), \\ n + (x - a_n) & \text{if } x \in [a_n, a_n + 1/(2n)), \\ a_n + (x - n) & \text{if } x \in [n, n + 1/(2n)). \end{cases}$$

Then both $(f_n : n \in \mathbb{N})$ and $(g_n : n \in \mathbb{N})$ converge pointwise to id, the identity mapping; however,

$$\lim_{n\to\infty} (f_n \circ g_n)(1) = \lim_{n\to\infty} f_n(g_n(1)) = \infty \neq (\mathrm{id} \circ \mathrm{id})(1) = \mathrm{id}(1) = 1.$$

This shows that the multiplication operation is not continuous.

On the other hand, sometimes, Example 2.6.18 does give rise to topological groups.

Example 2.6.20. Let (M, d) be a metric space, and let G be the set of all isometries on (M, d). Then $(G, \circ, \tau_p)$ is a topological group.

Proof. Firstly, recall that a local subbase for the topology τ_p at an element $f \in G$ consists of all sets of the form $W(f, x, \varepsilon) := \{g \in G : d(f(x), g(x)) < \varepsilon\}$, where $x \in M$ and $\varepsilon > 0$. Using this, we will show that "$\circ$" is continuous on $G \times G$. To this end, let $(f, g) \in G \times G$ and suppose that $x \in M$ and $\varepsilon > 0$ are given. We claim that $W(f, g(x), \varepsilon/2) \circ W(g, x, \varepsilon/2) \subseteq W(f \circ g, x, \varepsilon)$; which is sufficient to show that "$\circ$" is continuous at (f, g). To prove the claim, suppose that $f' \in W(f, g(x), \varepsilon/2)$ and $g' \in W(g, x, \varepsilon/2)$. Then

$$\begin{aligned} 0 &\leq d((f' \circ g')(x), (f \circ g)(x)) \\ &\leq d((f' \circ g')(x), (f' \circ g)(x)) + d((f' \circ g)(x), (f \circ g)(x)) \\ &= d(g'(x), g(x)) + d(f'(g(x)), f(g(x))) < \varepsilon/2 + \varepsilon/2 = \varepsilon. \end{aligned}$$

Next, we need to show that inversion is continuous. To this end, let $f \in G$ and suppose that $x \in M$ and $\varepsilon > 0$ are given. We claim that $[W(f, f^{-1}(x), \varepsilon)]^{-1} \subseteq W(f^{-1}, x, \varepsilon)$, which is sufficient to show that the inversion is continuous at f. To prove the claim, suppose that $f' \in W(f, f^{-1}(x), \varepsilon)$. Then

$$0 \leq d((f')^{-1}(x), f^{-1}(x)) = d(x, f'(f^{-1}(x)) = d(f(f^{-1}(x)), f'(f^{-1}(x))) < \varepsilon.$$

Thus $(G, \circ, \tau_p)$ is a topological group. □

The key notion in this section on semitopological groups is a Δ-Baire space.

Let (X, τ) be a topological space. We say that a subset $W \subseteq X \times X$ is *separately open in the second variable* if for each $x \in X$, $\{z \in X : (x, z) \in W\} \in \tau$.

We say that a topological space (X,τ) is a Δ-*Baire space* if for each separately open, in the second variable, neighbourhood W of $\Delta_X := \{(x,y) \in X\times X : x = y\}$ in $X \times X$, there exists a nonempty open subset U of X such that $U\times U \subseteq \overline{W}^{\tau\times\tau}$.

Our interest in Δ-Baire spaces is revealed in the next lemma.

Lemma 2.6.21 ([226, 267, 301]). *Let $(G,\cdot,\tau)$ be a semitopological group, let W be an open neighbourhood of e, and let $\varphi : G\times G \to G$ be defined by $\varphi(h,g) := h^{-1}\cdot g$. If (G,τ) is a Δ-Baire space, then there exists an open neighbourhood U of e such that $\varphi(U\times U) = U^{-1}\cdot U \subseteq \overline{W\cdot W}$.*

Proof. Let W be an open neighbourhood of e. Since (G,τ) is a Δ-Baire space and $\varphi^{-1}(W)$ is a separately open, in the second variable, neighbourhood of Δ_G, there exists a nonempty open subset U of G such that $U\times U \subseteq \overline{\varphi^{-1}(W)}^{\tau\times\tau}$. We claim that

$$\varphi(U\times U) = U^{-1}\cdot U \subseteq \overline{W\cdot W}.$$

So let us suppose, to obtain a contradiction, that there exists $(x,y) \in U\times U$ such that $\varphi(x,y) = x^{-1}\cdot y \notin \overline{W\cdot W}$ (i. e., $e \notin x\cdot \overline{W\cdot W}\cdot y^{-1}$). From this it follows that there exists an open neighbourhood N of e such that $N \cap (x\cdot W\cdot W\cdot y^{-1}) = \varnothing$ or, equivalently, $(x^{-1}\cdot N\cdot y)\cap W\cdot W = \varnothing$. This in turn implies that $(W^{-1}\cdot x^{-1}\cdot N\cdot y)\cap W = \varnothing$. Thus we get that

$$\varphi(x\cdot W\times N\cdot y)\cap W = \left[(x\cdot W)^{-1}\cdot(N\cdot y)\right]\cap W = \left(W^{-1}\cdot x^{-1}\cdot N\cdot y\right)\cap W = \varnothing.$$

However, this is impossible since

$$(x\cdot W)\times(N\cdot y)\cap(U\times U)\cap\varphi^{-1}(W) \neq \varnothing.$$

Hence $\varphi(U\times U) \subseteq \overline{W\cdot W}$. □

Note that if $(G,\cdot,\tau)$ is a semitopological group and the multiplication operation is quasicontinuous at (e,e), then for each neighbourhood N of e, there exist an open neighbourhood V of e and an element $n \in N$ such that $V\cdot V\cdot n \subseteq N$. To see this, note that by quasicontinuity there exist nonempty open subsets U' and V' such that $V'\cdot U' \subseteq N$. Choose $v \in V'$ and let $V := V'\cdot(v^{-1})$ and $U := v\cdot U'$. Note that V is an open neighbourhood of e and $V\cdot U = V'\cdot U' \subseteq N$. Choose $n \in U$ and note that since $e \in V$, $n \in N$. By possibly making V smaller we can assume that $V\cdot n \subseteq U$. Then $V\cdot V\cdot n \subseteq N$.

Theorem 2.6.22 ([226, Theorem 1]). *Let $(G,\cdot,\tau)$ be a semitopological group, and let (G,τ) be a regular Δ-Baire space. If the multiplication operation on $G\times G$ is quasicontinuous at (e,e), then $(G,\cdot,\tau)$ is a topological group.*

Proof. Given that $(G,\cdot,\tau)$ is a semitopological group and (G,τ) is regular, to prove that $(G,\cdot,\tau)$ is a topological group, it is sufficient to show that for each open neighbourhood W of e, there exists a nonempty open subset U of G such that $U^{-1}\cdot U \subseteq \overline{W}$. Let

$\varphi : G \times G \to G$ be defined by $\varphi(h,g) := h^{-1} \cdot g$, and let W be an arbitrary open neighbourhood of e. Since (G,ι) is a Δ-Baire space and $\varphi^{-1}(W)$ is separately open, in the second variable, neighbourhood of Δ_G, there exists a nonempty open set U such that $U \times U \subseteq \overline{\varphi^{-1}(W)}^{\tau\times\tau}$. We claim that $\varphi(U \times U) = U^{-1} \cdot U \subseteq \overline{W}$. So let us suppose, to obtain a contradiction, that there exists $(x,y) \in U \times U$ such that $\varphi(x,y) = x^{-1} \cdot y \notin \overline{W}$ (i. e., $e \notin x \cdot \overline{W} \cdot y^{-1}$). Then we may choose an open neighbourhood N of e such that
(i) $N \cdot y \subseteq U$ and
(ii) $\overline{N} \cap (x \cdot \overline{W} \cdot y^{-1}) = \varnothing$ or, equivalently, $(x^{-1} \cdot \overline{N} \cdot y) \cap \overline{W} = \varnothing$.

Now since multiplication is quasicontinuous at (e,e), there exist an open neighbourhood V of e and an element $n \in N$ such that $V \cdot V \cdot n \subseteq N$. Therefore $\overline{V \cdot V \cdot n} \subseteq \overline{N}$. By Lemma 2.6.21 there exists an open neighbourhood A of e such that $A^{-1} \cdot A \subseteq \overline{V \cdot V}$. Therefore $A^{-1} \cdot A \cdot n \subseteq \overline{N}$, and so by (ii) we have that

$$\varnothing = x^{-1} \cdot (A^{-1} \cdot A \cdot n) \cdot y \cap \overline{W} = (A \cdot x)^{-1} \cdot A \cdot (n \cdot y) \cap \overline{W}.$$

Let $y' := n \cdot y$. Then, by (i), $y' \in U$, and $\varphi(A \cdot x \times A \cdot y') \cap \overline{W} = \varnothing$. However, this is impossible since

$$(A \cdot x \times A \cdot y') \cap (U \times U) \cap \varphi^{-1}(W) \neq \varnothing.$$

Hence $\varphi(U \times U) = U^{-1} \cdot U \subseteq \overline{W}$. □

The significance of Theorem 2.6.22 depends upon the prevalence of Δ-Baire spaces.

Proposition 2.6.23 ([226, 267, 301]). *Every Baire metric space (X,d) is a Δ-Baire space.*

Proof. Let (X,d) be a Baire metric space, and let W be a separately open, in the second variable, neighbourhood W of Δ_X. For each $n \in \mathbb{N}$, let

$$F_n := \{x \in X : \{x\} \times B(x,1/n)\} \subseteq W.$$

Then, since W is a separately open, in the second variable, neighbourhood of Δ_X, $X = \bigcup_{n\in\mathbb{N}} F_n$. Now, because (X,d) is a Baire space, there exists $k \in \mathbb{N}$ such that $\mathrm{int}(\overline{F_k}) \neq \varnothing$. Next, let us choose $x_0 \in \mathrm{int}(\overline{F_k})$ and $0 < r < 1/(2k)$ such that $B(x_0,r) \subseteq \mathrm{int}(\overline{F_k})$. Let $U := B(x_0,r)$ and $x \in U \cap F_k$. Then $U = B(x_0,r) \subseteq B(x,1/k)$, and so

$$\{x\} \times U = \{x\} \times B(x_0,r) \subseteq \{x\} \times B(x,1/k) \subseteq W.$$

Since $(U \cap F_k)$ is dense in U, we have that $U \times U \subseteq \overline{W}$, which shows that (X,d) is a Δ-Baire space. □

Proposition 2.6.24 ([226, 267]). *Every regular countably Čech-complete space (X,τ) is a Δ-Baire space.*

Proof. Let (X,τ) be a regular countably Čech-complete space and suppose, for obtaining a contradiction, that (X,τ) is not a Δ-Baire space. Let $(\mathscr{A}_n : n \in \mathbb{N})$ be a sequence of open covers of X such that every decreasing sequence $(F_n : n \in \mathbb{N})$ of nonempty closed subsets of X has $\bigcap_{n\in\mathbb{N}} F_n \neq \varnothing$, provided that each F_n is $\mathscr{A}_n$-small, that is., for each $n \in \mathbb{N}$, there exists $A_n \in \mathscr{A}_n$ such that $F_n \subseteq A_n$.

Since (X,τ) is not a Δ-Baire space, there exists a separately open, in the second variable, neighbourhood W of Δ_X such that for each nonempty open subset U of X, $U \times U \nsubseteq \overline{W}^{\tau\times\tau}$.

We will inductively define a sequence of pairs $((U_n, y_n) \in \tau \times X : n \in \mathbb{N})$ such that

(a$_n$) $\overline{U_n} \cup \{y_n\} \subseteq U_{n-1}$, and $\overline{U_n}$ is $\mathscr{A}_n$-small;

(b$_n$) $U_n \times \{y_n\} \cap W = \varnothing$ for each $n \in \mathbb{N}$.

For notational reasons, set $U_0 := V_0 := X$.

Step 1. Since $U_0 \times U_0 \nsubseteq \overline{W}^{\tau\times\tau}$, there exist points $x_1, y_1 \in X$ such that $(x_1, y_1) \notin \overline{W}^{\tau\times\tau}$. Then, since $\overline{W}^{\tau\times\tau}$ is closed, there exist open neighbourhoods U_1 and x_1 and V_1 of y_1 such that $(U_1 \times V_1) \cap \overline{W}^{\tau\times\tau} = \varnothing$. By possibly making U_1 smaller we may assume that $\overline{U_1}$ is $\mathscr{A}_1$-small. Then (a_1) and (b_1) are satisfied.

Now suppose that $(x_j, y_j) \in X \times X$ and $(U_j, V_j) \in \tau \times \tau$ satisfying (a_j) and (b_j) are defined for all $1 \leq j \leq n$.

Step $n+1$. Since $U_n \times U_n \nsubseteq \overline{W}^{\tau\times\tau}$, there exist points $x_{n+1}, y_{n+1} \in X$ such that $(x_{n+1}, y_{n+1}) \notin \overline{W}^{\tau\times\tau}$. Then, since $\overline{W}^{\tau\times\tau}$ is closed, there exist open neighbourhoods U_{n+1} and x_{n+1} and V_{n+1} of y_{n+1} such that $(U_{n+1} \times V_{n+1}) \cap \overline{W}^{\tau\times\tau} = \varnothing$. By possibly making U_{n+1} smaller we may assume that $\overline{U_{n+1}} \subseteq U_n$ and $\overline{U_{n+1}}$ is $\mathscr{A}_{n+1}$-small. Then (a_{n+1}) and (b_{n+1}) are satisfied.

This completes the induction. Since $\overline{\{y_k : n < k\}} \subseteq \overline{U_n}$ for all $n \in \mathbb{N}$,

$$\varnothing \neq \bigcap_{n\in\mathbb{N}} \overline{\{y_k : n < k\}},$$

and so $(y_k : k \in \mathbb{N})$ has a cluster point $y_\infty \in X$. Moreover, since

$$y_\infty \in \overline{\{y_k : n < k\}} \subseteq \overline{U_n} \subseteq U_{n-1} \quad \text{for all } n \in \mathbb{N}, y_\infty \in \bigcap_{n\in\mathbb{N}} \overline{U_n} = \bigcap_{n\in\mathbb{N}} U_{n-1} = \bigcap_{n\in\mathbb{N}} U_n.$$

Now $(y_\infty, y_\infty) \in \Delta \subseteq W$. So there exists an open neighbourhood V of y_∞ such that $\{y_\infty\} \times V \subseteq W$. Since y_∞ is a cluster point of $(y_n : n \in \mathbb{N})$, there exists $k \in \mathbb{N}$ such that $y_k \in V$. Thus $(y_\infty, y_k) \in \{y_\infty\} \times V \subseteq W$. On the other hand, by property (b_k), $(y_\infty, y_k) \in U_k \times \{y_k\} \subseteq X \setminus W$, which is inconsistent with our earlier observation that $(y_\infty, y_k) \in W$. Therefore our original assumption that (X,τ) is not a Δ-Baire space must be false. Thus (X,τ) is a Δ-Baire space. □

We can now easily establish that many semitopological groups are in fact topological groups, provided that their underlying topology is respectable.

Corollary 2.6.25 ([165]). *Let $(G, \cdot, \tau)$ be a semitopological group, and let (G, τ) be a metrisable Baire space. Then $(G, \cdot, \tau)$ is a topological group.*

Proof. It follows from the Proposition 2.6.23 and Theorem 2.6.22 that it is sufficient to show that the multiplication operation on G is quasicontinuous at (e, e). However, this follows directly from Theorem 2.2.7. □

Corollary 2.6.26 ([165]). *Let $(G, \cdot, \tau)$ be a semitopological group, and let (G, τ) be a regular countably Čech-complete space. Then $(G, \cdot, \tau)$ is a topological group.*

Proof. It follows from the Proposition 2.6.24 and Theorem 2.6.22, that it is sufficient to show that the multiplication operation on G is quasicontinuous at (e, e). However, this follows directly from Theorem 2.2.15. □

If $(M, +)$ is an Abelian group endowed with a topology τ and $(R, +, *)$ is a ring endowed with a topology τ', then we say that $(M, +, \cdot, \tau)$ is a *semitopological R-module* (over R) if $(M, +, \cdot)$ is an R-module (over R), $(M, +, \tau)$ is a semitopological group, and the mapping $(r, x) \mapsto r \cdot x$, is separately continuous $(R, \tau') \times (M, \tau)$.

Corollary 2.6.27. *Let $(M, +, \cdot, \tau)$ be a semitopological R-module over a ring R. If R is a q-space and (M, τ) is a regular countably Čech-complete space, then $(M, +, \cdot, \tau)$ is a topological R-module, that is, $(M, +, \tau)$ is a topological group, and $(r, x) \mapsto r \cdot x$ is jointly continuous.*

Proof. By Corollary 2.6.26, $(M, +, \tau)$ is a topological group, and for each fixed $r \in R$, the mapping, $x \mapsto r \cdot x$, is a group homomorphism from $(M, +, \tau)$ into $(M, +, \tau)$. Therefore the result follows from Theorem 2.6.6. □

The last application we consider here is to CHART groups, which are an important class of "almost" compact topological groups that admit invariant measures [224, 229, 242].

We call a triple $(G, \cdot, \tau)$ a *right topological group* (*left topological group*) if $(G, \cdot)$ is a group, (G, τ) is a topological space, and for each $g \in G$, the mapping $x \mapsto x \cdot g$ ($x \mapsto g \cdot x$) is τ-continuous on G. Of course, if $(G, \cdot, \tau)$ is both a right topological group and a left topological group, then it is a semitopological group. Let $(G, \cdot, \tau)$ be a right topological group, and let $\Lambda(G, \tau)$ be the set of all $x \in G$ such that the mapping $y \mapsto x \cdot y$ is τ-continuous. If $\Lambda(G, \tau)$ is τ-dense in G, then $(G, \cdot, \tau)$ is said to be *admissible*. A compact Hausdorff admissible right topological group $(G, \cdot, \tau)$ is called a *CHART group*. As mentioned earlier, CHART groups appear in the study of topological dynamics and, in particular, in the study of distal flows on compact spaces.

The following results are from [225].

Lemma 2.6.28. *If $(G, \cdot, \tau)$ is a CHART group and N is an open neighbourhood of e, then for any $g \in G$, $g \cdot N^{-1} \cdot N = [g \cdot N^{-1} \cap \Lambda(G)] \cdot N \in \tau$.*

Proof. We will first show that for any dense subset D of G and any nonempty open subsets A and B of (G, τ), $A^{-1} \cdot B = (A \cap D)^{-1} \cdot B$. Let $x \in A^{-1} \cdot B$. Then for some $a \in A$, $a \cdot x \in B$. Since B is open and $A \cap D$ is dense in A, there is $c \in A \cap D$ such that $c \cdot x \in B$. Hence $x \in c^{-1} \cdot B \subseteq (A \cap D)^{-1} \cdot B$. Thus $A^{-1} \cdot B \subseteq (A \cap D)^{-1} \cdot B$. The reverse inclusion is obvious. We now prove the statement given in the lemma. Let N be an open neighbourhood of e, and let $g \in G$. Let $D := \Lambda(G) \cdot g$; which is dense in (G, τ) since $x \mapsto x \cdot g$ is a homeomorphism on G. Note also that $[\Lambda(G)]^{-1} = \Lambda(G)$. Therefore

$$\begin{aligned} g \cdot N^{-1} \cdot N &= g \cdot [N \cap \Lambda(G) \cdot g]^{-1} \cdot N \quad \text{by above} \\ &= g \cdot [N^{-1} \cap g^{-1} \cdot \Lambda(G)] \cdot N \\ &= [g \cdot N^{-1} \cap \Lambda(G)] \cdot N. \end{aligned}$$

This completes the proof. □

Proposition 2.6.29 ([225]). *If $(G, \cdot, \tau)$ is a CHART group and multiplication is quasicontinuous, then $(G, \cdot, \tau)$ is a topological group.*

Proof. Let $\pi : G \times G \to G$ be defined by $\pi(g, h) := g \cdot h$ for all $(g, h) \in G \times G$. We will first show that π is continuous at (e, e). To this end, let W be an open neighbourhood of e, and let

$$W^* := \{g \in G : (g, e) \in \operatorname{int}(\pi^{-1}(W))\}.$$

We claim that $e \in \overline{W^*}$. To justify this claim, consider an arbitrary open neighbourhood N of e. Since multiplication is quasicontinuous, there exists nonempty open subsets U and V of (G, τ) such that $\pi(U \times V) = U \cdot V \subseteq N \cap W$. Let $\lambda \in V \cap \Lambda(G)$. Then

$$(U \cdot \lambda) \cdot (\lambda^{-1} \cdot V) = U \cdot V \subseteq W \cap N \subseteq W.$$

Now $e \in \lambda^{-1} \cdot V$, and so $U \cdot \lambda \subseteq W^*$. On there other hand,

$$U \cdot \lambda = (U \cdot \lambda) \cdot e \subseteq (U \cdot \lambda) \cdot (\lambda^{-1} \cdot V) \subseteq W \cap N \subseteq N.$$

Therefore $\varnothing \neq U \cdot \lambda \subseteq W^* \cap N$. This completes the proof of the claim. Note also that since W^* is an open set, $e \in \overline{W^* \cap \Lambda(G)}$. Since (G, τ) is compact and Hausdorf,f to show that π is continuous at $(e, e) \in G \times G$, it is sufficient to show that $\bigcap_{N \in \mathcal{N}(e)} \overline{\pi(N \times N)} \subseteq \{e\}$, where $\mathcal{N}(e)$ denotes the set of all open neighbourhoods of e in G. So let $z \in \bigcap_{N \in \mathcal{N}(e)} \overline{\pi(N \times N)}$. Since (G, τ) is regular and Hausdorf,f to show that $z = e$, it is sufficient to show that $z \in \overline{V}$ for each $V \in \mathcal{N}(e)$. Let $V \in \mathcal{N}(e)$, and let $\lambda \in V^* \cap \Lambda(G)$. Then there exists $N \in \mathcal{N}(e)$ such that $(\lambda \cdot N) \cdot N = \lambda \cdot \pi(N \times N) \subseteq V$, and so $\lambda \cdot z \in \lambda \cdot \overline{\pi(N \times N)} \subseteq \overline{V}$. Now, as $e \in \overline{V^* \cap \Lambda(G)}$, $z = e \cdot z \in \overline{V}$. Thus π is continuous at (e, e). In fact, it is not hard to see that π is continuous at each point of $\Lambda(G) \times G$.

Next, we will show that the inversion is continuous at e. As with multiplication, it is sufficient to show that $\bigcap_{N\in\mathcal{N}(e)}\overline{N^{-1}} \subseteq \{e\}$. Let $y \in \bigcap_{N\in\mathcal{N}(e)}\overline{N^{-1}}$. Since (G,τ) is T_1, to show that $y = e$, it is sufficient to show that $e \in V$ for every open neighbourhood V of y. Thus let V be an open neighbourhood of $y = e \cdot y$. By the continuity of π at $(e, y) \in \Lambda(G) \times G$ there exist a neighbourhood N of e and a neighbourhood U of y such that $\pi(N \times U) \subseteq V$. Since $y \in \overline{N^{-1}}$, $N^{-1} \cap U \neq \varnothing$. Thus $e \in N \cdot U \subseteq V$. This shows that the inversion is continuous at e. Furthermore, since $N^{-1} \cdot N = (N \cap \Lambda(G))^{-1} \cdot N$ for each $N \in \mathcal{N}(e)$ (see Lemma 2.6.28), $\{N^{-1} \cdot N : N \in \mathcal{N}(e)\}$ is a local base for τ at e comprising of τ-open sets.

To show that the inversion is continuous on (G,τ), we simply need to note that for each $g \in G$:

(i) $\{N^{-1} \cdot N \cdot g : N \in \mathcal{N}(e)\}$ is a local base for τ at g;

(ii) $(N^{-1} \cdot N \cdot g)^{-1} = g^{-1} \cdot N^{-1} \cdot N = (g^{-1} \cdot N^{-1} \cap \Lambda(G)) \cdot N \in \tau$ (see Lemma 2.6.28).

From this and the fact that $(G,\cdot,\tau)$ is a right topological group it follows that $(G,\cdot,\tau)$ is also a left topological group, that is, $(G,\cdot,\tau)$ is a semitopological group. However, since π is continuous at (e,e), it now follows that π is continuous on $G \times G$. Therefore $(G,\cdot,\tau)$ is a topological group. □

2.7 Ascoli-type theorem for quasicontinuous functions

In what follows, let (X,τ) be a Hausdorff topological space, and let (Y,d) be a nontrivial metric space.

First, we will prove an Ascoli-type theorem for quasicontinuous subcontinuous functions (see [125]).

Recall that a function $f : (X,\tau) \to (Y,\tau')$ acting between topological spaces (X,τ) and (Y,τ') is *subcontinuous at* $x \in X$, [87, 184] if for each net $(x_\alpha : \alpha \in A)$ in (X,τ) that converges to $x \in X$, the net $(f(x_\alpha) : \alpha \in A)$ has a cluster point in (Y,τ'). A mapping $f : X \to Y$ is *subcontinuous* if it is subcontinuous at all points of X.

We will use the fact that if f is a subcontinuous function from a topological space (X,τ) into a regular space (Y,τ'), then for every compact set K in (X,τ), $\overline{f(K)}$ is a compact set in (Y,τ') (see [134, Proposition 3.13]).

An easy example of a quasicontinuous function that is subcontinuous but not continuous is the following one. Let $X := Y := \mathbb{R}$ equipped with the usual Euclidean topology. Define the function $f : X \to Y$ as follows: $f(x) := 0$ for $x \leq 0$ and $f(x) := 1$ for $x > 0$.

We denote by $F(X,Y)$ the set of all functions from a set X to a set Y. Then we will consider the set X with topology τ and the set Y with topology τ'. By $C(X,Y)$ we denote the set of all functions $f \in F(X,Y)$ such that $f : (X,\tau) \to (Y,\tau')$ is continuous, and by $S(X,Y)$ we denote the set of all functions $f \in F(X,Y)$ such that $f : (X,\tau) \to (Y,\tau')$ is subcontinuous. By $Q(X,Y)$ we denote the set of all functions $f \in F(X,Y)$ such that

$f : (X,\tau) \to (Y,\tau')$ is quasicontinuous, and, finally, by $QS(X,Y)$ we denote the set of all functions $f \in F(X,Y)$ such that $f : (X,\tau) \to (Y,\tau')$ is quasicontinuous and subcontinuous.

By τ_p, τ_{UC} and τ_U we denote the topology of pointwise convergence on $F(X,Y)$, the topology of uniform convergence on compact sets on $F(X,Y)$ and the topology of uniform convergence on $F(X,Y)$, respectively.

Let X be a set, and let (Y,d) be a metric space. The topology τ_p of *pointwise convergence on* $F(X,Y)$ is induced by the uniformity $\mathfrak{U}_p$ of pointwise convergence that has a base consisting of sets of the form

$$W(A,\varepsilon) := \{(f,g) : d(f(x),g(x)) < \varepsilon \text{ for all } x \in A\},$$

where A is a finite subset of X and $\varepsilon > 0$. A general τ_p-basic neighbourhood of $f \in F(X,Y)$ will be denoted by $W(f,A,\varepsilon)$, that is, $W(f,A,\varepsilon) := W(A,\varepsilon)[f]$. In other words,

$$W(f,A,\varepsilon) = \{g \in F(X,Y) : \ d(f(x),g(x)) < \varepsilon \text{ for all } x \in A\}.$$

Next, we define the topology τ_{UC} of *uniform convergence on compact sets* on $F(X,Y)$. This topology is induced by the uniformity $\mathfrak{U}_{UC}$ that has a base consisting of sets of the form

$$W(K,\varepsilon) := \{(f,g) : d(f(x),g(x)) < \varepsilon \text{ for all } x \in K\},$$

where K is a compact subset of X and $\varepsilon > 0$. A general τ_{UC}-basic neighbourhood of $f \in F(X,Y)$ will be denoted by $W(f,K,\varepsilon)$, that is, $W(f,K,\varepsilon) := W(K,\varepsilon)[f]$. In other words,

$$W(f,K,\varepsilon) = \{g \in F(X,Y) : \ d(f(x),g(x)) < \varepsilon \text{ for all } x \in K\}.$$

Finally, we define the topology τ_U of *uniform convergence* on $F(X,Y)$. Let ϱ be the (extended-valued) metric on $F(X,Y)$ defined by

$$\varrho(f,g) := \sup\{d(f(x),g(x)) : x \in X\}$$

for $f,g \in F(X,Y)$. Then the topology of uniform convergence for the space $F(X,Y)$ is the topology generated by the metric ϱ.

It is well known that a uniform limit of a net of quasicontinuous functions is quasicontinuous [249]. Thus the following proposition is easy to verify.

Proposition 2.7.1. *Let (X,τ) be a locally compact topological space, and let (Y,d) be a metric space. Then $Q(X,Y)$ is a closed set in $(F(X,Y),\tau_{UC})$.*

We even have the following characterisation of local compactness in the class of metric spaces.

Theorem 2.7.2. *Let (X,d) and (Y,e) be metric spaces. The following are equivalent:*

(i) *(X, d) is locally compact;*
(ii) *The space $Q(X, Y)$ is a closed subspace of $(F(X, Y), \tau_{UC})$.*

Proof. (i)⇒(ii) is trivial. (ii)⇒(i). Suppose that $x_0 \in X$ fails to have a local base of compact sets. Let $\delta_1 = 1$. There is a sequence $\{x_i^1 : i \in \mathbb{N}\}$ of different points of $\{z \in X : 0 < d(x_0, z) < \delta_1\}$ with no cluster point in X. There exists $\varepsilon_1 > 0$ such that $\varepsilon_1 < d(x_0, x_i^1)$ for every $i \in \mathbb{N}$. Next, let $\delta_2 = \min\{\frac{1}{2}, \frac{\varepsilon_1}{2}\}$, and let $\{x_i^2 : i \in \mathbb{N}\}$ be a sequence of different points of $\{z \in X : 0 < d(x_0, z) < \delta_2\}$ with no cluster point in (X, d). Choose $\varepsilon_2 > 0$ such that $\varepsilon_2 < d(x_0, x_i^2)$ for every $i \in \mathbb{N}$ and set $\delta_3 = \min\{\frac{1}{3}, \frac{\varepsilon_2}{2}\}$. Continuing, we can produce for each $n \in \mathbb{N}$ a sequence $\{x_i^n : i \in \mathbb{N}\}$ of different points with no cluster point in (X, d) such that $\delta_n = \min\{\frac{1}{n}, \frac{\varepsilon_{n-1}}{2}\}$, $0 < \varepsilon_n < \delta_n$ and $\{x_i^n : i \in \mathbb{N}\} \subseteq \{z \in X : \varepsilon_n < d(x_0, z) < \delta_n\}$. Let $B(x_i^n, \varepsilon_i^n)$ be the open ball with centre x_i^n and radius $\varepsilon_i^n < 1/i$ for all $i \in \mathbb{N}$ and $n \in \mathbb{N}$ such that the family $\{\overline{B(x_i^n, \varepsilon_i^n)} : i \in \mathbb{N}\}$ is pairwise disjoint and $\overline{B(x_i^n, \varepsilon_i^n)} \subseteq \{z \in X : \varepsilon_n < d(x_0, z) < \delta_n\}$. Let $\mathbb{N}^{\mathbb{N}}$ be partially ordered by the product order, that is, $h \le g$ if, $h(n) \le g(n)$ for every $n \in \mathbb{N}$. Let y_0, y_1 be two different points in Y. For every $g \in \mathbb{N}^{\mathbb{N}}$, define a function f_g as follows:

$$f_g(x) := \begin{cases} y_1 & \text{if } x \in \bigcup_{n\in\mathbb{N}} \overline{B(x^n_{g(n)}, \varepsilon^n_{g(n)})}, \\ y_1 & \text{if } x = x_0, \\ y_0 & \text{otherwise.} \end{cases}$$

Evidently, f_g is a quasicontinuous function. Now define the function f as follows:

$$f(x) := \begin{cases} y_1 & \text{if } x = x_0, \\ y_0 & \text{otherwise.} \end{cases}$$

Then $f \notin Q(X, Y)$, and the net $(f_g : g \in \mathbb{N}^{\mathbb{N}})$ converges in $(F(X, Y), \tau_{UC})$ to f. Thus $Q(X, Y)$ is not a closed subspace of $(F(X, Y), \tau_{UC})$. (Let K be a compact set in (X, d), and let $\varepsilon > 0$. Then for every $n \in \mathbb{N}$, there are only finitely many i such that $\overline{B(x_i^n, \varepsilon_i^n)} \cap K \neq \emptyset$. For every $n \in \mathbb{N}$, let $k_n \in \mathbb{N}$ be such that $\overline{B(x_k^n, \varepsilon_k^n)} \cap K = \emptyset$ for every $k_n \le k$. Define $g : \mathbb{N} \to \mathbb{N}$ as $g(n) := k_n$. Then for every $g \le h$, we have $e(f_h(x), f(x)) = 0 < \varepsilon$ for every $x \in K$.) □

If (X, τ) is a locally compact space and (Y, d) is a metric space, then the Ascoli theorem [159] says that a subset $\mathcal{E}$ of $(C(X, Y), \tau_{UC})$ is compact if and only if it is closed in $(C(X, Y), \tau_{UC})$, $\{f(x) \in Y; f \in \mathcal{E}\}$ has a compact closure in (Y, d) for each $x \in X$, and $\mathcal{E}$ is equicontinuous, where a subset $\mathcal{E}$ of $C(X, Y)$ is *equicontinuous*, provided that for all $x \in X$ and $\varepsilon > 0$, there is a neighbourhood U of x in (X, τ) with $d(f(x), f(z)) < \varepsilon$ for all $z \in U$ and $f \in \mathcal{E}$.

We will use the following definition introduced in [139] for the space of locally bounded functions. A function f from a topological space (X, τ) into a metric space (Y, d) is *locally bounded* if for every $x \in X$, there is an open neighbourhood U_x of x in (X, τ) such that $f(U_x)$ is a bounded set in (Y, d).

Let (X, τ) be a topological space, and let (Y, d) be a metric space. We say that a subset $\mathcal{E}$ of $F(X, Y)$ is *densely equiquasicontinuous at a point* x of (X, τ) if for every $\varepsilon > 0$, there exists a finite family $\mathcal{B}$ of nonempty subsets of (X, τ) that are either quasi-open or nowhere dense such that $\bigcup \mathcal{B}$ is a neighbourhood of x in (X, τ) and such that for every $f \in \mathcal{E}$, for every $B \in \mathcal{B}$ and for every $p, q \in B$, $d(f(p), f(q)) < \varepsilon$. We say that $\mathcal{E}$ is *densely equiquasicontinuous* if it is densely equiquasicontinuous at every point of (X, τ).

Note that if $\mathcal{E}$ is densely equiquasicontinuous (at x), then all its subsets are also densely equiquasicontinuous (at x).

Remarks 2.7.3. In the previous definition, replacing "quasi-open" by "open" produces the same concept.

Proposition 2.7.4. *Let (X, τ) be a topological space, and let (Y, d) be a metric space. Let $\mathcal{E} \subseteq F(X, Y)$ be densely equiquasicontinuous at $x \in X$, and let every $f \in \mathcal{E}$ be quasicontinuous at x. Then for every $\varepsilon > 0$ and every open neighbourhood U of $x \in X$, there exists a finite family $\mathcal{D}$ of nonempty subsets of U that are either open or nowhere dense such that $\bigcup \mathcal{D}$ is a neighbourhood of x in (X, τ) and for every $f \in \mathcal{E}$, every $D \in \mathcal{D}$, and every $p, q \in D$, $d(f(p), f(q)) < \varepsilon$ and for every $f \in \mathcal{E}$, there is an open set $U_f \in \mathcal{D}$ with $d(f(p), f(x)) < \varepsilon$ for every $p \in U_f$.*

Proof. Let $\varepsilon > 0$, and let U be an open neighbourhood of x in (X, τ). The dense equiquasicontinuity of $\mathcal{E}$ at x implies that there exists a finite family $\mathcal{B}$ of nonempty subsets of (X, τ) that are either open or nowhere dense such that $\bigcup \mathcal{B}$ is a neighbourhood of x in (X, τ) and such that for every $f \in \mathcal{E}$, every $B \in \mathcal{B}$ and every $p, q \in B$, $d(f(p), f(q)) < \varepsilon/2$. Put $\mathcal{D} := \{B \cap U : B \cap U \neq \emptyset, B \in \mathcal{B}\}$. Then $\mathcal{D}$ is a finite family of nonempty subsets of U that are either open or nowhere dense such that $\bigcup \mathcal{D}$ is a neighbourhood of x. Let $f \in \mathcal{E}$. The quasicontinuity of f at x implies that there is a nonempty open set $G \subseteq \cup \mathcal{D}$ such that $d(f(x), f(q)) < \varepsilon/2$ for every $q \in G$. There is an open set $U_f \in \mathcal{D}$ such that $U_f \cap G \neq \emptyset$. Let $p \in U_f$. There is some $q \in U_f \cap G$. We have

$$d(f(x), f(p)) \leq d(f(x), f(q)) + d(f(q), f(p)) < \varepsilon/2 + \varepsilon/2 = \varepsilon.$$

This completes the proof. □

Proposition 2.7.5. *Let (X, τ) be a topological space, and let (Y, d) be a metric space. Let $(f_\sigma : \sigma \in \Sigma)$ be a net of quasicontinuous functions pointwise convergent to a function $f : X \to Y$. If $\{f_\sigma : \sigma \in \Sigma\}$ is densely equiquasicontinuous, then f is quasicontinuous too.*

Proof. To prove that f is also quasicontinuous, let $x \in X$, let $\varepsilon > 0$, and let U be an open neighbourhood of x in (X, τ). By Proposition 2.7.4 there is a finite family $\mathcal{D}$ of nonempty subsets of U that are either open or nowhere dense such that $\bigcup \mathcal{D}$ is a neighbourhood of x and for every $\sigma \in \Sigma$, every $D \in \mathcal{D}$ and every $p, q \in D$, $d(f_\sigma(p), f_\sigma(q)) < \varepsilon/2$ and there is a nonempty open set $G_{f_\sigma} \in \mathcal{D}$ with $d(f_\sigma(x), f_\sigma(p)) < \varepsilon/2$ for every $p \in G_{f_\sigma}$.

Since $\mathcal{D}$ is a finite family, without loss of generality, we can suppose that there is $G \in \tau, G \neq \emptyset$, $G \subseteq U$, such that $d(f_\sigma(x), f_\sigma(p)) < \varepsilon/2$ for every $\sigma \in \Sigma$ and every $p \in G$. The pointwise convergence of $(f_\sigma : \sigma \in \Sigma)$ to f implies that $d(f(x), f(p)) < \varepsilon$ for every $p \in G$, and thus f is quasicontinuous at x. □

Remarks 2.7.6. Let $\mathcal{E}$ be a densely equiquasicontinuous subset of $F(X, Y)$. It is easy to see that the closure of $\mathcal{E}$ with respect to the topology τ_p is also densely equiquasicontinuous.

Theorem 2.7.7. *Let (X, τ) be a topological space, and let (Y, d) be a metric space. If $\mathcal{E}$ is a densely equiquasicontinuous subset of $F(X, Y)$, then the topologies τ_p and τ_{UC} on $\mathcal{E}$ coincide.*

Proof. Let $(f_\sigma : \sigma \in \Sigma)$ be a net in $\mathcal{E}$ that τ_p-converges to $f \in \mathcal{E}$. We will show that $(f_\sigma : \sigma \in \Sigma)$ converges to f also in $(F(X, Y), \tau_{UC})$.

Suppose that $(f_\sigma : \sigma \in \Sigma)$ fails to converge to f in $(F(X, Y), \tau_{UC})$. There exist a compact set A in (X, τ) and $\varepsilon > 0$ such that $(f_\sigma : \sigma \in \Sigma)$ is not cofinally in $W(f, A, 2\varepsilon)$.

For every $\sigma \in \Sigma$, there are $\beta_\sigma \geq \sigma$ and $a_\sigma \in A$ such that $\varepsilon < d(f(a_\sigma), f_{\beta_\sigma}(a_\sigma))$. Let a be a cluster point of $(a_\sigma : \sigma \in \Sigma)$ in (X, τ).

Since the family $\mathcal{E}$ is densely equiquasicontinuous, there is a finite family $\mathcal{B}$ of nonempty subsets of (X, τ) that are either open or nowhere dense such that $\bigcup \mathcal{B}$ is a neighbourhood of a and for every $g \in \mathcal{E}$, every $B \in \mathcal{B}$ and every $p, q \in B$, $d(g(p), g(q)) < \varepsilon/3$.

Without loss of generality, we can suppose that $a_\sigma \in \bigcup \mathcal{B}$ for every $\sigma \in \Sigma$, and since $\mathcal{B}$ is finite, we can suppose that there is $B \in \mathcal{B}$ such that $a_\sigma \in B$ for every $\sigma \in \Sigma$.

Choose a_η. There must exist $\sigma \in \Sigma$ such that $d(f(a_\eta), f_\delta(a_\eta)) < \varepsilon/3$ for every $\sigma \leq \delta$. There is $\sigma \leq \beta_\sigma$ such that $\varepsilon < d(f(a_\sigma), f_{\beta_\sigma}(a_\sigma))$. Then

$$d(f(a_\sigma), f_{\beta_\sigma}(a_\sigma)) \leq d(f(a_\sigma), f(a_\eta)) + d(f(a_\eta), f_{\beta_\sigma}(a_\eta)) + d(f_{\beta_\sigma}(a_\eta), f_{\beta_\sigma}(a_\sigma)) < \varepsilon,$$

a contradiction. Thus $(f_\sigma : \sigma \in \Sigma)$ converges to f in $(F(X, Y), \tau_{UC})$. □

Note that the above theorem improves Theorem 2.7 in [139].

If (Z, τ) is a topological space, then denote by $K(Z)$ the space of all nonempty compact subsets of (Z, τ).

We say that a subset $\mathcal{E}$ of $F(X, Y)$ is *pointwise bounded* if for every $x \in X$, $\overline{\{f(x) : f \in \mathcal{E}\}}$ is compact in (Y, d). We say that a subset $\mathcal{E}$ of $F(X, Y)$ is *compactly bounded* if for every $K \in K(X)$, $\overline{\bigcup \{f(K) : f \in \mathcal{E}\}}$ is compact in (Y, d).

If (Y, d) is a metric space, then the open d-ball, with centre $z_0 \in Y$ and radius $\varepsilon > 0$ is denoted by $B(z_0, \varepsilon)$, and the *ε-parallel body* $\bigcup_{a \in A} B(a, \varepsilon)$ for a subset A of (Y, d) is denoted by $B(A, \varepsilon)$.

Let H_d be the Hausdorff metric induced by d on $K(Y)$. It is known that for $A, C \in K(Y)$,

$$H_d(A, C) := \inf\{0 < \varepsilon : A \subseteq B(C, \varepsilon) \text{ and } C \subseteq B(A, \varepsilon)\}.$$

Lemma 2.7.8. *Let (X,τ) be a topological space, and let (Y,d) be a metric space. If $K \in \mathcal{K}(X)$, then the mapping*

$$M_K : (S(X,Y), \tau_{UC}) \to (\mathcal{K}(Y), H_d)$$

defined by $M_K(f) := \overline{f(K)}$ is continuous.

Proof. Of course, if f is a subcontinuous function, then $\overline{f(K)}$ is a compact set in (Y,d). Let $(f_\sigma : \sigma \in \Sigma)$ converge to f in $(S(X,Y), \tau_{UC})$. We will show that $(\overline{f_\sigma(K)} : \sigma \in \Sigma)$ converges to $\overline{f(K)}$ in $(\mathcal{K}(Y), H_d)$. Let $\varepsilon > 0$. There is $\sigma_0 \in \Sigma$ such that

$$d(f_\sigma(x), f(x)) < \varepsilon/2 \quad \text{for every } x \in K \text{ and every } \sigma_0 \le \sigma.$$

We claim that $H_d(\overline{f_\sigma(K)}, \overline{f(K)}) < \varepsilon$ for every $\sigma_0 \le \sigma$. Let $\sigma_0 \le \sigma$, and let $y \in \overline{f_\sigma(K)}$. There is $k \in K$ such that $d(y, f_\sigma(k)) < \varepsilon/2$. Then $d(y, f(k)) < \varepsilon$, that is, $\overline{f_\sigma(K)} \subseteq B(f(K), \varepsilon)$. The opposite inclusion is similar. □

Theorem 2.7.9. *Let (X,τ) be a locally compact topological space, and let (Y,d) be a metric space. A subset $\mathcal{E}$ of $(QS(X,Y), \tau_{UC})$ is compact if and only if $\mathcal{E}$ is closed, densely equiquasicontinuous and compactly bounded.*

Proof. Let $\mathcal{E}$ be compact in $(QS(X,Y), \tau_{UC})$. Let $K \in \mathcal{K}(X)$. By Lemma 2.7.8 the mapping

$$M_K : (QS(X,Y), \tau_{UC}) \to (\mathcal{K}(Y), H_d)$$

defined by $M_K(f) := \overline{f(K)}$ is continuous. Thus $M_K(\mathcal{E})$ is compact in $(\mathcal{K}(Y), H_d)$. Then $\bigcup_{f \in \mathcal{E}} \overline{f(K)}$ is compact in (Y,d), [206]. (The Vietoris topology and the Hausdorff metric topology coincide on $\mathcal{K}(Y)$.)

Now we show that $\mathcal{E}$ is densely equiquasicontinuous. Let $x \in (X,\tau)$ and $\varepsilon > 0$. Define a finite family $\mathcal{U}$ of open subsets of (X,τ) as follows. There is $U(x) \in \tau$ containing x such that $A := \overline{U(x)}$ is compact. There are $f_1, f_2, \ldots, f_n \in \mathcal{E}$ such that

$$\mathcal{E} \subseteq \bigcup_{1 \le j \le n} W(f_j, A, \varepsilon/3).$$

Since f_j is subcontinuous, $\overline{f_j(A)}$ is compact for every $1 \le j \le n$. Let $\mathcal{V} := \{V_1, \ldots, V_m\}$ be an open finite cover of $\bigcup_{1 \le j \le n} \overline{f_j(A)}$ such that each V_i has diameter less than $\varepsilon/3$. For all $i = 1, \ldots, m$ and $j = 1, \ldots, n$, let

$$D_i^j := \{z \in U(x) \cap C(f_j) : f_j(z) \in V_i\},$$

where $C(f_j) := \{z \in X : f_j \text{ is continuous at } z\}$. Since f_j is quasicontinuous, $C(f_j)$ is dense in (X,τ). For each $z \in D_i^j$, let $W_i^j(z)$ be an open neighbourhood of z in (X,τ) such that $f_j(W_i^j(z)) \subseteq V_i$, and let

$$W_i^j := \bigcup \{W_i^j(z) : z \in D_i^j\}.$$

Define

$$\mathcal{U} := \{U(x) \cap W^1_{i_1} \cap \cdots \cap W^n_{i_n} : i_j \in \{1, \ldots, m\} \text{ for each } j \in \{1, \ldots, n\}\}.$$

Let $U \in \mathcal{U}$, and let

$$U^j_i = \{z \in \overline{U} \setminus U : f_j(z) \in V_i\}.$$

Finally, define

$$\mathcal{R} := \{A \cap U^1_{i_1} \cap \cdots \cap U^n_{i_n} : i_j \in \{1, \ldots, m\} \text{ for } j \in \{1, \ldots, n\}, U \in \mathcal{U}\}.$$

Now we can check that $\bigcup \mathcal{U} \cup \bigcup \mathcal{R} = \overline{U(x)} = A$ and thus is a neighbourhood of x in (X, τ). Denote by $\mathcal{B}$ the family containing all nonempty sets from $\mathcal{U}$ and $\mathcal{R}$. It is easy to see that for every $f \in \mathcal{E}$, every $B \in \mathcal{B}$ and every $p, q \in B$, $d(f(p), f(q)) < \varepsilon$.

Let $\mathcal{E} \subseteq (QS(X, Y), \tau_{UC})$ be closed, densely equiquasicontinuous and compactly bounded. For every $x \in X$, put

$$\mathcal{E}_x := \overline{\{f(x) : f \in \mathcal{E}\}}.$$

Then $\mathcal{E}_x$ is a compact subset of (Y, d), and so the space $(\mathcal{E}_x, d)$ is a compact metric space. Put

$$H := \Pi_{x \in X} \mathcal{E}_x.$$

The space H is a compact subset of $Y^X = \Pi_{x \in X} Y_x$ $(Y_x = Y)$ with the relative product topology, and $\mathcal{E}$ is a subset of H. Let $(f_\sigma : \sigma \in \Sigma)$ be a net in $\mathcal{E}$. We would like to show that there is a function from $\mathcal{E}$ that is a cluster point of $(f_\sigma : \sigma \in \Sigma)$ in $(QS(X, Y), \tau_{UC})$.

There is $f \in H$ such that f is a cluster point of $(f_\sigma : \sigma \in \Sigma)$ in H. Without loss of generality, we can suppose that $(f_\sigma : \sigma \in \Sigma)$ converges to f in H. By Remark 2.7.6 the family $\mathcal{E}' := \mathcal{E} \cup \{f\}$ is densely equiquasicontinuous. Thus by Theorem 2.7.7 $(f_\sigma : \sigma \in \Sigma)$ converges to f also in $(F(X, Y), \tau_{UC})$. Since all functions f_σ $(\sigma \in \Sigma)$ are quasicontinuous, by Proposition 2.7.5, f is quasicontinuous too.

The function $f : X \to Y$ is subcontinuous. Let $x \in X$, and let $(x_i : i \in I)$ be a net in (X, τ) converging to x. Let O_x be an open neighbourhood of x such that $\overline{O_x}$ is compact. Without loss of generality, we can suppose that $\{x_i : i \in I\}$ is contained in O_x. The compact boundedness of $\mathcal{E}$ implies that

$$M := \overline{\bigcup \{h(\overline{O_x}) : h \in \mathcal{E}\}}$$

is compact. We show that $f(x_i) \in M$ for every $i \in I$. Let $i \in I$ and $\varepsilon > 0$. There is $\sigma_0 \in \Sigma$ such that $d(f(x_i), f_\sigma(x_i)) < \varepsilon$ for every $\sigma_0 \leq \sigma$. Since $f_{\sigma_0}(x_i) \in M$, we are done. Thus $(f(x_i) : i \in I)$ has a cluster point in M. We proved that $f : X \to Y$ is subcontinuous. This proves that $\mathcal{E}$ is compact in $(QS(X, Y), \tau_{UC})$. □

If (Y, d) is a complete metric space, then we have the following variant of the Ascoli theorem for $(QS(X, Y), \tau_{UC})$.

Theorem 2.7.10. *Let (X, τ) be a locally compact topological space, and let (Y, d) be a complete metric space. A subset $\mathcal{E}$ of $(QS(X, Y), \tau_{UC})$ is compact if and only if $\mathcal{E}$ is closed, densely equiquasicontinuous and pointwise bounded.*

Proof. Let $\mathcal{E}$ be compact in $(QS(X, Y), \tau_{UC})$. By Theorem 2.7.9 $\mathcal{E}$ must be closed, densely equiquasicontinuous and compactly bounded, and thus $\mathcal{E}$ is also pointwise bounded.

Suppose now that $\mathcal{E} \subseteq (QS(X, Y), \tau_{UC})$ is closed, densely equiquasicontinuous and pointwise bounded. We will proceed in the same way as in the proof of Theorem 2.7.9, except of the proof of the subcontinuity of $f : X \to Y$. We will show that the function $f : X \to Y$ is subcontinuous. Let $x \in X$, and let $(x_i : i \in I)$ be a net in (X, τ) converging to x. To prove that $(f(x_i) : i \in I)$ has a cluster point, it is sufficient to verify that $\bigcap_{i \in I} \overline{\{f(x_j) : i \le j\}} \neq \emptyset$. Note that for every $\varepsilon > 0$, there is $i \in I$ such that the set $\overline{\{f(x_j) : i \le j\}}$ can be covered by a finite family of sets of diameter smaller than ε. (Let $\varepsilon > 0$. The family $\mathcal{E}'$ is densely equiquasicontinuous at x, and thus there exists a finite family $\mathcal{B}$ of nonempty subsets of X such that $\bigcup \mathcal{B}$ is a neighbourhood of x and such that for every $g \in \mathcal{E}'$, every $B \in \mathcal{B}$ and every $p, q \in B$, $d(g(p), g(q)) < \varepsilon/2$. Let $i \in I$ be such that $\{x_j : i \le j\} \subseteq \bigcup \mathcal{B}$.

Since $f \in \mathcal{E}'$, the set $\{f(x_j) : i \le j\}$ can be covered by a finite family of sets of diameter smaller than $\varepsilon/2$, that is, the set $\overline{\{f(x_j) : i \le j\}}$ can be covered by a finite family of sets of diameter smaller than ε.) Thus by Corollary 1.5.3 we have $\bigcap_{i \in I} \overline{\{f(x_j) : i \le j\}} \neq \emptyset$. □

We even have the following characterisation of complete metric spaces.

Theorem 2.7.11. *Let (Y, d) be a metric space. The following are equivalent:*
(i) *(Y, d) is complete;*
(ii) *For every locally compact space (X, τ), a subset $\mathcal{E}$ of $(QS(X, Y), \tau_{UC})$ is compact if and only if $\mathcal{E}$ is closed, densely equiquasicontinuous and pointwise bounded.*

Proof. Of course, it is sufficient to prove that (ii)⇒(i). Suppose that (ii) holds and (Y, d) is not complete. There is a Cauchy sequence $(y_n : n \in \mathbb{N})$ in (Y, d) that has no cluster point in (Y, d). Let $X := (-1, 1)$ with the usual Euclidean topology. For every $n \in \mathbb{N}$, we define the function $f_n : X \to Y$ as follows:

$$f_n(x) := \begin{cases} y_1 & \text{for } x \in (-1, 0], \\ y_i & \text{for } x \in [\frac{1}{(i+1)}, \frac{1}{i}), i \le n, \\ y_n & \text{otherwise.} \end{cases}$$

Of course, for every $n \in \mathbb{N}$, the function f_n is quasicontinuous and subcontinuous. It is also easy to verify that the sequence $\{f_n : n \in \mathbb{N}\} \subseteq QS(X, Y)$ is densely equiquasicontinuous, pointwise bounded and closed in $(QS(X, Y), \tau_{UC})$. However, $\{f_n : n \in \mathbb{N}\}$ is not compact in $(QS(X, Y), \tau_{UC})$, a contradiction. □

Theorem 2.7.10 is a big generalisation of Theorem 2.8 from [139] stated for locally bounded quasicontinuous functions $Q^*(X,Y)$ from a locally compact topological space (X,τ) to a boundedly compact metric space (Y,d). We say that a metric space (Y,d) is *boundedly compact* [24] if every closed bounded subset is compact. Therefore (Y,d) is a locally compact separable metric space, and d is complete. In fact, any locally compact separable metric space has a compatible metric d such that (Y,d) is a boundedly compact space [305]. Boundedly compact metric spaces are called in [102, 139] *spaces with the Heine–Borel property.*

It is easy to verify that every subcontinuous function from a topological space (X,τ) into a metric space (Y,d) is locally bounded. If (Y,d) is a boundedly compact metric space, then the notion of local boundedness of a function is very strong. In fact, if (Y,d) is boundedly compact and $f : X \to Y$ is locally bounded, then for every $x \in X$, there is a neighbourhood U_x of x such that $\overline{f(U_x)}$ is compact. Thus every locally bounded function f from a topological space (X,τ) into a boundedly compact metric space is subcontinuous.

Corollary 2.7.12 (Theorem 2.8 in [139]). *Let (X,τ) be a locally compact space, and let (Y,d) be a boundedly compact metric space. A subset $\mathcal{E}$ of $Q^*(X,Y)$ is compact in $(Q^*(X,Y),\tau_{UC})$ if and only if $\mathcal{E}$ is closed, pointwise bounded and densely equiquasicontinuous.*

In the next part, we will prove Ascoli-type theorem for quasicontinuous functions (not necessarily subcontinuous) with values in boundedly compact metric spaces.

Our extension of dense equiquasicontinuity property to $F(X,Y)$ involves *Lebesgue covers* of (Y,d), which are covers for which there is a positive (Lebesgue) number λ such that every set of diameter less than λ is contained in an element of the cover.

Let (X,τ) be a topological space, and let (Y,d) be a boundedly compact metric space. We say that a subset $\mathcal{E}$ of $F(X,Y)$ is *densely equiquasicontinuous* at a point* x of (X,τ), provided that for every finite open Lebesgue cover $\mathcal{V}$ of (Y,d) that contains the complement of some nonempty compact set, there exists a finite family $\mathcal{B}$ of nonempty subsets of (X,τ) that are either open or nowhere dense such that $\bigcup\mathcal{B}$ is a neighbourhood of x and such that for every $f \in \mathcal{E}$, every $B \in \mathcal{B}$ and every $p,q \in B$, there exists $V \in \mathcal{V}$ with $f(p), f(q) \in V$. Then $\mathcal{E}$ is *densely equiquasicontinuous**, provided that it is densely equiquasicontinuous* at every point of (X,τ).

Note that if (X,τ) is a topological space, (Y,d) is a boundedly compact metric space, and a family $\mathcal{E} \subseteq F(X,Y)$ is densely equiquasicontinuous at $x \in (X,\tau)$, then $\mathcal{E}$ is also densely equiquasicontinuous* at x. The opposite implication does not hold. Let $X := Y := \mathbb{R}$ with the usual Euclidean topology, and let $f : X \to Y$ be a function defined as follows:

$$f(x) := \begin{cases} 0 & \text{if } x \le 0, \\ \frac{1}{x} & \text{otherwise.} \end{cases}$$

Put $\mathcal{E} := \{f\}$. Then $\mathcal{E}$ is densely equiquasicontinuous* at the point 0. However, it is not densely equiquasicontinuous at 0.

Remarks 2.7.13. It is easy to prove that if $\mathcal{E} \subseteq F(X,Y)$ is a densely equiquasicontinuous⋆ at $x \in X$, then the closure of $\mathcal{E}$ with respect to the topology τ_p is also densely equiquasicontinuous⋆ at x.

Let $H \subseteq F(X,Y)$, and let $x \in X$. Denote by $H[x]$ the set $\{f(x) \in Y : f \in H\}$. We say that a subset $\mathcal{E}$ of $F(X,Y)$ is *locally bounded at* $x \in X$ if there is an open neighbourhood U of x in (X,τ) such that $\bigcup\{f(U) : f \in \mathcal{E}\}$ is bounded in (Y,d).

Remarks 2.7.14. Note that if $\{A_1, A_2, \dots, A_n\}$ is a family of quasi-open sets of (X,τ), then the set $A := A_1 \cap \dots \cap A_n$ is nowhere dense, or it is a union of a nowhere dense set and an open set. In fact, consider first the case $\mathrm{int}(A) = \emptyset$. If $x \in A$, then there is $i \in \{1,2,\dots,n\}$ such that $x \in A_i \setminus \mathrm{int}(A_i)$, and hence $A \subseteq A_1 \setminus \mathrm{int}(A_1) \cup \dots \cup A_n \setminus \mathrm{int}(A_n)$. Since A_i is quasi-open for every $i \in \{1,2,\dots,n\}$, A is nowhere dense. Now let $\mathrm{int}(A) \neq \emptyset$. Then we can show as before that the set $A \setminus \mathrm{int}(A)$ is nowhere dense, and so $A = (A \setminus \mathrm{int}(A)) \cup \mathrm{int}(A)$ is a union of a nowhere dense set $A \setminus \mathrm{int}(A)$ and an open set $\mathrm{int}(A)$.

Lemma 2.7.15. *Let (X,τ) be a locally compact topological space, and let (Y,d) be a boundedly compact metric space. If $\mathcal{E}$ is a compact subset of $(Q(X,Y),\tau_{UC})$, then $\mathcal{E}$ is closed, pointwise bounded and densely equiquasicontinuous*.*

Proof. The set $\mathcal{E}$ is closed because $(Q(X,Y),\tau_{UC})$ is a Hausdorff space. The evaluation at x defined by $e_x(f) := f(x)$ for all $f \in Q(X,Y)$ is continuous with respect to τ_p topology on $Q(X,Y)$ [159], and hence it is continuous also with respect to τ_{UC} topology on $Q(X,Y)$, so the image $\mathcal{E}[x]$ of $\mathcal{E}$ is compact, and therefore $\mathcal{E}$ is pointwise bounded.

To prove that $\mathcal{E}$ is densely equiquasicontinuous*, we use an idea from the proof of [102, Theorem 5.7]. Let $x \in X$, and let $\mathcal{V} := \{V_1,\dots,V_m\}$ be a finite open Lebesgue cover that contains the complement of some nonempty compact set and with Lebesgue number $\varepsilon > 0$. Let O be an open neighbourhood of x in (X,τ) such that $A := \overline{O}$ is compact. For each $j \in \{1,\dots,m\}$, let $V_j^* := \{y \in V_j : \frac{\varepsilon}{3} < d(y, Y \setminus V_j)\}$. Then $\mathcal{V}^* := \{V_1^*,\dots,V_m^*\}$ is a Lebesgue cover of (Y,d) with Lebesgue number $\frac{\varepsilon}{3}$. We define a finite family $\mathcal{B}$ of quasi-open subsets of (X,τ) as follows. Since $\mathcal{E}$ is compact in $(Q(X,Y),\tau_{UC})$, there are $f_1,\dots,f_n \in \mathcal{E}$ such that

$$\mathcal{E} \subseteq W\left(f_1, A, \frac{\varepsilon}{3}\right) \cup \dots \cup W\left(f_n, A, \frac{\varepsilon}{3}\right).$$

For any $i \in \{1,2,\dots,n\}$ and $j \in \{1,2,\dots,m\}$, put $B_j^i := f_i^{-1}(V_j^*)$, which is a quasi-open set in (X,τ) since f_i is quasicontinuous. Denote by $\mathcal{F}$ the set of all functions from $\{1,2,\dots,n\}$ to $\{1,2,\dots,m\}$. For every $h \in \mathcal{F}$, put

$$P_h := (O \cap B_{h(1)}^1 \cap \dots \cap B_{h(n)}^n) \setminus \mathrm{int}(O \cap B_{h(1)}^1 \cap \dots \cap B_{h(n)}^n)$$

and

$$R_h := \operatorname{int}(O \cap B^1_{h(1)} \cap \cdots \cap B^n_{h(n)}).$$

By Remark 2.7.14, for every $h \in \mathcal{F}$, the set P_h is a nowhere dense set. Denote by $\mathcal{B}$ the family containing all nonempty sets P_h and R_h, where $h \in \mathcal{F}$. We show that $\bigcup \mathcal{B}$ is a neighbourhood of x in (X, τ). Let $z \in O$. Then there is $h \in \mathcal{F}$ such that $f_i(z) \in V^*_{h(i)}$ for every $i \in \{1, 2, \ldots, n\}$. So $z \in B^i_{h(i)}$ for every $i \in \{1, 2, \ldots, n\}$. Then $z \in P_h$ or $z \in R_h$, and thus $z \in \bigcup \mathcal{B}$. Now let $f \in \mathcal{E}$, let $B \in \mathcal{B}$, and let $p, q \in B$. Of course, there is $h \in \mathcal{F}$ such that $B \subseteq B^1_{h(1)} \cap \cdots \cap B^n_{h(n)}$. Because

$$\mathcal{E} \subseteq W\left(f_1, A, \frac{\varepsilon}{3}\right) \cup \cdots \cup W\left(f_n, A, \frac{\varepsilon}{3}\right),$$

there exists $i \in \{1, 2, \ldots, n\}$ such that $f \in W(f_i, A, \frac{\varepsilon}{3})$. So $d(f(p), f_i(p)) < \frac{\varepsilon}{3}$ and $d(f(q), f_i(q)) < \frac{\varepsilon}{3}$. Because $p, q \in B^i_{h(i)}$, we have that $f_i(p) \in V^*_{h(i)}$ and $f_i(q) \in V^*_{h(i)}$, and so $f(p), f(q)$ are elements of $B(V^*_{h(i)}, \frac{\varepsilon}{3})$, and hence they are elements of $V_{h(i)}$. Hence $\mathcal{E}$ is densely equiquasicontinuous*. □

The following result presents an example of a non-compact subset $\mathcal{E}$ of $(Q((-1,1), \mathbb{R}), \tau_{UC})$ that is closed, pointwise bounded and densely equiquasicontinuous*.

Example 2.7.16. Consider $X := (-1, 1)$ endowed with the usual topology and consider the function $f : X \to \mathbb{R}$ defined by,

$$f(x) := \begin{cases} 0 & \text{if } x \in (-1, 0], \\ \frac{1}{x} & \text{if } x \in (0, 1). \end{cases}$$

For each $n \in \mathbb{N}$, define $f_n : X \to \mathbb{R}$ as follows:

$$f_n(x) := \begin{cases} 0 & \text{if } x \in (-1, 0], \\ n & \text{if } x \in (0, \frac{1}{n}), \\ \frac{1}{x} & \text{if } x \in [\frac{1}{n}, 1). \end{cases}$$

The sequence $(f_n : n \in \mathbb{N})$ of quasicontinuous functions pointwise converges to the quasicontinuous function f but fails to converge to f with respect to the topology of uniform convergence on compact sets. It is easy to see that the subset $\{f_n : n \in \mathbb{N}\}$ of $(Q((-1,1), \mathbb{R}), \tau_{UC})$ is closed and pointwise bounded. It is clear that $\{f_n : n \in \mathbb{N}\}$ is densely equiquasicontinuous* at every point $x \neq 0$. We show that $\{f_n : n \in \mathbb{N}\}$ is densely equiquasicontinuous* at $x = 0$.

Let $\mathcal{V}$ be a finite open Lebesgue cover of $\mathbb{R}$ that contains the complement of a nonempty compact set A. There is $n_0 \in \mathbb{N}$ such that $\{n \in \mathbb{N} : n_0 \leq n\} \subseteq Y \setminus A$. Denote by $\mathcal{B}$ the family $\{(-1, 0), \{0\}, (0, \frac{1}{n_0})\}$. Then for every $n_0 \leq n$ and every $p, q \in (0, \frac{1}{n_0})$, we have $f_n(p), f_n(q) \in Y \setminus A$. For every $n < n_0$ and every $p \in (0, \frac{1}{n_0})$, we have $f_n(p) = n$. For every $n \in \mathbb{N}$ and every $p \in (-1, 0]$, we have $f_n(p) = 0$. Hence $\{f_n : n \in \mathbb{N}\}$ is densely equiquasicontinuous* at $x = 0$, but $\{f_n : n \in \mathbb{N}\}$ is not compact in $(Q((-1,1), \mathbb{R}), \tau_{UC})$.

Lemma 2.7.17. *Let (X,τ) be a topological space, and let (Y,d) be a boundedly compact metric space. Let $x \in X$, and let $\mathcal{E}$ be densely equiquasicontinuous* at x and locally bounded at x. Then $\mathcal{E}$ is densely equiquasicontinuous at x.*

Proof. Let $\varepsilon > 0$. The local boundedness of $\mathcal{E}$ at x implies that there are $n \in \mathbb{N}$ and an open neighbourhood U_x of x in (X,τ) such that $\bigcup\{f(U_x) : f \in \mathcal{E}\} \subseteq B(y_0, n)$, where $y_0 \in Y$. Let $m \in \mathbb{N}$ be such that $n < m$ and $1/m < \varepsilon$. Let $\mathcal{V}$ be a finite open Lebesgue cover of (Y,d) such that $\mathcal{V}$ contains $Y \setminus \overline{B(y_0, m)}$ and all other elements of $\mathcal{V}$ have diameters less than $\frac{1}{m}$. Since $\mathcal{E}$ is densely equiquasicontinuous* at x, there exists a finite family $\mathcal{B}$ of nonempty subsets of (X,τ) that are either open or nowhere dense such that $\bigcup \mathcal{B}$ is a neighbourhood of x in (X,τ) and such that for every $f \in \mathcal{E}$, every $B \in \mathcal{B}$ and every $p, q \in B$, there exists $V \in \mathcal{V}$ with $f(p), f(q) \in V$. Put $\mathcal{B}' := \{B \cap U_x : B \in \mathcal{B}\}$. Then $\mathcal{B}'$ is a finite family of subsets of (X,τ) that are either open or nowhere dense such that $\bigcup \mathcal{B}'$ is a neighbourhood of x in (X,τ) and such that for every $f \in \mathcal{E}$, every $B \in \mathcal{B}'$ and every $p, q \in B$, $d(f(p), f(q)) < \varepsilon$. Thus $\mathcal{E}$ is densely equiquasicontinuous at x. □

Proposition 2.7.18. *Let (X,τ) be a locally compact topological space, and let (Y,d) be a boundedly compact metric space. Let $\mathcal{E}$ be densely equiquasicontinuous* and pointwise bounded. Then there exists*
(i) *a dense G_δ-set G in (X,τ) such that $\mathcal{E}$ is equicontinuous at each point $x \in G$ and*
(ii) *a dense open set M in (X,τ) such that $\mathcal{E}$ is locally bounded at every $x \in M$ and $\mathcal{E}$ is densely equiquasicontinuous at each point $x \in M$.*

Proof. We first prove (i). We use an idea from the proof of [102, Theorem 5.4]. Choose any $y_0 \in (Y,d)$. Let $(\mathcal{V}_n : n \in \mathbb{N})$ be a sequence of finite open Lebesgue covers of (Y,d) such that each $\mathcal{V}_{n+1}$ strongly refines $\mathcal{V}_n$ (i. e. every element of $\mathcal{V}_{n+1}$ is a proper subset of an element from $\mathcal{V}_n$) and such that each $\mathcal{V}_n$ contains $Y \setminus \overline{B(y_0, n)}$ and all other elements of $\mathcal{V}_n$ have diameters less than $\frac{1}{n}$. By using Zorn's lemma we obtain, for each positive integer n, a maximal collection $\mathbf{U}_n$ of nonempty finite families of nonempty open subsets of (X,τ) satisfying:
(a) for each $\mathcal{U}_1, \mathcal{U}_2 \in \mathbf{U}_n$ with $\mathcal{U}_1 \neq \mathcal{U}_2$, $\overline{\bigcup \mathcal{U}_1} \cap \overline{\bigcup \mathcal{U}_2} = \emptyset$;
(b) for all $f \in \mathcal{E}$, $\mathcal{U} \in \mathbf{U}_n$, $U \in \mathcal{U}$ and $p, q \in U$, there exists $V \in \mathcal{V}_n$ such that $f(p)$, $f(q) \in V$.

Define $G := \bigcap_{n \in \mathbb{N}} \bigcup\{\bigcup \mathcal{U} : \mathcal{U} \in \mathbf{U}_n\}$, which is a G_δ subset of (X,τ) and is dense because of the maximality of each $\mathbf{U}_n$. Let $x \in G$, and let $\varepsilon > 0$. Because $\mathcal{E}[x]$ is bounded, there is an integer $\frac{1}{\varepsilon} < n$ such that $\mathcal{E}[x] \subseteq \overline{B(y_0, n)}$. There is $U_x \in \mathcal{U}$, where $\mathcal{U} \in \mathbf{U}_n$ is such that $x \in U_x$. Then for every $f \in \mathcal{E}$ and every $y \in U_x$, $d(f(x), f(y)) < \varepsilon$, and so $\mathcal{E}$ is equicontinuous at x. This completes the proof of part (i). We now prove part (ii). From before we have that $\bigcup\{f(U_x) : f \in \mathcal{E}\}$ is bounded in (Y,d). Put $M := \bigcup\{U_x : x \in G\}$. Since $\mathcal{E}$ is densely equiquasicontinuous*, and because for every $z \in M$, there is a neighbourhood V of z in (X,τ) such that $\bigcup\{f(V) : f \in \mathcal{E}\}$ is bounded in (Y,d), we have by Lemma 2.7.17 that $\mathcal{E}$ is densely equiquasicontinuous at each $x \in M$. □

We say that a system $\mathcal{E} \subseteq F(X,Y)$ is *supported at* $x \in X$ if for every $\varepsilon > 0$, there exist a neighbourhood $U(x)$ of x in (X,τ) and a finite family $\{\mathcal{E}_1, \mathcal{E}_2, \ldots, \mathcal{E}_n\}$ of nonempty subsets of $\mathcal{E}$ such that $\bigcup_{i=1}^{n} \mathcal{E}_i = \mathcal{E}$ and for every $z \in U(x)$, every $i \in \{1,2,\ldots,n\}$ and every $f,g \in \mathcal{E}_i$, $d(f(z),g(z)) < \varepsilon$.

Lemma 2.7.19. *Let (X,τ) be a locally compact topological space, and let (Y,d) be a metric space. If $\mathcal{E}$ is a compact subset of $(Q(X,Y),\tau_{UC})$, then $\mathcal{E}$ is supported at each point $x \in X$.*

Proof. Let $x \in (X,\tau)$, and let $\varepsilon > 0$. Let $U(x)$ be an open neighbourhood of x in (X,τ) such that $\overline{U(x)}$ is compact. Since $\mathcal{E}$ is compact in $(Q(X,Y),\tau_{UC})$, there are $f_1,\ldots,f_n \in \mathcal{E}$ such that

$$\mathcal{E} \subseteq W\left(f_1, \overline{U(x)}, \frac{\varepsilon}{2}\right) \cup \cdots \cup W\left(f_n, \overline{U(x)}, \frac{\varepsilon}{2}\right).$$

Put $\mathcal{E}_i := \mathcal{E} \cap W(f_i, \overline{U(x)}, \frac{\varepsilon}{2})$ for every $i \in \{1,2,\ldots,n\}$. So $\bigcup_{i=1}^{n} \mathcal{E}_i = \mathcal{E}$, and for every $z \in U(x)$, every $i \in \{1,2,\ldots,n\}$ and every $f,g \in \mathcal{E}_i$, $d(f(z),g(z)) < \varepsilon$. □

Theorem 2.7.20. *Let (X,τ) be a locally compact topological space, and let (Y,d) be a boundedly compact metric space. A subset $\mathcal{E}$ of $(Q(X,Y),\tau_{UC})$ is compact if and only if $\mathcal{E}$ is closed, pointwise bounded, densely equiquasicontinuous* and supported at the points of non-local boundedness of $\mathcal{E}$.*

Proof. Let $\mathcal{E}$ be a compact set in $(Q(X,Y),\tau_{UC})$. By Lemma 2.7.15, $\mathcal{E}$ is closed, pointwise bounded and densely equiquasicontinuous*. By Lemma 2.7.19, $\mathcal{E}$ is supported at each point of X.

For the converse, suppose that $\mathcal{E}$ is closed, pointwise bounded, densely equiquasicontinuous* and supported at the points of non-local boundedness of $\mathcal{E}$. The product $\prod_{x\in X} \overline{\mathcal{E}[x]}$ is a compact subset of $Y^X = \prod_{x\in X} Y_x$ with the relative product topology, where $Y_x := Y$ for every $x \in X$. Let $(f_\sigma : \sigma \in \Sigma)$ be a net in $\mathcal{E}$. There is $f \in \prod_{x\in X} \overline{\mathcal{E}[x]}$ such that f is a cluster point of $(f_\sigma : \sigma \in \Sigma)$ in $\prod_{x\in X} \overline{\mathcal{E}[x]}$. Without loss of generality, we can suppose that $(f_\sigma : \sigma \in \Sigma)$ converges to f in $\prod_{x\in X} \overline{\mathcal{E}[x]}$. We will show that $(f_\sigma : \sigma \in \Sigma)$ converges uniformly to f on every compact set K in X. Then, by Proposition 2.7.1, f is quasicontinuous, and thus $f \in \mathcal{E}$. So suppose, to obtaining a contradiction, that $(f_\sigma : \sigma \in \Sigma)$ fails to converge uniformly on compact sets to f. There exist a compact set A in X and $\varepsilon > 0$ such that for every $\sigma \in \Sigma$, there are $\sigma \leq \beta_\sigma$ and $a_\sigma \in A$ such that $\varepsilon < d(f(a_\sigma), f_{\beta_\sigma}(a_\sigma))$. Let a be a cluster point of $(a_\sigma : \sigma \in \Sigma)$. Without loss of generality, we can suppose that $(a_\sigma : \sigma \in \Sigma)$ converges to a.

Suppose first that $\mathcal{E}$ is locally bounded at a. Then, by Lemma 2.7.17, $\mathcal{E}$ is densely equiquasicontinuous at a. The dense equiquasicontinuity of $\mathcal{E} \cup \{f\}$ at the point a (see Remark 2.7.6) implies that there is a finite family $\mathcal{B}$ of nonempty subsets of (X,τ) that are either open or nowhere dense such that $\bigcup \mathcal{B}$ is a neighbourhood of a and such that for every $g \in \mathcal{E} \cup \{f\}$, every $B \in \mathcal{B}$ and every $p,q \in B$, $d(g(p),g(q)) < \varepsilon/3$. Without loss of generality, we can suppose that $a_\sigma \in \bigcup \mathcal{B}$ for every $\sigma \in \Sigma$, and since $\mathcal{B}$ is finite, we can suppose that there is $B \in \mathcal{B}$ such that $a_\sigma \in B$ for every $\sigma \in \Sigma$.

Choose a_η. Since the net $(f_\sigma : \sigma \in \Sigma)$ pointwise converges to f, there must exist $\gamma \in \Sigma$ such that $d(f(a_\eta), f_\delta(a_\eta)) < \varepsilon/3$ for every $\gamma \le \delta$. There is $\gamma \le \beta_\gamma$ such that $\varepsilon < d(f(a_\gamma), f_{\beta_\gamma}(a_\gamma))$. Then

$$d(f(a_\gamma), f_{\beta_\gamma}(a_\gamma)) \le d(f(a_\gamma), f(a_\eta)) + d(f(a_\eta), f_{\beta_\gamma}(a_\eta)) + d(f_{\beta_\gamma}(a_\eta), f_{\beta_\gamma}(a_\gamma)) < \varepsilon,$$

a contradiction.

Suppose now that $\mathcal{E}$ is not locally bounded at a. Then $\mathcal{E}$ is supported at a. There exist a neighbourhood $U(a)$ of a in (X, τ) and a finite family $\{\mathcal{E}_1, \mathcal{E}_2, \dots, \mathcal{E}_n\}$ of nonempty subsets of $\mathcal{E}$ such that $\bigcup_{i=1}^n \mathcal{E}_i = \mathcal{E}$ and such that for every $z \in U(a)$, every $i \in \{1, 2, \dots, n\}$ and every $f, g \in \mathcal{E}_i$, $d(f(z), g(z)) < \varepsilon$. Without loss of generality, we can suppose that $a_\sigma \in U(a)$ for every $\sigma \in \Sigma$ and since $\{\mathcal{E}_1, \mathcal{E}_2, \dots, \mathcal{E}_n\}$ is finite, we can suppose that there is $\mathcal{E}_i \in \{\mathcal{E}_1, \mathcal{E}_2, \dots, \mathcal{E}_n\}$ such that $f_{\beta_\sigma} \in \mathcal{E}_i$ for every $\sigma \in \Sigma$. The net $(f_{\beta_\sigma} : \sigma \in \Sigma)$ converges pointwise to f, and so $d(f_{\beta_\sigma}(x), f(x)) \le \varepsilon$ for every $x \in U(a)$, a contradiction. □

Theorem 2.7.21. *Let (X, τ) be a locally compact topological space, and let (Y, d) be a boundedly compact metric space. A subset $\mathcal{E}$ of $(Q(X, Y), \tau_{UC})$ is compact if and only if $\mathcal{E}$ is closed, pointwise bounded, there is a dense open set M in (X, τ) such that $\mathcal{E}$ is densely equiquasicontinuous at each $x \in M$ and $\mathcal{E}$ is supported at each point $x \in X \setminus M$.*

Proof. Let $\mathcal{E}$ be a compact set in $(Q(X, Y), \tau_{UC})$. The proof that $\mathcal{E}$ is closed, pointwise bounded and densely equiquasicontinuous* follows from Lemma 2.7.15. By Proposition 2.7.18 there is a dense open set M in (X, τ) such that $\mathcal{E}$ is equicontinuous at each $x \in M$. By Lemma 2.7.19, $\mathcal{E}$ is supported at each point $x \in X \setminus M$.

The proof of the converse is similar to the proof of Theorem 2.7.20, because if a cluster point a of the net $(a_\sigma : \sigma \in \Sigma)$ in (X, τ) is an element of the set M, then $\mathcal{E}$ is densely equiquasicontinuous at a, and if $a \notin M$, then $\mathcal{E}$ is supported at a. □

Theorem 2.7.22. *Let (X, τ) be a locally compact topological space and let (Y, d) be a boundedly compact metric space. A subset $\mathcal{E}$ of $(Q(X, Y), \tau_{UC})$ is compact if and only if $\mathcal{E}$ is closed and pointwise bounded, there is a dense G_δ-set G in (X, τ) such that $\mathcal{E}$ is equicontinuous at each point $x \in G$, and $\mathcal{E}$ is supported at each point $x \in X \setminus G$.*

Proof. Let $\mathcal{E}$ be a compact set in $(Q(X, Y), \tau_{UC})$. The proof that $\mathcal{E}$ is closed, pointwise bounded and densely equiquasicontinuous* follows from Lemma 2.7.15. By Proposition 2.7.18 there is a dense G_δ-set G in (X, τ) such that $\mathcal{E}$ is equicontinuous at each $x \in G$. By Lemma 2.7.19, $\mathcal{E}$ is supported at each point $x \in X \setminus G$.

The proof of the converse is similar to that of Theorem 2.7.20, because if a cluster point a of the net $(a_\sigma : \sigma \in \Sigma)$ in (X, τ) from the proof of Theorem 2.7.20 is an element of the set G, then $\mathcal{E}$ is densely equiquasicontinuous at a, and if $a \notin G$, then $\mathcal{E}$ is supported at a. □

Exercises

1. Show that every subcontinuous function from a topological space (X, τ) into a metric space (Y, d) is locally bounded.
2. Let (X, τ) be a topological space, and let (Y, d) be boundedly compact. If $f : X \to Y$ is locally bounded, then for every $x \in X$, there is a neighbourhood U_x of x in (X, τ) such that $\overline{f(U_x)}$ is compact. Thus every locally bounded function f from a topological space (X, τ) into a boundedly compact metric space (Y, d) is subcontinuous.
3. Show that notions of equiquasicontinuity and dense equiquasicontinuity are independent.
4. Let (X, τ) be a topological space, and let (Y, d) be a boundedly compact metric space. Prove that if $\mathcal{E} \subseteq F(X, Y)$ is densely equiquasicontinuous* at $x \in X$, then the closure of $\mathcal{E}$, with respect to the topology τ_p, is also densely equiquasicontinuous* at x.

2.8 Metrisability of quasicontinuous functions

In this section, let (X, τ) be a Hausdorff topological space, and let (Y, d) be a non-trivial metric space. A topological space (X, τ) is *hemicompact* [78] if in the family of all compact subspaces of (X, τ), ordered by set inclusion, there exists a countable cofinal subfamily. Every hemicompact space is σ-compact but not vice versa. The space of rationals with usual topology is a σ-compact space that is not hemicompact. On the other hand, a locally compact σ-compact space is hemicompact.

Theorem 2.8.1. *Let (X, τ) be a topological space, and let (Y, d) be a metric space. Then the following are equivalent:*

(i) *The uniformity $\mathfrak{U}_{UC}$ on $Q(X, Y)$ is induced by a metric;*
(ii) *$(Q(X, Y), \tau_{UC})$ is metrisable;*
(iii) *$(Q(X, Y), \tau_{UC})$ is first countable;*
(iv) *(X, τ) is hemicompact.*

Proof. (iv)⇒(i) Let $\{K_n : n \in \mathbb{N}\}$ be a countable cofinal subfamily in $K(X)$ with respect to the inclusion. The family

$$W(K, \varepsilon) := \{(f, g) : d(f(x), g(x)) < \varepsilon \text{ for all } x \in K\},$$

where $K \in K(X)$ and $\varepsilon > 0$, is a base of $\mathfrak{U}_{UC}$ on $Q(X, Y)$. Since for every $K \in K(X)$, there is $n \in \mathbb{N}$ with $K \subseteq K_n$, the family

$$\left\{W\left(K_n, \frac{1}{m}\right) : n, m \in \mathbb{N}\right\}$$

is a countable base of $\mathfrak{U}_{UC}$. Thus, by the metrisation theorem in [159], $(Q(X, Y), \mathfrak{U}_{UC})$ is metrisable.

(i) ⇒ (ii) and (ii) ⇒ (iii) are obvious.

(iii) ⇒ (iv) Let $y_0 \in Y$ and let f be the constant function on X mapping each point to y_0. By the assumption f has a countable base $\{W(f, K_n, \varepsilon_n) : n \in \mathbb{N}\}$. We claim that $\{K_n : n \in \mathbb{N}\}$ is a countable cofinal family in $K(X)$ with respect to the inclusion. Suppose that this is not true. Thus there is $K \in K(X)$ such that for each $n \in \mathbb{N}$, there is $k_n \in K \setminus K_n$. For every $n \in \mathbb{N}$, there is an open neighbourhood U_n of k_n such that $\overline{U_n} \cap K_n = \emptyset$. Let $y_1 \in Y$ be a point different from y_0. For every $n \in \mathbb{N}$, let $f_n : X \to Y$ be the function defined by

$$f_n(x) := \begin{cases} y_1 & \text{if } x \in \overline{U_n}, \\ y_0 & \text{otherwise.} \end{cases}$$

It is easy to see that the function f_n is quasicontinuous for every $n \in \mathbb{N}$. Since $f_n \notin W(f, K, d(y_0, y_1))$ for every $n \in \mathbb{N}$, the family $\{W(f, K_n, \varepsilon_n) : n \in \mathbb{N}\}$ fails to be a base of neighbourhoods of f (otherwise, there is $n \in \mathbb{N}$ such that $f_n \in W(f, K_n, \varepsilon_n) \subseteq W(f, K, d(y_0, y_1))$, a contradiction). □

Let (X, τ) be a hemicompact space, and let $\{K_n : n \in \mathbb{N}\}$ be a fixed countable cofinal subfamily in $K(X)$ with respect to set inclusion. We define a metric on the space $Q(X, Y)$. For every $K \in K(X)$, let p_K be the (extended-valued) pseudometric on $Q(X, Y)$ defined by

$$p_K(f, g) := \sup\{d(f(x), g(x)) : x \in K\}.$$

Then for every $K \in K(X)$, we have a real-valued pseudometric h_K defined by

$$h_K(f, g) := \min\{1, p_K(f, g)\}.$$

We define the function $\rho : Q(X, Y) \times Q(X, Y) \to \mathbb{R}$ as follows:

$$\rho(f, g) := \sum_{n=1}^{\infty} \frac{1}{2^n} h_{K_n}(f, g).$$

It is easy to see that ρ is a metric on $Q(X, Y)$ and that the uniformity $\mathfrak{U}_{UC}$ is generated by ρ. We say that a Hausdorff space (X, τ) is of *pointwise countable type* [78] if for every point $x \in X$, there exists a compact subset C of (X, τ) such that $x \in C$ and $\chi(C, X) \leq \aleph_0$. Another more general property is that of being a q-space (see page 66). This is a space such that for each point, there exists a sequence $(U_n : n \in \mathbb{N})$ of neighbourhoods of that point in (X, τ) such that if $x_n \in U_n$ for each $n \in \mathbb{N}$, then $(x_n : n \in \mathbb{N})$ has a cluster point in (X, τ) [211].

Theorem 2.8.2. *Let (X, τ) be a topological space. Then the following are equivalent:*
(i) *The uniformity $\mathfrak{U}_{UC}$ on $Q(X, \mathbb{R})$ is induced by a metric;*
(ii) *$(Q(X, \mathbb{R}), \tau_{UC})$ is metrisable;*
(iii) *$(Q(X, \mathbb{R}), \tau_{UC})$ is first countable;*

(iv) $(Q(X,\mathbb{R}),\tau_{UC})$ *is of pointwise countable type;*
(v) $(Q(X,\mathbb{R}),\tau_{UC})$ *is a q-space;*
(vi) (X,τ) *is hemicompact.*

Proof. (vi)⇒(i) It is the same as in Theorem 2.8.1. (i)⇒ (ii)⇒(iii)⇒(iv)⇒(v) are obvious.

(v)⇒(vi) Suppose that $(Q(X,\mathbb{R}),\tau_{UC})$ is a q-space. Let f be the zero function on X. By assumption there is a sequence $(W(f,K_n,\varepsilon_n) : n \in \mathbb{N})$ such that if $f_n \in W(f,K_n,\varepsilon_n)$ for each $n \in \mathbb{N}$, then $(f_n : n \in \mathbb{N})$ has a cluster point in $(Q(X,\mathbb{R}),\tau_{UC})$. Without loss of generality, we may assume that the sequence $(\varepsilon_n : n \in \mathbb{N})$ converges to 0. We claim that $X = \bigcup\{K_n : n \in \mathbb{N}\}$. So let us suppose, to obtain a contradiction, that there is $x \in X\setminus\bigcup\{K_n : n \in \mathbb{N}\}$. For each $n \in \mathbb{N}$, there is an open neighbourhood U_n of x in (X,τ) such that $\overline{U_n} \cap K_n = \varnothing$. For each $n \in \mathbb{N}$, let $f_n \in Q(X,\mathbb{R})$ be defined as follows:

$$f_n(z) := \begin{cases} n & \text{if } z \in \overline{U_n}, \\ 0 & \text{otherwise.} \end{cases}$$

Evidently, f_n is quasicontinuous for every $n \in \mathbb{N}$, and $f_n \in W(f,K_n,\varepsilon_n)$ for every $n \in \mathbb{N}$, but $(f_n : n \in \mathbb{N})$ does not have a cluster point in $Q(X,\mathbb{R})$. Now for every $n \in \mathbb{N}$, put $C_n := \bigcup\{K_i : i \le n\}$. We claim that $\{C_n : n \in \mathbb{N}\}$ is a cofinal family in $K(X)$ with respect to set inclusion. Suppose that this is not true. Then there is $K \in K(X)$ such that for each $n \in \mathbb{N}$, there is $k_n \in K \setminus C_n$. For every $n \in \mathbb{N}$, there is an open neighbourhood V_n of k_n such that $\overline{V_n} \cap C_n = \varnothing$. For every $n \in \mathbb{N}$, let $g_n : X \to \mathbb{R}$ be a function defined by

$$g_n(z) := \begin{cases} 1 & \text{if } z \in \overline{V_n}, \\ 0 & \text{otherwise.} \end{cases}$$

Of course, $g_n \in Q(X,\mathbb{R})$ for every $n \in \mathbb{N}$, and $g_n \in W(f,C_n,\varepsilon_n) \subseteq W(f,K_n,\varepsilon_n)$. By assumption $(g_n : n \in \mathbb{N})$ has a cluster point $g \in Q(X,\mathbb{R})$. Then $g(x) = 0$ for every $x \in X$, a contradiction, since $g_n \notin W(g,K,d(0,1))$ for every $n \in \mathbb{N}$. □

Remarks 2.8.3. It is easy to verify that the previous theorem also works for a non-compact metric space (Y,d).

Theorem 2.8.4. *Let (X,τ) be a regular topological space with a countable pseudocharacter, and let (Y,d) be a metric space. Then the following are equivalent:*
(i) *The uniformity $\mathfrak{U}_{UC}$ on $Q(X,Y)$ is induced by a metric;*
(ii) $(Q(X,Y),\tau_{UC})$ *is metrizable;*
(iii) $(Q(X,Y),\tau_{UC})$ *is first countable;*
(iv) $(Q(X,Y),\tau_{UC})$ *is of pointwise countable type;*
(v) $(Q(X,Y),\tau_{UC})$ *is a q-space;*
(vi) (X,τ) *is hemicompact.*

Proof. (vi)⇒(i) It is the same as in Theorem 2.8.1. (i)⇒(ii)⇒(iii)⇒(iv)⇒(v) are obvious.

(v)⇒(vi) Let $y_0 \in Y$, and let f be the constant function on X mapping each point to y_0. Suppose that $(Q(X,Y), \tau_{UC})$ is a q-space. By the assumption there is a sequence $(W(f, K_n, \varepsilon_n) : n \in \mathbb{N})$ such that if $f_n \in W(f, K_n, \varepsilon_n)$ for each $n \in \mathbb{N}$, then $(f_n : n \in \mathbb{N})$ has a cluster point in $(Q(X,Y), \tau_{UC})$. Without loss of generality, we may assume that the sequence $(\varepsilon_n : n \in \mathbb{N})$ converges to 0. We claim that $X = \bigcup\{K_n : n \in \mathbb{N}\}$. So let us suppose, to obtain a contradiction, that there is $x \in X \setminus \bigcup\{K_n : n \in \mathbb{N}\}$. Then there is a sequence $(U_n : n \in \mathbb{N})$ of open neighbourhoods of x such that for each $n \in \mathbb{N}$, $\overline{U}_{n+1} \subseteq U_n$, $\overline{U_n} \cap K_n = \varnothing$ and $\{x\} = \bigcap\{\overline{U_n} : n \in \mathbb{N}\}$. Let $y_1 \in Y$ be a point different from y_0. For each $n \in \mathbb{N}$, let $f_n \in Q(X,Y)$ be defined as follows:

$$f_n(z) := \begin{cases} y_1 & \text{if } z \in \overline{U_n}, \\ y_0 & \text{otherwise.} \end{cases}$$

For every $n \in \mathbb{N}$, $f_n \in W(f, K_n, \varepsilon)$, but $(f_n : n \in \mathbb{N})$ does not have a cluster point in $Q(X,Y)$. Now for every $n \in \mathbb{N}$, put $C_n := \bigcup\{K_i : i \leq n\}$. We claim that $\{C_n : n \in \mathbb{N}\}$ is a cofinal family in $K(X)$ with respect to set inclusion. Suppose that this is not true. Then there is $K \in K(X)$ such that for each $n \in \mathbb{N}$, there is $k_n \in K \setminus C_n$. For every $n \in \mathbb{N}$, there is an open neighbourhood V_n of k_n such that $\overline{V_n} \cap C_n = \varnothing$. For every $n \in \mathbb{N}$, let $g_n : X \to Y$ be a function defined by,

$$g_n(x) := \begin{cases} y_1 & \text{if } x \in \overline{V_n}, \\ y_0 & \text{otherwise.} \end{cases}$$

Of course, $g_n \in Q(X,Y)$ for every $n \in \mathbb{N}$, and $g_n \in W(g, C_n, \varepsilon_n) \subseteq W(f, K_n, \varepsilon_n)$. By assumption $(g_n : n \in \mathbb{N})$ has a cluster point $g \in Q(X,Y)$. Then $g(x) = y_0$ for every $x \in X$, a contradiction, since $g_n \notin W(g, K, d(y_0, y_1))$ for every $n \in \mathbb{N}$. □

To study the complete metrisability of $(Q(X,Y), \tau_{UC})$, we will start with the following theorem.

Theorem 2.8.5. *Let (X, τ) be a locally compact topological space, and let (Y, d) be a complete metric space. Then $(Q(X,Y), \mathfrak{U}_{UC})$ is a complete uniform space.*

Proof. By [159, Theorem 7.10] the uniformity $\mathfrak{U}_{UC}$ on $F(X,Y)$ is complete. By Proposition 2.7.1, $Q(X,Y)$ is a closed set in $(F(X,Y), \tau_{UC})$. Thus $(Q(X,Y), \mathfrak{U}_{UC})$ is a complete uniform space. □

We have the following characterisation of local compactness in the class of metric spaces.

Proposition 2.8.6. *Let (X, d) be a metric space, and let (Y, e) be a complete metric space. The following are equivalent:*
(i) *The uniform space $(Q(X,Y), \mathfrak{U}_{UC})$ is complete;*
(ii) *(X, d) is locally compact.*

Proof. (ii)⇒(i) is by Theorem 2.8.5. (i)⇒(ii) We will prove the contrapositive statement. Suppose that (X, d) is not locally compact. Choose $x_0 \in X$ so that x_0 fails to have a local base of compact sets. We will use the construction from Theorem 2.7.2. The net $(f_g : g \in \mathbb{N}^{\mathbb{N}})$ is $\mathfrak{U}_{UC}$-Cauchy. However, the net $(f_g : g \in \mathbb{N}^{\mathbb{N}})$ fails to have a limit point in $(Q(X, Y), \mathfrak{U}_{UC})$. Indeed, the net $(f_g : g \in \mathbb{N}^{\mathbb{N}})$ converges in $(F(X, Y), \tau_{UC})$ to the function $f : X \to Y$ from Theorem 2.7.2, which is not quasicontinuous. Thus $(Q(X, Y), \mathfrak{U}_{UC})$ is not complete. □

We have the following results concerning complete metrisability of $Q(X, Y)$.

Theorem 2.8.7. *Let (X, τ) be a locally compact topological space, and let (Y, d) be a complete metric space. Then the following are equivalent:*
(i) *The uniformity $\mathfrak{U}_{UC}$ on $Q(X, Y)$ is induced by a complete metric;*
(ii) *$(Q(X, Y), \tau_{UC})$ is completely metrisable;*
(iii) *$(Q(X, Y), \tau_{UC})$ is first countable;*
(iv) *(X, τ) is hemicompact.*

Proof. Of course, it is sufficient to prove that (iv)⇒(i). By Theorem 2.8.5 the uniform space $(Q(X, Y), \mathfrak{U}_{UC})$ is complete. By Theorem 2.8.1 the uniformity $\mathfrak{U}_{UC}$ on $Q(X, Y)$ is induced by a metric, and thus $\mathfrak{U}_{UC}$ on $Q(X, Y)$ is induced by a complete metric. □

Remarks 2.8.8. If (X, τ) is a locally compact and hemicompact topological space and (Y, d) is a complete metric space, then the metric ρ defined above is a complete metric on $Q(X, Y)$.

We will now apply our previous results to characterise sequentially compact subsets of $(Q(X, Y), \tau_{UC})$.

Theorem 2.8.9. *Let (X, τ) be a locally compact hemicompact space, let (Y, d) be a boundedly compact metric space, and let $Q^*(X, Y)$ be the set of all locally bounded quasicontinuous functions from (X, τ) to (Y, d). A subset $\mathcal{E}$ of $(Q^*(X, Y), \tau_{UC})$ is sequentially compact if and only if it is closed, pointwise bounded and densely equiquasicontinuous.*

Proof. If (X, τ) is hemicompact, then by Theorem 2.8.1, $(Q^*(X, Y), \tau_{UC})$ is metrisable. In metrisable spaces, compactness and sequential compactness are equivalent. Thus we can apply Corollary 2.7.12. □

Corollary 2.8.10. *Let (X, τ) be a locally compact hemicompact space, and let (Y, d) be a boundedly compact metric space. If $(f_n : n \in \mathbb{N})$ is a pointwise bounded densely equiquasicontinuous sequence in $Q^*(X, Y)$, then there is a τ_{UC} convergent subsequence $(f_{n_k} : k \in \mathbb{N})$.*

Proof. Put $\mathcal{E} := \overline{\{f_n : n \in \mathbb{N}\}}$ in $(Q^*(X, Y), \tau_{UC})$. Then $\mathcal{E}$ is closed, pointwise bounded and densely equiquasicontinuous, [139]. By Theorem 2.8.9, $\mathcal{E}$ is sequentially compact in $(Q^*(X, Y), \tau_{UC})$. Thus there is a τ_{UC} convergent subsequence $(f_{n_k} : k \in \mathbb{N})$. □

3 Applications of usco mappings

This chapter contains applications of usco mappings that demonstrate the utility of this class of set-valued mappings. In particular, this chapter contains applications of usco mappings to topology, approximation theory, differentiability theory of convex functions, variational analysis and the differentiation theory of Lipschitz functions.

3.1 Usco mappings in topology

In this section, we consider some applications of uscos in topology.

Let (X,τ) be a completely regular topological space. Then we say that an ordered pair $((Y,\tau'),e)$ is a *compactification of* (X,τ) if (i) (Y,τ') is a compact Hausdorff space, (ii) $e : X \to Y$ is a homeomorphic embedding of (X,τ) into (Y,τ'), and (iii) $e(X)$ is dense in (Y,τ').

We say that a compactification $((Y,\tau'),e)$ of a completely regular topological space (X,τ) is a *Stone–Čech compactification* if for every bounded real-valued continuous function $g : X \to \mathbb{R}$, there exists a continuous function $\widehat{g} : Y \to \mathbb{R}$ such that $g(x) = \widehat{g}(e(x))$ for all $x \in X$, that is, $g = \widehat{g} \circ e$.

It can be shown that every completely regular topological space admits a Stone–Čech compactification and that, up to homeomorphism, this compactification is unique in the sense that if $((Y,\tau'),e)$ and $((Y',\tau''),e')$ are Stone–Čech compactifications of a completely regular space (X,τ), then there exists a homeomorphism $h : Y \to Y'$ such that $h|_{e(X)} : e(X) \to e'(X)$ is a homeomorphism; see [78]. Hence we use the notation $((\beta X,\tau_\beta),e)$ to denote the Stone–Čech compactification of (X,τ). When there is no confusion, we will just write βX for the Stone–Čech compactification of (X,τ).

3.1.1 Extensions of usco mappings

Our first extension theorem concerns Čech-complete spaces. We say that a topological space (X,τ) is *Čech-complete* if (i) (X,τ) is completely regular and (ii) $e(X)$ is a G_δ subset of $(\beta X,\tau_\beta)$, where $((\beta X,\tau_\beta),e)$ is the Stone–Čech compactification of (X,τ).

Theorem 3.1.1 ([111]). *Let (Y,τ') be a completely regular topological space. Then (Y,τ') is Čech-complete if and only if for every usco mapping $\Phi : D \to 2^Y$ defined on a dense subset D of a topological space (X,τ), there exists a G_δ-subset G of (X,τ) containing D and an usco $\Phi^* : G \to 2^Y$ such that $\Phi^*|_D = \Phi$.*

Proof. Suppose that (Y,τ') is Čech-complete. Then there exists a sequence $(O_n : n \in \mathbb{N})$ of open subsets of $(\beta Y,\tau'_\beta)$ (the Stone–Čech compactification of (Y,τ')) such that $Y = \bigcap_{n\in\mathbb{N}} O_n$. Let $\Phi : D \to 2^Y$ be an usco mapping defined on a dense subset D

https://doi.org/10.1515/9783110750188-003

of a topological space (X, τ). Consider $\Phi_E : X \to 2^{\beta Y}$ defined by $\Phi_E(x) := \Phi(x)$ if $x \in D$ and $\Phi_E(x) := \varnothing$ if $x \notin D$. Note that Φ_E is a τ'-usco at each point of D. We may define $\mathrm{USC}(\Phi_E) : X \to 2^{\beta Y}$ by

$$\mathrm{USC}(\Phi_E)(x) := \bigcap \{\overline{\Phi_E(U)} : U \text{ is an open neighbourhood of } x\}.$$

Then $\mathrm{USC}(\Phi_E)$ is densely defined and has a closed graph (see Exercise 1.1.16). Therefore, by Corollary 1.1.15, $\mathrm{USC}(\Phi_E)$ is an usco on (X, τ). Furthermore, by Exercise 1.1.16, $\mathrm{USC}(\Phi_E)|_D = \Phi_E|_D = \Phi$. Now if

$$G := \{x \in X : \mathrm{USC}(\Phi_E)(x) \subseteq Y\} = \bigcap_{n \in \mathbb{N}} \{x \in X : \mathrm{USC}(\Phi_E)(x) \subseteq O_n\},$$

then G is a G_δ-subset of (X, τ) that contains D. Hence $\Phi^* := \mathrm{USC}(\Phi_E)|_G$ is the desired usco extension of Φ.

Let (Y, τ') be a completely regular topological space such that for every usco mapping $\Phi : D \to 2^Y$ defined on a dense subset D of a topological space (X, τ), there exists a G_δ-subset G of (X, τ) containing D and an usco $\Phi^* : G \to 2^Y$ such that $\Phi^*|_D = \Phi$. Let $X := \beta Y$, τ be the topology on βY, and let $\Phi : Y \to 2^Y$ be defined by $\Phi(x) := \{x\}$. Then Φ is an usco defined on (Y, τ). Therefore by the hypotheses there exists a G_δ-subset G of (X, τ) containing Y and an usco mapping $\Phi^* : G \to 2^Y$ such that $\Phi^*|_Y = \Phi$. (Note that we also may and do view Φ^* as a mapping into 2^G). We claim that $Y = G$. Let $f : G \to G$ be defined by $f(x) := x$. Then, by Corollary 1.3.14, $f(x) \in \Phi^*(x)$ for all $x \in G$, since $f(x) \in \{x\} = \Phi(x) = \Phi^*(x)$ for all $x \in Y$. Therefore

$$G = f(G) \subseteq \Phi^*(G) \subseteq Y \subseteq G,$$

that is, $Y = G$. This completes the proof. □

Remarks 3.1.2. Note that in the previous proof, we in fact showed that if (Y, τ') is a dense subspace of a Hausdorff space (Z, τ'') and (Y, τ') has the property that for every usco mapping $\Phi : D \to 2^Y$ defined on a dense subset D of a topological space (X, τ), there exists a G_δ-subset G of (X, τ) containing D and an usco $\Phi^* : G \to 2^Y$ such that $\Phi^*|_D = \Phi$, then Y is a dense G_δ-subset of (Z, τ''). Therefore if (Y, τ') is Čech-complete and dense in a Hausdorff space (Z, τ''), then Y is a G_δ-subset of (Z, τ'').

Corollary 3.1.3. *Let D be a subset of a metric space (M, d). If D, with the relative topology inherited from (M, d), is Čech-complete, then D is a G_δ-subset of (M, d).*

Proof. Let $X := \overline{D}$. From Remark 3.1.2 we see that D is a dense G_δ-subset of (X, d_X), where d_X denotes the restriction of the metric d to the set X. Now, as X is a closed subset of (M, d), it is a G_δ-subset of (M, d). Therefore D, as a G_δ-subset of a G_δ-subset of (M, d), is itself a G_δ-subset of (M, d). □

3.1.2 Spaces with a G_δ-diagonal

Lemma 3.1.4. *Let (X,τ) and (Y,τ') be topological spaces, and let $\Phi : X \to 2^Y$ be an usco on (X,τ). If (Y,τ') has a G_δ-diagonal, then $\{x \in X : \Phi(x)$ is a singleton$\}$ is a G_δ-subset of (X,τ).*

Proof. Since (Y,τ') has a G_δ-diagonal, there exist open neighbourhoods $(W_n : n \in \mathbb{N})$ of $\Delta_Y := \{(x,y) \in Y \times Y : x = y\}$ such that $\bigcap_{n\in\mathbb{N}} W_n = \Delta_Y$. For each $n \in \mathbb{N}$, let

$$O_n := \bigcup\{U \in \tau : \Phi(U) \times \Phi(U) \subseteq W_n\}.$$

Clearly, each set O_n is open in (X,τ). Therefore $G := \bigcap_{n\in\mathbb{N}} O_n$ is a G_δ-subset of (X,τ). If $x \in G$, then $\Phi(x) \times \Phi(x) \subseteq \bigcap_{n\in\mathbb{N}} W_n = \Delta_Y$, and so $\Phi(x)$ is a singleton. On the other hand, if $x \in X$ and $\Phi(x)$ is a singleton, then for each $n \in \mathbb{N}$, there exists an open neighbourhood V_n of $\Phi(x)$ such that $V_n \times V_n \subseteq W_n$. Fix $n \in \mathbb{N}$. Then by the upper semicontinuity of Φ at x there exists an open neighbourhood U_n of x such that $\Phi(U_n) \subseteq V_n$. Therefore $\Phi(U_n) \times \Phi(U_n) \subseteq V_n \times V_n \subseteq W_n$. Hence $x \in U_n \subseteq O_n$. Thus $G = \{x \in X : \Phi(x)$ is a singleton$\}$. □

Corollary 3.1.5. *Let (Y,τ') be a topological space. Then (Y,τ') has the property that every usco mapping $\Phi : X \to 2^Y$ acting from a topological space (X,τ) into a subset of (Y,τ') is single-valued at the points of a G_δ-subset of (X,τ) if and only if (Y,τ') has a G_δ-diagonal.*

Proof. Suppose that (Y,τ') has the property that every usco mapping $\Phi : X \to 2^Y$ acting from a topological space (X,τ) into a subset of (Y,τ') is single-valued at the points of a G_δ-subset of (X,τ). Consider the mapping $\Phi : Y \times Y \to 2^Y$ defined by $\Phi(x,y) := \{x,y\}$. From Example (v) in Section 1.1.2, we see that Φ is an usco on $(Y \times Y, \tau' \times \tau')$. Now

$$\{(x,y) \in Y \times Y : \Phi(x,y) \text{ is a singleton}\} = \Delta_Y.$$

Therefore Δ_Y is a G_δ-subset of $(Y \times Y, \tau' \times \tau')$. The converse direction follows directly from Lemma 3.1.4. □

3.1.3 Extensions of continuous functions

Theorem 3.1.6 ([50, 111, 116]). *Let D be a dense subset of a topological space (X,τ), and let (Y,τ') be a Čech-complete space with a G_δ-diagonal. If $f : D \to Y$ is a continuous function, then there exist a G_δ-subset G of (X,τ) containing D and a continuous function $\tilde{f} : G \to Y$ such that $\tilde{f}|_D = f$.*

Proof. Let $\Phi : D \to 2^Y$ be defined by $\Phi(x) := \{f(x)\}$ for $x \in D$. Then by Theorem 3.1.1 there exists a G_δ-subset G_1 of (X,τ) containing D and an usco $\Phi^* : G_1 \to 2^Y$ such that

$\Phi^*|_D = \Phi$. By Lemma 3.1.4 there exists a G_δ-subset G of (G_1, τ'') containing D such that Φ^* is single-valued on G, where τ'' denotes the relative τ-topology on the set G_1. As the set G is a G_δ-subset of a G_δ-subset of (X, τ), it is itself a G_δ-subset of (X, τ). Finally, the mapping $\Phi^*|_G$ may be identified with a continuous extension of the function f. □

It is interesting to see that the extension property considered above in fact provides a characterisation of a completely regular space being Čech-complete and possessing a G_δ-diagonal.

Lemma 3.1.7. *Let (Y, τ') be a completely regular topological space such that for every continuous function $f : D \to Y$ defined on a dense subset D of a topological space (X, τ), there exist a G_δ-subset G of (X, τ) containing D and a continuous function $\tilde{f} : G \to 2^Y$ such that $\tilde{f}|_D = f$. Then $(Y \times Y, \tau' \times \tau')$ also has the same extension property.*

Proof. Let $f : D \to Y \times Y$ be a continuous function defined on a dense subset D of a topological space (X, τ). By the definition of the product topology on $Y \times Y$ it follows that there exist continuous functions $f_1 : D \to Y$ and $f_2 : D \to Y$ such that $f(x) = (f_1(x), f_2(x))$ for all $x \in D$. Thus by the assumption on (Y, τ'), there exist G_δ-subsets G_1 and G_2 of (X, τ) containing D and continuous functions $\tilde{f}_1 : G_1 \to Y$ and $\tilde{f}_2 : G_2 \to Y$ such that $\tilde{f}_1|_D = f_1$ and $\tilde{f}_2|_D = f_2$. Let $G := G_1 \cap G_2$ and define $\tilde{f} : G \to Y \times Y$ by $\tilde{f}(x) := (\tilde{f}_1(x), \tilde{f}_2(x))$ for $x \in G$. Then G is a G_δ-subset of (X, τ) containing D, and $\tilde{f}$ is the desired extension of f. □

Theorem 3.1.8 ([50]). *Let (Y, τ') be a completely regular topological space. Then (Y, τ') is Čech-complete and possesses a G_δ-diagonal if and only if for every continuous function $f : D \to Y$ defined on a dense subset D of a topological space (X, τ), there exist a G_δ-subset G of (X, τ) containing D and a continuous function $\tilde{f} : G \to Y$ such that $\tilde{f}|_D = f$.*

Proof. The fact that the extension property holds if (Y, τ') is Čech-complete and possesses a G_δ-diagonal is simply a restatement of Theorem 3.1.6. So we consider the converse. Suppose that (Y, τ') has the property that for every continuous function $f : D \to Y$ defined on a dense subset D of a topological space (X, τ), there exist a G_δ-subset G of (X, τ) containing D and a continuous function $\tilde{f} : G \to Y$ such that $\tilde{f}|_D = f$. We will first show that (Y, τ') is Čech-complete. Let $X := \beta Y$, let τ be the topology on βY, and let $f : Y \to Y$ be defined by $f(x) := x$ for $x \in Y$. Then f is continuous on (Y, τ'). Therefore by the hypothesis there exist a G_δ-subset G of (X, τ) containing Y and a continuous function $\tilde{f} : G \to Y$ such that $\tilde{f}|_Y = f$. We claim that $Y = G$. Let $g : G \to G$ be defined by $g(x) := x$ for $x \in G$. Then, since $\tilde{f}|_Y = f = g|_Y$, we have that $\tilde{f} = g$. Therefore

$$G = g(G) = \tilde{f}(G) \subseteq Y \subseteq G,$$

that is, $Y = G$. Next, we show that (Y, τ') has a G_δ-diagonal. By Lemma 3.1.7 we know that $(Y \times Y, \tau' \times \tau')$ also has this extension property. Let $S := \{1/n : n \in \mathbb{N}\} \cup \{0\}$

(endowed with the relative topology inherited from $\mathbb{R}$), and let $X := Y^2 \times S$ endowed with the product topology τ. Furthermore, let

$$D := Y^2 \times [S \setminus \{0\}] \cup \Delta_Y \times \{0\},$$

where $\Delta_Y := \{(y, y) : y \in Y\}$. Note that D is dense in (X, τ). Let $\mathscr{E} := \{1/2n : n \in \mathbb{N}\}$ and $\mathscr{O} := \{1/(2n-1) : n \in \mathbb{N}\}$. Let us now define $f : D \to Y \times Y$ by

$$f(x,y,t) := \begin{cases} (x,y) & \text{if } (x,y,t) \in Y^2 \times \mathscr{E}, \\ (y,x) & \text{if } (x,y,t) \in Y^2 \times \mathscr{O}, \\ (x,x) & \text{if } (x,y,t) \in \Delta_Y \times \{0\}. \end{cases}$$

Then f is continuous on the open subspaces $Y^2 \times \mathscr{E}$ and $Y^2 \times \mathscr{O}$ of D, and f is continuous at each point of $(x, x, 0) \in D$ since $f^{-1}(U \times U) = (U \times U \times S) \cap D$ for each open neighbourhood U of x in (Y, τ'). It is easy to check that the function f cannot be extended to a continuous function on any subset of X that is strictly larger than D. Hence D is itself a G_δ-subset of (X, τ). Now since $Y^2 \times \{0\}$ is a G_δ-subset of (X, τ), we see that $\Delta_Y \times \{0\} = (Y^2 \times \{0\}) \cap D$ is a G_δ-subset of (X, τ). Finally, since the mapping $g : Y^2 \to Y^2 \times \{0\}$ defined by $g(x, y) := (x, y, 0)$ for $(x, y) \in Y^2$ is continuous, we see that $\Delta_Y = g^{-1}(\Delta_Y \times \{0\})$ is a G_δ-subset of $(Y \times Y, \tau' \times \tau')$. □

3.1.4 Extensions of functions on compact sets

Our theorem in this subsection shows that the Stone–Čech compactification enjoys a very strong extension property.

Theorem 3.1.9. *Let (X, τ) be a completely regular space. Then for every continuous function $f : (X, \tau) \to (Y, \tau')$ into a compact Hausdorff space (Y, τ'), there exists a continuous function $\tilde{f} : (\beta X, \tau_\beta) \to (Y, \tau')$, such that $f(x) = \tilde{f}(e(x))$ for all $x \in X$, that is, $f = \tilde{f} \circ e$, where $((\beta X, \tau_\beta), e)$ denotes the Stone–Čech compactification of (X, τ).*

Proof. Let $f : (X, \tau) \to (Y, \tau')$ be a continuous function into a compact Hausdorff space (Y, τ'), and let $\sigma : e(X) \to Y$ be defined by $\sigma := f \circ e^{-1}$. Note that σ is continuous on $e(X)$ and $\sigma \circ e = f$. Let $F : \beta X \to 2^Y$ be defined by $F := \mathrm{USC}(\sigma)$. It follows from Corollary 1.3.11 that F is a minimal usco on $(\beta X, \tau_\beta)$. Furthermore, by Corollary 1.3.12, $F(z) = \{\sigma(z)\}$ for every $z \in e(X)$. Therefore,

$$(F \circ e)(x) = F(e(x)) = \{\sigma(e(x))\} = \{f(x)\} \quad \text{for all } x \in X. \quad (*)$$

We will show that F is single-valued on βX. Let $g \in C(Y)$ (note that since (Y, τ') is compact, g is bounded on Y). Then $(g \circ F) : \beta X \to 2^{\mathbb{R}}$ is a minimal usco (see Proposition 1.3.6). By the defining property of the Stone–Čech compactification there exists a continuous function $h : \beta X \to \mathbb{R}$ such that $h(e(x)) = (g \circ f)(x)$ for all $x \in X$. Now, since

$$h(e(x)) = (g \circ f)(x) = g(f(x)) \in g(F(e(x))) = (g \circ F)(e(x)) \quad \text{for all } x \in X,$$

we have that $h(z) \in (g \circ F)(z)$ for all $z \in e(X)$. Thus, by Corollary 1.3.14, $h(z) \in (g \circ F)(z)$ for all $z \in \beta X$. Therefore, by the minimality of $(g \circ F)$, $\{h(z)\} = (g \circ F)(z)$ for all $z \in \beta X$, that is, $(g \circ F)$ is single-valued on βX. Since $g \in C(Y)$ was arbitrary, F must be single-valued everywhere on βX. Let $\tilde{f} : \beta X \to Y$ be the unique selection of F on βX. Then $\tilde{f}$ is continuous and by equation $(*)$, $\tilde{f} \circ e = f$. □

Exercise 3.1.10. Use the previous theorem to show that if $((Y,\tau'),e)$ and $((Y',\tau''),e')$ are Stone–Čech compactifications of a completely regular space (X,τ), then there exists a homeomorphism $h : Y \to Y'$ such that $h|_{e(X)} : e(X) \to e'(X)$ is a homeomorphism.

Henceforth, we will identify $e(X)$ with X, that is, we will suppose that X is a subset of βX and that the topology τ on X is simply the relative τ_β on X.

3.1.5 $\mathcal{K}$-countably determined spaces

In this subsection, we investigate a class of topological spaces that may be defined in terms of usco mappings.

A topological space (X,τ) is called a $\mathcal{K}$*-countably determined space* (see [6, 304]) if it is completely regular and there exists a countable family $(C_n : n \in \mathbb{N})$ of closed subsets of βX such that for every pair of points $x, y \in \beta X$ with $x \in X$ and $y \in \beta X \setminus X$, there exists $n \in \mathbb{N}$ such that $x \in C_n$ and $y \notin C_n$.

To explore some basic properties of $\mathcal{K}$-countably determined spaces, we need the following fact.

Proposition 3.1.11 ([175]). *Let (Y,τ) be a completely regular topological space. Then the following are equivalent:*

(i) *(Y,τ) is a $\mathcal{K}$-countably determined space;*

(ii) *there exist a separable metric space (M,d) and a surjective usco mapping $\Phi : M \to 2^Y$ onto (Y,τ);*

(iii) *there exist a second countable topological space (X,τ') and a surjective usco mapping $\Phi : X \to 2^Y$ onto (Y,τ).*

Proof. (i) ⇒ (ii). Let $(C_n : n \in \mathbb{N})$ be a family of closed subsets of $(\beta Y, \tau_\beta)$ such that for every $x, y \in \beta Y$, if $x \in Y$ and $y \in \beta Y \setminus Y$, then there exists $n \in \mathbb{N}$ such that $x \in C_n$ and $y \notin C_n$. Consider $\mathbb{N}^{\mathbb{N}}$ endowed with the Baire metric d, that is, if $s \neq t \in \mathbb{N}^{\mathbb{N}}$, then $d(s,t) := 1/k$, where $k := \min\{i \in \mathbb{N} : s_i \neq t_i\}$, and if $s = t$, then $d(s,t) := 0$. Now $(\mathbb{N}^{\mathbb{N}}, d)$ is a separable metric space. Let

$$\Sigma := \left\{ s \in \mathbb{N}^{\mathbb{N}} : \varnothing \neq \bigcap_{i \in \mathbb{N}} C_{s_i} \subseteq Y \right\},$$

and let d^Σ be the restriction of the metric d to Σ. Then (Σ, d^Σ) is a separable metric space. We define $\Phi : \Sigma \to 2^Y$ by $\Phi(s) := \bigcap_{i\in\mathbb{N}} C_{s_i}$ for $s \in \Sigma$. We claim that Φ is a surjective usco onto Y. Firstly, by the definition of Σ, Φ has nonempty compact images. Next, suppose that $s \in \Sigma$ and W is an open neighbourhood of $\Phi(s)$. For each $n \in \mathbb{N}$, let $K_n := \bigcap_{1\le i\le n} C_{s_i}$. Then $\bigcap_{n\in\mathbb{N}} K_n = \bigcap_{i\in\mathbb{N}} C_{s_i} = \Phi(s) \subseteq W$. Let W' be any open subset of βY such that $W' \cap Y = W$. Since $(K_n : n \in \mathbb{N})$ is a decreasing sequence of compact subsets of $(\beta Y, \tau_\beta)$, there exists $n \in \mathbb{N}$ such that $K_n \subseteq W'$, since otherwise, $\varnothing \neq \bigcap_{n\in\mathbb{N}}(K_n \setminus W') = (\bigcap_{n\in\mathbb{N}} K_n) \setminus W'$, which is impossible since $\bigcap_{n\in\mathbb{N}} K_n \subseteq W'$. Thus $\Phi(B(s,1/n)) \subseteq W$, where,

$$B(s,1/n) := \{s' \in \Sigma : d(s,s') < 1/n\} = \{s' \in \Sigma : s'_i = s_i \text{ for all } 1 \le i \le n\}$$

since $\Phi(B(s,1/n)) \subseteq K_n \cap Y \subseteq W$. Hence Φ is an usco on (Σ, d^Σ). To show that Φ maps onto Y, we consider the following. Let $x \in Y$, and let $J := \{n \in \mathbb{N} : x \in C_n\}$. Furthermore, let $s := (s_i : i \in \mathbb{N})$ be an enumeration of the elements of J. Then $x \in \bigcap_{i\in\mathbb{N}} C_{s_i} \subseteq Y$ by the definition of $(C_n : n \in \mathbb{N})$, and so $s \in \Sigma$. Thus $x \in \Phi(s) \subseteq \Phi(\Sigma)$.

(ii) $\Rightarrow$ (iii) follows from the fact that every separable metric space has a countable base.

(iii) $\Rightarrow$ (i). Suppose that (X, τ') is a second countable topological space and $\Phi : X \to 2^Y$ is a surjective usco onto (Y, τ). Let $\mathcal{B} := \{B_n : n \in \mathbb{N}\}$ be a base for τ', and for each $n \in \mathbb{N}$, let $C_n := \overline{\Phi(B_n)}$, where the closure is taken in βY. We claim that $(C_n : n \in \mathbb{N})$ has the desired separation properties. To see this, suppose that $y \in Y$ and $z \in \beta Y \setminus Y$. Then $y \in \Phi(x)$ for some $x \in X$. Note that $z \notin \Phi(x)$ since $\Phi(x) \subseteq Y$. Now, as βY is Hausdorff, there exist disjoint open subsets U and V of $(\beta Y, \tau_\beta)$ such that $y \in \Phi(x) \subseteq U$ and $z \in V$. By the upper semicontinuity of Φ there exists $n \in \mathbb{N}$ such that $x \in B_n$ and $\Phi(B_n) \subseteq U$. Thus $\overline{\Phi(B_n)} \subseteq \overline{U} \subseteq \beta Y \setminus V$. Therefore $y \in C_n$ and $z \notin C_n$. □

Since it is not hard to show that the usco image of a Lindelöf space is again Lindelöf (see Exercise 8 in Section 1.6), we see that every $\mathcal{K}$-countably determined space is Lindelöf.

To obtain some further facts about $\mathcal{K}$-countably determined spaces, we need the following technical result.

Proposition 3.1.12 ([175]). *Let (X, τ) be a topological space, and let (Y, τ') be a completely regular topological space. Then the following conditions are equivalent:*
(i) *there exists a surjective usco mapping $\Phi : X \to 2^Y$ onto (Y, τ');*
(ii) *there exists a closed subset G of $X \times \beta Y$ such that $X = \pi_1(G)$ and $Y = \pi_2(G)$, where $\pi_1 : X \times \beta Y \to X$ is defined by $\pi_1(x,y) := x$ for $(x,y) \in X \times \beta Y$, and $\pi_2 : X \times \beta Y \to \beta Y$ is defined by $\pi_2(x,y) := y$ for $(x,y) \in X \times \beta Y$;*
(iii) *there exist a topological space (Z, τ'') and mappings $f : Z \to X$ and $g : Z \to Y$ such that f is a perfect mapping and g is a continuous surjection onto (Y, τ').*

Proof. (i) $\Rightarrow$ (ii). Suppose that Φ is an usco from (X, τ) onto (Y, τ'). We may view Φ as an usco from X into $2^{\beta Y}$ such that $\Phi(X) = Y$. Let $G := \mathrm{Gr}(\Phi)$. By Proposition 1.1.8, G is a closed subset of $X \times \beta Y$, and since Φ has nonempty images, $\pi_1(G) = X$. Finally, since Φ maps onto Y, $\pi_2(G) = Y$.

(ii) $\Rightarrow$ (iii). Let $Z := G$, and let τ'' be the relative product topology on G. Let $f := \pi_1|_G$ and $g := \pi_2|_G$ (note that $f : Z \to X$ and $g : Z \to Y$). Then g is a continuous surjection onto Y, and f is a perfect mapping onto X. To see that f is a perfect mapping, consider the following. Define $\Phi : X \to 2^{\beta Y}$ by $\Phi(x) := \{y \in \beta Y : (x, y) \in G\}$ for $x \in X$. Then Φ has nonempty images since $\pi_1(G) = X$. Note also that $\mathrm{Gr}(\Phi) = G$. Therefore, by Corollary 1.1.15, Φ is an usco on (X, τ). The fact that f is a perfect mapping now follows from Exercise 11 in Section 1.6.

(iii) $\Rightarrow$ (i). By Example (iii) in Section 1.1.2, $f^{-1} : X \to 2^Z$ is an usco. Therefore, by Construction (iii) in Section 1.1.3, $\Phi := g \circ f^{-1}$ is an usco from (X, τ) into subsets of (Y, τ'). Since both f^{-1} and g are surjective, so is Φ. □

Let us now recall that a completely regular space (Y, τ') is called a *Lindelöf Σ-space* [239] if it is the continuous image of a topological space (Z, τ'') that can be perfectly mapped onto a second countable topological space (X, τ). Therefore it follows from Propositions 3.1.11 and 3.1.12 that a completely regular space (Y, τ') is $\mathcal{K}$-countably determined space if and only if it is a Lindelöf Σ-space [276, p. 96].

We can now apply some basic properties of usco mappings to establish the fundamental properties of $\mathcal{K}$-countably determined spaces/Lindelöf Σ-spaces.

As noted before (and not surprisingly, given their definition), Lindelöf Σ-spaces are Lindelöf. From Construction (xii) in Section 1.1.3, and Proposition 3.1.12, part (i), it follows that if $((X_n, \tau_n) : n \in \mathbb{N})$ are all Lindelöf Σ-spaces, then so is $\prod_{n\in\mathbb{N}} X_n$, endowed with the product topology. It also follows that if (X, τ) is a Lindelöf Σ-space and Y is a closed subset of (X, τ), then Y, endowed with the relative topology, is a Lindelöf Σ-space (see Exercise 9 in Section 1.6 and Proposition 3.1.12, part (i)). Furthermore, it follows from Construction (iii) in Section 1.1.3, and Proposition 3.1.12, part (i), that the continuous image of a Lindelöf Σ-space is again a Lindelöf Σ-space. In fact, we can equivalently define the class of all Lindelöf Σ-spaces as the smallest class of completely regular topological spaces that contains all compact Hausdorff spaces and all separable metric spaces and is closed under countable products, closed subspaces and continuous images.

Proposition 3.1.13 ([81, Theorem 7.1.3]). *Let (X, τ) be a completely regular topological space. If $(X_n : n \in \mathbb{N})$ are $\mathcal{K}$-countably determined subspaces of (X, τ), then so are $\bigcap_{n\in\mathbb{N}} X_n$ and $\bigcup_{n\in\mathbb{N}} X_n$.*

Proof. By the preceding we see that $\prod_{n\in\mathbb{N}} X_n$ endowed with the product topology is a $\mathcal{K}$-countably determined space. Let

$$D := \left\{(x_n)_{n\in\mathbb{N}} \in \prod_{n\in\mathbb{N}} X_n : x_1 = x_2 = x_3 = \cdots = x_n = \cdots\right\}.$$

Then D is a closed subset of $\prod_{n\in\mathbb{N}} X_n$ and thus a $\mathcal{K}$-countably determined space. Let $f : D \to X$ be defined by $f((x_n)_{n\in\mathbb{N}}) := x_1$ for $(x_n)_{n\in\mathbb{N}} \in D$. Then $f(D) = \bigcap_{n\in\mathbb{N}} X_n$ is a $\mathcal{K}$-countably determined space, as it is a continuous image of a $\mathcal{K}$-countably determined space.

Let us now consider $\mathbb{N}$ endowed with the discrete topology τ_d. Then $(\mathbb{N}, \tau_d)$ is a $\mathcal{K}$-countably determined space (since the sets $C_n := \{n\}$ for $n \in \mathbb{N}$ separate $\mathbb{N}$ from $\beta\mathbb{N} \setminus \mathbb{N}$), and therefore so is $\mathbb{N} \times (\prod_{n\in\mathbb{N}} X_n)$. Define $f : \mathbb{N} \times (\prod_{n\in\mathbb{N}} X_n) \to X$ by $f(m, (x_n)_{n\in\mathbb{N}}) := x_m$ for $(m, (x_n)_{n\in\mathbb{N}}) \in \mathbb{N} \times (\prod_{n\in\mathbb{N}} X_n)$. Then $f(\mathbb{N} \times (\prod_{n\in\mathbb{N}} X_n)) = \bigcup_{n\in\mathbb{N}} X_n$ is a $\mathcal{K}$-countably determined space as it is a continuous image of a $\mathcal{K}$-countably determined space. □

3.2 Uscos in approximation theory

In this section, we will see how minimal uscos (and, more generally, minimal mappings) appear in the study of nearest and farthest points in Banach spaces.

3.2.1 Nearest points

We begin with some basic definitions. Let $(X, \|\cdot\|)$ be a normed linear space, and let K be a nonempty subset of X. For any point $x \in X$, we define

$$d(x, K) := \inf_{y\in K} \|x - y\|$$

and call this the *distance from x to K*. We will also refer to the mapping $x \mapsto d(x, K)$ as the *distance function for K*.

We associate with the distance function the set-valued mapping $P_K : X \to 2^K$ defined by

$$P_K(x) := \{y \in K : \|x - y\| = d(x, K)\} \text{ if } K \text{ is nonempty and by } P_K(x) := \varnothing \text{ if } K = \varnothing.$$

We refer to the elements of $P_K(x)$ as the *best approximations of x in K* (or the *nearest points to x in K*), and we refer to P_K as the *metric projection mapping (for K)*.

A normed linear space $(X, \|\cdot\|)$ is said to be *strictly convex* (or, *rotund*) if for any $x, y \in S_X$,

$$\left\|\frac{x+y}{2}\right\| = 1 \quad \text{if and only if } x = y.$$

We are now in a position to formulate Stečkin's conjecture. In 1963, Stečkin [289] proved that for any nonempty subset K of a strictly convex normed linear space $(X, \|\cdot\|)$,

$$\{x \in X : P_K(x) \text{ is at most a singleton}\}$$

is dense in $(X, \|\cdot\|)$.

In the same paper, Stečkin also conjectured that for any nonempty subset K of a strictly convex space $(X, \|\cdot\|)$,

$$\{x \in X : P_K(x) \text{ is at most a singleton}\}$$

is *residual in* $(X, \|\cdot\|)$, that is, contains the intersection of a countable family of dense open subsets of $(X, \|\cdot\|)$. Stečkin's conjecture is still open. For the latest state of play, see [266].

We now present several partial solutions to Stečkin's conjecture. However, to accomplish this, we first need to establish some basic properties concerning distance functions and metric projections.

The following simple result regarding the distance function will be used repeatedly throughout the remainder of this subsection.

Proposition 3.2.1 ([83]). *Let K be a nonempty subset of a normed linear space $(X, \|\cdot\|)$. Then the distance function for K is nonexpansive (and hence continuous).*

Proof. For each $k \in K$, let $g_k : X \to [0, \infty)$ be defined by $g_k(x) := \|x - k\|$ for $x \in X$. Then each g_k is nonexpansive and bounded below by 0. Now for each $x \in X$, $d(x, K) = \inf\{g_k(x) : k \in K\}$. Thus, as the pointwise infimum of a family of pointwise bounded below nonexpansive mappings, the distance function is itself nonexpansive. □

Although the metric projection is not always a (norm)-usco (see [83, Example 2.54]), it does always have a closed graph.

Lemma 3.2.2 ([83]). *Let K be a nonempty closed subset of a normed linear space $(X, \|\cdot\|)$. Then P_K has a closed graph.*

Proof. Let $((x_n, y_n) : n \in \mathbb{N})$ be a sequence in $\mathrm{Gr}(P_K)$ converging to some $(x, y) \in X \times X$, that is, $y_n \in P_K(x_n)$ for all $n \in \mathbb{N}$, $x = \lim_{n\to\infty} x_n$, and $y = \lim_{n\to\infty} y_n$. First, note that $y \in K$, since K is closed. Furthermore,

$$\|x - y\| = \lim_{n\to\infty} \|x_n - y_n\| = \lim_{n\to\infty} d(x_n, K) = d(x, K),$$

since the distance function for K is continuous by Proposition 3.2.1. Thus $y \in P_K(x)$, that is, $(x, y) \in \mathrm{Gr}(P_K)$. □

Let $(X, \|\cdot\|)$ and $(Y, \|\cdot\|')$ be normed linear spaces, and let $\Phi : X \to 2^Y$ be a function. We say that Φ is *locally bounded on* $(X, \|\cdot\|)$ if for every $x_0 \in X$, there exist $r > 0$ and $M > 0$ such that $\Phi(B(x_0, r)) \subseteq \mathrm{MB}_Y$.

Lemma 3.2.3 ([83]). *Let K be a nonempty subset of a normed linear space $(X, \|\cdot\|)$. Then the metric projection mapping $x \mapsto P_K(x)$ is locally bounded on $(X, \|\cdot\|)$.*

Proof. Fix $x_0 \in X$ and let $r := 1$ and $M := d(x_0, K) + \|x_0\| + 2$. Suppose $y \in P_K(B(x_0, r))$, thats is, $y \in P_K(x)$ for some $x \in B(x_0, 1)$. Using the triangle inequality (twice) and the fact that the distance function for K is nonexpansive, we have that

$$\begin{aligned}\|y\| &= \|y - 0\| \\ &\le \|y - x_0\| + \|x_0 - 0\| \\ &\le \|y - x\| + \|x - x_0\| + \|x_0 - 0\| \\ &= d(x, K) + (\|x - x_0\| + \|x_0\|) \\ &\le (d(x_0, K) + \|x - x_0\|) + (\|x - x_0\| + \|x_0\|) \quad \text{(since } z \mapsto d(z, K) \text{ is nonexpansive)} \\ &< d(x_0, K) + 2 + \|x_0\| \\ &= M.\end{aligned}$$

Hence $P_K(B(x_0, r)) \subseteq MB_Y$, and so the metric projection mapping for K is locally bounded on $(X, \|\cdot\|)$. □

Let K be a subset of a normed linear space $(X, \|\cdot\|)$. We say that K is *boundedly compact* if for every $r > 0$, $rB_X \cap K$ is compact.

Exercise 3.2.4. Let K be a nonempty boundedly compact subset of a normed linear space $(X, \|\cdot\|)$. Show that P_K has nonempty compact images. *Hint:* For each $x \in X$,

$$B[x, d(x, K)] \cap K = \bigcap_{n \in \mathbb{N}} B[x, d(x, K) + 1/n] \cap K \subseteq B[0, d(x, K) + \|x\| + 1] \cap K.$$

Our reason for considering boundedly compact sets is revealed in the next theorem.

Theorem 3.2.5. *Let K be a nonempty boundedly compact subset of a normed linear space $(X, \|\cdot\|)$. Then the metric projection mapping $y \mapsto P_K(y)$ is an usco on $(X, \|\cdot\|)$.*

Proof. Since P_K has a closed graph (see Lemma 3.2.2) and nonempty images (see Exercise 3.2.4), it follows from Exercise 1.1.17 that it is sufficient to show that for each $x \in X$, there exists a neighbourhood U_x of x such that $\overline{P_K(U_x)}$ is compact. However, since P_K is locally bounded (see Lemma 3.2.3) for each $x \in X$, there exist a neighbourhood U_x of x and $M > 0$ such that $\overline{P_K(U_x)} \subseteq K \cap B[0, M]$, which is compact, since K is boundedly compact. □

A particular case of the previous theorem is the following folklore result.

Corollary 3.2.6. *The metric projection mapping for a nonempty closed set in a finite-dimensional normed linear space is an usco.*

Before going to establish our first result concerning Stečkin's conjecture, we need some basic facts about strictly convex (i. e., rotund) normed linear spaces. Recall that a function $\varphi : D \to \mathbb{R}$ defined on a nonempty convex subset D of a vector space $(V, +, \cdot)$ is a *convex function* if

$$\varphi(\lambda x + (1 - \lambda)y) \le \lambda\varphi(x) + (1 - \lambda)\varphi(y) \quad \text{for all } x, y \in D \text{ and all } 0 < \lambda < 1.$$

Exercise 3.2.7. Part I. Let $\varphi : [0,1] \to \mathbb{R}$ be a convex function such that $\varphi(0) = \varphi(1)$. Show that φ is a constant function if and only if there exists $x \in (0,1)$ such that

$$\varphi(x) = \varphi(0) = \varphi(1).$$

Exercise 3.2.8. Part II. Let $(X, \|\cdot\|)$ is a strictly convex normed linear space, let $x, y \in S_X := \{z \in X : \|z\| = 1\}$, and let $0 < \lambda < 1$. Show that if $\lambda x + (1-\lambda)y \in S_X$, then $x = y$. *Hint:* Consider $\varphi : [0,1] \to \mathbb{R}$ defined by $\varphi(t) := \|tx + (1-t)y\|$.

Proposition 3.2.9. *Let $(X, \|\cdot\|)$ be a rotund normed linear space, and let $x, y \in X \setminus \{0\}$. Then $\|x+y\| = \|x\| + \|y\|$ if and only if $y = (\|y\|/\|x\|)x$.*

Proof. Suppose that $y = (\|y\|/\|x\|)x$. Then

$$\|x+y\| = \left\| \frac{\|x\|x + \|y\|x}{\|x\|} \right\| = \left\| \frac{(\|x\| + \|y\|)x}{\|x\|} \right\| = \frac{\|x\| + \|y\|}{\|x\|}\|x\| = \|x\| + \|y\|.$$

On the other hand, suppose that $\|x+y\| = \|x\| + \|y\|$. Then

$$\left\| \frac{\|x\|\widehat{x}}{\|x\| + \|y\|} + \frac{\|y\|\widehat{y}}{\|x\| + \|y\|} \right\| = 1, \quad \text{where } \widehat{x} := x/\|x\| \text{ and } \widehat{y} := y/\|y\|.$$

Let $\lambda := \|x\|/(\|x\| + \|y\|)$. Then $0 < \lambda < 1$ and $\|\lambda\widehat{x} + (1-\lambda)\widehat{y}\| = 1$. Thus, by Exercise 3.2.8, $\widehat{x} = \widehat{y}$. The result now follows. □

Exercise 3.2.10. Let $(X, \|\cdot\|)$ be a normed linear space, and let $x, y, z \in X$. Show that if $y \in (x,z)$, then $\|z - x\| = \|z - y\| + \|y - x\|$.

Lemma 3.2.11. *Let x and z be distinct points of a rotund normed linear space $(X, \|\cdot\|)$. Then for any $y \in (x,z)$, $B[y, \|z-y\|] \subseteq B(x, \|z-x\|) \cup \{z\}$.*

Proof. Clearly,

$$B[y, \|z-y\|] \subseteq B[x, \|z-x\|] = B(x, \|z-x\|) \cup S(x, \|z-x\|).$$

Indeed, if $w \in B[y, \|z-y\|]$, then

$$\begin{aligned} \|w-x\| &\le \|w-y\| + \|y-x\| \quad \text{by the triangle inequality} \\ &\le \|z-y\| + \|y-x\| \quad \text{since } w \in B[y, \|z-y\|] \\ &= \|z-x\| \quad \text{by Exercise 3.2.10, since } y \in (x,z). \end{aligned}$$

On the other hand, if $w \in B[y, \|z-y\|] \cap S(x, \|z-x\|)$, then

$$\begin{aligned} \|z-x\| &= \|w-x\| \quad \text{since } w \in S(x, \|z-x\|) \\ &\le \|w-y\| + \|y-x\| \quad \text{by the triangle inequality} \\ &\le \|z-y\| + \|y-x\| \quad \text{since } w \in B[y, \|z-y\|] \\ &= \|z-x\| \quad \text{by Exercise 3.2.10, since } y \in (x,z). \end{aligned}$$

From this we can deduce that:

(i) $\|w - x\| = \|w - y\| + \|y - x\|$,
(ii) $\|w - y\| + \|y - x\| = \|z - y\| + \|y - x\|$, and
(iii) $\|z - x\| = \|z - y\| + \|y - x\|$.

Therefore by properties (i) and (iii), respectively, and Proposition 3.2.9 we have that

$$w - y = \frac{\|w - y\|}{\|y - x\|}(y - x) \quad \text{and} \quad z - y = \frac{\|z - y\|}{\|y - x\|}(y - x).$$

Now, by property (ii) it follows that $\|w - y\| = \|z - y\|$. Therefore $w - y = z - y$, thats is, $w = z$. This completes the proof. □

Corollary 3.2.12. *Let K be a nonempty subset of a rotund normed linear space $(X, \|\cdot\|)$. Suppose that $x \in X \setminus K$ and $z \in P_K(x)$. Then for any $y \in (x, z]$, we have that $P_K(y) = \{z\}$.*

Proof. The result is clearly true if $y = z$, so we only consider the case where $y \in (x, z)$. Now

$$z \in B[y, \|z - y\|] \cap K \subseteq (B(x; \|z - x\|) \cup \{z\}) \cap K = \{z\} \cap K = \{z\},$$

that is, $B[y, \|z - y\|] \cap K = \{z\}$. Hence $P_K(y) = \{z\}$. □

From this result it follows that for any nonempty subset K of a rotund space $(X, \|\cdot\|)$,

$$\{x \in X : P_K(x) \text{ is at most a singleton}\}$$

is dense in $(X, \|\cdot\|)$.

Theorem 3.2.13. *Let K be a nonempty boundedly compact subset of a rotund normed linear space $(X, \|\cdot\|)$. Then the metric projection mapping $y \mapsto P_K(y)$ is a minimal usco on $(X, \|\cdot\|)$. In particular, $\{x \in X : P_K(x) \text{ is at most a singleton}\}$ is a dense G_δ-subset of $(X, \|\cdot\|)$.*

Proof. As already shown in Theorem 3.2.5, P_K is an usco. So it remains to show that it is a minimal usco. To accomplish this, we appeal to Proposition 1.3.5. Let U and W be a pair of open subsets of $(X, \|\cdot\|)$ such that $P_K(U) \cap W \neq \varnothing$. Choose x and z so that $x \in U$ and $z \in P_K(x) \cap W$. If $x \in K$, then $P_K(x) = \{x\} = \{z\}$, and so by the upper semicontinuity of P_K there exists an open neighbourhood V of x, contained in U such that $P_K(V) \subseteq W$. So let us suppose that $x \notin K$. Then $x \neq z$ (as $z \in K$). We may now choose $y \in (x, z)$ such that $y \in U$. By Corollary 3.2.12, $P_K(y) = \{z\} \subseteq W$, and so by the upper semicontinuity of P_K there exists an open neighbourhood V of y, contained in U, such that $P_K(V) \subseteq W$. Therefore, in either case, we have established the existence of a nonempty open subset V of U such that $P_K(V) \subseteq W$. This completes the proof that P_K is a minimal usco on $(X, \|\cdot\|)$. The fact that $\{x \in X : P_K(x) \text{ is at most a singleton}\}$ is a dense G_δ-subset of $(X, \|\cdot\|)$ now follows directly from Corollary 1.5.18. □

Corollary 3.2.14. *Let K be a nonempty closed subset of a rotund finite-dimensional normed linear space $(X, \|\cdot\|)$. Then the metric projection mapping, $y \mapsto P_K(y)$ is a minimal usco on $(X, \|\cdot\|)$. In particular, $\{x \in X : P_K(x)$ is at most a singleton$\}$ is a dense G_δ-subset of $(X, \|\cdot\|)$.*

Thus Stečkin's conjecture has a positive solution for closed sets in finite-dimensional normed linear spaces.

By working a bit harder we can extend this result beyond finite-dimensional spaces. In fact, we can show that there is a positive answer to Stečkin's conjecture whenever $(X, \|\cdot\|)$ has a weakly locally uniformly rotund norm (weak LUR).

A normed linear space $(X, \|\cdot\|)$ is called *weakly locally uniformly rotund* if for every sequence $(y_n : n \in \mathbb{N})$ in S_X and every point $x \in S_X$, $(y_n : n \in \mathbb{N})$ converges to x with respect to the weak topology on $(X, \|\cdot\|)$ whenever $\lim_{n\to\infty} \|x + y_n\| = 2$. It is easy to check that all weakly locally uniformly rotund normed linear spaces are rotund and that, in the realm of finite-dimensional normed linear spaces, all rotund spaces are indeed weakly locally uniformly rotund [70]. However, outside finite-dimensional spaces, there are many examples of rotund normed linear spaces that are not weakly locally uniformly rotund; see [285].

Outside finite-dimensional spaces, it is unreasonable to expect that the metric projection mapping is always a minimal usco, essentially due to the lack of compactness in infinite-dimensional spaces. However, as we will see, the metric projection may retain the minimality property expressed in Proposition 1.3.5.

Let (X, τ) and (Y, τ') be topological spaces, and let $\Phi : X \to 2^Y$. Recall that, by Proposition 2.3.21, Φ is a *minimal mapping* if for each pair of open subsets U of (X, τ) and W of (Y, τ'), with $\Phi(U) \cap W \neq \varnothing$, there exists a nonempty τ-open subset V of U such that $\Phi(V) \subseteq W$.

Exercise 3.2.15. Let $0 < \lambda < 1$, and let $\varphi_n : [0,1] \to \mathbb{R}$ be a sequence of convex functions such that (i) $\varphi_n(0) = \varphi_n(1) = 1$ for all $n \in \mathbb{N}$ and (ii) $\lim_{n\to\infty} \varphi_n(\lambda) = 1$. Show that $\lim_{n\to\infty} \varphi_n(1/2) = 1$. *Hint:* Use the convexity to show that if $0 < \lambda < 1/2$, then

$$1 - \frac{1}{2}\left[\frac{1 - \varphi_n(\lambda)}{\lambda}\right] \leq \varphi_n(1/2) \leq 1$$

and if $1/2 < \lambda < 1$, then

$$1 - \frac{1}{2}\left[\frac{1 - \varphi_n(\lambda)}{1 - \lambda}\right] \leq \varphi_n(1/2) \leq 1.$$

Exercise 3.2.16. Let $(X, \|\cdot\|)$ be a weakly locally uniformly rotund normed linear space, let $\{x\} \cup \{y_n : n \in \mathbb{N}\} \subseteq S_X$, and let $0 < \lambda < 1$. Show that if $\lim_{n\to\infty} \|\lambda x + (1-\lambda)y_n\| = 1$, then $(y_n : n \in \mathbb{N})$ converges to x with respect to the weak topology on $(X, \|\cdot\|)$. *Hint:* For each $n \in \mathbb{N}$, define $\varphi_n : [0,1] \to \mathbb{R}$ by $\varphi_n(t) := \|tx + (1-t)y_n\|$.

Proposition 3.2.17. *Let $(X, \|\cdot\|)$ be a weakly locally uniformly rotund normed linear space, let $\{x\} \cup \{y_n : n \in \mathbb{N}\} \subseteq X \setminus \{0\}$, and let $0 \le r$. If*

$$\lim_{n\to\infty} \|y_n\| = r \quad \textit{and} \quad \lim_{n\to\infty} \|x + y_n\| = \|x\| + r,$$

then $(y_n : n \in \mathbb{N})$ converges to $(r/\|x\|)x$ with respect to the weak topology on $(X, \|\cdot\|)$.

Proof. If $r = 0$, then the result is obvious. So assume that $r > 0$. Now

$$\lim_{n\to\infty} \|x + y_n\| = \|x\| + r \Longrightarrow \lim_{n\to\infty} \left\| \frac{\|x\|\widehat{x}}{\|x\| + r} + \frac{\|y_n\|\widehat{y_n}}{\|x\| + r} \right\| = 1,$$

where $\widehat{x} = x/\|x\|$ and $\widehat{y_n} = y_n/\|y_n\|$. Let $\lambda := \|x\|/(\|x\| + r)$. Then $0 < \lambda < 1$, and

$$\begin{aligned} \left\| \frac{\|x\|\widehat{x}}{\|x\| + r} + \frac{\|y_n\|\widehat{y_n}}{\|x\| + r} \right\| - \left| \frac{r - \|y_n\|}{\|x\| + r} \right| &\le \left\| \frac{\|x\|\widehat{x}}{\|x\| + r} + \frac{r\widehat{y_n}}{\|x\| + r} \right\| \\ &\le \left\| \frac{\|x\|\widehat{x}}{\|x\| + r} + \frac{\|y_n\|\widehat{y_n}}{\|x\| + r} \right\| + \left| \frac{r - \|y_n\|}{\|x\| + r} \right|. \end{aligned}$$

Therefore, by the squeeze theorem, $\lim_{n\to\infty} \|\lambda\widehat{x} + (1 - \lambda)\widehat{y_n}\| = 1$. Hence, by Exercise 3.2.16, $(\widehat{y_n} : n \in \mathbb{N})$ converges to $\widehat{x}$ with respect to the weak topology on $(X, \|\cdot\|)$. Thus $(y_n : n \in \mathbb{N})$ converges to $(r/\|x\|)x$ with respect to the weak topology on $(X, \|\cdot\|)$. □

Theorem 3.2.18. *Let K be a nonempty subset of a weakly locally uniformly rotund normed linear space $(X, \|\cdot\|)$. Then the metric projection mapping $y \mapsto P_K(y)$ is a (weak)-minimal set-valued mapping on $(X, \|\cdot\|)$ (i. e. minimal with respect to the weak topology on $(X, \|\cdot\|)$).*

Proof. Let U be a (norm)-open subsets of $(X, \|\cdot\|)$, and let W be a weak-open subset of $(X, \|\cdot\|)$ such that $P_K(U) \cap W \ne \varnothing$. Choose x and z such that $x \in U$ and $z \in P_K(x) \cap W$. We consider two cases.

(i) Suppose that $x \in K$. Then $P_K(x) = \{x\} = \{z\}$. Choose $r > 0$ such that $B[x, 2r] \subseteq U \cap W$. Let $V := B(x, r)$. Then $\varnothing \ne V \subseteq U$. We claim that $P_K(V) \subseteq W$. To see this, suppose, to obtain a contradiction, that $P_K(V) \not\subseteq W$. Then there exist $x' \in V$ and $z' \in P_K(x') \setminus W$. In particular, $r < \|x' - z'\| = d(x', K)$, since, by the triangle inequality, $B[x', r] \subseteq B[x, 2r] \subseteq W$. However, $x = z \in K$ and $\|x' - x\| \le r < d(x', K)$, which contradicts the fact that z' is the closest point in K to x'.

(ii) Suppose $x \notin K$. Then $x \ne z$ (as $z \in K$). We may now choose $y \in (x, z)$ such that $y \in U$. Note that, by Corollary 3.2.12, $P_K(y) = \{z\}$ and, furthermore, $z = y + \frac{\|z-y\|}{\|y-x\|}(y - x)$. We now claim that for some $r > 0$, $P_K(B[y, r]) \subseteq W$. Indeed, if this is not the case, then for each $n \in \mathbb{N}$, $P_K(B[y, 1/n]) \not\subseteq W$. Hence for each $n \in \mathbb{N}$, there exist $x_n \in X$ and $z_n \in X$ such that $\|x_n - y\| \le 1/n$ and $z_n \in P_K(x_n) \setminus W$. Now,

$$\|z_n - x_n\| - \|x_n - y\| \le \|z_n - y\| \le \|z_n - x_n\| + \|x_n - y\|.$$

However, as $\lim_{n\to\infty}\|z_n - x_n\| = \lim_{n\to\infty} d(x_n, K) = d(y, K)$ and $\lim_{n\to\infty}\|x_n - y\| = 0$, we have that $r := \lim_{n\to\infty}\|z_n - y\| = d(y, K) = \|z - y\|$ and

$$\begin{aligned} r + \|y - x\| &= \|z - y\| + \|y - x\| \\ &= \|z - x\| \quad \text{(by Exercise 3.2.10, since } y \in (x, z)) \\ &= d(x, K) \le \|z_n - x\| \quad (\text{since } z_n \in K) \\ &= \|(z_n - y) + (y - x)\| \le \|z_n - y\| + \|y - x\|. \end{aligned}$$

Therefore $\lim_{n\to\infty}\|(z_n - y) + (y - x)\| = r + \|y - x\|$, and so, by Proposition 3.2.17, $(z_n - y : n \in \mathbb{N})$ converges weakly to $(r/\|y - x\|)(y - x)$. Thus $(z_n : n \in \mathbb{N})$ converges weakly to

$$y + \frac{r}{\|y - x\|}(y - x) = y + \frac{\|z - y\|}{\|y - x\|}(y - x) = z.$$

However, this contradicts the fact that $z \in W$ and $z_n \notin W$ for all $n \in \mathbb{N}$. Hence there must exist $r > 0$ such that $V := B(y, r) \subseteq U$ and $P_K(V) \subseteq W$.

This completes the proof. □

The relevance of this result to Stečkin's conjecture is revealed in the next result.

Proposition 3.2.19. *Let $(X, \|\cdot\|)$ be a rotund normed linear space, and let $\Phi : A \to 2^X$ be a locally bounded (weak)-minimal set-valued mapping defined on a topological space (A, τ). Then $\{x \in A : \Phi(x)$ is at most a singleton$\}$ is residual in (A, τ).*

Proof. Define $d : X \times X \to [0, \infty)$ by

$$d(x, y) := \max\{\|x\|, \|y\|\} - \left\|\frac{1}{2}x + \frac{1}{2}y\right\|.$$

Then, for any $x, y \in X$,

$$\left\|\frac{1}{2}x + \frac{1}{2}y\right\| \le \frac{1}{2}\|x\| + \frac{1}{2}\|y\| \le \frac{1}{2}\max\{\|x\|, \|y\|\} + \frac{1}{2}\max\{\|x\|, \|y\|\} = \max\{\|x\|, \|y\|\},$$

and so $0 \le d(x, y)$. Moreover, from the above inequality we can see that $d(x, y) = 0$ if and only if
(i) $\|x\| = \max\{\|x\|, \|y\|\}$ and $\|y\| = \max\{\|x\|, \|y\|\}$, that is, $\|x\| = \|y\|$, and
(ii) $\|\frac{1}{2}x + \frac{1}{2}y\| = \frac{1}{2}\|x\| + \frac{1}{2}\|y\|$, that is, $\|x + y\| = \|x\| + \|y\|$.

Since $(X, \|\cdot\|)$ is rotund, this can only happen if $x = y$ (see Proposition 3.2.9). Hence $d(x, y) = 0$ if and only if $x = y$. Note also that $d(x, y) = d(y, x)$ for all $x, y \in X$. However, in general, the triangle inequality does not hold.

For each $\varepsilon > 0$, let $O_\varepsilon := \bigcup\{U \in \tau : d\text{-}\mathrm{diam}(\Phi(U)) < \varepsilon\}$. Then each O_ε is open, and Φ is at most single-valued at each point of $\bigcap_{n\in\mathbb{N}} O_{1/n}$. So it remains to show that each O_ε is dense in (A, τ).

To this end, fix $\varepsilon > 0$. Let W be a nonempty open subset of (A, τ). We may assume, by possibly making W smaller, that $\Phi(W)$ is bounded in $(X, \|\cdot\|)$. We need to show that $O_\varepsilon \cap W \neq \emptyset$. If $\Phi(W) = \emptyset$, then $\emptyset \neq W \subseteq O_\varepsilon \cap W$. So let us suppose that $\Phi(W) \neq \emptyset$. Let $s := \sup\{\|x\| : x \in \Phi(W)\}$. If $s = 0$, then $\Phi(W) = \{0\}$, and so $\emptyset \neq W \subseteq O_\varepsilon \cap W$. So we will assume that $s > 0$.

Choose $x_0 \in \Phi(W)$ such that $s - \varepsilon/2 < \|x_0\|$. Then we may apply the Hahn–Banach extension theorem [75, p. 62] to deduce the existence of a continuous linear functional $x^* \in X^*$ such that $\|x^*\| = 1$ and $x^*(x_0) = \|x_0\|$. Let $V := \{x \in X : s - \varepsilon/2 < x^*(x)\}$. Then V is weakly open, and $\Phi(W) \cap V \neq \emptyset$, since $x_0 \in \Phi(W) \cap V$. Therefore by the (weak)-minimality property of Φ there exists a nonempty τ-open subset U of W such that $\Phi(U) \subseteq \Phi(W) \cap V$.

We claim that d-diam$(\Phi(W) \cap V) < \varepsilon$. Indeed, suppose that $x, y \in \Phi(W) \cap V$. Then $\max\{\|x\|, \|y\|\} \leq s$ and $s - \varepsilon/2 < \|\frac{x+y}{2}\|$, since $\frac{x+y}{2} \in V$ and

$$s - \varepsilon/2 < x^*(z) \leq \|x^*\|\|z\| = \|z\| \quad \text{for all } z \in V.$$

Therefore

$$d(x, y) = \max\{\|x\|, \|y\|\} - \left\|\frac{x+y}{2}\right\| \leq s - (s - \varepsilon/2) = \varepsilon/2.$$

Hence it follows that d-diam$(\Phi(U)) < \varepsilon$. This shows that $\emptyset \neq U \subseteq O_\varepsilon \cap W$, which completes the proof. □

Thus Theorem 3.2.18 and Proposition 3.2.19 together, imply a positive solution to Stečkin's conjecture in the case of weakly locally uniformly rotund normed linear spaces.

Corollary 3.2.20. *Let K be a nonempty subset of a weakly locally uniformly rotund normed linear space $(X, \|\cdot\|)$. Then $\{x \in X : P_K(x)$ is at most a singleton$\}$ is residual in $(X, \|\cdot\|)$.*

3.2.2 Farthest points

Let $(X, \|\cdot\|)$ be a normed linear space, and let K be a nonempty bounded subset of X. For any point $x \in X$, we define $r(x, K) := \sup_{y \in K} \|x - y\|$. We refer to the mapping $x \mapsto r(x, K)$ as the *radial function for K*.

Proposition 3.2.21. *Let K be a nonempty bounded subset of a normed linear space $(X, \|\cdot\|)$. Then the radial function for K is convex and nonexpansive (and hence continuous).*

Proof. For each $k \in K$, let $g_k : X \to [0, \infty)$ be defined by $g_k(x) := \|x - k\|$ for $x \in X$. Then each g_k is convex and nonexpansive. Now, for each $x \in X$, $r(x, K) = \sup\{g_k(x) : k \in K\}$. Thus, as the pointwise supremum of a family of pointwise bounded above and convex nonexpansive mappings, the radial function is itself convex and nonexpansive. □

Let $(X, \|\cdot\|)$ be a normed linear space, and let K be a subset of X. We define the set-valued mapping $F_K : X \to 2^K$ by $F_K(x) := \{y \in K : \|x - y\| = r(x, K)\}$ if K is nonempty and bounded and by $F_K(x) := \varnothing$ otherwise. We refer to the elements of $F_K(x)$ as the *farthest points from x in K*.

We say that K is a *remotal set* if $F_K(x)$ is a nonempty for each $x \in X$. Furthermore, we say that K is a *uniquely remotal set* if $F_K(x)$ is a singleton for each $x \in X$.

Remarks 3.2.22. It follows from the definition that every remotal set is nonempty and bounded.

Lemma 3.2.23. *Let K be a nonempty subset of a rotund normed linear space $(X, \|\cdot\|)$. Suppose that $x \in X$ and $z \in F_K(x)$. Then for any $y \in X$ such that $x \in (z, y)$, we have that $F_K(y) = \{z\}$.*

Proof. Suppose $x \in (z, y)$. Then by Lemma 3.2.11 we have that

$$K \subseteq B[x, \|x - z\|] \subseteq B(y, \|y - z\|) \cup \{z\} \subseteq B[y, \|y - z\|].$$

Therefore $K \setminus \{z\} \subseteq B(y, \|y - z\|)$. Hence $F_K(y) = \{z\}$. □

Proposition 3.2.24. *Let K be a remotal subset of a rotund normed linear space $(X, \|\cdot\|)$. Then $F_K(x) = F_{\overline{\text{co}}(K)}(x)$ for all $x \in X$. In particular, if K is a uniquely remotal set, then so is $\overline{\text{co}}(K)$.*

Proof. Without loss of generality, we may assume that K is non-trivial, that is, it is not a singleton set. Firstly, let us note that if $A \subseteq B \subseteq X$ (and B is bounded), then $r(x, A) \le r(x, B)$ for all $x \in X$, and thus $r(x, K) \le r(x, \overline{\text{co}}(K))$ for all $x \in X$. On the other hand, if $x \in X$, then $K \subseteq B[x, r(x, K)]$, and so $\overline{\text{co}}(K) \subseteq B[x, r(x, K)]$ since $B[x, r(x, K)]$ is closed and convex. Therefore $r(x, \overline{\text{co}}(K)) \le r(x, K)$. Thus $r(x, \overline{\text{co}}(K)) = r(x, K)$ for all $x \in X$. It now follows that $F_K(x) \subseteq F_{\overline{\text{co}}(K)}(x)$ for all $x \in X$. We will now show that $F_{\overline{\text{co}}(K)}(x) \subseteq F_K(x)$ for all $x \in X$. Let $x \in X$ and $z \in F_{\overline{\text{co}}(K)}(x)$. To show that $z \in F_K(x)$, it is sufficient to show that $z \in K$. To this end, choose $y \in X$ such that $x \in (z, y)$. Then, by Lemma 3.2.23, $F_{\overline{\text{co}}(K)}(y) = \{z\}$. However, as

$$\varnothing \ne F_K(y) \subseteq F_{\overline{\text{co}}(K)}(y) = \{z\},$$

we must have that $\{z\} = F_K(y) \subseteq K$, which completes the proof. □

In this section, we also consider the following "quantitative" version of these notions. Let $\varepsilon > 0$. Then we say that K is an *ε-uniquely remotal set* if K is remotal and $\text{diam}(F_K(x)) \le \varepsilon$ for each $x \in X$.

We will consider the following slightly weaker notion of upper semicontinuity. A set-valued mapping Φ from a topological space (A, τ') into nonempty subsets of a linear topological space $(X, +, \cdot, \tau)$ is *τ-Hausdorff upper semicontinuous at $t_0 \in A$* if for every τ-open neighbourhood W of 0 in $(X, +, \cdot, \tau)$, there exists a neighbourhood U of t_0 such that $\Phi(U) \subseteq \Phi(t_0) + W$. If Φ is τ-Hausdorff semicontinuous at each point of

(A, τ'), then we say that Φ is *τ-Hausdorff upper semicontinuous on (A, τ')*. In general, τ-Hausdorff upper semicontinuity is a weaker notion than τ-upper semicontinuity, but these notions coincide if the mapping Φ has nonempty compact images.

A useful result connecting τ-Hausdorff upper semicontinuity to τ-upper semicontinuity is the following.

Proposition 3.2.25. *Let $\Phi : A \to 2^X$ be a τ-Hausdorff upper semicontinuous set-valued mapping from a topological space (A, τ') into nonempty subsets of a Hausdorff locally convex space $(X, +, \cdot, \tau)$. If for each $t \in A$, $\overline{co}^{\tau}(\Phi(t))$ is a compact subset of (X, τ), then the mapping $\Psi : A \to 2^X$ defined by $\Psi(t) := \overline{co}^{\tau}(\Phi(t))$ for $t \in A$ is a τ-cusco on (A, τ').*

Proof. The proof of this result follows the lines of that of Proposition 3.3.31. Therefore the interested reader is directed to look at the proof of Proposition 3.3.31. □

Theorem 3.2.26. *Let $(X, \|\cdot\|)$ be a Banach space, let $\varepsilon > 0$, and let K be an ε-uniquely remotal set in $(X, \|\cdot\|)$. If $\overline{co}(K)$ is weakly compact and the farthest point mapping $x \mapsto F_K(x)$ is weak-Hausdorff upper semicontinuous on $(\overline{co}(K), \text{weak})$, then $\operatorname{diam}(K) \leq 2\varepsilon$.*

Proof. By assumption $C := \overline{co}(K)$ is nonempty and weakly compact. Let $G : C \to 2^C$ be defined by $G(x) := \overline{co}(F_K(x))$ for each $x \in C$. Then, by Proposition 3.2.25, G is a weak cusco on (C, weak). Hence by the Kakutani–Glicksberg–Fan fixed point theorem (Theorem 1.2.9) there exists $x_0 \in C$ such that $x_0 \in G(x_0)$, that is, $x_0 \in \overline{co}(F_K(x_0))$. Since

$$\operatorname{diam}(G(x_0)) = \operatorname{diam}(\overline{co}(F_K(x_0))) = \operatorname{diam}(F_K(x_0)) \leq \varepsilon,$$

we have that $F_K(x_0) \subseteq G(x_0) \subseteq B[x_0, \varepsilon]$. Thus $K \subseteq B[x_0, \varepsilon]$, and so $\operatorname{diam}(K) \leq 2\varepsilon$. □

Lemma 3.2.27 ([83]). *Let K be a nonempty closed and bounded subset of a normed linear space $(X, \|\cdot\|)$. Then F_K has a closed graph.*

Proof. Let $((x_n, y_n) : n \in \mathbb{N})$ be a sequence in $\operatorname{Gr}(F_K)$ converging to some $(x, y) \in X \times X$, that is, $y_n \in F_K(x_n)$ for all $n \in \mathbb{N}$, $x = \lim_{n\to\infty} x_n$, and $y = \lim_{n\to\infty} y_n$. Firstly, note that $y \in K$, since K is closed. Furthermore,

$$\|x - y\| = \lim_{n\to\infty} \|x_n - y_n\| = \lim_{n\to\infty} r(x_n, K) = r(x, K),$$

since the radial function for K is continuous by Proposition 3.2.21. Thus $y \in F_K(x)$, that is, $(x, y) \in \operatorname{Gr}(F_K)$. □

Corollary 3.2.28. *Let $(X, \|\cdot\|)$ be a Banach space, and let $\varepsilon > 0$. If K is a compact ε-uniquely remotal set in $(X, \|\cdot\|)$, then $\operatorname{diam}(K) \leq 2\varepsilon$. In particular, every closed ε-uniquely remotal subset K of a finite-dimensional Banach space $(X, \|\cdot\|)$ has $\operatorname{diam}(K) \leq 2\varepsilon$.*

Proof. Since K is compact, $\overline{co}(K)$ is norm compact (see [75, p. 416] or [195]) and hence weakly compact. Furthermore, from Lemma 3.2.27 and the fact that F_K has nonempty images (since K is remotal), combined with the fact that a set-valued mapping with a closed graph and nonempty images that maps into a compact space is an usco mapping, (see Corollary 1.1.15), we have that the farthest point mapping $x \mapsto F_K(x)$ is an usco on $\overline{\text{co}}(K)$. The result then follows directly from Theorem 3.2.26. □

From Corollary 3.2.28 we can immediately obtain the following classical result.

Corollary 3.2.29. *Every closed uniquely remotal subset of a finite-dimensional normed linear space is a singleton.*

By appealing to Proposition 3.2.24 we can further refine Corollary 3.2.29.

Corollary 3.2.30. *Every uniquely remotal subset of a finite-dimensional rotund normed linear space is a singleton.*

For further information on Chebyshev sets and uniquely remotal sets, see [14, 83, 227]

3.3 Differentiability of convex functions

The key concept, which runs throughout this section, is the notion of a convex set. A subset C of a vector space $(V, +, \cdot)$ over the real numbers is called *convex* if for every pair of points $x, y \in C$ and $0 < \lambda < 1$, we have $\lambda x + (1 - \lambda)y \in C$. For functions, there is a corresponding notion. A real-valued function $\varphi : C \to \mathbb{R}$ defined on a nonempty convex subset C of a vector space V is said to be a *convex function* if for every pair of points $x, y \in C$ and $0 < \lambda < 1$, we have

$$\varphi(\lambda x + (1 - \lambda)y) \leq \lambda\varphi(x) + (1 - \lambda)\varphi(y).$$

A closely related notion is that of being sublinear. A real-valued function $s : V \to \mathbb{R}$ is said to be *sublinear* if for every pair of points $x, y \in V$ and $0 < \lambda < \infty$, we have that $s(x + y) \leq s(x) + s(y)$ (i. e., s is *subadditive*) and $s(\lambda x) = \lambda s(x)$ (i. e., s is *positively homogeneous*).

Exercise 3.3.1. Let $(V, +, \cdot)$ be a vector space over the real numbers, and let $f : V \to \mathbb{R}$. Show that if f is sublinear, then f is convex. On the other hand, show that if f is convex and positively homogeneous, then f is sublinear.

Proposition 3.3.2. *Let $(V, +, \cdot)$ be a vector space over the real numbers, and let $\varphi : C \to \mathbb{R}$ be a convex function defined on a nonempty convex subset C of V. Suppose that $x_0 \in C$, t_1, t_2 are positive real numbers, and $y \in V$. If $0 < t_1 < t_2$ and $x_0 + t_2 y \in C$, then*

$$\frac{\varphi(x_0 + t_1 y) - \varphi(x_0)}{t_1} \leq \frac{\varphi(x_0 + t_2 y) - \varphi(x_0)}{t_2} \leq \frac{\varphi(x_0 + t_2 y) - \varphi(x_0 + t_1 y)}{t_2 - t_1}.$$

Proof. First, note that $x_0 + t_1 y = (1 - t_1/t_2)x_0 + (t_1/t_2)(x_0 + t_2 y) \in C$. Therefore

$$\varphi(x_0 + t_1 y) \le (1 - t_1/t_2)\varphi(x_0) + (t_1/t_2)\varphi(x_0 + t_2 y). \qquad (*)$$

Thus

$$\varphi(x_0 + t_1 y) - \varphi(x_0) \le (t_1/t_2)(\varphi(x_0 + t_2 y) - \varphi(x_0)).$$

The first inequality now follows by multiplying both sides by t_1^{-1}. Now by going back to equation $(*)$ we get that

$$\varphi(x_0 + t_1 y) \le (t_1/t_2)\varphi(x_0 + t_2 y) + (1 - t_1/t_2)\varphi(x_0).$$

By subtracting $\varphi(x_0 + t_2 y)$ from both sides we get

$$\begin{aligned}\varphi(x_0 + t_1 y) - \varphi(x_0 + t_2 y) &\le (t_1/t_2 - 1)\varphi(x_0 + t_2 y) - (t_1/t_2 - 1)\varphi(x_0)\\ &= \left(\frac{t_1 - t_2}{t_2}\right)\varphi(x_0 + t_2 y) - \left(\frac{t_1 - t_2}{t_2}\right)\varphi(x_0).\end{aligned}$$

The second inequality now follows by multiplying both sides by $-(t_2 - t_1)^{-1}$. □

Proposition 3.3.3. *Let $(V, +, \cdot)$ be a vector space over the real numbers, and let $\varphi : C \to \mathbb{R}$ be a convex function defined on a nonempty convex subset C of V. Suppose that $x_0 \in C$, $s, t \in \mathbb{R}$ and $y \in V$. If $s < 0 < t$ and both $x_0 + sy$ and $x_0 + ty$ are elements of C, then*

$$\frac{\varphi(x_0 + sy) - \varphi(x_0)}{s} \le \frac{\varphi(x_0 + ty) - \varphi(x_0)}{t}.$$

Proof. To prove this, we can simply apply Proposition 3.3.2 with $x_0' := x_0 + sy$, $t_1' := (-s)$ and $t_2' := (-s) + t$. □

Exercise 3.3.4. Let $(V, +, \cdot)$ be a vector space over the real numbers, and let $s : V \to \mathbb{R}$. Show that if s is sublinear and $s(-x) = -s(x)$ for all $x \in V$, then s is linear.

Theorem 3.3.5 (Hahn–Banach theorem [75, p. 62]). *Let Y be a subspace of a vector space $(V, +, \cdot)$ over the real numbers, and let $p : V \to \mathbb{R}$ be a sublinear functional on V. If f is a linear functional on Y and $f(y) \le p(y)$ for all $y \in Y$, then there exists a linear functional $F : V \to \mathbb{R}$ such that $F|_Y = f$ and $F(x) \le p(x)$ for all $x \in V$.*

We need the following very important fact regarding the continuity of convex functions.

Proposition 3.3.6 ([257, Proposition 1.6]). *Let U be a nonempty open convex subset of a normed linear space $(X, \|\cdot\|)$, and let $\varphi : U \to \mathbb{R}$ be a convex function. If φ is locally bounded above on U, that is, for every $x_0 \in U$, there exist $M > 0$ and $\delta > 0$ such that $B(x_0, \delta) \subseteq U$ and $\varphi(x) \le M$ for all $x \in B(x_0, \delta)$, then φ is locally Lipschitz on U, that is, for every $x_0 \in U$, there exist $L > 0$ and $\delta' > 0$ such that $B(x_0, \delta') \subseteq U$ and*

$$|\varphi(x) - \varphi(y)| \le L\|x - y\|$$

for all $x, y \in B(x_0, \delta')$.

Proof. Let $x_0 \in U$. Choose $M^* > 0$ and $\delta > 0$ such that $B(x_0, 2\delta) \subseteq U$ and $\varphi(x) \le M^*$ for all $x \in B(x_0, 2\delta)$. Then for all $x \in B(x_0, \delta)$, we have that $2x_0 - x = x_0 - (x - x_0) \in B(x_0, \delta)$ and $x_0 = (1/2)(2x_0 - x) + (1/2)x$. Hence

$$\varphi(x_0) \le \frac{\varphi(2x_0 - x) + \varphi(x)}{2} \le \frac{M^* + \varphi(x)}{2},$$

so $-\varphi(x) \le M^* + 2|\varphi(x_0)|$, that is, $|\varphi(x)| \le (M^* + 2|\varphi(x_0)|) =: M'$ for all $x \in B(x_0, \delta)$. So $|\varphi|$ is bounded by M' on $B(x_0, \delta)$. Let $\delta' := \delta/2$. We will now show that φ is Lipschitz on $B(x_0, \delta')$. To this end, let x and y be distinct points in $B(x_0, \delta')$. Set $\alpha := \|x - y\|$ and $z := y + (\delta'/\alpha)(y - x)$. Note that $z \in B(x_0, 2\delta')$. Since $y = [\alpha/(\alpha + \delta')]z + [\delta'/(\alpha + \delta')]x$ is a convex combination lying in $B(x_0, 2\delta')$, we have that $\varphi(y) \le [\alpha/(\alpha + \delta')]\varphi(z) + [\delta'/(\alpha + \delta')]\varphi(x)$, and so

$$\begin{aligned}\varphi(y) - \varphi(x) &\le [\alpha/(\alpha + \delta')](\varphi(z) - \varphi(x)) + [\delta'/(\alpha + \delta')](\varphi(x) - \varphi(x)) \\ &\le (\alpha/\delta')2M' = (2M'/\delta')\|x - y\|.\end{aligned}$$

Interchanging x and y gives the desired result, with $M := 2M'/\delta'$. □

Suppose that $\varphi : C \to \mathbb{R}$ is a convex function defined on a nonempty convex subset C of a normed linear space $(X, \|\cdot\|)$ and $x \in C$. Then we define the *subdifferential* $\partial\varphi(x)$ by

$$\partial\varphi(x) := \{x^* \in X^* : x^*(y) - x^*(x) \le \varphi(y) - \varphi(x) \text{ for all } y \in C\}.$$

We can also define the subdifferential in terms of the right-hand derivative of φ. Suppose that $f : U \to \mathbb{R}$ is a real-valued function defined on a nonempty open subset U of a normed linear space $(X, \|\cdot\|)$. Let $x_0 \in U$ and $v \in X$. Then the *right-hand directional derivative of f at the point $x_0 \in U$ in the direction v* is defined as

$$f'_+(x_0; v) := \lim_{\lambda \to 0^+} \frac{f(x_0 + \lambda v) - f(x_0)}{\lambda}.$$

Similarly, the *left-hand directional derivative of f at the point $x_0 \in U$ in the direction v* is defined as

$$f'_-(x_0; v) := \lim_{\lambda \to 0^-} \frac{f(x_0 + \lambda v) - f(x_0)}{\lambda}.$$

For convex functions, the right-hand and left-hand directional derivatives always exist. To see this, let us suppose that f is convex and x_0 and $v \neq 0$ are as in the definition of $f'_+(x_0; v)$. Suppose also that $\delta > 0$ is such that $x_0 + tv \in U$ whenever $-\delta \le t \le \delta$. Then, by Proposition 3.3.2, $t \mapsto \frac{f(x_0+tv)-f(x_0)}{t}$ is an increasing function over $(0, \delta)$. Furthermore, by Proposition 3.3.3

$$\frac{f(x_0 + sv) - f(x_0)}{s} \le \frac{f(x_0 + tv) - f(x_0)}{t} \quad \text{for all } -\delta < s < 0 \text{ and } 0 < t < \delta.$$

Therefore the right-hand quotients are bounded from below. Hence the right-hand directional derivative of f at x_0 in the direction v exists. For the left-hand directional derivative, we simply note that $f'_-(x_0; v) = (-1)f'_+(x_0; -v)$. In this way, we see that both directional derivatives always exist for convex functions.

We can now give the basic properties of the subdifferential mapping $x \mapsto \partial\varphi(x)$.

Lemma 3.3.7 ([257, Proposition 1.8]). *Let U be a nonempty open and convex subset of a normed linear space $(X, \|\cdot\|)$, and let $\varphi : U \to \mathbb{R}$ be a continuous convex function. If $x_0 \in U$, then $\partial\varphi(x_0) = \{x^* \in X^* : x^*(y) \le \varphi'_+(x_0; y)$ for all $y \in X\}$.*

Proof. Let $x_0 \in U$. We will first show that

$$\partial\varphi(x_0) \subseteq \{x^* \in X^* : x^*(y) \le \varphi'_+(x_0; y) \text{ for all } y \in X\}.$$

To this end, let $x_0^* \in \partial\varphi(x_0)$. If $y \in X$ and $\lambda > 0$, then we have that

$$x_0^*(y) = \frac{x_0^*(x_0 + \lambda y) - x_0^*(x_0)}{\lambda} \le \frac{\varphi(x_0 + \lambda y) - \varphi(x_0)}{\lambda}.$$

By taking the limit as $\lambda \to 0^+$ we get that $x_0^*(y) \le \varphi'_+(x_0; y)$. Since $y \in X$ was arbitrary, we have that $x_0^*(y) \le \varphi'_+(x_0; y)$ for all $y \in X$. This shows that

$$\partial\varphi(x_0) \subseteq \{x^* \in X^* : x^*(y) \le \varphi'_+(x_0; y) \text{ for all } y \in X\}.$$

Next, we show that

$$\{x^* \in X^* : x^*(y) \le \varphi'_+(x_0; y) \text{ for all } y \in X\} \subseteq \partial\varphi(x_0).$$

So let $x_0^* \in \{x^* \in X^* : x^*(y) \le \varphi'_+(x_0; y)$ for all $y \in X\}$. Fix $x \in U$. Then

$$\begin{aligned} x_0^*(x - x_0) &\le \lim_{\lambda \to 0^+} \frac{\varphi(x_0 + \lambda(x - x_0)) - \varphi(x_0)}{\lambda} \\ &\le \frac{\varphi(x_0 + 1(x - x_0)) - \varphi(x_0)}{1} = \varphi(x) - \varphi(x_0), \end{aligned}$$

since $\lambda \mapsto \frac{\varphi(x_0+\lambda(x-x_0))-\varphi(x_0)}{\lambda}$ is increasing on $(0, 1]$ by Proposition 3.3.2. Therefore, since $x \in U$ was arbitrary, we have that $x_0^* \in \partial\varphi(x_0)$. This completes the proof. □

Lemma 3.3.8 ([257, Lemma 1.2]). *Let U be a nonempty open and convex subset of a normed linear space $(X, \|\cdot\|)$, and let $\varphi : U \to \mathbb{R}$ be a continuous convex function. If $x_0 \in U$, then the mapping $y \mapsto \varphi'_+(x_0; y)$ from X into $\mathbb{R}$ is sublinear.*

Proof. Let $x_0 \in U$ and define $p : X \to \mathbb{R}$ by $p(x) := \varphi'_+(x_0; x)$ for $x \in X$. Note that p is well-defined. Let $0 < \mu < \infty$, and let $x \in X$. Then

$$\begin{aligned} p(\mu x) &= \lim_{\lambda \to 0^+} \frac{\varphi(x_0 + \lambda(\mu x)) - \varphi(x_0)}{\lambda} \\ &= \mu \lim_{\lambda \to 0^+} \frac{\varphi(x_0 + (\lambda\mu)x) - \varphi(x_0)}{\lambda\mu} \end{aligned}$$

$$= \mu \lim_{\lambda' \to 0^+} \frac{\varphi(x_0 + \lambda' x) - \varphi(x_0)}{\lambda'} \quad \text{(where } \lambda' := \lambda\mu)$$
$$= \mu p(x).$$

So p is positively homogeneous on X. Next, choose $\delta > 0$ such that $B[x_0, \delta] \subseteq U$. We claim that p is convex on $B[0, \delta]$. Fix $n \in \mathbb{N}$ and define $p_n : B[0, \delta] \to \mathbb{R}$ by

$$p_n(x) := \frac{\varphi(x_0 + (1/n)x) - \varphi(x_0)}{(1/n)} \quad \text{for } x \in B[0, \delta].$$

Since $x \mapsto x_0 + (1/n)x$ is an affine mapping, $x \mapsto \varphi(x_0 + (1/n)x)$ is convex, and so p_n is also convex. Now $p(x) = \lim_{n\to\infty} p_n(x)$ for each $x \in B[0, \delta]$. Therefore $p|_{B[0,\delta]}$ is convex, as the pointwise limit of convex functions is again convex. Since p is also positively homogeneous on X, it is an easy exercise to show that p is sublinear on X. □

Corollary 3.3.9. *Let U be a nonempty open and convex subset of a normed linear space $(X, \|\cdot\|)$, and let $\varphi : U \to \mathbb{R}$ be a continuous convex function. If $x_0 \in U$, then for all $y \in X$, $-\varphi'_+(x_0; -y) \le \varphi'_+(x_0; y)$.*

Proof. Let $x_0 \in U$ and $y_0 \in X$. Then since $y \mapsto \varphi'_+(x_0; y)$ is sublinear, we have that

$$0 = \varphi'_+(x_0; 0) = \varphi'_+(x_0; -y_0 + y_0) \le \varphi'_+(x_0; -y_0) + \varphi'_+(x_0; y_0).$$

Therefore $-\varphi'_+(x_0; -y_0) \le \varphi'_+(x_0; y_0)$. □

Lemma 3.3.10 ([257, Proposition 1.11]). *Let U be a nonempty open and convex subset of a normed linear space $(X, \|\cdot\|)$, and let $\varphi : U \to \mathbb{R}$ be a continuous convex function. If $x_0 \in U$, then $\partial\varphi(x_0) \neq \varnothing$.*

Proof. Let y_0 be any element of S_X. By Corollary 3.3.9, $-\varphi'_+(x_0; -y_0) \le \varphi'_+(x_0; y_0)$. Choose $\alpha \in [-\varphi'_+(x_0; -y_0), \varphi'_+(x_0; y_0)]$. Then define $f_\alpha : \operatorname{span}\{y_0\} \to \mathbb{R}$ by

$$f_\alpha(\lambda y_0) := \lambda\alpha \quad \text{for all } \lambda \in \mathbb{R}.$$

Now fix $0 < \lambda < \infty$. Then by the positive homogeneity of $y \mapsto \varphi'_+(x_0; y)$ we have

$$-\varphi'_+(x_0; (-\lambda)y_0) \le \lambda\alpha \le \varphi'_+(x_0; \lambda y_0).$$

Therefore $f_\alpha(\lambda y_0) \le \varphi'_+(x_0; \lambda y_0)$, and $f_\alpha((-\lambda)y_0) \le \varphi'_+(x_0; (-\lambda)y_0)$. Hence we obtain $f_\alpha(\lambda y_0) \le \varphi'_+(x_0; \lambda y_0)$ for all $\lambda \in \mathbb{R}$. Thus by the Hahn–Banach Theorem (see Theorem 3.3.5) there exists a linear functional $F_\alpha : X \to \mathbb{R}$ such that $F_\alpha(y) \le \varphi'_+(x_0; y)$ for all $y \in X$ and $F_\alpha(y_0) = \alpha$. Note also that by Proposition 3.3.6 and the definition of $\varphi'_+(x_0; y)$ there exists $L > 0$ such that $F_\alpha(y) \le \varphi'_+(x_0; y) \le L\|y\|$ for all $y \in X$. Thus $F_\alpha \in X^*$. Furthermore, by Lemma 3.3.7 we have that $F_\alpha \in \partial\varphi(x_0)$. □

Proposition 3.3.11 ([257, Proposition 1.11]). *Let U be a nonempty open and convex subset of a normed linear space $(X, \|\cdot\|)$, and let $\varphi : U \to \mathbb{R}$ be a continuous convex function.*

If $x_0 \in U$, then $\partial\varphi(x_0)$ is a weak*-compact convex subset of X^*. Moreover, the mapping $x \mapsto \partial\varphi(x)$ is locally bounded at x_0, that is, there exist $L > 0$ and $\delta > 0$ such that $B(x_0, \delta) \subseteq U$ and $\|x^*\| \leq L$ whenever $x \in B(x_0, \delta)$ and $x^* \in \partial\varphi(x)$.

Proof. For each $x \in U$, let

$$F_x := \{x^* \in X^* : x^*(x - x_0) \leq \varphi(x) - \varphi(x_0)\} = (\widehat{x - x_0})^{-1}(-\infty, \varphi(x) - \varphi(x_0)].$$

Thus each set F_x is weak* closed and convex. Now $\partial\varphi(x_0) = \bigcap_{x \in U} F_x$. Therefore $\partial\varphi(x_0)$ is weak* closed and convex. Let us now show that $x \mapsto \partial\varphi(x)$ is locally bounded at x_0. (Note that this will then automatically show that $\partial\varphi(x)$ is weak* compact by the Banach–Alaoglu theorem [1].) By Proposition 3.3.6 there exists $L > 0$ and $\delta > 0$ such that $B(x_0, \delta) \subseteq U$ and $|\varphi(x) - \varphi(y)| \leq L\|x - y\|$ for all $x, y \in B(x_0, \delta)$. We claim that $\|x^*\| \leq L$ whenever $x \in B(x_0, \delta)$ and $x^* \in \partial\varphi(x)$. To this end, let $x \in B(x_0, \delta)$ and $x^* \in \partial\varphi(x)$. Let $v \in S_X$ and choose $0 < \mu$ such that $x + \mu v \in B(x_0, \delta)$. Then

$$x^*(v) = \frac{x^*((x + \mu v) - x)}{\mu} \leq \frac{\varphi(x + \mu v) - \varphi(x)}{\mu} \leq \frac{L\|\mu v\|}{\mu} = L.$$

Thus $\|x^*\| \leq L$. Note that here we used the simple fact that if $x^*(v) \leq L$ for all $v \in S_X$, then $\|x^*\| \leq L$. □

Proposition 3.3.12 ([257, Proposition 2.5]). *If $\varphi : U \to \mathbb{R}$ is a continuous convex function defined on a nonempty open convex subset U of a normed linear space $(X, \|\cdot\|)$, then the subdifferential mapping $x \mapsto \partial\varphi(x)$ is a weak*-cusco on U.*

Proof. It follows from Lemma 3.3.10 and Proposition 3.3.11 that we need only show that $x \mapsto \partial\varphi(x)$ is weak*-upper semicontinuous on U. So suppose, to obtain a contradiction, that $\partial\varphi$ is not weak* upper semicontinuous at some point $x_0 \in U$. Then there exists a weak* open subset W of X^*, containing $\partial\varphi(x_0)$, such that, for every $\delta > 0$, $\partial\varphi(B(x_0, \delta)) \nsubseteq W$. Therefore, in particular, there exist sequences $(x_n : n \in \mathbb{N})$ in U and $(x_n^* : n \in \mathbb{N})$ in X^* such that $x_0 = \lim_{n\to\infty} x_n$ and $x_n^* \in \partial\varphi(x_n) \setminus W$ for all $n \in \mathbb{N}$. Furthermore, by Proposition 3.3.11 we can assume that the sequence $(x_n^* : n \in \mathbb{N})$ is norm bounded in X^*, that is, there exists $L > 0$ such that $\|x_n^*\| \leq L$ for all $n \in \mathbb{N}$. Hence by the Banach–Alaoglu theorem [1] the sequence $(x_n^* : n \in \mathbb{N})$ has a weak* cluster-point x_∞^*, which must lie in $X^* \setminus W$. We will obtain our desired contradiction by showing that $x_\infty^* \in \partial\varphi(x_0) \subseteq W$. To this end, fix $x \in U$ and $\varepsilon > 0$. Since φ is continuous at x_0, there exists $N \in \mathbb{N}$ such that $\|x_n - x_0\| < \varepsilon/L$ and $|\varphi(x_n) - \varphi(x_0)| < \varepsilon$ for all $n > N$. Let $n > N$. Then

$$\begin{aligned}(\widehat{x - x_0})(x_n^*) = x_n^*(x - x_0) &= x_n^*(x - x_n) + x_n^*(x_n - x_0)\\ &\leq \varphi(x) - \varphi(x_n) + \|x_n^*\|\|x_n - x_0\|\\ &\leq [\varphi(x) - \varphi(x_0)] + [\varphi(x_0) - \varphi(x_n)] + L\|x_n - x_0\|\\ &\leq [\varphi(x) - \varphi(x_0)] + \varepsilon + \varepsilon.\end{aligned}$$

Therefore $x_\infty^*(x-x_0) = \widehat{(x-x_0)}(x_\infty^*) \le [\varphi(x)-\varphi(x_0)]+2\varepsilon$. Since $\varepsilon > 0$ was arbitrary, we have that $x_\infty^*(x-x_0) \le \varphi(x)-\varphi(x_0)$. Since $x \in U$ was arbitrary, we have that $x_\infty^* \in \partial\varphi(x_0)$, as desired. □

One of the most important features of the subdifferential mapping of a convex function is that it belongs to a much studied class of set-valued mappings called "monotone operators".

Let $T : X \to 2^{X^*}$ be a set-valued mapping acting from a normed linear space $(X, \|\cdot\|)$ into subsets of its dual X^*. T is said to be a *monotone operator* if $0 \le (x^* - y^*)(x-y)$ whenever $x, y \in X$ and $x^* \in T(x)$, $y^* \in T(y)$.

Proposition 3.3.13 ([257, Example 2.2]). *If $\varphi : U \to \mathbb{R}$ is a continuous convex function defined on a nonempty open convex subset U of a normed linear space $(X, \|\cdot\|)$, then $T : X \to 2^{X^*}$ defined by*

$$T(x) := \begin{cases} \partial\varphi(x) & \text{if } x \in U, \\ \varnothing & \text{if } x \notin U \end{cases}$$

is a monotone operator on $(X, \|\cdot\|)$.

Proof. Let $x^*, y^* \in X^*$ and suppose that $x^* \in T(x)$ and $y^* \in T(y)$ for some $x, y \in X$. Then $x, y \in U$ since $T(x) \ne \varnothing$ and $T(y) \ne \varnothing$. In fact, $T(x) = \partial\varphi(x)$ and $T(y) = \partial\varphi(y)$. Therefore

$$x^*(y-x) \le \varphi(y) - \varphi(x) \quad \text{and} \quad (-y^*)(y-x) = y^*(x-y) \le \varphi(x) - \varphi(y).$$

By adding these inequalities we get $(x^* - y^*)(y-x) \le 0$, and so $0 \le (x^* - y^*)(x-y)$. Hence, T is indeed a monotone operator. □

Suppose that (X, τ) is a topological space, $(Y, +, \cdot, \tau')$ is a linear topological space, and $\Phi : X \to 2^Y$ is a cusco on (X, τ). Then we say that Φ is a *minimal cusco* if the graph of Φ does not contain, as a proper subset, the graph of any other cusco defined on (X, τ).

Proposition 3.3.14. *Let $\Phi : U \to 2^{X^*}$ be a* weak**-cusco defined on a nonempty open subset U of a normed linear space $(X, \|\cdot\|)$. If the mapping $T : X \to 2^{X^*}$ defined by $T(x) := \Phi(x)$ if $x \in U$ and $T(x) := \varnothing$ if $x \in X \setminus U$ is a monotone operator, then Φ is a minimal weak*-cusco on U.*

Proof. Suppose, to obtain a contradiction, that Φ is not a minimal weak*-cusco on U. Then there exists a weak*-cusco $\Psi : U \to 2^{X^*}$ such that $\Psi(x) \subseteq \Phi(x)$ for all $x \in U$ but $\Psi(x_0) \ne \Phi(x_0)$ for some $x_0 \in U$. Choose $x_0^* \in \Phi(x_0) \setminus \Psi(x_0) = T(x_0) \setminus \Psi(x_0)$. By the separation theorem [75, p. 418] applied in (X^*, weak^*) there exists $y \in X$ such that $\sup_{y^* \in \Psi(x_0)} \hat{y}(y^*) < \hat{y}(x_0^*)$. Let $W := \{x^* \in X^* : \hat{y}(x^*) < \hat{y}(x_0^*)\}$. Then W is a weak*-open subset of X^* containing $\Psi(x_0)$. Therefore there exists an open neighbourhood $V \subseteq U$

of x_0 such that $\Psi(V) \subseteq W$. Choose $0 < t < \infty$ such that $x_0 + ty \in V$. Let $y^* \in \Psi(x_0+ty) \subseteq \Phi(x_0+ty) = T(x_0+ty)$. Since T is a monotone operator, $x_0^* \in T(x_0)$, and $y^* \in T(x_0+ty)$, we have that

$$0 \leq (y^* - x_0^*)((x_0 + ty) - x_0) = t(y^* - x_0^*)(y),$$

which implies that $x_0^*(y) \leq y^*(y)$. However, this contradicts the fact that $y^* \in W$, that is, $\widehat{y}(y^*) < \widehat{y}(x_0^*)$. Thus Φ must be a minimal weak*-cusco on U. □

Corollary 3.3.15. *If $\varphi : U \to \mathbb{R}$ is a continuous convex function defined on a nonempty open convex subset U of a normed linear space $(X, \|\cdot\|)$, then the subdifferential mapping $x \mapsto \partial\varphi(x)$ is a minimal weak*-cusco on U.*

Proof. By Proposition 3.3.12 we have that $x \mapsto \partial\varphi(x)$ is a weak*-cusco on U. So the result follows from Propositions 3.3.13 and 3.3.14. □

Thus we have established a close connection between the differentiability theory of continuous convex functions and the study of minimal cuscos. We will explore this relationship further by showing that a continuous convex function $\varphi : U \to \mathbb{R}$ defined on a nonempty open subset U of a normed linear space $(X, \|\cdot\|)$ is Gâteaux differentiable at a point $x_0 \in U$ if and only if $\partial\varphi(x_0)$ is a singleton.

3.3.1 Gâteaux differentiability of convex functions

Suppose that $f : U \to \mathbb{R}$ is a real-valued function defined on a nonempty open subset U of a normed linear space $(X, \|\cdot\|)$. Let $x_0 \in U$ and $v \in X$. Then the *directional derivative of f at the point x_0 in the direction v* is defined as

$$f'(x_0; v) := \lim_{\lambda \to 0} \frac{f(x_0 + \lambda v) - f(x_0)}{\lambda}.$$

Note that $f'(x_0; v)$ exists if and only if both $f'_+(x_0; v)$ and $f'_-(x_0; v)$ exist and $f'_+(x_0; v) = f'_-(x_0; v)$. So, if f is a convex function, then $f'(x_0; v)$ exists if and only if $f'_+(x_0; v) = f'_-(x_0; v)$. Let us also recall again the general fact that for any function, $f'_-(x_0; v) = (-1)f'_+(x_0; -v)$. Thus, if f is a continuous convex function, then $f'(x_0; v)$ exists if and only if $f'_+(x_0; -v) = (-1)f'_+(x_0; v)$. We now consider the stronger notion of a Gâteaux derivative. We will say that f is *Gateaux differentiable at a point $x_0 \in U$* if there exists a continuous linear functional $x^* \in X^*$ such that

$$x^*(v) = \lim_{\lambda \to 0} \frac{f(x_0 + \lambda v) - f(x_0)}{\lambda} \quad \text{for all } v \in X. \qquad (*)$$

It follows from this definition that if f is Gâteaux differentiable at $x_0 \in U$, then the mapping $v \mapsto f'(x_0; v)$ is a continuous linear functional on X. On the other hand, if the mapping $v \mapsto f'(x_0; v)$ is well-defined, continuous and linear, then f is Gâteaux differentiable at $x_0 \in U$ (simply set $x^*(v) := f'(x_0, ; v)$ for $v \in X$ and check that it satisfies equation $(*)$).

Exercise 3.3.16. Let U be a nonempty open convex subset of a normed linear space $(X, \|\cdot\|)$, and let $\varphi : U \to \mathbb{R}$ be a continuous convex function. Show that if φ is Gâteaux differentiable at $x_0 \in U$ with Gâteaux derivative $x^* \in X^*$, then $x^* \in \partial\varphi(x_0)$.

Theorem 3.3.17. *Let U be a nonempty open convex subset of a normed linear space $(X, \|\cdot\|)$, and let $\varphi : U \to \mathbb{R}$ be a continuous convex function. If $x_0 \in U$, then φ is Gâteaux differentiable at x_0 if and only if $y \mapsto \varphi'_+(x_0; y)$ is a linear function on $(X, \|\cdot\|)$.*

Proof. Suppose that $y \mapsto \varphi'_+(x_0; y)$ is linear. By Proposition 3.3.6 and the definition of $\varphi'_+(x_0; y)$ there exists $L > 0$ such that $\varphi'_+(x_0; y) \le L\|y\|$ for all $y \in X$. Furthermore, if $y \in X$, then

$$\varphi'_-(x_0; y) = (-1)\varphi'_+(x_0; -y) = (-1)(-1)\varphi'_+(x_0; y) = \varphi'_+(x_0; y).$$

Therefore $\varphi'(x_0; y)$ exists. Since $y \in X$ was arbitrary, we have that for every $y \in X$, (i) $\varphi'(x_0; y)$ exists, and (ii) $\varphi'(x_0; y) = \varphi'_+(x_0; y) \le L\|y\|$. Hence $y \mapsto \varphi'(x_0; y)$, is a continuous linear functional on X, and so φ is Gâteaux differentiable at x_0.

Conversely, if φ is Gâteaux differentiable at x_0, then by definition $y \mapsto \varphi'(x_0; y)$ is linear. Thus $y \mapsto \varphi'_+(x_0; y)$, is linear. □

Corollary 3.3.18. *Let U be a nonempty open convex subset of a normed linear space $(X, \|\cdot\|)$, and let $\varphi : U \to \mathbb{R}$ be a continuous convex function. If $x_0 \in U$ and $\varphi'(x_0; y)$ exists for all $y \in X$, then φ is Gâteaux differentiable at x_0.*

Proof. By Theorem 3.3.17 it is sufficient to show that $y \mapsto \varphi'_+(x_0; y)$ is linear. Indeed, by Exercise 3.3.4 it is sufficient to show that

$$\varphi'_+(x_0; -y) = -\varphi'_+(x_0; y) \quad \text{for all } y \in X.$$

However, this follows from the assumption that $\varphi'_+(x_0; y) = \varphi'_-(x_0; y) = (-1)\varphi'_+(x_0; -y)$ for all $y \in X$. □

Exercise 3.3.19. Let $(V, +, \cdot)$ be a vector space over the real numbers, and let $s : V \to \mathbb{R}$ be a linear functional. Show that if $t : V \to \mathbb{R}$ is any linear functional on V such that $t(v) \le s(v)$ for all $v \in V$, then $s = t$.

Next we give what is arguably the most important result in this section. It says that the problem of showing that a continuous convex function is Gâteaux differentiable reduces to the problem of showing that a certain set-valued mapping, namely the subdifferential mapping, is single-valued.

Theorem 3.3.20. *Let U be a nonempty open convex subset of a normed linear space $(X, \|\cdot\|)$, and let $\varphi : U \to \mathbb{R}$ be a continuous convex function. If $x_0 \in U$, then φ is Gâteaux differentiable at x_0 if and only if $\partial\varphi(x_0)$ is a singleton.*

Proof. Suppose that φ is Gâteaux differentiable at x_0. Then $y \mapsto \varphi'_+(x_0;y)$ is linear on X, and so by Lemma 3.3.7 and Exercise 3.3.19 $\partial\varphi(x_0)$ is a singleton.

Conversely (and this is the more important implication), suppose that $\partial\varphi(x_0)$ is a singleton and suppose also, to obtaing a contradiction, that φ is not Gâteaux differentiable at $x_0 \in U$. Then by Theorem 3.3.17 $y \mapsto \varphi'_+(x_0;y)$ is not linear on X. In fact, by Exercise 3.3.4 there must exist $y_0 \in X$ such that $\varphi'_+(x_0;-y_0) \neq -\varphi'_+(x_0;y_0)$. Now, because of Corollary 3.3.9, it must be the case that $-\varphi'_+(x_0;-y_0) < \varphi'_+(x_0;y_0)$. For each $\alpha \in [-\varphi'_+(x_0;-y_0), \varphi'_+(x_0;y_0)]$, we will construct a continuous linear functional $F_\alpha \in \partial\varphi(x_0)$ such that $F_\alpha(y_0) = \alpha$ (note that $F_\alpha(y_0) = \alpha$ implies that all F_α are distinct). To this end, choose $\alpha \in [-\varphi'_+(x_0;-y_0), \varphi'_+(x_0;y_0)]$. Then define $f_\alpha : \operatorname{span}\{y_0\} \to \mathbb{R}$ by

$$f_\alpha(\lambda y_0) := \lambda\alpha \quad \text{for } \lambda \in \mathbb{R}.$$

Now fix $0 < \lambda < \infty$. Then by the positive homogeneity of $y \mapsto \varphi'_+(x_0;y)$ we have

$$-\varphi'_+(x_0;(-\lambda)y_0) \le \lambda\alpha \le \varphi'_+(x_0;\lambda y_0).$$

Therefore $f_\alpha(\lambda y_0) \le \varphi'_+(x_0;\lambda y_0)$ and $f_\alpha((-\lambda)y_0) \le \varphi'_+(x_0;(-\lambda)y_0)$. Hence we obtain $f_\alpha(\lambda y_0) \le \varphi'_+(x_0;\lambda y_0)$ for all $\lambda \in \mathbb{R}$. Thus by the Hahn–Banach theorem (see Theorem 3.3.5) there exists a linear functional $F_\alpha : X \to \mathbb{R}$ such that $F_\alpha(y) \le \varphi'_+(x_0;y)$ for all $y \in X$ and $F_\alpha(y_0) = \alpha$. Note also that by Proposition 3.3.6 and the definition of $\varphi'_+(x_0;y)$ there exists $L > 0$ such that $F_\alpha(y) \le \varphi'_+(x_0;y) \le L\|y\|$ for all $y \in X$. Thus $F_\alpha \in X^*$. Furthermore, by Lemma 3.3.7 we have that $F_\alpha \in \partial\varphi(x_0)$. □

Exercise 3.3.21. Show that every nonempty open subset of a Baire space is itself a Baire space with the relative topology.

It follows from Exercise 3.3.21 and Corollary 1.4.24 that every nonempty open subset of a complete metric space is a Baire space with the relative topology. However, in this case, we can do even better. Although some open subsets of a compete metric space are not necessarily complete metric spaces under their given metrics, they can be "re-metrized" to become a complete metric space under a new metric while retaining the same (relative) topology. Indeed, suppose that A is a nonempty open subset of a complete metric space (M, d). Then $(M \times \mathbb{R}, \rho)$ is also a complete metric space under the metric

$$\rho((x_1,r_1),(x_2,r_2)) := d(x_1,x_2) + |r_1 - r_2| \quad \text{for all } (x_1,r_1),(x_2,r_2) \in M \times \mathbb{R}.$$

Let $f : A \to \mathbb{R}$ be defined by $f(x) := \inf\{d(x,y) \in \mathbb{R} : y \in M \setminus A\} = d(x, M \setminus A)$. Note that f is continuous on A, see Proposition 3.2.1. Let $G := \{(x,r) \in M \times \mathbb{R} : x \in A \text{ and } r = 1/f(x)\}$. Then G is a closed subset of $(M \times \mathbb{R}, \rho)$ and hence is a complete metric space with respect to the restriction of the metric ρ to G. Finally, let us note that G is homeomorphic to A. Indeed, the mapping $\pi : G \to A$ defined by $\pi(x,r) := x$ for $(x,r) \in G$ is such a homeomorphism. Thus a nonempty open subset of a complete metric space is "completely metrisable".

Thus, if A is a nonempty open subset of a complete metric space (M,d) and F is a first category subset of A with respect to the relative topology (or equivalently, first category with respect to (M,d)), then $A \setminus F$ contains a dense G_δ-subset of A.

We say that a Banach space $(X, \|\cdot\|)$ is a *weak Asplund space* if every continuous convex function defined on a nonempty open convex subset U of X is Gâteaux differentiable at each point of a dense G_δ-subset of U.

Proposition 3.3.22. *Let (X,τ) be a topological space, let $(Y,+,\cdot,\tau')$ be a Hausdorff locally convex space, and let $\Phi : X \to 2^Y$ be a cusco. Then Φ is a minimal cusco if and only if for each open subset U of (X,τ) and closed and convex subset K of $(Y,+,\cdot,\tau')$ such that $\Phi(U) \not\subseteq K$, there exists a nonempty τ-open subset V of U such that $\Phi(V) \cap K = \varnothing$.*

Proof. The proof is left to the reader: it follows very closely the proof of Proposition 1.3.5. □

Proposition 3.3.23. *Let Φ be a minimal cusco acting from a topological space (X,τ) into subsets of a Hausdorff locally convex space $(Y,+,\cdot,\tau')$. If $\{C_n : n \in \mathbb{N}\}$ is a countable family of closed and convex subsets of $(Y,+,\cdot,\tau')$, then there exists a first Baire category subset F of (X,τ) such that for each $x_0 \in X \setminus F$ and each $n_0 \in \mathbb{N}$, if $\Phi(x_0) \cap C_{n_0} \neq \varnothing$, then there exists an open neighbourhood N of x_0 such that $\Phi(N) \subseteq C_{n_0}$. In particular, if the sets $\{C_n : n \in \mathbb{N}\}$ separate the points of Y (i. e. for each pair of distinct points $x, y \in Y$, there exists $n_0 \in \mathbb{N}$ such that $\{x,y\} \cap C_{n_0}$ is a singleton), then Φ is single-valued at each point of $X \setminus F$.*

Proof. For each $n \in \mathbb{N}$, let $D_n := \{x \in X : \Phi(x) \cap C_n \neq \varnothing\}$. Then each set D_n is closed in (X,τ). Further, for each $n \in \mathbb{N}$, we let $F_n := \mathrm{Bd}(D_n) = D_n \setminus \mathrm{int}(D_n)$. Then each F_n is closed and has no interior. Let $F := \bigcup_{n\in\mathbb{N}} F_n$. Then F is of the first Baire category. Let $x_0 \in X \setminus F$ and $n_0 \in \mathbb{N}$ and suppose that $\Phi(x_0) \cap C_{n_0} \neq \varnothing$. Therefore $x_0 \in D_{n_0} \setminus F_{n_0} = \mathrm{int}(D_{n_0})$. Since $\Phi(x) \cap C_{n_0} \neq \varnothing$ for all $x \in \mathrm{int}(D_{n_0})$, it follows from Proposition 3.3.22 that $\Phi(\mathrm{int}(D_{n_0})) \subseteq C_{n_0}$. Thus $\Phi(N) \subseteq C_{n_0}$, where $N := \mathrm{int}(D_{n_0})$. This completes the first part of the proof.

Next, suppose that the sets $\{C_n : n \in \mathbb{N}\}$ separate the points of Y. Let $x_0 \in X \setminus F$ and suppose, to obtain a contradiction, that there exist distinct points $y, y' \in \Phi(x_0)$. Then, after possibly re-labelling, we may assume that there exists $n_0 \in \mathbb{N}$ such that $y \in C_{n_0}$ and $y' \notin C_{n_0}$. On the other hand, by the preceding, $y' \in \Phi(x_0) \subseteq C_{n_0}$. This gives us our desired contradiction. □

Corollary 3.3.24. *For every minimal cusco Φ acting from a Baire space (A,τ) into subsets of $\mathbb{R}$, endowed with the usual topology, $\{x \in A : \Phi(x)$ is a singleton$\}$ is a dense G_δ-subset of subset of (A,τ).*

Proof. Let $\{r_n : n \in \mathbb{N}\}$ be an enumeration of the rational numbers of $\mathbb{R}$. For each $n \in \mathbb{N}$, let $C_n := (-\infty, r_n]$. Then the sets $\{C_n : n \in \mathbb{N}\}$ are closed and convex and separate the points of $\mathbb{R}$. Therefore by Proposition 3.3.23 $\{x \in A : \Phi(x)$ is a singleton$\}$ is dense in

(A, τ). By Lemma 3.1.4 $\{x \in A : \Phi(x)$ is a singleton$\}$ is a G_δ-subset of (A, τ), since $\mathbb{R}$ has a G_δ-diagonal. □

Proposition 3.3.25. *Let Φ be a minimal cusco acting from a topological space (X, τ) into subsets of a Hausdorff locally convex space $(Y, +, \cdot, \tau')$, and let f be a continuous linear mapping from $(Y, +, \cdot, \tau')$ into a Hausdorff locally convex space $(Z, +, \cdot, \tau'')$. Then the mapping $(f \circ \Phi) : X \to 2^Z$ is a minimal cusco on (X, τ).*

Proof. Clearly, $(f \circ \Phi) : X \to 2^Z$ is a cusco on (X, τ) (see Construction (iii) in Section 1.1.3), so it remains to show that it is a minimal cusco on (X, τ). To do this, we will appeal to Proposition 3.3.22. Consider a closed convex subset K of $(Z, +, \cdot, \tau'')$ and an open subset U of (X, τ) such that $(f \circ \Phi)(U) \not\subseteq K$. Since f is continuous and linear on $(Y, +, \cdot, \tau')$, $f^{-1}(K)$ is a closed and convex subset of $(Y, +, \cdot, \tau')$. Since Φ is a minimal cusco and $\Phi(U) \not\subseteq f^{-1}(K)$, there exists, due to Proposition 3.3.22, a nonempty τ-open subset V of U such that $\Phi(V) \cap f^{-1}(K) = \varnothing$. Hence $(f \circ \Phi)(V) \cap K = \varnothing$. □

Theorem 3.3.26 (Mazur's theorem [196]). *Every separable Banach space is a weak Asplund space.*

Proof. Let $(X, \|\cdot\|)$ be a separable Banach space, and let $\varphi : U \to \mathbb{R}$ be a continuous convex function defined on a nonempty open convex subset U of $(X, \|\cdot\|)$. Let $\{x_n : n \in \mathbb{N}\}$ be any subset of $(X, \|\cdot\|)$ such that $\overline{\text{span}}\{x_n : n \in \mathbb{N}\} = X$. Fix $n \in \mathbb{N}$. Then by Corollary 3.3.15 and Proposition 3.3.25 the mapping $x \mapsto (\widehat{x_n} \circ \partial\varphi)(x)$ is a minimal cusco from U into subsets of $\mathbb{R}$. Hence by Corollary 3.3.24 there exists a dense G_δ-subset G_n of U such that $(\widehat{x_n} \circ \partial\varphi)$ is single-valued at each point of G_n. Let $G := \bigcap_{n \in \mathbb{N}} G_n$. Then G is a dense G_δ-subset of U. We claim that $\partial\varphi$ is single-valued at each point of G. To this end, let $x_0 \in G$. Now suppose that $x^* \in \partial\varphi(x_0)$ and $y^* \in \partial\varphi(x_0)$. Then

$$x^*(x_n) = \widehat{x_n}(x^*) = \widehat{x_n}(y^*) = y^*(x_n) \quad \text{for all } n \in \mathbb{N}.$$

Therefore $\{x_n : n \in \mathbb{N}\} \subseteq \text{Ker}(x^* - y^*)$. Since $\text{Ker}(x^* - y^*)$ is a closed linear subspace of $(X, \|\cdot\|)$, we must have that $X = \overline{\text{span}}\{x_n : n \in \mathbb{N}\} \subseteq \text{Ker}(x^* - y^*)$, that is, $x^* = y^*$. This shows that $\partial\varphi(x_0)$ is a singleton. It now follows from Theorem 3.3.20 that φ is Gâteaux differentiable at each point of G. Thus $(X\|\cdot\|)$ is a weak Asplund space. □

Let $f : X \to (-\infty, \infty]$ be a function on a topological space (X, τ). Then we say that f is *lower semicontinuous* if for each $r \in \mathbb{R}$, $\{x \in X : f(x) \le r\}$ is a closed subset of (X, τ). Note that this is equivalent to saying that the *epigraph* of f, which is the set: $\{(x, r) \in X \times \mathbb{R} : f(x) \le r\}$, is closed in $X \times \mathbb{R}$ endowed with the product topology.

Lemma 3.3.27 (Kenderov's theorem [162]). *Suppose that $\Phi : X \to 2^Y$ is an usco mapping acting from a topological space (X, τ) into subsets of a Hausdorff locally convex space $(Y, +, \cdot, \tau')$. Suppose that $f : Y \to \mathbb{R}$ is a lower semicontinuous function. Then the function $\varphi : X \to \mathbb{R}$ defined by*

$$\varphi(x) := \min\{f(y) : y \in \Phi(x)\} \quad \textit{for } x \in X$$

is well-defined and lower semicontinuous on (X, τ). If, in addition, f is convex, Φ is a minimal cusco, and φ is continuous at a point $x_0 \in X$, then $\Phi(x_0) \subseteq f^{-1}(\varphi(x_0))$.

Proof. Let $x_0 \in X$. Since $\Phi(x_0)$ is nonempty and compact and f is lower semicontinuous, the minimum of f over $\Phi(x_0)$ exists, see Lemma 3.4.13. Hence φ is well-defined. To see that φ is lower semicontinuous, let $r \in \mathbb{R}$ be an arbitrary real number. Let $K := \{y \in Y : f(y) \leq r\}$. Then K is closed, and so $\{x \in X : \Phi(x) \cap K \neq \varnothing\}$ is closed. However, $\{x \in X : \varphi(x) \leq r\} = \{x \in X : \Phi(x) \cap K \neq \varnothing\}$. Thus φ is lower semicontinuous.

Next, suppose that f is convex (in addition to being lower semicontinuous), Φ is a minimal cusco, and φ is continuous at a point $x_0 \in X$. Clearly, $\Phi(x_0) \subseteq f^{-1}([\varphi(x_0), \infty))$. So we need only show that for each $\varepsilon > 0$, $\Phi(x_0) \subseteq f^{-1}((-\infty, \varphi(x_0) + \varepsilon])$. To this end, fix $\varepsilon > 0$. Since φ is continuous at x_0, there exists an open neighbourhood U of x_0 such that $\varphi(x) < \varphi(x_0) + \varepsilon$ for all $x \in U$. Let $K := \{y \in Y : f(y) \leq \varphi(x_0) + \varepsilon\}$. Then K is closed and convex and $\Phi(x) \cap K \neq \varnothing$ for all $x \in U$. Hence, by Proposition 3.3.22, $\Phi(x_0) \subseteq \Phi(U) \subseteq K$. This completes the proof. □

Lemma 3.3.28 ([78]). *Let $f : A \to \mathbb{R}$ be a lower semicontinuous function defined on a Baire space (A, τ). Then there exists a dense G_δ-subset G of (A, τ) such that f is continuous at each point of G.*

Proof. Let $\{r_n : n \in \mathbb{N}\}$ be an enumeration of the rational numbers of $\mathbb{R}$. For each $n \in \mathbb{N}$, let $D_n := \{x \in A : f(x) \leq r_n\}$. Then each D_n is closed. Further, for each $n \in \mathbb{N}$, let $F_n := \mathrm{Bd}(D_n) = D_n \setminus \mathrm{int}(D_n)$. Then each F_n is closed and has no interior. Let $F = \bigcup_{n \in \mathbb{N}} F_n$. Then $G := A \setminus F$ is a dense G_δ-subset of (A, τ). We claim that f is continuous at each point of G. To this end, let $x_0 \in G$. To show that f is continuous at x_0, it suffices to show that for each $\varepsilon > 0$, there exists a neighbourhood N of x_0 such that $f(x) < f(x_0) + \varepsilon$ for all $x \in N$. So let us fix $\varepsilon > 0$ and choose $f(x_0) < r_n < f(x_0) + \varepsilon$. Then $x_0 \in D_n \setminus F_n = \mathrm{int}(D_n)$. Let $N := \mathrm{int}(D_n)$. Then $f(x) \leq r_n < f(x_0) + \varepsilon$ for all $x \in N$. □

Exercise 3.3.29. Let $(X, \|\cdot\|)$ be a normed linear space. Show that the dual norm on X^* is weak*-lower semicontinuous on (X^*, weak^*). *Hint:* For each $0 < r < \infty$, $\{x^* \in X^* : \|x^*\| \leq r\} = rB_{X^*}$.

Theorem 3.3.30 ([10, Theorem 2]). *Let $(X, \|\cdot\|)$ be a Banach space and suppose that $(X, \|\cdot\|)$ can be equivalently renormed so that its dual norm is rotund. Then $(X, \|\cdot\|)$ is a weak Asplund space.*

Proof. Let $\varphi : U \to \mathbb{R}$ be a continuous convex function defined on a nonempty open convex subset U of $(X, \|\cdot\|)$. Let $\Psi : U \to \mathbb{R}$ be defined by $\Psi(x) := \min\{\|x^*\| : x^* \in \partial\varphi(x)\}$ for all $x \in U$. Since the dual norm is weak*-lower semicontinuous on X^*, we may apply Lemmas 3.3.27 and 3.3.28. Hence there exists a dense G_δ-subset G of U such that $\partial\varphi(x) \subseteq \Psi(x)S_{X^*}$ for each $x \in G$. Let $x \in G$. Then, as $\partial\varphi(x)$ is a convex subset of $\Psi(x)S_{X^*}$, from the rotundity of the dual norm we have that $\partial\varphi(x)$ is a singleton. It now follows from Theorem 3.3.20 that φ is Gâteaux differentiable at each point of G. Hence $(X, \|\cdot\|)$ is a weak Asplund space. □

An important property of a topological space related to the study of weak Asplund spaces is the following. We say that a topological space (X,τ) has *property* (S) (named after Stegall [292]) if every minimal usco $\Phi : A \to 2^X$ acting from a Baire space (A,τ') into subsets of (X,τ) is single-valued at the points of a dense G_δ- subset of (A,τ'). Note that Stegall's initial description of this class of spaces was somewhat more convoluted than the definition given here, which seems to have first been formulated in the Rainwater Seminar [243]. It can be shown that a Hausdorff topological space (X,τ) has property (S) if and only if it is a Stegall space; see Proposition 2.3.17 or [81, Theorem 3.2.6].

To see the connection between weak Asplund spaces and topological spaces possessing property (S), we need the following result.

Proposition 3.3.31 ([152, 257]). *Let $\Phi : A \to 2^X$ be a τ-usco acting from a topological space (A,τ') into nonempty subsets of a locally convex space $(X,+,\cdot,\tau)$. If for each $t \in A$, $\overline{co}^\tau\Phi(t)$ is a compact subset of $(X,+,\cdot,\tau)$, then the mapping $\Psi : A \to 2^X$ defined by $\Psi(t) := \overline{co}^\tau\Phi(t)$ for $t \in A$ is a τ-cusco on (A,τ').*

Proof. Clearly, Ψ has nonempty compact convex images. So it is sufficient to show that Ψ is τ-upper semicontinuous on (A,τ'). Let $x_0 \in A$, and let W be a τ-open subset of X containing $\Psi(x_0)$. Since vector addition is continuous, for each $x \in \Psi(x_0)$, there exist τ-open convex neighbourhoods U_x of x and V_x of 0 such that $x = x + 0 \subseteq U_x + V_x \subseteq W$. Since linear topological spaces are also regular, we can assume, by possibly making V_x smaller, that $U_x + \overline{V_x}^\tau \subseteq W$. Now $\{U_x : x \in \Psi(x_0)\}$ is an open cover of $\Psi(x_0)$. Therefore there exists a finite subcover $\{U_{x_k} : 1 \le k \le n\}$ of $\{U_x : x \in \Psi(x_0)\}$. Let $V := \bigcap_{1\le k\le n} V_{x_k}$. Then V is a convex open neighbourhood of 0, and, furthermore,

$$\Psi(x_0) + \overline{V}^\tau \subseteq \left(\bigcup_{1\le k\le n} U_{x_k}\right) + \overline{V}^\tau = \bigcup_{1\le k\le n} (U_{x_k} + \overline{V}^\tau) \subseteq \bigcup_{1\le k\le n} (U_{x_k} + \overline{V_{x_k}}^\tau) \subseteq W.$$

Since $\Phi(x_0) \subseteq \Psi(x_0) + V$, which is τ-open, there exists an open neighbourhood U of x_0 such that $\Phi(U) \subseteq \Psi(x_0) + V$. Let $x \in U$. Then

$$\begin{aligned}\Psi(x) &= \overline{co}^\tau\Phi(x) \subseteq \overline{co}^\tau(\Psi(x_0) + V)\\ &\subseteq \Psi(x_0) + \overline{V}^\tau \subseteq W \quad \text{as, } \Psi(x_0) + \overline{V}^\tau \text{ is closed and convex.}\end{aligned}$$

Here we used the fact that the sum of a closed set and a compact set is closed. □

Corollary 3.3.32. *Let $\Phi : A \to 2^X$ be a minimal τ-usco acting from a topological space (A,τ') into nonempty subsets of a Hausdorff locally convex space $(X,+,\cdot,\tau)$. If for each $t \in A$, $\overline{co}^\tau\Phi(t)$ is a compact subset of $(X,+,\cdot,\tau)$, then the mapping $\Psi : A \to 2^X$ defined by $\Psi(t) := \overline{co}^\tau\Phi(t)$ for $t \in A$ is a minimal τ-cusco on (A,τ').*

Proof. It follows from Proposition 3.3.31 that Ψ is a τ-cusco. So it remains to show that it is minimal. To achieve this, we appeal to Proposition 3.3.22. So let U be an open subset of (A, τ'), and let K be a closed convex subset of $(X, +, \cdot, \tau)$ such that $\Psi(U) \not\subseteq K$. By the definition of Ψ it follows that $\Phi(U) \not\subseteq K$. Choose $x_0 \in \Phi(U) \setminus K$. Then choose a convex τ-open neighbourhood W of x_0 such that $\overline{W}^\tau \cap K = \varnothing$. Now by Proposition 1.3.5 there exists a nonempty τ'-open subset V of U such that $\Phi(V) \subseteq W$, since $\Phi(U) \cap W \neq \varnothing$. Again by appealing to the definition of Ψ we get that

$$\Psi(x) = \overline{\text{co}}^\tau \Phi(x) \subseteq \overline{W}^\tau \subseteq X \setminus K \quad \text{for each } x \in V,$$

that is, $\Phi(V) \cap K = \varnothing$. This proves that Ψ is a minimal τ-cusco. □

Theorem 3.3.33 ([243, 292]). *Let $(X, \|\cdot\|)$ be a Banach space. If $(X^*, weak^*)$ possesses property (S), then $(X, \|\cdot\|)$ is a weak Asplund space.*

Proof. Suppose that (X^*, weak^*) possesses property (S) and let $\varphi : U \to \mathbb{R}$ be a continuous convex function defined on a nonempty open convex subset U of $(X, \|\cdot\|)$. Now, by Proposition 3.3.12, $x \mapsto \partial\varphi(x)$ is a weak*-cusco. Therefore by Proposition 1.3.3 there exists a minimal weak*-usco $\Phi : U \to 2^{X^*}$ such that $\Phi(x) \subseteq \partial\varphi(x)$ for all $x \in U$. Let $\Psi : U \to 2^{X^*}$ be defined by $\Psi(x) := \overline{\text{co}}^{w^*} \Phi(x)$ for $x \in U$. Then, by Proposition 3.3.31, Ψ is a weak*-cusco on U. Moreover, since $\partial\varphi$ has weak* closed and convex images, $\Psi(x) \subseteq \partial\varphi(x)$ for all $x \in U$. However, since $x \mapsto \partial\varphi(x)$ is a minimal weak*-cusco on U (see Corollary 3.3.15), we have that

$$\overline{\text{co}}^{w^*} \Phi(x) = \Psi(x) = \partial\varphi(x) \quad \text{for all } x \in U. \qquad (*)$$

Since (X, weak^*) possesses property (S), there exists a dense G_δ-subset G of U such that Φ is single-valued at each point of G. It now follows from equation $(*)$ that $\partial\varphi$ is single-valued at each point of G. As we have done before, we can apply Theorem 3.3.20 to deduce that φ is Gâteaux differentiable at each point of G. This shows that $(X, \|\cdot\|)$ is a weak Asplund space. □

Although the next result does not feature in our studies of weak Asplund spaces, it again highlights the similarity between the theory of minimal uscos and the theory of minimal cuscos.

Proposition 3.3.34. *Let $(X, +, \cdot, \tau)$ be a Hausdorff locally convex space and suppose that $\overline{\text{co}}^\tau(K)$ is compact for each compact subset K of $(X, +, \cdot, \tau)$. Then (X, τ) possesses property (S) if and only if every minimal cusco $\Psi : A \to 2^X$ defined on a Baire space (A, τ') is single-valued at the points of a dense G_δ-subset of (A, τ').*

Proof. Suppose that (X, τ) possesses property (S). Let $\Phi' : A \to 2^X$ be a minimal cusco defined on a Baire space (A, τ'). Then by Proposition 1.3.3 there exists a minimal usco $\Phi : A \to 2^X$ such that $\Phi(x) \subseteq \Phi'(x)$ for all $x \in A$. Let $\Psi : A \to 2^X$ be defined by $\Psi(x) := \overline{\text{co}}^\tau \Phi(x)$ for $x \in A$. Then, by Proposition 3.3.31, Ψ is a cusco on (A, τ'). Moreover, since Φ' has closed and convex images, $\Psi(x) \subseteq \Phi'(x)$ for all $x \in A$. However, since $x \mapsto \Phi'(x)$, is a minimal cusco on (A, τ'), we must have that

$$\overline{\text{co}}^\tau \Phi(x) = \Psi(x) = \Phi'(x) \quad \text{for all } x \in A. \qquad (*)$$

Since (X,τ) possesses property (S), there exists a dense G_δ-subset G of (A,τ') such that Φ is single-valued at each point of G. It now follows from equation $(*)$ that Φ' is single-valued at each point of G. This completes the proof of the first implication.

Now suppose that $(X,+,\cdot,\tau)$ has the property that every minimal cusco acting from a Baire space into subsets of $(X,+,\cdot,\tau)$ is single-valued at the points of a dense G_δ-subset of its domain. Let $\Phi : A \to 2^X$ be a minimal usco defined on a Baire space (A,τ'). Let $\Psi : A \to 2^X$ be defined by $\Psi(x) := \overline{\text{co}}^\tau \Phi(x)$ for $x \in A$. Then, by Corollary 3.3.32, Ψ is a minimal cusco on (A,τ'). Hence by assumption there exists a dense G_δ-subset G of (A,τ') such that Ψ is single-valued at each point of G. Then, clearly, Φ is single-valued at each point of G. Therefore (X,τ) possesses property (S). □

There are some interesting examples of spaces where the hypotheses of the previous proposition hold. For example, if $(X,\|\cdot\|)$ is a Banach space, then, by the Krein–Phillips theorem [174, 258], $\overline{\text{co}}(K)$ is weakly compact for each weakly compact subset K of $(X,\|\cdot\|)$. Also, by the uniform boundedness theorem [75, p. 66], $\overline{\text{co}}^{w^*}(K)$ is weak* compact for each weak* compact subset K of X^*. Finally, $\overline{\text{co}}^\tau(K)$ is compact for each compact subset K of a Fréchet space [75]. Recall that a linear topological space $(X,+,\cdot,\tau)$ is a *Fréchet space* if it is locally convex and the topology is generated by a translation-invariant complete metric d on X. By *translation-invariant* we mean that $d(x,y) = d(x+z,y+z)$ for all $x,y,z \in X$.

3.3.2 Fréchet differentiability of convex functions

Let $f : U \to \mathbb{R}$ be a real-valued function on a nonempty open subset U of a normed linear space $(X,\|\cdot\|)$. We say that f is *Fréchet differentiable at a point* $x_0 \in U$ if there exists a continuous linear functional $x^* \in X^*$ such that for every $\varepsilon > 0$, there exists $\delta > 0$ such that

$$|f(x+x_0) - f(x_0) - x^*(x)| \le \varepsilon\|x\| \quad \text{for all } \|x\| < \delta.$$

Note that (i) if f is Fréchet differentiable at x_0, then f is also Gâteaux differentiable at x_0 (to see this, replace x with λv, fix $v \in X$, and let $\lambda \to 0$); and (ii) if in the definition of the Gâteaux derivative the limit is uniform with respect to all $v \in S_X$, then f is Fréchet differentiable, that is, f is Fréchet differentiable at $x_0 \in U$ if there exists a continuous linear functional $x^* \in X^*$ such that for every $\varepsilon > 0$, there exists $\delta > 0$ such that

$$\left|\frac{f(x_0+\lambda v) - f(x_0)}{\lambda} - x^*(v)\right| < \varepsilon \quad \text{for all } 0 < |\lambda| < \delta \text{ and } v \in S_X.$$

Like with Gâteaux differentiability of continuous convex functions, Frêchet differentiability may also be characterised in terms of the subdifferential mapping. Of course, then we must require more than just single-valuedness of the subdifferential mapping.

A set-valued mapping $\Phi : A \to 2^X$ acting between a topological space (A, τ) and a normed linear space $(X, \|\cdot\|)$ is said to be *single-valued and norm upper semicontinuous at* $a_0 \in A$ if $\Phi(a_0) =: \{x_0\}$ is a singleton and for each $\varepsilon > 0$, there exists an open neighbourhood U of a_0 such that $\Phi(U) \subseteq B(x_0, \varepsilon)$.

Theorem 3.3.35. *Let U be a nonempty open convex subset of a normed linear space $(X, \|\cdot\|)$, and let $\varphi : U \to \mathbb{R}$ be a continuous convex function. If $x_0 \in U$, then φ is Fréchet differentiable at x_0 if and only if $\partial\varphi$ is single-valued and norm upper semicontinuous at x_0.*

Proof. Let $x_0 \in U$ and suppose that $\partial\varphi$ is single-valued and norm upper semicontinuous at x_0. Since $\partial\varphi$ is single-valued at x_0, there exists $x_0^* \in X^*$ such that $\partial\varphi(x_0) = \{x_0^*\}$. We claim that x_0^* is the Fréchet derivative of φ at x_0. To verify this claim, let us consider an arbitrary $\varepsilon > 0$. Since $\partial\varphi$ is norm upper semicontinuous at x_0, there exists $\delta > 0$ such that $B(x_0, \delta) \subseteq U$ and $\partial\varphi(B(x_0, \delta)) \subseteq B(x_0^*, \varepsilon)$. Now since $x_0^* \in \partial\varphi(x_0)$, we have that

$$0 \le \varphi(x + x_0) - \varphi(x_0) - x_0^*(x) \qquad (*)$$

for all $\|x\| < \delta$. Let x be any element of $B(0, \delta)$, and let $x^* \in \partial\varphi(x + x_0)$. Then

$$x^*(-x) \le \varphi(x_0) - \varphi(x + x_0)$$

or, equivalently,

$$\varphi(x + x_0) - \varphi(x_0) \le x^*(x). \qquad (**)$$

Substituting inequality $(**)$ into inequality $(*)$, we get that

$$0 \le \varphi(x + x_0) - \varphi(x_0) - x_0^*(x) \le (x^* - x_0^*)(x) \le \|x^* - x_0^*\| \cdot \|x\| \le \varepsilon\|x\|.$$

Therefore $|\varphi(x + x_0) - \varphi(x_0) - x_0^*(x)| \le \varepsilon\|x\|$. Thus φ is Fréchet differentiable at x_0.

Conversely, suppose that φ is Fréchet differentiable at $x_0 \in U$ with Fréchet derivative x_0^*. From Exercise 3.3.16 and Theorem 3.3.20 we see that $\partial\varphi(x_0) = \{x_0^*\}$. So it is sufficient to show that $\partial\varphi$ is norm upper semicontinuous at x_0. Let $\varepsilon > 0$, and let $0 < \varepsilon' < \varepsilon/2$. Then there exists $\delta > 0$ such that (i) $B(x_0, 2\delta) \subseteq U$; (ii) φ is L-Lipschitz on $B(x_0, 2\delta)$, that is, there exists $L > 0$ such that $|\varphi(x) - \varphi(y)| \le L\|x - y\|$ for all $x, y \in B(x_0, 2\delta)$; and (iii) $\varphi(x + x_0) - \varphi(x_0) - x_0^*(x) \le \varepsilon'\|x\|$ for all $\|x\| \le \delta$. Let $\delta' := \min\{\delta, \varepsilon'\delta/2L\}$. We claim that $\partial\varphi(B(x_0, \delta')) \subseteq B(x_0^*, \varepsilon)$. To this end, let $x \in B(x_0, \delta')$ and $x^* \in \partial\varphi(x)$. Let $v \in S_X$. Then

$$\begin{aligned}
\delta x^*(v) = x^*(\delta v) &\le \varphi(x + \delta v) - \varphi(x) \quad (\text{because } x^* \in \partial\varphi(x)) \\
&\le \varphi(x_0 + \delta v) - \varphi(x_0) + 2L\|x - x_0\| \quad (\text{since } \varphi \text{ is } L\text{-Lipschitz}) \\
&\le \varphi(x_0 + \delta v) - \varphi(x_0) + \varepsilon'\delta \quad (\text{since } \|x - x_0\| < \delta' \le \varepsilon'\delta/2L)
\end{aligned}$$

$$< x_0^*(\delta v) + \varepsilon'\|\delta v\| + \varepsilon'\delta \quad \text{(by (iii) above)}$$
$$= \delta(x_0^*(v) + \varepsilon' + \varepsilon').$$

Hence $(x^* - x_0^*)(v) \leq 2\varepsilon' < \varepsilon$. Since $v \in S_X$ was arbitrary, $\|x^* - x_0^*\| < \varepsilon$. Therefore $x^* \in B(x_0^*, \varepsilon)$. □

We say that a Banach space $(X, \|\cdot\|)$ is an *Asplund space* if every continuous convex function defined on a nonempty open convex subset U of $(X, \|\cdot\|)$ is Fréchet differentiable at each point of a dense G_δ-subset of U. Therefore every Asplund space is a weak Asplund space.

Theorem 3.3.36 ([10]). *Let $(X, \|\cdot\|)$ be a Banach space. If $(X^*, \|\cdot\|)$ is separable (with respect to the norm topology), then $(X, \|\cdot\|)$ is an Asplund space.*

Proof. Let $\varphi : U \to \mathbb{R}$ be a continuous convex function defined on a nonempty open convex subset U of $(X, \|\cdot\|)$. For $n \in \mathbb{N}$, let

$$O_n := \bigcup \{V \in 2^U : V \text{ is open and } \|\cdot\| - \operatorname{diam}[\partial\varphi(V)] \leq 1/n\}.$$

Clearly, each set O_n is open, as it is a union of open sets. We claim that each set O_n is dense as well. To justify this claim, we fix $n \in \mathbb{N}$ and consider a nonempty open subset W of U. Let $D := \{x_k : k \in \mathbb{N}\}$ be a norm dense subset of X^*, and let $C_k := B[x_k, 1/2n]$ for $k \in \mathbb{N}$. By Proposition 3.3.23 there exists a first category set F such that for every $x \in U \setminus F$ and every $k \in \mathbb{N}$, if $\partial\varphi(x) \cap C_k \neq \varnothing$, then there exists an open neighbourhood N of x such that $\partial\varphi(N) \subseteq C_k$. Choose $x_0 \in W \setminus F$. Note that this is possible since W is not a first category subset of U. Now, since $\partial\varphi(x_0) \neq \varnothing$, there exists $n_0 \in \mathbb{N}$ such that $\partial\varphi(x_0) \cap C_{n_0} \neq \varnothing$. Therefore there exists an open neighbourhood N of x_0 contained in W such that $\partial\varphi(N) \subseteq C_{n_0}$. Thus $\|\cdot\| - \operatorname{diam}[\partial\varphi(N)] \leq 1/n$. Hence $\varnothing \neq N \subseteq W \cap O_n$. This shows that O_n is dense in U. Next, we observe that $\partial\varphi$ is single-valued and norm upper semicontinuous at each point of $\bigcap_{n\in\mathbb{N}} O_n$. It now only remains to apply Theorem 3.3.35 to deduce that φ is Fréchet differentiable at each point of $\bigcap_{n\in\mathbb{N}} O_n$; which shows that $(X, \|\cdot\|)$ is an Asplund space. □

Note that if $(X, \|\cdot\|)$ is a Banach space whose dual space $(X^*, \|\cdot\|)$ is separable, then $(X, \|\cdot\|)$ must also be separable. To see this, suppose that $\{x_n^* : n \in \mathbb{N}\}$ is a countable dense subset of S_{X^*}. For each $n \in \mathbb{N}$, choose $x_n \in S_X$ such that $1/2 < x_n^*(x_n)$. Set $Y := \overline{\operatorname{span}}\{x_n : n \in \mathbb{N}\}$. Then Y is a separable subspace of $(X, \|\cdot\|)$. In fact, $Y = X$. This is because if $Y \neq X$, then by the separation theorem [75, p. 418] there would exist $x^* \in S_{X^*}$ such that $x^*(y) = 0$ for all $y \in Y$. On the other hand, there must exist $x_n^* \in S_{X^*}$ such that $\|x^* - x_n^*\| < 1/4$. However,

$$1/4 = 1/2 - 1/4 \leq 1/2 - \|x_n^* - x^*\| \leq x_n^*(x_n) - [x_n^*(x_n) - x^*(x_n)] = x^*(x_n),$$

which is impossible, since we assumed that $x^*(x_n) = 0$.

Our first characterisation of Asplund spaces involves the definition of a "slice". Suppose that K is a nonempty bounded subset of a normed linear space $(X, \|\cdot\|)$, $x^* \in X^*$ and $\delta > 0$. Then a *slice* of K is the set

$$S(K, x^*, \delta) := \{x \in K : \sup\{x^*(k) : k \in K\} - \delta < x^*(x)\}.$$

If K is a nonempty bounded subset of X^*, then we define a *weak** *slice* analogously, but with the functional coming from $\widehat{X}$, rather than from X^{**}.

Theorem 3.3.37 (Asplund–Namioka–Phelps theorem [10, 245]). *Let $(X, \|\cdot\|)$ be a Banach space. Then the following properties are equivalent:*
(i) *for every nonempty weak* compact subset K of X^* and $\varepsilon > 0$, there exist $x \in S_X$ and $\delta > 0$ such that the norm diameter of $S(K, \widehat{x}, \delta)$ is less than ε;*
(ii) *for every nonempty weak* compact subset K of X^* and $\varepsilon > 0$, there exists a weak* open subset W of X^* such that $K \cap W \neq \varnothing$ and the norm diameter of $K \cap W$ is less than ε;*
(iii) *every locally bounded minimal weak*-usco $\Phi : A \to 2^{X^*}$ acting from a Baire space (A, τ) into subsets of X^* is single-valued and norm upper semicontinuous at the points of a dense G_δ-subset of (A, τ);*
(iv) *$(X, \|\cdot\|)$ is an Asplund space.*

Proof. (i)⇒(ii). Follows directly from the definition of $S(K, \widehat{x}, \delta)$.

(ii)⇒(iii) Let $\Phi : A \to 2^{X^*}$ be a locally bounded minimal weak*-usco from a Baire space (A, τ) into subsets of X^*. For each $n \in \mathbb{N}$, let

$$O_n := \bigcup\{V' \in \tau : \|\cdot\| - \operatorname{diam}[\Phi(V')] < 1/n\}.$$

Clearly, each set O_n is open, as it is a union of open sets. We claim that each set O_n is dense as well. To justify this claim, fix $n \in \mathbb{N}$ and let U be a nonempty open subset of (A, τ). By possibly making U smaller we may assume that Φ is bounded on U. Let $K := \overline{\Phi(U)}^{\text{weak}^*}$. Then K is a nonempty weak* compact subset of X^*. Therefore by assumption there exists a weak* open subset W of X^* such that $K \cap W \neq \varnothing$ and the norm diameter of $K \cap W$ is less than $1/n$. Note that $\Phi(U) \cap W \neq \varnothing$. Hence by Proposition 1.3.5 there exists a nonempty τ-open subset V of U such that $\Phi(V) \subseteq W$, that is,

$$\Phi(V) \subseteq \Phi(U) \cap W \subseteq K \cap W.$$

Hence $\|\cdot\| - \operatorname{diam}[\Phi(V)] < 1/n$. Thus $\varnothing \neq V \subseteq U \cap O_n$. This shows that O_n is dense in (A, τ). It now only remains to observe that Φ is single-valued and norm upper semicontinuous at each point of $\bigcap_{n \in \mathbb{N}} O_n$.

(iii)⇒(iv). Let $\varphi : U \to \mathbb{R}$ be a continuous convex function defined on a nonempty open convex subset U of $(X, \|\cdot\|)$. Now, by Proposition 3.3.12, $x \mapsto \partial\varphi(x)$ is a weak*-cusco. Therefore by Proposition 1.3.3 there exists a minimal weak*-usco

$\Phi : U \to 2^{X^*}$ such that $\Phi(x) \subset \partial\varphi(x)$ for all $x \in U$. Let $\Psi : U \to 2^{X^*}$ be defined by $\Psi(x) := \overline{\text{co}}^{w^*}\Phi(x)$ for $x \in U$. Then, by Proposition 3.3.31, Ψ is a weak*-cusco on U. Moreover, since $\partial\varphi$ has weak* closed and convex images, $\Psi(x) \subseteq \partial\varphi(x)$ for all $x \in U$. However, since $x \mapsto \partial\varphi(x)$, is a minimal weak*-cusco on U (see, Corollary 3.3.15), we must have that

$$\overline{\text{co}}^{w^*}\Phi(x) = \Psi(x) = \partial\varphi(x) \quad \text{for all } x \in U. \qquad (*)$$

Finally, note that since $x \mapsto \partial\varphi(x)$ is locally bounded on U (see Proposition 3.3.11), Φ is locally bounded on U as well. Hence by assumption there exists a dense G_δ-subset G of U such that Φ is single-valued and norm upper semicontinuous at each point of G. We claim that $\partial\varphi$ is also single-valued and norm upper semicontinuous at each point of G. So let us consider $x_0 \in G$. It follows from equation $(*)$ that $\partial\varphi$ is single-valued at x_0, that is, $\Phi(x_0) = \partial\varphi(x_0) =: \{x_0^*\}$. So we must show that $\partial\varphi$ is norm upper semicontinuous at x_0. Let $\varepsilon > 0$. Since Φ is norm upper semicontinuous at x_0, there exists $\delta > 0$ such that $\Phi(B(x_0, \delta)) \subseteq B[x_0^*, \varepsilon]$. Thus, if $x \in B(x_0, \delta)$, then $\Phi(x) \subseteq B[x_0^*, \varepsilon]$, which is weak* closed and convex. Therefore $\partial\varphi(x) = \overline{\text{co}}^{w^*}\Phi(x) \subseteq B[x_0^*, \varepsilon]$. Thus $\partial\varphi(B(x_0, \delta)) \subseteq B[x_0^*, \varepsilon]$. Hence $\partial\varphi$ is single-valued and norm upper semicontinuous at x_0. We now apply Theorem 3.3.35 to deduce that φ is Fréchet differentiable at x_0. This shows that $(X, \|\cdot\|)$ is an Asplund space.

(iv)⇒(i). Let K be a nonempty weak* compact subset of X^*. By the uniform boundedness theorem [75, p. 66] there exists a positive real number L such that $\|x^*\| \leq L$ for all $x^* \in K$. Let $p : X \to \mathbb{R}$ be defined by

$$p(x) := \sup\{x^*(x) : x^* \in K\} \quad \text{for } x \in X.$$

Then, as the pointwise supremum of a pointwise bounded above family of linear L-Lipschitz functions, p is itself convex and L-Lipschitz. (Recall the general fact that the pointwise supremum of a pointwise bounded above family of convex functions is convex and the pointwise supremum of a pointwise bounded above family of L-Lipschitz functions is again L-Lipschitz.) Of course, in this case, p is also positively homogeneous, which means that p is in fact sublinear. Now, as $(X, \|\cdot\|)$ is an Asplund space, there exists a point $x_0 \in X \setminus \{0\}$ such that p is Fréchet differentiable at x_0 with Fréchet derivative x_0^*. We next show that

$$x_0^*(x) \leq p(x) \quad \text{for all } x \in X \qquad (*)$$

and $x_0^*(x_0) = p(x_0)$. So consider any $x \in X$. Then since $x_0^* \in \partial p(x_0)$ (see Exercise 3.3.16),

$$x_0^*(x) \leq p(x + x_0) - p(x_0) \leq p(x) \quad \text{by the triangle inequality.}$$

Furthermore, if we set $x := -x_0$, then we get that

$$-x_0^*(x_0) = x_0^*(-x_0) \leq p(-x_0 + x_0) - p(x_0) = p(0) - p(x_0) = -p(x_0),$$

and so $p(x_0) \le x_0^*(x_0)$. By combining this equality with inequality $(*)$ we obtain $x_0^*(x_0) = p(x_0)$. Let us now show that $x_0^* \in K$. Indeed, if $x_0^* \notin K$, then we may apply a separation argument in (X, weak^*) to obtain $x \in X$ such that

$$p(x) = \sup\{x^*(x) : x^* \in K\} = \sup\{\widehat{x}(x^*) : x^* \in K\} < \widehat{x}(x_0^*) = x_0^*(x),$$

but this contradicts inequality $(*)$. Hence $x_0^* \in K$. In fact, $x_0^* \in \operatorname{argmax}(\widehat{x_0}|_K)$ since $\widehat{x}(x_0^*) = p(x_0) = \sup\{\widehat{x_0}(x^*) : x^* \in K\}$.

Let $\varepsilon > 0$ and $0 < \varepsilon' < \varepsilon/4$. Since p is Fréchet differentiable at x_0, there exists $\delta > 0$ such that

$$0 \le p(x + x_0) - p(x_0) - x_0^*(x) \le \varepsilon'\|x\| \quad \text{for all } \|x\| \le \delta.$$

We claim that the norm diameter of $S(K, \widehat{x_0}, \varepsilon'\delta)$ is less than ε. To verify this claim, it is sufficient to show that for every $x^* \in S(K, \widehat{x_0}, \varepsilon'\delta)$, $\|x^* - x_0^*\| \le 2\varepsilon'$. To this end, let $x^* \in S(K, \widehat{x_0}, \varepsilon'\delta)$. Then, for any $x \in X$, $x^*(x - x_0) \le p(x) - p(x_0) + \varepsilon'\delta$, since for any $x \in X$, $x^*(x) \le p(x)$ as $x^* \in K$ and $p(x_0) - \varepsilon'\delta < x^*(x_0)$. Let $v \in S_X$. Then

$$\begin{aligned}(x^* - x_0^*)(\delta v) &\le [p(\delta v + x_0) - p(x_0) + \varepsilon'\delta] - x_0^*(\delta v)\\ &= [p(\delta v + x_0) - p(x_0) - x_0^*(\delta v)] + \varepsilon'\delta\\ &\le \varepsilon'\|\delta v\| + \varepsilon'\delta = 2\varepsilon'\delta.\end{aligned}$$

Thus $(x^* - x_0^*)(v) \le 2\varepsilon'$. Since $v \in S_X$ was arbitrary, $\|x^* - x_0^*\| \le 2\varepsilon'$. □

Lemma 3.3.38. *If $(X, \|\cdot\|)$ is a separable Banach space, then (B_{X^*}, weak^*) is a compact metrizable topological space.*

Proof. The fact that (B_{X^*}, weak^*) is a compact follows directly from the Banach–Alaoglu theorem [1]. So we need only show that (B_{X^*}, weak^*) is metrizable. Let $\{x_n : n \in \mathbb{N}\}$ be any dense subset of B_X and define $d : B_{X^*} \times B_{X^*} \to [0, 2]$ by

$$d(x^*, y^*) := \sum_{n\in\mathbb{N}} \frac{|\widehat{x_n}(x^* - y^*)|}{2^n} \quad \text{for } x^*, y^* \in B_{X^*}.$$

Firstly, d is well-defined since for each fixed $x^*, y^* \in B_{X^*}$,

$$0 \le |\widehat{x_n}(x^* - y^*)|/2^n \le 1/2^{n-1} \quad \text{for all } n \in \mathbb{N} \text{ and } \sum_{n\in\mathbb{N}} 1/2^{n-1} = 2 < \infty.$$

Therefore by the comparison test the series in the definition of d is convergent. It is also obvious that (i) $0 \le d(x^*, y^*) \le 2$ for all $x^*, y^* \in B_{X^*}$ and (ii) d is symmetric, that is, $d(x^*, y^*) = d(y^*, x^*)$. The fact that d satisfies the triangle inequality is standard and left as an exercise for the reader. Next, we show that d 'separates' the points of B_{X^*}. So let $x^*, y^* \in B_{X^*}$ and suppose that $d(x^*, y^*) = 0$. Then

$$|(x^* - y^*)(x_n) = |\widehat{x_n}(x^* - y^*)| = 0 \quad \text{for all } n \in \mathbb{N}.$$

Thus $\{x_n : n \in \mathbb{N}\} \subseteq \mathrm{Ker}(x^* - y^*)$. Since $\mathrm{Ker}(x^* - y^*)$ is a closed subspace of $(X, \|\cdot\|)$,

$$X = \overline{\mathrm{span}}\{x_n : n \in \mathbb{N}\} \subseteq \mathrm{Ker}(x^* - y^*),$$

that is, $x^* = y^*$. Thus d separates the points of B_{X^*}. Next, we show that d generates the relative weak* topology on B_{X^*}. Fix $y^* \in B_{X^*}$. Then: (i) the mapping $x^* \mapsto |\widehat{x_n}(x^* - y^*)|/2^n$ is weak* continuous on B_{X^*}; (ii) $0 \le |\widehat{x_n}(x^* - y^*)|/2^n \le 1/2^{n-1}$ for all $x^* \in B_{X^*}$ and $n \in \mathbb{N}$; and (iii) $\sum_{n\in\mathbb{N}} 1/2^{n-1} = 2 < \infty$. Therefore by the Weierstrass M-test the function $x^* \mapsto d(x^*, y^*)$ is weak* continuous on B_{X^*}. Hence, for all $y^* \in B_{X^*}$ and $\varepsilon > 0$, $B_d(y^*, \varepsilon) := \{x^* \in B_{X^*} : d(x^*, y^*) < \varepsilon\}$ is weak* open in B_{X^*}. So the metric topology τ_d is weaker (or equal to) the relative weak* topology on B_{X^*}. Therefore the mapping $I : (B_{X^*}, \text{weak}^*) \to (B_{X^*}, \tau_d)$ defined by $I(x^*) := x^*$ for $x^* \in B_{X^*}$ is one-to-one, continuous and onto. Since (B_{X^*}, weak^*) is compact and (B_{X^*}, τ_d) is Hausdorff (and so every compact subset of (B_{X^*}, τ_d) is closed), $I(C)$ is closed in (B_{X^*}, τ_d) for each weak* closed subset C of B_{X^*}, that is, I maps closed sets to closed sets. This shows that I is a homeomorphism. Thus, (B_{X^*}, weak^*) is a compact metrizable topological space. □

Theorem 3.3.39 ([245, Corollary 10]). *Let $(X, \|\cdot\|)$ be a separable Asplund space. Then $(X^*, \|\cdot\|)$ is norm separable.*

Proof. Suppose, to obtain a contradiction, that $(X^*, \|\cdot\|)$ is not separable. Then it follows that B_{X^*} is not norm separable. Now recall that for $\varepsilon > 0$, an *ε-net* in a metric space (M, d) is any subset S of M such that for every pair of distinct elements $x, y \in S$, $\varepsilon < d(x, y)$. By Zorn's lemma, for each $\varepsilon > 0$ and each metric space (M, d), there exists a maximal ε-net in M. Note that if S is a maximal ε-net in (M, d), then $M \subseteq \bigcup_{x\in S} B[x, \varepsilon]$. Let us now return to our particular situation. For each $k \in \mathbb{N}$, let F_k be a maximal $(1/k)$-net in B_{X^*}. If for each $k \in \mathbb{N}$, F_k is at most countable, then $F := \bigcup_{k\in\mathbb{N}} F_k$ is an at most countable dense subset of B_{X^*}. Hence there must exist $n \in \mathbb{N}$ for which F_n is uncountable. Let $L := F_n$, and let $\varepsilon := 1/n$. Furthermore, let

$$\mathscr{U} := \{U : U \text{ is a relatively weak}^* \text{ open subset of } B_{X^*}, \text{ and } L \cap U \text{ is at most countable}\}.$$

Let $U_0 := \bigcup\{U : U \in \mathscr{U}\}$. Then U_0 is a relatively weak* open subset of B_{X^*} and $L \cap U_0$ is at most countable. To see this last statement, note that since (B_{X^*}, weak^*) is compact and metrisable, it is hereditarily Lindelöf (since every compact metric space is separable and every separable metric space is hereditarily Lindelöf). Therefore there exists an at most countable subset $C \subseteq \mathscr{U}$ such that $U_0 = \bigcup\{U : U \in C\}$. Therefore $L \cap U_0 = \bigcup\{L \cap U : U \in C\}$, which is at most countable as a countable union of countable sets. Thus we have shown that $U_0 \in \mathscr{U}$. In fact, U_0 is the largest element in the partially ordered set $(\mathscr{U}, \subseteq)$. Let $M := L \setminus U_0$. Then M is an uncountable ε-net in B_{X^*}, and for every relatively weak* open subset W of B_{X^*} such that $\varnothing \ne M \cap W$, $M \cap W$ is uncountable, as $W \not\subseteq U_0$ and so $W \notin \mathscr{U}$. Therefore, in particular, $\varepsilon < \|\cdot\|\text{-diam}[M \cap W]$. If we set $K := \overline{M}^{w^*}$, then K is a weak* compact subset of X^* such that $\varepsilon < \|\cdot\| - \mathrm{diam}[K \cap W]$ for every weak*

open subset of X^* with $K \cap W \neq \varnothing$. However, this contradicts part (ii) of Theorem 3.3.37. Hence $(X^*, \|\cdot\|)$ must be separable. □

Corollary 3.3.40 ([245, Corollary 10]). *Let $(X, \|\cdot\|)$ be a separable Banach space. Then $(X, \|\cdot\|)$ is an Asplund space if and only if $(X^*, \|\cdot\|)$ is separable.*

Proof. To get this result, we simply combine Theorems 3.3.36 and 3.3.39. □

Proposition 3.3.41 ([245, Theorem 12]). *Every closed subspace of an Asplund space is again an Asplund space. In particular, every closed separable subspace of an Asplund space has a separable dual space.*

Proof. Let $(Y, \|\cdot\|)$ be a closed subspace of an Asplund space $(X, \|\cdot\|)$. To show that $(Y, \|\cdot\|)$ is an Asplund space, we appeal to part (ii) of Theorem 3.3.37. To this end, let K be a nonempty weak* compact subset of Y^*, and let $\varepsilon > 0$. By the uniform boundedness theorem [75, p. 66] there exists $L >$ such that $\|y^*\| \leq L$ for all $y^* \in K$. Let $R : X^* \to Y^*$ be defined by $R(x^*) := x^*|_Y$ for $x^* \in X^*$. Then R is weak*-to-weak* continuous. Let $M := R^{-1}(K) \cap L \cdot B_{X^*}$. Then M is a weak* compact subset of X^*. Furthermore, $R|_M : M \to K$ is a weak*-to-weak* continuous surjection. The fact that $R|_M$ is surjective follows from the Hahn–Banach theorem (Theorem 3.3.5), since every $y^* \in K$ has a norm-preserving extension to a continuous linear functional of $(X, \|\cdot\|)$. In fact, $R|_M$ is a perfect mapping. Let $\Psi : K \to 2^{X^*}$ be defined by $\Psi(y^*) := (R|_M)^{-1}(y^*)$ for $y^* \in K$. Then Ψ is a weak*-cusco on (K, weak^*); see Example (iii) in Section 1.1.2. Therefore by Proposition 1.3.3 there exists a minimal weak*-usco $\Phi : K \to 2^{X^*}$ such that $\Phi(y^*) \subseteq \Psi(y^*)$ for all $y^* \in K$. Now, since $(X, \|\cdot\|)$ is an Asplund space, from part (iii) of Theorem 3.3.37 and the fact that (K, weak^*) is a Baire space it follows that there exists a dense G_δ-subset G of (K, weak^*) such that Φ is single-valued and norm upper semicontinuous at each point of G. In particular, there exists a nonempty relatively weak* open subset U of K such that $\|\cdot\| - \text{diam}[\Phi(U)] < \varepsilon$. Next, we note that for any $y_1^*, y_2^* \in K$ and any $x_1^* \in \Phi(y_1^*)$ and $x_2^* \in \Phi(y_2^*)$, we have

$$\|y_2^* - y_1^*\| = \|(R|_M)(x_2^*) - (R|_M)(x_1^*)\| \leq \|x_2^* - x_1^*\|.$$

Therefore, $\|\cdot\| - \text{diam}[U] \leq \|\cdot\| - \text{diam}[\Phi(U)] < \varepsilon$. This completes the proof. □

In the following part of this section, we will show that Asplund spaces are intimately related to spaces with the Radon–Nikodým property. We say that a Banach space $(X, \|\cdot\|)$ possesses the *Radon–Nikodým property* if for each nonempty bounded subset K of $(X, \|\cdot\|)$ and each positive real number $\varepsilon > 0$, there exist $x^* \in X^*$ and $\delta > 0$ such that the norm diameter of the slice $S(K, x^*, \delta)$ is less than ε. Our investigations concerning spaces with the Radon–Nikodým property involves the study of "trees".

For the remainder of this section, we denote by T the set of all finite sequences of 0s and 1s, including the sequence of length 0, which will be denoted by $\emptyset$. It $t \in T$ (i. e., $t := (t_1, t_2, \ldots, t_n)$ for some $n \in \mathbb{N}$, or $t = \emptyset$), then we define the *length* of t, denoted $|t|$, to be n if $t := (t_1, t_2, \ldots, t_n)$ for some $n \in \mathbb{N}$ or 0 if $t = \emptyset$. We will also write $t1$ for the sequence $(t_1, t_2, \dots, t_n, 1)$ and $t0$ for the sequence $(t_1, t_2, \dots, t_n, 0)$. Of course, if $|t| = 0$, then $\emptyset 1 = (1)$ and $\emptyset 0 = (0)$.

An *infinite tree* in a Banach space $(X, \|\cdot\|)$ is any function $f : T \to X$ such that

$$f(t) = \frac{1}{2}(f(t0) + f(t1)) \quad \text{for all } t \in T.$$

An infinite tree for which there exists $\delta > 0$ such that $2\delta < \|f(t0) - f(t1)\|$ for all $t \in T$ is called an *infinite δ-tree.*

Note that if $f : T \to X$ is a bounded infinite δ-tree for some $\delta > 0$, then every slice of $A := \{f(t) : t \in T\}$ has a diameter of at least δ. Indeed, if $S(A, x^*, \alpha)$ is a slice of A, then there exists $t \in T$ such that $f(t) \in S(A, x^*, \alpha)$. It then follows by a simple convexity argument that either $f(t0)$ or $f(t1)$ (or both) is an element of $S(A, x^*, \alpha)$. However, $\delta < \|f(t0) - f(t)\|$ and $\delta < \|f(t1) - f(t)\|$.

Thus there are no bounded infinite δ-trees in Banach spaces with the Radon–Nikodým property. The following lemma is based upon [257, Lemma 5.5].

Lemma 3.3.42. *Let $\{K_t : t \in T\}$ be a family of nonempty compact convex subsets of a linear topological space $(X, +, \cdot, \tau)$ such that $K_{t0} \cup K_{t1} \subseteq K_t$ for all $t \in T$. Then there exists an infinite tree $f : T \to X$ such that $f(t) \in K_t$ for all $t \in T$.*

Proof. Let $K := \prod_{t \in T} K_t$ be the Tychonoff product of all the sets $\{K_t : t \in T\}$. By Tychonoff's theorem (see [78]) K is nonempty and compact and consists of all functions $f : T \to X$ such that $f(t) \in K_t$ for all $t \in T$. Define, for each $n \in \{0, 1, 2, \dots\}$,

$$B_n := \{f \in K : f(t) = (1/2)(f(t0) + f(t1)) \text{ for all } t \in T \text{ with } |t| = n\}.$$

Each set B_n is closed and hence compact, and a function $f \in K$ is an infinite tree if $f \in \bigcap_{n=0}^{\infty} B_n$. Thus we need to show that $\bigcap_{n=0}^{\infty} B_n \neq \emptyset$. By compactness it suffices to prove that $B_0 \cap B_1 \cap \dots \cap B_k \neq \emptyset$ for each $k \in \{0, 1, 2, \dots, \}$. To this end, fix $k \in \{0, 1, 2, \dots, \}$ and define $f \in K$ as follows. If $t \in T$ and $k < |t|$, then let $f(t)$ be any element of K_t. If $t \in T$ and $|t| = k$, then we define $f(t) := (1/2)(f(t0) + f(t1)) \in K_t$. Now this is possible since both $f(t0)$ and $f(t1)$ are already defined and $(1/2)K_{t0} + (1/2)K_{t1} \subseteq K_t$. Next, consider $t \in T$ with $|t| = k - 1$. Then we define $f(t) := (1/2)(f(t0) + f(t1)) \in K_t$. Again, this is possible since $f(t0)$ and $f(t1)$ were previously defined and $(1/2)K_{t0} + (1/2)K_{t1} \subseteq K_t$. Continue on in this fashion until we reach the case where $|t| = 0$. Then $t = \emptyset$, and we define $f(\emptyset) := (1/2)(f((0)) + f((1))) \in K_\emptyset$. Then $f \in B_0 \cap B_1 \cap \dots \cap B_k$. This completes the proof. □

Lemma 3.3.43. *Let $(X, \|\cdot\|)$ be a Banach space and suppose that there exist a nonempty weak* compact (and hence bounded) subset K of X^* and $\varepsilon > 0$ such that $\varepsilon < \|\cdot\|$ – diam$[U]$ whenever U is a nonempty relatively weak* open subset of K. Then there exists an infinite $(\varepsilon/2)$-tree contained in $\overline{\text{co}}^{w^*}(K)$.*

Proof. We will inductively (on the length $|t|$ of $t \in T$) define the nonempty relatively weak* open subsets U_t of K and elements $x_t \in S_X$ that fulfil the following properties:

(a) $U_{t0} \cup U_{t1} \subseteq U_t$;
(b) if $x^* \in \overline{\text{co}}^{w^*}(U_{t0})$ and $y^* \in \overline{\text{co}}^{w^*}(U_{t1})$, then $\varepsilon \leq (x^* - y^*)(x_t)$.

First we set $U_\emptyset := K$.

Base Step. By assumption $\varepsilon < \|\cdot\| - \text{diam}[U_\emptyset]$, and hence there exist $x^*, y^* \in U_\emptyset$ such that $\varepsilon < \|x^* - y^*\|$. Choose $x_\emptyset \in S_X$ such that $(x^* - y^*)(x_\emptyset) = \varepsilon + \delta$ for some $\delta > 0$. Let

$$U_{(1)} := \{z^* \in U_\emptyset : z^*(x_\emptyset) > x^*(x_\emptyset) - \delta/2\} \quad \text{and} \quad U_{(0)} := \{z^* \in U_\emptyset : z^*(x_\emptyset) < y^*(x_\emptyset) + \delta/2\}.$$

Then

$$\overline{\text{co}}^{w^*}(U_{(1)}) \subseteq \{z^* \in X^* : z^*(x_\emptyset) \geq x^*(x_\emptyset) - \delta/2\}$$

and

$$\overline{\text{co}}^{w^*}(U_{(0)}) \subseteq \{z^* \in X^* : z^*(x_\emptyset) \leq y^*(x_\emptyset) + \delta/2\}.$$

Clearly, $U_{(0)}$ and $U_{(1)}$ are nonempty relatively weak* open subsets of K, and $x_\emptyset \in S_X$, which together satisfy properties (a) and (b).

Now suppose that U_t and $x_{t'}$ are defined for all $t \in T$ with $|t| \leq n$ and all $t' \in T$ with $|t'| < n$.

Inductive step. Consider $t \in T$ of length n. By assumption $\varepsilon < \|\cdot\| - \text{diam}[U_t]$, and hence there exist $x^*, y^* \in U_t$ such that $\varepsilon < \|x^* - y^*\|$. Choose $x_t \in S_X$ such that $(x^* - y^*)(x_t) = \varepsilon + \delta$ for some $\delta > 0$. Let

$$U_{t1} := \{z^* \in U_t : z^*(x_t) > x^*(x_t) - \delta/2\} \quad \text{and} \quad U_{t0} := \{z^* \in U_t : z^*(x_t) < y^*(x_t) + \delta/2\}.$$

Then

$$\overline{\text{co}}^{w^*}(U_{t1}) \subseteq \{z^* \in X^* : z^*(x_t) \geq x^*(x_t) - \delta/2\}$$

and

$$\overline{\text{co}}^{w^*}(U_{t0}) \subseteq \{z^* \in X^* : z^*(x_t) \leq y^*(x_t) + \delta/2\}.$$

Clearly, U_{t0} and U_{t1} are nonempty relatively weak* open subsets of K, and $x_t \in S_X$, which together satisfy properties (a) and (b).

Now, for each $t \in T$, let K_t denote the weak* closed convex hull of U_t. Each K_t is nonempty, weak* compact and convex, and (a) implies that $K_{t0} \cup K_{t1} \subseteq K_t$ for all $t \in T$. Thus by Lemma 3.3.42 there exists an infinite tree in X^* such that $f(t) \in K_t$ for all $t \in T$. Note that by (b) $\varepsilon \leq \|f(t0) - f(t1)\|$ for all $t \in T$, and so f is an infinite $(\varepsilon/2)$-tree contained in the weak* closed convex hull of K. □

Theorem 3.3.44 (Stegall's Asplund space characterisation [290]). *Let* $(X, \|\cdot\|)$ *be a Banach space. Then* $(X, \|\cdot\|)$ *is an Asplund space if and only if* $(X^*, \|\cdot\|)$ *has the Radon–Nikodým property.*

Proof. Suppose first that $(X, \|\cdot\|)$ is an Asplund space. Suppose also that A is a nonempty bounded subset of X^* and $\varepsilon > 0$. Set $K := \overline{A}^{w^*}$. Then K is a nonempty weak* compact subset of X^*. Hence by part (i) of Theorem 3.3.37 there exist $x \in S_X$ and $\delta > 0$ such that the norm diameter of $S(K, \widehat{x}, \delta)$ is less than ε. Now since A is weak* dense in K,

$$\sup\{\widehat{x}(x^*) : x^* \in A\} = \sup\{\widehat{x}(x^*) : x^* \in K\} = \max\{\widehat{x}(x^*) : x^* \in K\}.$$

Hence $S(A, \widehat{x}, \delta) \subseteq S(K, \widehat{x}, \delta)$, and so the norm diameter of $S(A, \widehat{x}, \delta)$ is less than ε too. Therefore $(X^*, \|\cdot\|)$ possesses the Radon–Nikodým property.

Next, we consider the converse. Suppose that $(X^*, \|\cdot\|)$ possesses the Radon–Nikodým property, but, to obtain a contradiction, let us also suppose that $(X, \|\cdot\|)$ is not an Asplund space. Then by Theorem 3.3.37, part (ii), there exist a nonempty weak* compact subset K of X^* and $\varepsilon > 0$ such that every nonempty relatively weak* open subset of K has diameter greater than ε. Thus by Lemma 3.3.43 there exists a bounded infinite $(\varepsilon/2)$-tree in X^*. However, as we deduced in the discussion just prior to Lemma 3.3.42, this implies that $(X^*, \|\cdot\|)$ does not possess the Radon–Nikodým property. Hence our assumption that $(X, \|\cdot\|)$ is not an Asplund space is false, that is, $(X, \|\cdot\|)$ is indeed an Asplund space. □

Theorem 3.3.45. *Let* $(X, \|\cdot\|)$ *be a Banach space. Then* $(X, \|\cdot\|)$ *is an Asplund space if every closed separable subspace of* $(X, \|\cdot\|)$ *is an Asplund space, that is,* $(X, \|\cdot\|)$ *is an Asplund space if every closed separable subspace of* $(X, \|\cdot\|)$ *has a separable dual space.*

Proof. Let us suppose, to obtain a contradiction, that every closed separable subspace of $(X, \|\cdot\|)$ is an Asplund space but $(X, \|\cdot\|)$ is not an Asplund space. Then by Theorem 3.3.37, part (ii), there exist a nonempty weak* compact subset K of X^* and $\varepsilon > 0$ such that every nonempty relatively weak* open subset of K has diameter greater than ε. Thus by Lemma 3.3.43 there exists a bounded infinite $(\varepsilon/2)$-tree $f : T \to \overline{\text{co}}^{w^*}(K)$. By re-examining the proof of Lemma 3.3.43 we see that in addition to the infinite $(\varepsilon/2)$-tree f, we also constructed the elements $\{x_t : t \in T\}$ in S_X such that $0 < \varepsilon < (f(t1) - f(t0))(x_t)$ for all $t \in T$. Let $Y := \overline{\text{span}}\{x_t : t \in T\}$. Since T is countable, $(Y, \|\cdot\|)$ is a separable closed subspace of $(X, \|\cdot\|)$. Let $R : X^* \to Y^*$ be defined by $R(x^*) := x^*|_Y$. Then R is a bounded linear operator from X^* into Y^*. Therefore $(R \circ f) : T \to Y^*$ is a bounded infinite tree in Y^*. Moreover, since

$$0 < \varepsilon < (f(t1) - f(t0))(x_t) = ((R \circ f)(t1) - (R \circ f)(t0))(x_t) \quad \text{for all } t \in T,$$

we have that $\varepsilon < \|(R \circ f)(t1) - (R \circ f)(t0)\|$ for all $t \in T$, that is, $(R \circ f) : T \to Y^*$ is a bounded infinite $(\varepsilon/2)$-tree in Y^*. Hence by the discussion just prior to Lemma 3.3.42

this implies that $(Y^*, \|\cdot\|)$ does not possess the Radon–Nikodým property. Therefore, by Theorem 3.3.44, $(Y, \|\cdot\|)$ is not an Asplund space. However, this contradicts our assumption that every closed separable subspace of $(X, \|\cdot\|)$ is an Asplund space. Thus $(X, \|\cdot\|)$ is indeed an Asplund space. □

Corollary 3.3.46 ([290, Theorem 2]). *Let $(X, \|\cdot\|)$ be a Banach space. Then $(X, \|\cdot\|)$ is an Asplund space if and only if every closed separable subspace of $(X, \|\cdot\|)$ has a separable dual space.*

Proof. Suppose that $(X, \|\cdot\|)$ is an Asplund space. Then by Proposition 3.3.41 every closed separable subspace of $(X, \|\cdot\|)$ has a separable dual. The converse statement follows from Theorem 3.3.45. □

Lemma 3.3.47. *Let $\Phi : M \to 2^{X^*}$ be a weak*-usco acting from a metric space (M, d) into subsets of the dual of a Banach space $(X, \|\cdot\|)$. Then Φ is locally bounded on (M, d).*

Proof. Let $\Phi : M \to 2^{X^*}$ be a weak*-usco from a metric space (M, d) into subsets of the dual of a Banach space $(X, \|\cdot\|)$. Let $x_0 \in M$, and let us suppose, to obtain a contradiction, that Φ is not locally bounded at x_0. Then for every $n \in \mathbb{N}$, $\Phi(B(x_0, 1/n)) \not\subseteq nB_{X^*}$. Hence there exist sequences $(x_n : n \in \mathbb{N})$ in M and $(x_n^* : n \in \mathbb{N})$ in X^* such that $\|x_n - x_0\| < 1/n$ and $x_n^* \in \Phi(x_n) \setminus nB_{X^*}$ for all $n \in \mathbb{N}$. Note that (i) $(x_n : n \in \mathbb{N})$ converges to x_0 and (ii) $(x_n^* : n \in \mathbb{N})$ is unbounded. Let $K := \{x_0\} \cup \{x_n : n \in \mathbb{N}\}$. Then K is compact. Therefore, by Proposition 1.1.7, $\Phi(K)$ is weak*-compact and hence, by the uniform boundedness theorem [75, p. 66], bounded. However, this is impossible since $\{x_n^* : n \in \mathbb{N}\} \subseteq \Phi(K)$. Thus Φ is locally bounded at x_0. Since the point $x_0 \in M$ was arbitrary, we have the desired result that Φ is locally bounded on (M, d). □

Theorem 3.3.48. *Let $(X, \|\cdot\|)$ be an Asplund space. Then every weak*-usco $\Phi : M \to 2^{X^*}$ from a complete metric space (M, d) into subsets of X^* admits a Baire one selection.*

Proof. To prove this theorem, we appeal to the Jayne–Rogers selection theorem (see Theorem 1.4.28). To this end, let (A, ρ) be a nonempty complete metric space, and let $\Phi : A \to 2^{X^*}$ be a minimal weak*-usco. By Lemma 3.3.47, Φ is locally bounded on (A, ρ). Hence, by Theorem 3.3.37, part (iii), Φ is single-valued and norm upper semicontinuous at the points of a dense G_δ-subset of (A, ρ). (Here we used the fact that every nonempty complete metric space is a Baire space; see Corollary 1.4.24.) The result now follows from the Jayne–Rogers selection theorem. □

3.4 Variational analysis

Let X be a nonempty set. We will consider the vector space $\mathbb{R}^X$ of all real-valued functions defined on X endowed with the operations of pointwise addition and pointwise scalar multiplication. If A is a nonempty subset of X, then we call the weak topology on $\mathbb{R}^X$, generated by $\{\delta_a : a \in A\}$ the *topology of pointwise convergence on A*, where for each $a \in A$, $\delta_a : \mathbb{R}^X \to \mathbb{R}$ is defined by

$$\delta_a(f) := f(a) \quad \text{for } f \in \mathbb{R}^X.$$

We denote the topology of pointwise convergence on A by $\tau_p(A)$.

Let X be any nonempty set, and let $Y \subseteq \mathbb{R}^X$. Throughout this section, the family Y of real-valued functions (i. e. the set of all perturbations) will be considered with pointwise addition, pointwise scalar multiplication and with a complete norm $\|\cdot\|$ whose topology on Y is at least as strong as the $\tau_p(X)$ topology on Y. We emphasise here, right at the outset, that there is at most one norm $\|\cdot\|$, up to equivalence of norms, that can be placed on Y with this property. In this way, we see that the norm on the set Y of all perturbations is completely determined by the set Y itself.

Proposition 3.4.1 ([235, Proposition 3.1]). *Let X be a nonempty set, and let $Y \subseteq \mathbb{R}^X$. If $(Y, \|\cdot\|_1)$ and $(Y, \|\cdot\|_2)$ are both Banach spaces under pointwise addition and pointwise scalar multiplication and if the topologies of both $\|\cdot\|_1$ and $\|\cdot\|_2$ are at least as strong as the $\tau_p(X)$-topology on Y, then $\|\cdot\|_1$ and $\|\cdot\|_2$ are equivalent norms (i. e. there is at most one complete norm on Y, up to equivalence of norms, whose topology is at least as strong as $\tau_p(X)$).*

Proof. Suppose that $(Y, \|\cdot\|_1)$ and $(Y, \|\cdot\|_2)$ are Banach spaces and that the topologies of both norms $\|\cdot\|_1$ and $\|\cdot\|_2$ are at least as strong as $\tau_p(X)$. Then consider the identity mapping $I : (Y, \|\cdot\|_1) \to (Y, \|\cdot\|_2)$. Note that since the $\tau_p(X)$-topology on Y is Hausdorff,

$$\mathrm{Gr}(I) = \Delta_Y = \{(x, y) \in Y \times Y : x = y\}$$

is closed in $Y \times Y$ when $Y \times Y$ is equipped with the product topology generated by $\tau_p(X)$. Hence $\mathrm{Gr}(I)$ is closed in $Y \times Y$ when $Y \times Y$ is endowed with the product topology generated by $(Y, \|\cdot\|_1)$ and $(Y, \|\cdot\|_2)$. Therefore, since $I : (Y, \|\cdot\|_1) \to (Y, \|\cdot\|_2)$ is linear, we have by the closed graph theorem [75, p. 57] that $I : (Y, \|\cdot\|_1) \to (Y, \|\cdot\|_2)$ is bounded. Of course, the same argument applies to $I^{-1} : (Y, \|\cdot\|_2) \to (Y, \|\cdot\|_1)$, and so the norms $\|\cdot\|_1$ and $\|\cdot\|_2$ are equivalent. This completes the proof. □

In the setting described above, it is possible to consider the canonical mapping $\widehat{\cdot} : X \to Y^*$ defined by $\widehat{x}(y) := y(x)$ for $y \in Y$. Note that this mapping is well-defined since for each $x \in X$, the mapping $\widehat{x} : Y \to \mathbb{R}$ is linear, $\tau_p(X)$-continuous and hence norm continuous, since we are assuming that the norm topology on Y is at least as strong as the $\tau_p(X)$-topology on Y.

Next, we consider the function $\rho_Y : X \times X \to [0, \infty)$ defined by

$$\rho_Y(x, z) := \|\widehat{x} - \widehat{z}\| = \sup\{|y(x) - y(z)| : y \in B_Y\} \quad \text{for } x, z \in X.$$

It is not hard to see that ρ_Y is a pseudometric on X, and if Y separates the points on X, then it is a metric on X. Furthermore, the topology induced by the pseudometric ρ_Y

on X is always at least as strong as the $\sigma(X, Y)$-topology on X, that is, the weak topology on X generated by Y.

We shall say that a function $f : X \to [-\infty, \infty)$ attains (or has) a *strong maximum at* $x_0 \in X$ *with respect to* ρ_Y if

$$f(x_0) = \sup_{x \in X} f(x) \quad \text{and} \quad \lim_{n\to\infty} \rho_Y(x_n, x_0) = 0$$

whenever $(x_n : n \in \mathbb{N})$ is a sequence in X such that

$$\lim_{n\to\infty} f(x_n) = \sup_{x \in X} f(x) = f(x_0).$$

To expedite our results on maxima, we will utilise the theory of conjugates, but first let us recall that: (i) for any function $f : X \to [-\infty, \infty]$ defined on a nonempty set X,

$$\mathrm{Dom}(f) := \{x \in X : f(x) \in \mathbb{R}\};$$

(ii) a function $f : X \to [-\infty, \infty]$ is said to be a *proper function* if $\mathrm{Dom}(f) \neq \varnothing$; (iii) a function $f : X \to (-\infty, \infty]$ defined on a vector space $(V, +, \cdot)$ over $\mathbb{R}$ is said to be a *convex function* if for all $x, y \in \mathrm{Dom}(f)$ and $0 < \lambda < 1$, $f(\lambda x + (1-\lambda)y) \leq \lambda f(x) + (1-\lambda) f(y)$; and (iv) for any function $f : X \to (-\infty, \infty]$ defined on a nonempty set X,

$$\mathrm{argmax}(f) := \{x \in X : f(y) \leq f(x) \text{ for all } y \in X\}.$$

With these preliminary definitions under our belt, we proceed to the definition of a conjugate mapping. Let X be an arbitrary set, and let $Y \subseteq \mathbb{R}^X$. For a proper function $f : X \to (-\infty, \infty]$, the *conjugate* $f_Y^* : Y \to (-\infty, \infty]$ is defined by

$$f_Y^*(y) := \sup_{x \in X} [y(x) - f(x)] = \sup_{x \in \mathrm{Dom}(f)} [y(x) - f(x)] \quad \text{for } y \in Y.$$

In the classical situation where $(X, \|\cdot\|)$ is a Banach space and $(Y, \|\cdot\|)$ is the dual space of $(X, \|\cdot\|)$, we simply write f^* for the "Fenchel conjugate/Fenchel transform" [35, 42, 272, 274] instead of the more complicated notation $f_{X^*}^*$.

Our results on conjugate mappings depend upon the notion of ε-subderivatives. So we next give the definition of an ε-subderivative.

Let $f : X \to (-\infty, \infty]$ be a proper function on a normed linear space $(X, \|\cdot\|)$ and $x \in \mathrm{Dom}(f)$. Then for any $\varepsilon > 0$, we define the *ε-subdifferential* $\partial_\varepsilon f(x)$ by

$$\partial_\varepsilon f(x) := \{x^* \in X^* : x^*(y - x) \leq f(y) - f(x) + \varepsilon \text{ for all } y \in \mathrm{Dom}(f)\}.$$

Proposition 3.4.2. *Let $f : X \to (-\infty, \infty]$ be a proper function defined on a nonempty set X, and let $(Y, \|\cdot\|)$ be a Banach space of real-valued functions defined on X under pointwise addition and pointwise scalar multiplication. Suppose that the norm-topology on Y is at least as strong as the $\tau_p(X)$-topology on Y. Then the conjugate f_Y^* satisfies the following properties:*

(i) *f_Y^* is a convex, and $\tau_p(X)$-lower semicontinuous function on $Dom(f_Y^*)$;*
(ii) *f_Y^* is continuous on $int(Dom(f_Y^*))$;*
(iii) *If $y \in Dom(f_Y^*)$ and $x \in argmax(y - f)$, then $\widehat{x} \in \partial f_Y^*(y)$;*
(iv) *If $\varepsilon > 0$, $y \in Dom(f_Y^*)$, $x \in X$, and $f_Y^*(y) - \varepsilon < y(x) - f(x)$, then $\widehat{x} \in \partial_\varepsilon f_Y^*(y)$;*
(v) *If $y_0 \in int(Dom(f_Y^*))$, $x_0 \in argmax(y_0 - f)$, and $y \mapsto \partial f_Y^*(y)$ is single-valued and norm-upper semicontinuous at y_0, then $y_0 - f$ has a strong maximum at x_0 with respect to ρ_Y.*

Proof. People familiar with the Fenchel conjugate may skip the proofs of (i)–(iv).
(i) For each $x \in \mathrm{Dom}(f)$, define $g_x : Y \to \mathbb{R}$ by

$$g_x(y) := \widehat{x}(y) - f(x) \quad \text{for } y \in Y.$$

Then each function g_x is $\tau_p(X)$-continuous and affine. Now for each $y \in Y$,

$$f_Y^*(y) = \sup_{x \in \mathrm{Dom}(f)} g_x(y).$$

Thus, as the pointwise supremum of a family of $\tau_p(X)$-continuous affine mappings, the conjugate of f is itself convex and $\tau_p(X)$-lower semicontinuous. (Recall the general fact that the pointwise supremum of a family of convex functions is convex and the pointwise supremum of a family of lower semicontinuous mappings is again lower semicontinuous.)
(ii) Since this statement is vacuously true when $\mathrm{int}(\mathrm{Dom}(f_Y^*)) = \varnothing$, we will assume that $\mathrm{int}(\mathrm{Dom}(f_Y^*))$ is nonempty. Let us first recall that by Proposition 3.3.6 and the fact that f_Y^* and $\mathrm{int}(\mathrm{Dom}(f_Y^*))$ are convex it is sufficient to show that f_Y^* is locally bounded above on $\mathrm{int}(\mathrm{Dom}(f_Y^*))$. In fact, as we will now show, it is sufficient to show that f_Y^* is locally bounded above at a single point $y_0 \in \mathrm{int}(\mathrm{Dom}(f_Y^*))$. To this end, suppose that f_Y^* is locally bounded above at $y_0 \in \mathrm{int}(\mathrm{Dom}(f_Y^*))$. Then there exist $M > 0$ and $\delta > 0$ such that $f_Y^*(y) \le M$ for all $y \in B[y_0, \delta]$. Let x be any point in $\mathrm{int}(\mathrm{Dom}(f_Y^*))$. Since $\mathrm{int}(\mathrm{Dom}(f_Y^*))$ is an open convex set, there exist a point $y \in \mathrm{int}(\mathrm{Dom}(f_Y^*))$ and $0 < \lambda < 1$ such that $x = \lambda y + (1-\lambda) y_0$. Let $M^* := \max\{M, f_Y^*(y)\}$ and note that

$$x \in B[x, (1-\lambda)\delta] = \lambda y + (1-\lambda) B[y_0, \delta] \subseteq \mathrm{int}(\mathrm{Dom}(f_Y^*)), \quad \text{as } \mathrm{int}(\mathrm{Dom}(f_Y^*)) \text{ is convex.}$$

We claim that f_Y^* is bounded above by M^* on $B[x, (1-\lambda)\delta]$. To see this, let z be any element of $B[x, (1-\lambda)\delta]$. Then $z = \lambda y + (1-\lambda) w$ for some $w \in B[y_0, \delta]$ since

$$B[x, (1-\lambda)\delta] = x + (1-\lambda) B[0, \delta] = (\lambda y + (1-\lambda) y_0) + (1-\lambda) B[0, \delta] = \lambda y + (1-\lambda) B[y_0, \delta].$$

Therefore

$$f_Y^*(z) = f_Y^*(\lambda y + (1-\lambda) w) \le \lambda f_Y^*(y) + (1-\lambda) f_Y^*(w) \le \lambda M^* + (1-\lambda) M^* = M^*.$$

Next, we will use the fact that since $\mathrm{int}(\mathrm{Dom}(f_Y^*))$ is a nonempty open subset of a complete metric space, it is itself a Baire space with the relative topology. Now for $n \in \mathbb{N}$, let

$$F_n := \{x \in \mathrm{int}(\mathrm{Dom}(f_Y^*)) : f_Y^*(x) \leq n\}.$$

Since f_Y^* is $\tau_p(X)$-lower semicontinuous, it is lower semicontinuous with respect to the norm topology too. Therefore each set F_n is closed with respect to the relative norm topology on $\mathrm{int}(\mathrm{Dom}(f_Y^*))$. Since $\mathrm{int}(\mathrm{Dom}(f_Y^*)) = \bigcup_{n\in\mathbb{N}} F_n$, there exists $n_0 \in \mathbb{N}$ such that $\mathrm{int}(F_{n_0}) \neq \varnothing$. Hence f_Y^* is locally bounded above at each point of $\mathrm{int}(F_{n_0})$. This completes the proof of part (ii).

(iii) Let $y \in \mathrm{Dom}(f_Y^*)$ and suppose that $x \in \mathrm{argmax}(y - f)$. Let z be any element of $\mathrm{Dom}(f_Y^*)$. Then

$$\begin{aligned}\widehat{x}(z) - \widehat{x}(y) &= z(x) - y(x)\\ &= [z(x) - f(x)] - [y(x) - f(x)]\\ &= [z(x) - f(x)] - f_Y^*(y)\\ &\leq f_Y^*(z) - f_Y^*(y).\end{aligned}$$

Therefore $\widehat{x} \in \partial f_Y^*(y)$.

(iv) Suppose that $\varepsilon > 0$, $y \in \mathrm{Dom}(f_Y^*)$, and $(f_Y^*(y) - \varepsilon) < y(x) - f(x)$. Let z be any element of $\mathrm{Dom}(f_Y^*)$. Then

$$\begin{aligned}\widehat{x}(z) - \widehat{x}(y) &= z(x) - y(x)\\ &= [z(x) - f(x)] - [y(x) - f(x)]\\ &\leq [z(x) - f(x)] - [f_Y^*(y) - \varepsilon]\\ &\leq f_Y^*(z) - f_Y^*(y) + \varepsilon.\end{aligned}$$

Therefore $\widehat{x} \in \partial_\varepsilon f_Y^*(y)$.

(v) Let $(x_n : n \in \mathbb{N})$ be a sequence in X such that

$$\lim_{n\to\infty} (y_0 - f)(x_n) = \sup_{x\in X}(y_0 - f)(x) = f_Y^*(y_0).$$

We will show that $(x_n : n \in \mathbb{N})$ converges to x_0 with respect to the metric ρ_Y. Let $\varepsilon > 0$, and let $0 < \varepsilon' < \varepsilon/2$. By (iii) and the assumption that $\partial f_Y^*(y_0)$ is a singleton we have that $\partial f_Y^*(y_0) = \{\widehat{x_0}\}$. Since $y \mapsto \partial f_Y^*(y)$ is norm-upper semicontinuous at y_0, there exists $\delta > 0$ such that if $\|y - y_0\| \leq \delta$, then $\|F - \widehat{x_0}\| < \varepsilon'$ for all $F \in \partial f_Y^*(y)$. Choose $N \in \mathbb{N}$ such that $f_Y^*(y_0) - \varepsilon'\delta < (y_0 - f)(x_n)$ for all $N < n$. Then, by (iv), $\widehat{x_n} \in \partial_{\varepsilon'\delta} f_Y^*(y_0)$ for all $n > N$. Fix $n > N$ and let $v \in S_Y$. Then

$$\widehat{x_n}(\delta v) - \widehat{x_0}(\delta v) \leq [f_Y^*(y_0 + \delta v) - f_Y^*(y_0) + \varepsilon'\delta] - \widehat{x_0}(\delta v). \qquad (*)$$

Let $F \in \partial f_Y^*(y_0 + \delta v)$. Then

$$F(-\delta v) \le f_Y^*(y_0) - f_Y^*(y_0 + \delta v)$$

or, equivalently,

$$f_Y^*(y_0 + \delta v) - f_Y^*(y_0) \le F(\delta v). \qquad (**)$$

Substituting inequality $(**)$ into inequality $(*)$, we get that

$$\widehat{x_n}(\delta v) - \widehat{x_0}(\delta v) \le (F - \widehat{x_0})(\delta v) + \varepsilon'\delta \le \|F - \widehat{x_0}\| \cdot \delta + \varepsilon'\delta < \varepsilon'\delta + \varepsilon'\delta.$$

Hence $(\widehat{x_n} - \widehat{x_0})(v) = \widehat{x_n}(v) - \widehat{x_0}(v) \le 2\varepsilon' < \varepsilon$. Since $v \in S_Y$ was arbitrary, $\|\widehat{x_n} - \widehat{x_0}\| < \varepsilon$, and so $\rho_Y(x_n, x_0) = \|\widehat{x_n} - \widehat{x_0}\| < \varepsilon$.

This completes the proof. □

Theorem 3.4.3 (First variational theorem). *Let $f : X \to (-\infty, \infty]$ be a proper function defined on a nonempty set X, and let $(Y, \|\cdot\|)$ be a Banach space of real-valued functions defined on X under pointwise addition and pointwise scalar multiplication. Suppose that the norm-topology on Y is at least as strong as the $\tau_p(X)$-topology on Y and that (X, ρ_Y) is separable. If there exist a nonempty open subset A of $\mathrm{Dom}(f_Y^*)$ and a dense G_δ-subset R of A such that $\mathrm{argmax}(y - f) \ne \varnothing$ for each $y \in R$, then there exists a dense G_δ-subset R' of A such that $(y - f) : X \to [-\infty, \infty)$ has a strong maximum with respect to ρ_Y for each $y \in R'$. In addition, if $0 \in A$ and $\varepsilon > 0$, then there exists $y_0 \in Y$ with $\|y_0\| < \varepsilon$ such that $(y_0 - f) : X \to [-\infty, \infty)$ has a strong maximum with respect to ρ_Y.*

Proof. Let $n \in \mathbb{N}$, and let

$$O_n := \bigcup\{U \in 2^A : U \text{ is open and } \|\cdot\| - \mathrm{diam}[\partial f_Y^*(U)] \le 1/n\}.$$

Clearly, O_n is open, as it is a union of open sets. We will show that O_n is also dense in A. To this end, let W be a nonempty open subset of A, and let F be a first Baire category subset of Y such that $\mathrm{argmax}(y - f) \ne \varnothing$ for all $y \in A \setminus F$. Note that by Proposition 3.4.2, part (iii), $\partial f_Y^*(y) \cap \widehat{X} \ne \varnothing$ for all $y \in A \setminus F$. Since (X, ρ_Y) is separable, there exists a countable dense subset $\{x_n : n \in \mathbb{N}\}$ of (X, ρ_Y). Then it follows that $\{\widehat{x_n} : n \in \mathbb{N}\}$ is dense in $(\widehat{X}, \|\cdot\|)$. For each $n \in \mathbb{N}$, let $C_n := B[\widehat{x_n}, 1/2n]$. Then

$$\varnothing \ne \partial f_Y^*(y) \cap \widehat{X} \subseteq \partial f_Y^*(y) \cap \bigcup_{n \in \mathbb{N}} C_n \quad \text{for all } y \in A \setminus F.$$

Let us also recall from Corollary 3.3.15 that $y \mapsto \partial f_Y^*(y)$ is a minimal weak* cusco on A. Therefore by Proposition 3.3.23 there exists a first Baire category set F' such that for all $y \in A \setminus F'$ and $n \in \mathbb{N}$, if $\partial f_Y^*(y) \cap C_n \ne \varnothing$, then there exists an open neighbourhood N of y such that $\partial f_Y^*(N) \subseteq C_n$. Let $y_0 \in W \setminus (F \cup F')$. Since $\partial f_Y^*(y_0) \cap \bigcup_{n \in \mathbb{N}} C_n \ne \varnothing$, there exists $n_0 \in \mathbb{N}$ such that $\partial f_Y^*(y_0) \cap C_{n_0} \neq \varnothing$. Now, because $y_0 \notin F'$, there exists an open neighbourhood N of y_0 contained in W such that $\partial f_Y^*(N) \subseteq C_{n_0}$. Thus $\varnothing \neq N \subseteq W \cap O_n$. Hence O_n is dense in A. It now follows that $y \mapsto \partial f_Y^*(y)$ is single-valued and norm upper semicontinuous at each point of $\bigcap_{n\in\mathbb{N}} O_n$. Let $R' := R \cap \bigcap_{n\in\mathbb{N}} O_n$. Then R' contains a dense G_δ-subset of A, and by Proposition 3.4.2, part (v), $(y - f) : X \to [-\infty, \infty)$ has a strong maximum with respect to ρ_Y at each $y \in R'$. □

Our second variational theorem applies to dual differentiation spaces [93]. Recall that a Banach space $(X, \|\cdot\|)$ is called a *dual differentiability space (or DD-space for short)* if every continuous convex function $\varphi : A \to \mathbb{R}$ defined on a nonempty open convex subset A of $(X^*, \|\cdot\|)$ such that $\{x^* \in A : \partial\varphi(x^*) \cap \widehat{X} \neq \varnothing\}$ contains a dense G_δ-subset of A has the property that its subdifferential mapping $\partial\varphi : A \to 2^{X^{**}}$ is single-valued and norm upper semicontinuous at each point of a dense G_δ-subset of A (or, equivalently, φ is Fréchet differentiable at the points of a dense G_δ-subset of A; see Theorem 3.3.35). Let us also observe that if $(X, \|\cdot\|)$ is a Banach space, then $\rho_{X^*}(x, y) = \|x - y\|$ for all $x, y \in X$. Therefore, when the context is clear, we will simply say that a function $f : X \to [-\infty, \infty)$ has *strong maximum*, rather than has a strong maximum with respect to ρ_{X^*}.

We also say that a function $f : X \to (-\infty, \infty]$ has *strong minimum at $x_0 \in X$ with respect to ρ_Y* if

$$f(x_0) = \inf_{x\in X} f(x) \quad \text{and} \quad \lim_{n\to\infty} \rho_Y(x_n, x_0) = 0$$

whenever $(x_n : n \in \mathbb{N})$ is a sequence in X such that

$$\lim_{n\to\infty} f(x_n) = \inf_{x\in X} f(x) = f(x_0).$$

Note that a function $f : X \to (-\infty, \infty]$ has strong minimum with respect to ρ_Y if and only if $(-f) : X \to [-\infty, \infty)$ has strong maximum with respect to ρ_Y.

Theorem 3.4.4 (Second variational theorem). *Let $f : X \to (-\infty, \infty]$ be a proper function on a dual differentiation space $(X, \|\cdot\|)$. If there exists a nonempty open subset A of $Dom(f^*)$ and a dense G_δ-subset R of A such that $argmax(x^* - f) \neq \varnothing$ for each $x^* \in R$, then there exists a dense G_δ-subset R' of A such that $(x^* - f) : X \to [-\infty, \infty)$ has a strong maximum for each $x^* \in R'$. In addition, if $0 \in A$ and $\varepsilon > 0$, then there exists $x_0^* \in X^*$ with $\|x_0^*\| < \varepsilon$ such that $(x_0^* - f) : X \to [-\infty, \infty)$ has a strong maximum.*

Proof. Consider $\partial f^* : A \to 2^{X^{**}}$. Then, by Proposition 3.4.2, part (iii),

$$R_1 := \{x^* \in A : \partial f^*(x^*) \cap \widehat{X} \neq \varnothing\}$$

contains a dense G_δ-subset of A. Since $(X, \|\cdot\|)$ is a dual differentiation space,

$$R_2 := \{x^* \in A : \partial f^* \text{ is single-valued and norm upper semicontinuous at } x^*\}$$

contains a dense G_δ-subset of A. Let $R' := R_1 \cap R_2$. Then R' contains a dense G_δ-subset of A, and by Proposition 3.4.2, part (v), $(x^* - f)$ has a strong maximum for each $x^* \in R'$. □

Remarks 3.4.5. There are two main weaknesses of this theorem: (i) although it is known that many Banach spaces (e. g. all spaces with the Radon–Nikodým property [93], all weakly Lindelöf spaces [169], all spaces that admit an equivalent locally uniformly rotund norm [94] and all spaces whose dual space X^* is weak Asplund [93]) are dual differentiation spaces, it is still an open question as to whether every Banach space is a dual differentiation space; (ii) it is not clear how we would go about showing that there exists a "large" subset R of int(Dom(f^*)) such that $\operatorname{argmax}(x^* - f) \neq \varnothing$ for each $x^* \in R$.

Next, we give an application of the second variational theorem to the geometry of Banach spaces. However, to do this, we need the notion of a strongly exposed point.

Let C be a nonempty closed bounded convex subset of a normed linear space $(X, \|\cdot\|)$. We say that a point $x_0 \in C$ is a *strongly exposed point of* C if there exists $x^* \in X^*$ such that $x^*|_C$ has a strong maximum at x_0, and we denote by Exp(C) the set of all strongly exposed points of C. Note that if $f : X \to (-\infty, \infty]$ is defined by $f(x) := 0$ if $x \in C$ and $f(x) := \infty$ otherwise, then we have the following:

If $x^* \in X^*$ and $x^* - f$ has a strong maximum at $x_0 \in X$, then $x_0 \in C$, and x_0 is in fact a strongly exposed point of C.

Theorem 3.4.6 ([236]). *If C is a nonempty closed bounded convex subset of a DD-space $(X, \|\cdot\|)$ and*

$$\{x^* \in X^* : x^* \text{ attains its supremum over } C\}$$

is residual in $(X^*, \|\cdot\|)$, then $C = \overline{\text{co}}(Exp(C))$.

Proof. Let $f : X \to (-\infty, \infty]$ be defined by $f(x) := 0$ if $x \in C$ and $f(x) := \infty$ otherwise. Then by assumption

$$R := \{x^* \in X^* : \operatorname{argmax}(x^*|_C) \neq \varnothing\} = \{x^* \in X^* : \operatorname{argmax}(x^* - f) \neq \varnothing\}$$

is residual in X^*. Therefore by Theorem 3.4.4 there exists a dense G_δ-subset R' of X^* such that $(x^* - f)$ has a strong maximum for each $x^* \in R'$. Now suppose, to obtain a contradiction, that $C \neq \overline{\text{co}}(\text{Exp}(C))$. Then there exist $x_0 \in C \setminus \overline{\text{co}}(\text{Exp}(C))$ and $x^* \in X^*$ such that

$$\sup\{x^*(c) : c \in \overline{\text{co}}(\text{Exp}(C))\} < x^*(x_0).$$

Since C is bounded and R' is dense in X^*, we can assume, without loss of generality, that $x^* \in R'$. But then $\operatorname{argmax}(x^*|_C) = \operatorname{argmax}(x^* - f) =: \{x\}$ is a strong maximum of

$(x^* - f)$, and hence a strongly exposed point of C. On the other hand,

$$\sup\{x^*(c) : c \in \overline{\text{co}}(\text{Exp}(C))\} < x^*(x_0) \leq x^*(x);$$

which implies that $x \notin \text{Exp}(C)$. Thus $C = \overline{\text{co}}(\text{Exp}(C))$. □

For our third variational theorem, we need to recall some basic facts.

Lemma 3.4.7. *Let A be a nonempty bounded subset of the dual of a normed linear space $(X, \|\cdot\|)$. Then*

$$\|\cdot\| - diam[\overline{co}^{w^*}(A)] = \|\cdot\| - diam[A].$$

Proof. Clearly, $\|\cdot\| - \text{diam}[A] \leq \|\cdot\| - \text{diam}[\overline{\text{co}}^{w^*}(A)]$. So it is sufficient to show that $\|\cdot\| - \text{diam}[\overline{\text{co}}^{w^*}(A)] \leq \|\cdot\| - \text{diam}[A] =: d$. So let x^* and y^* be fixed but arbitrary elements of $\overline{\text{co}}^{w^*}(A)$. Next, let a be any element of A. Then

$$x^* \in \overline{\text{co}}^{w^*}(A) \subseteq B[a, d] \quad \text{since } B[a, d] \text{ is weak}^* \text{ closed and convex and } A \subseteq B[a, d].$$

Therefore, by symmetry, $a \in B[x^*, d]$. Since $a \in A$ was arbitrary, $A \subseteq B[x^*, d]$. Thus

$$y^* \in \overline{\text{co}}^{w^*}(A) \subseteq B[x^*, d] \quad \text{since } B[a, d] \text{ is weak}^* \text{ closed and convex and } A \subseteq B[x^*, d].$$

This shows that $\|y^* - x^*\| \leq d$. Since x^* and y^* were arbitrary elements of $\overline{\text{co}}^{w^*}(A)$,

$$\|\cdot\| - \text{diam}[\overline{\text{co}}^{w^*}(A)] \leq d = \|\cdot\| - \text{diam}[A].$$

This completes the proof. □

If we impose some stronger assumptions upon the function $f : X \to (-\infty, \infty]$, then we may extract more information about the mapping $x^* \mapsto \text{argmax}(x^* - f)$.

For example, if $f : X \to (-\infty, \infty]$ is a proper lower semicontinuous function defined on a Banach space $(X, \|\cdot\|)$, $x^* \in \text{Dom}(f^*)$ and $(x_n : n \in \mathbb{N})$ is any sequence in X such that

$$\lim_{n\to\infty}(x^* - f)(x_n) = \sup_{x\in X}(x^* - f),$$

then $\text{cl}_{\|\cdot\|}(x_n : n \in \mathbb{N}) \subseteq \text{argmax}(x^* - f)$, that is, all the cluster points of $(x_n : n \in \mathbb{N})$ lie in $\text{argmax}(x^* - f)$.

Theorem 3.4.8 (Third variational theorem). *Let $f : X \to (-\infty, \infty]$ be a proper lower semicontinuous function on a Banach space $(X, \|\cdot\|)$ with the Radon–Nikodým property. If there exists a nonempty open subset A of $\text{Dom}(f^*)$ and a dense subset D of A such that $\text{argmax}(x^* - f) \neq \varnothing$ for each $x^* \in D$, then there exists a dense G_δ-subset R of A such that $(x^* - f) : X \to [-\infty, \infty)$ has a strong maximum for each $x^* \in R$. In addition, if $0 \in A$ and $\varepsilon > 0$, then there exists $x_0^* \in X^*$ with $\|x_0^*\| < \varepsilon$ such that $(x_0^* - f) : X \to [-\infty, \infty)$ has a strong maximum.*

Proof. Define $J : A \to 2^X$ by $J(x^*) := \{x \in X : \widehat{x} \in \partial f^*(x^*)\}$ for $x^* \in A$. Note that by Proposition 3.4.2, part (iii), $\operatorname{argmax}(x^* - f) \subseteq J(x^*)$ for each $x^* \in A$. Therefore $J(x^*) \neq \varnothing$ for all $x^* \in D$. It follows from Proposition 3.3.22 that for any open subset U of A, $x^* \in X^*$ and $\alpha \in \mathbb{R}$, if $J(U) \cap \{x \in X : \alpha < x^*(x)\} \neq \varnothing$, that is, $J(U) \not\subseteq K := \{x \in X : x^*(x) \leq \alpha\}$, then there exists a nonempty open subset V of U such that

$$J(V) \subseteq \{x \in X : \alpha < x^*(x)\}, \quad \text{that is, } J(V) \cap K = \varnothing.$$

Next, we define (as we have done several times before)

$$O_\varepsilon := \bigcup \{U \in 2^A : U \text{ is open and } \|\cdot\| - \operatorname{diam}[\partial f^*(U)] < \varepsilon\}.$$

Clearly, O_ε is open. We will show that O_ε is also dense in A. To this end, let W be a nonempty open subset of A. By possibly making W smaller we may assume by Proposition 3.3.11 that $\partial f^*(W)$ is a norm bounded subset of $(X^{**}, \|\cdot\|)$. Therefore $J(W)$ is also a norm bounded subset (of $(X, \|\cdot\|)$). Since $(X, \|\cdot\|)$ has the Radon–Nikodým property, there exist $x^* \in X^*$ and $\delta > 0$ such that $\|\cdot\| - \operatorname{diam}[S(J(W), x^*, \delta)] < \varepsilon$. From our earlier observation concerning the mapping J we see that there exists a nonempty open subset V of W such that

$$\begin{aligned} J(V) &\subseteq J(W) \cap \{x \in X : (s - \delta) < x^*(x)\} \quad \text{where } s := \sup\{x^*(y) : y \in J(W)\} \\ &= S(J(W), x^*, \delta). \end{aligned}$$

Thus $\|\cdot\| - \operatorname{diam}[J(V)] < \varepsilon$. We now claim that $\partial f^*(V) \subseteq \overline{\operatorname{co}}^{w^*}(\widehat{J(V)})$. To see this, let us suppose, to obtain a contradiction, that $\partial f^*(V) \not\subseteq \overline{\operatorname{co}}^{w^*}(\widehat{J(V)})$. Then by Proposition 3.3.22 there exists a nonempty open subset V' of V such that $\partial f^*(V') \cap \overline{\operatorname{co}}^{w^*}(\widehat{J(V)}) = \varnothing$. In particular,

$$\widehat{J(V')} = \widehat{J(V')} \cap \widehat{J(V')} \subseteq \widehat{J(V')} \cap \widehat{J(V)} \subseteq \partial f^*(V') \cap \widehat{J(V)} = \varnothing.$$

However, $J(V') \neq \varnothing$ since $V' \cap D \neq \varnothing$ and

$$\varnothing \neq \operatorname{argmax}(x^* - f) \subseteq J(x^*) \quad \text{for all } x^* \in V' \cap D.$$

Thus $\partial f^*(V) \subseteq \overline{\operatorname{co}}^{w^*}(\widehat{J(V)})$. Therefore, by Lemma 3.4.7, $\|\cdot\| - \operatorname{diam}[\partial f^*(V)] < \varepsilon$, and so $\varnothing \neq V \subseteq W \cap O_\varepsilon$. Hence O_ε is dense in A. It now follows that $x^* \mapsto \partial f^*(x^*)$ is single-valued and norm upper semicontinuous at each point of $\bigcap_{n\in\mathbb{N}} O_{1/n}$. We will now show that $\operatorname{argmax}(x^* - f) \neq \varnothing$ for all $x^* \in \bigcap_{n\in\mathbb{N}} O_{1/n}$. To accomplish this, consider an arbitrary element $x_0^* \in \bigcap_{n\in\mathbb{N}} O_{1/n}$. Choose $\{x_n^* : n \in \mathbb{N}\} \subseteq D$ such that $\lim_{n\to\infty} x_n^* = x_0^*$ and choose $x_n \in \operatorname{argmax}(x_n^* - f)$ for all $n \in \mathbb{N}$. Note that $\widehat{x_n} \in \partial f^*(x_n^*)$ for all $n \in \mathbb{N}$, and so $\lim_{n\to\infty} \widehat{x_n} = F$, where $\{F\} := \partial f^*(x_0^*)$. Since $\widehat{X}$ is norm closed, $F \in \widehat{X}$ and so $F = \widehat{x_0}$ for some $x_0 \in X$. Thus $\lim_{n\to\infty} x_n = x_0$. Now

$$f^*(x_n^*) - \|x_n^* - x_0^*\|\|x_n\| \leq (x_n^* - f)(x_n) - (x_n^* - x_0^*)(x_n) = (x_0^* - f)(x_n) \leq f^*(x_0^*), \qquad (*)$$

and $(x_n : n \in \mathbb{N})$ is bounded as (i) $(\widehat{x_n} : n \in \mathbb{N})$ is convergent and hence bounded, and (ii) $\|x_n\| = \|\widehat{x_n}\|$ for all $n \in \mathbb{N}$. Furthermore, by Proposition 3.4.2, part (ii), $x^* \mapsto f^*(x^*)$ is continuous at x_0^*. Therefore by the squeeze theorem applied to $(*)$ we get that

$$\lim_{n\to\infty} (x_0^* - f)(x_n) = f^*(x_0^*).$$

Moreover, since f is lower semicontinuous and $\lim_{n\to\infty} x_n = x_0$, we have, from the observation just before the statement of this theorem, that $x_0 \in \mathrm{argmax}(x_0^* - f)$. The result now follows from Proposition 3.4.2, part (v). □

One draw back to the third variational theorem is that it still contains the troublesome requirement that there exists a dense subset D of A such that $\mathrm{argmax}(x^* - f) \neq \varnothing$ for each $x^* \in D$. To overcome this impediment, we will apply the Brøndsted–Rockafellar theorem.

Theorem 3.4.9 (Brøndsted–Rockafellar theorem [48]). *Let $f : X \to (-\infty, \infty]$ be a proper convex lower semicontinuous function on a Banach space $(X, \|\cdot\|)$. Then, given any $x_0 \in \mathrm{Dom}(f)$, $\varepsilon > 0$ and $x_0^* \in \partial_\varepsilon f(x_0)$, there exist $x \in \mathrm{Dom}(f)$ and $x^* \in X^*$ such that $x^* \in \partial f(x)$, $\|x - x_0\| \leq \sqrt{\varepsilon}$ and $\|x^* - x_0^*\| \leq \sqrt{\varepsilon}$.*

Exercise 3.4.10. Let $f : X \to (-\infty, \infty]$ be a proper convex lower semicontinuous function on a Banach space $(X, \|\cdot\|)$. Use the Brøndsted–Rockafellar theorem to show that $\{x^* \in \mathrm{int}(\mathrm{Dom}(f)) : \mathrm{argmax}(x^* - f) \neq \varnothing\}$ is dense in $\mathrm{int}(\mathrm{Dom}(f))$. Hint: First recall that if $\varepsilon > 0$, $x_0^* \in \mathrm{Dom}(f^*)$ and $f^*(x_0^*) - \varepsilon < (x_0^* - f)(x)$, then $x_0^* \in \partial_\varepsilon f(x)$. Secondly, observe that $x^* \in \partial f(x_0)$ for some $x_0 \in \mathrm{Dom}(f)$ if and only if $x_0 \in \mathrm{argmax}(x^* - f)$.

To apply the Brøndsted–Rockafellar theorem to functions that are not necessarily convex, we introduce the following method of "convexifying" an arbitrary function that is bounded below by a continuous linear functional.

Let $(X, \|\cdot\|)$ be a normed linear space, and let $f : X \to (-\infty, \infty]$ be a proper function that is bounded below by at least one continuous linear functional. We define the function $\overline{\mathrm{co}}(f)$: $X \to (-\infty, \infty]$ by

$$\overline{\mathrm{co}}(f)(x) := \sup\{\psi(x) : \psi : X \to \mathbb{R} \text{ is continuous and linear, and } \psi(y) \leq f(y) \text{ for all } y \in X\}.$$

It is immediate from this definition that:
(i) $\overline{\mathrm{co}}(f)$ is convex and lower semicontinuous;
(ii) $\overline{\mathrm{co}}(f)(x) \leq f(x)$ for all $x \in X$, and
(iii) $\inf\{f(x) : x \in X\} = \inf\{\overline{\mathrm{co}}(f)(x) : x \in X\}$.

Note that from (ii) and (iii) it follows that $\mathrm{argmin}(f) \subseteq \mathrm{argmin}(\overline{\mathrm{co}}(f))$. A further useful observation is that if $x^* \in X^*$, then

$$\overline{\mathrm{co}}(f - x^*) = \overline{\mathrm{co}}(f) - x^*.$$

Exercise 3.4.11. Let $f : X \to (-\infty, \infty]$ be a proper lower semicontinuous function on a Banach space $(X, \|\cdot\|)$. Show that if $\overline{\text{co}}(f)$ has as strong minimum at $x_0 \in \text{Dom}(f)$, then $\text{argmin}(f) = \text{argmin}(\overline{\text{co}}(f)) = \{x_0\}$, and f has a strong minimum at x_0 as well.

We can now state and prove the well-known Stegall variational theorem.

Corollary 3.4.12 (Stegall variational theorem [291]). *Let $f : X \to (-\infty, \infty]$ be a proper lower semicontinuous function on a Banach space $(X, \|\cdot\|)$ with the Radon–Nikodým property. If A is a nonempty open subset of $\text{Dom}(f^*)$, then there exists a dense G_δ-subset R of A such that $(x^* - f) : X \to [-\infty, \infty)$ has a strong maximum for each $x^* \in R$. In addition, if $0 \in A$ and $\varepsilon > 0$, then there exists $x_0^* \in X^*$ with $\|x_0^*\| < \varepsilon$ such that $(x_0^* - f) : X \to [-\infty, \infty)$ has a strong maximum.*

Proof. Let $f : X \to (-\infty, \infty]$ be a proper lower semicontinuous function on $(X, \|\cdot\|)$, and let A be a nonempty open subset of $\text{int}(\text{Dom}(f^*))$. Then $\overline{\text{co}}(f) : X \to (-\infty, \infty]$ is also a proper lower semicontinuous convex function on $(X, \|\cdot\|)$. Furthermore, $\text{Dom}(f^*) = \text{Dom}(\overline{\text{co}}(f)^*)$. Now by Exercise 3.4.10 $\{x^* \in A : \text{argmax}(x^* - \overline{\text{co}}(f)) \neq \varnothing\}$ is dense in A. Hence we may apply the third variational theorem to obtain a dense G_δ-subset R of A such that $x^* - \overline{\text{co}}(f)$ has a strong maximum (or, equivalently, $\overline{\text{co}}(f) - x^* = \overline{\text{co}}(f - x^*)$ has a strong minimum) at each point $x^* \in R$. It now follows from Exercise 3.4.11 that $(f - x^*)$ has a strong minimum (or, equivalently, $(x^* - f)$ has a strong maximum) at each $x^* \in R$. □

We now consider a variational theorem in the setting of compact topological spaces.

Lemma 3.4.13. *Let $f : K \to (-\infty, \infty]$ be a proper lower semicontinuous function on a nonempty compact topological space (K, τ). Then argmin(f) is nonempty.*

Proof. Let us first show that f is bounded from below. Indeed, if this is not the case, then for every $n \in \mathbb{N}$, $C_n := \{x \in K : f(x) \leq -n\}$ is a nonempty closed and hence compact subset of (K, τ). Moreover, $C_{n+1} \subseteq C_n$ for all $n \in \mathbb{N}$. Therefore, by the finite intersection property, $\varnothing \neq \bigcap_{n\in\mathbb{N}} C_n$. However, if $x \in \bigcap_{n\in\mathbb{N}} C_n$, then $f(x) \leq -n$ for all $n \in \mathbb{N}$; which is impossible, as no real number has this property. Therefore f must be bounded from below. Let $I := \inf\{f(x) : x \in K\}$ (which is finite since f is a proper function), and for each $n \in \mathbb{N}$, let $D_n := \{x \in K : f(x) \leq I + 1/n\}$. Then each set D_n is nonempty and closed, and as with the sets $\{C_n : n \in \mathbb{N}\}$ earlier, the sets $\{D_n : n \in \mathbb{N}\}$ are decreasing, that is, $D_{n+1} \subseteq D_n$ for all $n \in \mathbb{N}$. Therefore $\varnothing \neq \bigcap_{n\in\mathbb{N}} D_n$. Let $x_0 \in \bigcap_{n\in\mathbb{N}} D_n$, and let $n_0 \in \mathbb{N}$. Then $I \leq f(x_0) \leq I + (1/n_0)$ as $x_0 \in D_{n_0}$. Since $n_0 \in \mathbb{N}$ was arbitrary, $I \leq f(x_0) \leq I$. Therefore $f(x_0) = \inf\{f(x) : x \in K\}$, that is, f attains its minimum value at x_0. □

Exercise 3.4.14. Let $f : X \to (-\infty, \infty]$ be a proper lower semicontinuous function defined on a topological space (X, τ). Show that if $g : X \to \mathbb{R}$ is a continuous function then (i) $\text{Dom}(f + g) = \text{Dom}(f)$, and (ii) $(f + g) : X \to (-\infty, \infty]$ is lower semicontinuous on (X, τ).

Proposition 3.4.15. *Let $f : K \to (-\infty, \infty]$ be a proper lower semicontinuous function on a nonempty compact Hausdorff space (K, τ). Then the set-valued mapping $M_f : C(K) \to 2^K$ defined by $M_f(g) := \operatorname{argmin}(f + g)$ for $g \in C(K)$ is a minimal τ-usco on $(C(K), \|\cdot\|_\infty)$.*

Proof. Let $g \in C(K)$. Then, by Exercise 3.4.14, $f + g$ is lower semicontinuous on (K, τ). Therefore, by Lemma 3.4.13, $M_f(g) = \operatorname{argmin}(f + g) \neq \varnothing$. Since

$$\operatorname{argmin}(f + g) = \{x \in K : (f + g)(x) \le r\}, \quad \text{where } r := \min\{(f + g)(k) : k \in K\},$$

$\operatorname{argmin}(f+g)$ is also closed and hence compact, that is, $M_f(g)$ is a nonempty compact subset of (K, τ). So to show that M_f is a τ-usco at g, we need only show that M_f is τ-upper semicontinuous at g. To this end, let W be a τ-open subset of K containing $M_f(g)$, that is, $M_f(g) \subseteq W$. If $W = K$, then $M_f(C(K)) \subseteq W$, and we are done. So we may assume that $W \neq K$. If $f|_{K\setminus W} \equiv \infty$ on $K \setminus W$, then $M_f(C(K)) \subseteq W$, and we are done. So we may assume that $f|_{K\setminus W}$ is a proper lower semicontinuous function on $K \setminus W$. Let $h := (f + g)|_{K\setminus W}$. Then, by Exercise 3.4.14, h is proper lower semicontinuous on $K \setminus W$. Therefore by Lemma 3.4.13 there exists $x_0 \in K \setminus W$ such that $h(x_0) \le h(x) = (f + g)(x)$ for all $x \in K \setminus W$. Since $x_0 \notin W$, $x_0 \notin \operatorname{argmin}(f + g)$. Therefore

$$I := \min\{(f + g)(k) : k \in K\} < (f + g)(x_0) = h(x_0) \le h(x) = (f + g)(x) \quad \text{for all } x \in K \setminus W.$$

Let $\varepsilon := (f + g)(x_0) - I$. Then $\varepsilon > 0$. Let $0 < \delta < \varepsilon/3$. We claim that $M_f(B(g, \delta)) \subseteq W$. To substantiate this claim, let us consider an arbitrary element $g' \in B(g, \delta)$ and arbitrary $k_0 \in \operatorname{argmin}(f + g)$. Then

$$\min\{(f + g')(k) : k \in K\} \le (f + g')(k_0) = (f + g)(k_0) + (g' - g)(k_0) < I + \delta.$$

On the other hand, if $x \in K \setminus W$, then

$$I + 3\delta < I + \varepsilon = (f + g)(x_0) \le (f + g)(x) = (f + g')(x) + (g - g')(x) < (f + g')(x) + \delta.$$

Therefore $I + 2\delta < (f + g')(x)$, and so

$$\min\{(f + g')(k) : k \in K\} < I + \delta < I + 2\delta < (f + g')(x).$$

Hence $x \notin \operatorname{argmin}(f + g')$. Since $x \in K \setminus W$ was arbitrary, $\operatorname{argmin}(f + g') \cap (K \setminus W) = \varnothing$, that is, $\operatorname{argmin}(f + g') \subseteq W$. This completes the proof that M_f is a τ-usco. Next, we show that M_f is in fact a minimal τ-usco on $(C(K), \|\cdot\|_\infty)$. To achieve this, we appeal to Proposition 1.3.5. In this direction, let U be an open subset of $C(K)$, and let W be an open subset of (K, τ) such that $M_f(U) \cap W \neq \varnothing$. Choose $k_0 \in M_f(U) \cap W$ and $f_0 \in U$ such that $k_0 \in M_f(f_0)$. Since U is an open set, there exists $\varepsilon > 0$ such that $B(f_0, 2\varepsilon) \subseteq U$. Since (K, τ) is compact and Hausdorff, (K, τ) is normal. Therefore there exists a continuous function $h : K \to [0, 1]$ such that $h(k_0) = 1$ and $h(k) = 0$ for all $k \in K \setminus W$ (see [78]). Let

$g := f_0 - \varepsilon h$. Then $\|g - f_0\|_\infty = \varepsilon$, and so $g \in B(f_0, 2\varepsilon) \subseteq U$. Furthermore,

$$\begin{aligned}\min\{(f+g)(k') : k' \in K\} &\leq (f+g)(k_0)\\ &= (f+f_0)(k_0) - \varepsilon h(k_0)\\ &= \min\{(f+f_0)(k') : k' \in K\} - \varepsilon.\end{aligned}$$

On the other hand, if $k \in K \setminus W$, then

$$\begin{aligned}\min\{(f+f_0)(k') : k' \in K\} &\leq (f+f_0)(k)\\ &= (f+f_0)(k) - \varepsilon h(k) \quad \text{since, } h(k) = 0.\\ &= (f+g)(k).\end{aligned}$$

Hence

$$\begin{aligned}\min\{(f+g)(k') : k' \in K\} &\leq \min\{(f+f_0)(k') : k' \in K\} - \varepsilon\\ &< \min\{(f+f_0)(k') : k' \in K\}\\ &\leq (f+g)(k).\end{aligned}$$

Thus $k \notin \operatorname{argmin}(f+g)$. Therefore $M_f(g) = \operatorname{argmin}(f+g) \subseteq W$. Since, as established earlier, M_f is τ-upper semicontinuous, there exists an open neighbourhood V of g, contained in U, such that $M_f(V) \subseteq W$. This shows that M_f is a minimal τ-usco on $(C(K), \|\cdot\|_\infty)$. □

To state our fourth variational theorem, we need to recall one further notion concerning minima of real-valued functions. We say that a function $f : X \to (-\infty, \infty]$ defined on a topological space (X, τ) has *strong minimum at $x_0 \in X$ with respect to τ* if

$$f(x_0) = \inf_{x \in X} f(x) \quad \text{and} \quad x_0 = \lim_{n \to \infty} x_n$$

whenever $(x_n : n \in \mathbb{N})$ is a sequence in X such that

$$\lim_{n \to \infty} f(x_n) = \inf_{x \in X} f(x) = f(x_0).$$

The origins of the next variational theorem may be traced back to [163].

Theorem 3.4.16 (Fourth variational theorem). *Let $f : K \to (-\infty, \infty]$ be a proper lower semicontinuous function on a nonempty compact Hausdorff Stegall space (K, τ). Then*

$$\{g \in C(K) : f + g \text{ has a strong minimum with respect to } \tau\}$$

contains a dense G_δ-subset of $(C(K), \|\cdot\|_\infty)$.

Proof. By Proposition 3.4.15, $g \mapsto \operatorname{argmin}(f + g)$ is a minimal τ-usco on $(C(K), \|\cdot\|_\infty)$, and by Corollary 1.4.24 $(C(K), \|\cdot\|_\infty)$ is a Baire space. Hence by Proposition 2.3.17 there exists a dense G_δ-subset G of $(C(K), \|\cdot\|_\infty)$ such that $\operatorname{argmin}(f+g)$ is a singleton for each $g \in G$. We claim that at each $g \in G$, $f + g$ has a strong minimum with respect to τ. To verify this claim, consider any $g \in G$. Let $k_0 \in K$ be defined so that $\operatorname{argmin}(f+g) =: \{k_0\}$, and let $(k_n : n \in \mathbb{N})$ be any sequence in K such that

$$\lim_{n\to\infty} (f + g)(k_n) = (f + g)(k_0) = \min\{(f + g)(k) : k \in K\}.$$

We need to show that $k_0 = \lim_{n\to\infty} k_n$ (with respect to the topology τ). So let W be a τ-open neighbourhood of k_0. If $W = K$, then $k_n \in W$ for all $n \in \mathbb{N}$, and we are done. So we may assume that $W \neq K$. If $f|_{K\setminus W} \equiv \infty$ on $K \setminus W$, then there exists $N \in \mathbb{N}$ such that $k_n \in \operatorname{Dom}(f) \subseteq W$ for all $n > N$, and we are done. So we may assume that $f|_{K\setminus W}$ is a proper lower semicontinuous function on $K \setminus W$. Let $h := (f + g)|_{K\setminus W}$. It follows from Exercise 3.4.14 that h is a proper lower semicontinuous on $K \setminus W$. Therefore by Lemma 3.4.13 there exists $x_0 \in K\setminus W$ such that $h(x_0) \le h(x) = (f+g)(x)$ for all $x \in K\setminus W$. Since $x_0 \notin W$, $x_0 \notin \operatorname{argmin}(f + g)$. Thus

$$\min\{(f + g)(k) : k \in K\} < (f + g)(x_0) = h(x_0) \le h(x) = (f + g)(x) \quad \text{for all } x \in K \setminus W.$$

Let $\varepsilon := (f + g)(x_0) - \min\{(f + g)(k) : k \in K\}$. Then $\varepsilon > 0$. Since

$$\lim_{n\to\infty} (f + g)(k_n) = \min\{(f + g)(k) : k \in K\},$$

there exists $N \in \mathbb{N}$ such that

$$(f + g)(k_n) < \min\{(f + g)(k) : k \in K\} + \varepsilon = (f + g)(x_0) = \min\{(f + g)(x) : x \in K \setminus W\}$$

for all $n > N$. Therefore $k_n \in W$ for all $n > N$. This shows that $k_0 = \lim_{n\to\infty} k_n$. Hence $(f + g)$ has a strong minimum at k_0 with respect to τ. □

Before we can state the next corollary, we need to recollect from Chapter 1 the following definition. A topological space (X, τ) is *fragmentable* if, there exists a metric ρ on X such that for every nonempty subset B of X and every $\varepsilon > 0$, there exists a τ-open subset W of X such that (i) $\varnothing \neq B \cap W$ and (ii) $\rho - \operatorname{diam}(B \cap W) < \varepsilon$.

Corollary 3.4.17. *Let $f : K \to (-\infty, \infty]$ be a proper lower semicontinuous function defined on a nonempty compact Hausdorff fragmentable space (K, τ). Then*

$$\{g \in C(K) : f + g \text{ has a strong minimum with respect to } \tau\}$$

contains a dense G_δ-subset of $(C(K), \|\cdot\|_\infty)$.

Proof. By Proposition 2.3.16 every fragmentable space is a Stegall space. The result then follows from Theorem 3.4.16. □

We end this section with one last variational theorem. Unfortunately, the proof of this last theorem is beyond the scope of this book, because although the proof relies heavily upon the careful handling of minimal cuscos, it also relies upon a deep understanding of James' weak compactness theorem [147, 148, 228]. We encourage the interested reader to consult the paper [235] for the proof.

Theorem 3.4.18 (Abstract variational theorem [235]). *Let $f : X \to [-\infty, \infty)$ be a proper function defined on a nonempty set X, and let $(Y, \|\cdot\|)$ be a Banach space of real-valued functions defined on X under pointwise addition and pointwise scalar multiplication. Suppose that the norm-topology on Y is at least as strong as the $\tau_p(X)$-topology on Y and that B_Y is $\tau_p(X)$-compact. If there exists a nonempty open subset of A of Y such that $\operatorname{argmax}(f+y) \neq \varnothing$ for each $y \in A$, then there exists a dense G_δ-subset of R of A such that $(f+y) : X \to [-\infty, \infty)$ has a strong maximum with respect to ρ_Y for each $y \in R$.*

An important particular case of this theorem is the case where X is a normed linear space and $Y = X^*$. In this case, we have the following theorem.

Corollary 3.4.19 ([228, Theorem 5.6]). *Let $f : X \to [-\infty, \infty)$ be a proper function on a normed linear space $(X, \|\cdot\|)$. If there exists a nonempty open subset A of X^* such that $\operatorname{argmax}(f + x^*) \neq \varnothing$ for each $x^* \in A$, then there exists a dense G_δ-subset R of A such that for each $x^* \in R$, $(f + x^*) : X \to [-\infty, \infty)$ has a strong maximum with respect to the norm-topology. In addition, if $0 \in A$ and $\varepsilon > 0$, then there exists $x_0^* \in X^*$ with $\|x_0^*\| < \varepsilon$ such that $(f + x_0^*) : X \to [-\infty, \infty)$ has a strong maximum with respect to the norm-topology.*

Proof. It follows from the Banach–Alaoglu theorem that $(B_Y, \tau_p(X)) = (B_{X^*}, \text{weak}^*)$ is compact. The result then follows from the Abstract variational theorem. □

3.5 James' weak compactness theorem

In this section, among other things, we prove James' weak compactness theorem in the setting of separable Banach spaces.

For any x in a normed linear space $(X, \|\cdot\|)$, we define $\hat{x} \in X^{**}$ by $\hat{x}(x^*) := x^*(x)$ for $x^* \in X^*$. Then $x \mapsto \hat{x}$ is a linear isometric embedding of X into X^{**}. In particular, if $(X, \|\cdot\|)$ is a Banach space, then $\widehat{X}$ is a closed linear subspace of X^{**}.

Let K be a weak* compact convex subset of the dual of a Banach space $(X, \|\cdot\|)$. A subset B of K is called a *boundary* of K if for every $\hat{x} \in \widehat{X}$, there exists $b^* \in B$ such that $\hat{x}(b^*) = \sup\{\hat{x}(y^*) : y^* \in K\}$. We say B *(I)-generates* K if for every countable cover $\{C_n : n \in \mathbb{N}\}$ of B by weak* compact convex subsets of K, the convex hull of $\bigcup_{n\in\mathbb{N}} C_n$ is norm dense in K [84].

James' theorem relies upon several prerequisite results.

Lemma 3.5.1 ([237]). *Let $\beta > 0$, $\beta' > 0$ and suppose that $\varphi : [0, \beta + \beta'] \to \mathbb{R}$ is a convex function. Then*

$$\frac{\varphi(\beta) - \varphi(0)}{\beta} \leq \frac{\varphi(\beta + \beta') - \varphi(\beta)}{\beta'}.$$

Proof. The inequality given in the statement of the lemma follows by rearranging the inequality $\varphi(\beta) \leq \frac{\beta}{\beta+\beta'}\varphi(\beta + \beta') + \frac{\beta'}{\beta+\beta'}\varphi(0)$. □

Lemma 3.5.2 (The moving away lemma [237]). *Let V be a vector space over $\mathbb{R}$, and let $p : V \to \mathbb{R}$ be a sublinear function. If $(A_n : n \in \mathbb{N})$ is a decreasing sequence of nonempty convex subsets of V, $(\beta_n : n \in \mathbb{N})$ is any sequence of strictly positive numbers and*

$$r < \inf_{a \in A_1} p(a) \quad \textit{for some } 0 < r,$$

then there exists a sequence $(a_n : n \in \mathbb{N})$ in V such that:
(i) *$a_n \in A_n$, and*
(ii) *$p(\sum_{i=1}^{n} \beta_i a_i) + \beta_{n+1} r < p(\sum_{i=1}^{n+1} \beta_i a_i)$ for all $n \in \mathbb{N}$.*

Proof. We proceed in two parts. Firstly, we prove that if $u \in V$ and

$$\beta_n r + p(u) < \inf_{a \in A_n} p(u + \beta_n a) \quad \text{for some } n \in \mathbb{N},$$

then there exists $a_n \in A_n$, such that

$$\beta_{n+1} r + p(u + \beta_n a_n) < \inf_{a \in A_n} p(u + \beta_n a_n + \beta_{n+1} a).$$

To see this, suppose that $u \in V$ and that $\beta_n r + p(u) < \inf_{a \in A_n} p(u + \beta_n a)$. Then there exists $\varepsilon > 0$ such that

$$r + 2\varepsilon < \frac{\inf_{a \in A_n} p(u + \beta_n a) - p(u)}{\beta_n}. \qquad (*)$$

So choose $a_n \in A_n$ such that $p(u + \beta_n a_n) < \inf_{a \in A_n} p(u + \beta_n a) + \beta_{n+1}\varepsilon$. Let $a \in A_n$. Then $v := (\beta_n a_n + \beta_{n+1} a)/(\beta_n + \beta_{n+1}) \in A_n$, and so

$$\begin{aligned} r + 2\varepsilon &< \frac{p(u + \beta_n v) - p(u + 0v)}{\beta_n} && \text{by } (*) \text{ and the fact that } v \in A_n \\ &\leq \frac{p(u + (\beta_n + \beta_{n+1})v) - p(u + \beta_n v)}{\beta_{n+1}} && \text{by Lemma 3.5.1.} \end{aligned}$$

Rearranging gives $\beta_{n+1}(r + \varepsilon) + [p(u + \beta_n v) + \beta_{n+1}\varepsilon] < p(u + \beta_n a_n + \beta_{n+1} a)$. Since $a \in A_n$ was arbitrary and $p(u + \beta_n a_n) < [p(u + \beta_n v) + \beta_{n+1}\varepsilon]$, the desired inequality follows.

From this we may inductively construct a sequence $(a_n : n \in \mathbb{N})$. For the first step, we set $u := 0$, and then by the hypothesis we have that $\beta_1 r + p(0) < \inf_{a \in A_1} p(\beta_1 a) =$

$\inf_{a\in A_1} p(0+\beta_1 a)$. So, by the above there exists $a_1 \in A_1$ such that $\beta_2 r + p(\beta_1 a_1) < \inf_{a\in A_1} p(\beta_1 a_1 + \beta_2 a)$.

For the nth step, set $u := \sum_{i=1}^{n-1} \beta_i a_i$. Since $A_n \subseteq A_{n-1}$ and by the way a_{n-1} was constructed, we have that $\beta_n r + p(u) < \inf_{a\in A_{n-1}} p(u+\beta_n a) \le \inf_{a\in A_n} p(u+\beta_n a)$. So, by the first result again, there exists $a_n \in A_n$ such that $\beta_{n+1} r + p(\sum_{i=1}^n \beta_i a_i) < \inf_{a\in A_n} p(\sum_{i=1}^n \beta_i a_i + \beta_{n+1} a)$, which completes the induction. The sequence $(a_n : n \in \mathbb{N})$ has the properties claimed above. □

We may now state and prove the main theorem in this section.

Theorem 3.5.3 ([84, 219, 237]). *Let K be a weak* compact convex subset of the dual of a Banach space $(X, \|\cdot\|)$, and let B be a boundary of K. Then B (I)-generates K.*

Proof. After possibly translating K, we may assume that $0 \in B$. Let $\{C_n : n \in \mathbb{N}\}$ be weak* compact convex subsets of K such that $B \subseteq \bigcup_{n\in\mathbb{N}} C_n$ and suppose, to obtain a contradiction, that $\mathrm{co}[\bigcup_{n\in\mathbb{N}} C_n]$ is not norm dense in K. Then there must exist $\varepsilon > 0$ and $y^* \in K$ such that

$$y^* \in K \setminus \left(\mathrm{co}\left[\bigcup_{n\in\mathbb{N}} C_n\right] + \varepsilon B_{X^*}\right)$$

Since, for each $n \in \mathbb{N}$, $\mathrm{co}[\bigcup_{j=1}^n C_j]$ is weak* compact and convex, there exist $(\widehat{x}_n : n \in \mathbb{N})$ in $S_{\widehat{X}}$ such that

$$\begin{aligned}
&\max\left\{\widehat{x}_n(x^*) : x^* \in \mathrm{co}\left[\bigcup_{j=1}^n C_j\right]\right\} + \varepsilon \\
&= \max\left\{\widehat{x}_n(x^*) : x^* \in \mathrm{co}\left[\bigcup_{j=1}^n C_j\right] + \varepsilon B_{X^*}\right\} < \widehat{x}_n(y^*) \qquad (*)
\end{aligned}$$

for all $n \in \mathbb{N}$. Now $(\widehat{x}_n(y^*) : n \in \mathbb{N})$ is a bounded sequence of real numbers and thus has a convergent subsequence $(\widehat{x}_{n_k}(y^*) : k \in \mathbb{N})$. Let $s := \lim_{k\to\infty} \widehat{x}_{n_k}(y^*)$. Then $\varepsilon \le s$ as $0 \in C_N$ for some $N \in \mathbb{N}$ and $\varepsilon = 0 + \varepsilon < \widehat{x}_n(y^*)$ for all $n \ge N$ by inequality $(*)$. After relabelling the sequence $(\widehat{x}_n : n \in \mathbb{N})$ if necessary, we may assume that $\lim_{n\to\infty} \widehat{x}_n(y^*) = s$ and $|\widehat{x}_n(y^*) - s| < \varepsilon/3$ for all $n \in \mathbb{N}$. Note that this relabelling does not disturb the inequality in $(*)$.

We define $A_n := \mathrm{co}\{\widehat{x}_k : n \le k\}$ for $n \in \mathbb{N}$ and note that (i) $(A_n : n \in \mathbb{N})$ is a decreasing sequence of nonempty convex subsets of $\widehat{X}$ and (ii) if $b^* \in C_N$ and $n > N$, then

$$g(b^*) < [g(y^*) - \varepsilon] \quad \text{for all } g \in A_n \qquad (**)$$

since $\{\widehat{x}_k : n \le k\} \subseteq \{\widehat{x} \in \widehat{X} : \widehat{x}(b^* - y^*) < -\varepsilon\}$, which is convex. Next, we define $p : \widehat{X} \to \mathbb{R}$ by

$$p(\widehat{x}) := \sup_{x^*\in K} \widehat{x}(x^*) \quad \text{for } \widehat{x} \in \widehat{X}.$$

Then p defines a sublinear functional on $\widehat{X}$. Moreover, for all $g \in A_1$, we have $(s-\varepsilon/3) < g(y^*)$ since $\{\widehat{x}_n : n \in \mathbb{N}\} \subseteq \{\widehat{x} \in \widehat{X} : (s-\varepsilon/3) < \widehat{x}(y^*)\}$, which is convex. Therefore $(s-\varepsilon/3) < p(g)$ for all $g \in A_1$ since $y^* \in K$.

Let $(\beta_n : n \in \mathbb{N})$ be any sequence of strictly positive real numbers such that $\lim_{n\to\infty}(\sum_{i=n+1}^{\infty}\beta_i)/\beta_n = 0$. Now

$$0 < (s-\varepsilon/2) < (s-\varepsilon/3) \leq \inf_{g\in A_1} p(g).$$

Therefore by Lemma 3.5.2 there exists a sequence $(g_n : n \in \mathbb{N})$ in $\widehat{X}$ such that $g_n \in A_n$ and

$$p\left(\sum_{i=1}^{n}\beta_i g_i\right) + \beta_{n+1}(s-\varepsilon/2) < p\left(\sum_{i=1}^{n+1}\beta_i g_i\right) \quad \text{for all } n \in \mathbb{N}. \qquad (***)$$

Note also that since $\lim_{n\to\infty}\widehat{x}_n(y^*) = s$, $\lim_{n\to\infty} g_n(y^*) = s$. As $\|g_n\| \leq 1$ for all $n \in \mathbb{N}$, we have that $\sum_{i=1}^{\infty}\|\beta_i g_i\| \leq \sum_{i=1}^{\infty}\beta_i < \infty$. Since X is a Banach space, this implies that $g := \sum_{i=1}^{\infty}\beta_i g_i \in \widehat{X}$. Because p is continuous, this implies that $(p(\sum_{i=1}^{n}\beta_i g_i) : n \in \mathbb{N})$ is a convergent and hence bounded sequence in $\mathbb{R}$. Moreover, by inequality $(***)$ we have that $(p(\sum_{i=1}^{n}\beta_i g_i) : n \in \mathbb{N})$ is an increasing sequence. Therefore, by the monotone convergence theorem, $(p(\sum_{i=1}^{n}\beta_i g_i) : n \in \mathbb{N})$ converges to its supremum, that is,

$$\sup_{n\in\mathbb{N}} p\left(\sum_{i=1}^{n}\beta_i g_i\right) = \lim_{n\to\infty} p\left(\sum_{i=1}^{n}\beta_i g_i\right) = p\left(\lim_{n\to\infty}\sum_{i=1}^{n}\beta_i g_i\right) = p(g). \qquad (****)$$

Since $g \in \widehat{X}$ and B is a boundary for K, there exists $b^* \in B$ such that

$$g(b^*) = \sup\{g(x^*) : x^* \in K\} = p(g).$$

Let $n \in \mathbb{N}$. Then

$$\begin{aligned}
\beta_n(s-\varepsilon/2) &< p\left(\sum_{i=1}^{n}\beta_i g_i\right) - p\left(\sum_{i=1}^{n-1}\beta_i g_i\right) && \text{by } (***)\\
&\leq p(g) - p\left(\sum_{i=1}^{n-1}\beta_i g_i\right) && \text{by } (****)\\
&= g(b^*) - p\left(\sum_{i=1}^{n-1}\beta_i g_i\right)\\
&\leq g(b^*) - \left(\sum_{i=1}^{n-1}\beta_i g_i\right)(b^*) = \sum_{i=n}^{\infty}\beta_i g_i(b^*).
\end{aligned}$$

Since $B \subseteq \bigcup_{n\in\mathbb{N}} C_n$, $b^* \in C_N$ for some $N \in \mathbb{N}$. Thus, if $n > N$, then

$$(s-\varepsilon/2) < \frac{1}{\beta_n}\left(\sum_{i=n+1}^{\infty}\beta_i g_i(b^*)\right) + g_n(b^*) < \frac{1}{\beta_n}\left(\sum_{i=n+1}^{\infty}\beta_i g_i(b^*)\right) + [g_n(y^*) - \varepsilon] \quad \text{by } (**)$$

since $g_n \in A_n$. By taking the limit as n tends to infinity we get that $(s - \varepsilon/2) \leq (s - \varepsilon)$, which is impossible. Therefore B (I)-generates K. □

Remarks 3.5.4. If $\beta_n := \frac{1}{n!}$ for all $n \in \mathbb{N}$ or $\beta_n := \frac{1}{2^{n^2}}$ for all $n \in \mathbb{N}$, then

$$\lim_{n\to\infty} \frac{\sum_{i=n+1}^{\infty} \beta_i}{\beta_n} = 0.$$

Corollary 3.5.5 ([98, Theorem I.2]). *Let K be a weak* compact convex subset of the dual of a Banach space $(X, \|\cdot\|)$, and let B be a norm separable boundary of K. Then $K = \overline{\mathrm{co}}(B)$. In particular, K is norm separable.*

Proof. To show that $K \subseteq \overline{\mathrm{co}}(B)$, it is sufficient to show that for every $\varepsilon > 0$, $K \subseteq \overline{\mathrm{co}}(B) + 2\varepsilon B_{X^*}$. To this end, let ε be an arbitrary positive real number. Let $\{x_n^* : n \in \mathbb{N}\}$ be a countable norm dense subset of B. Then

$$B \subseteq \bigcup_{n\in\mathbb{N}} B[x_n^*, \varepsilon] = \bigcup_{n\in\mathbb{N}} (x_n^* + \varepsilon B_{X^*}) \subseteq B + \varepsilon B_{X^*}.$$

Therefore

$$\begin{aligned}
K &\subseteq \overline{\mathrm{co}}\Big(\bigcup_{n\in\mathbb{N}} B[x_n^*, \varepsilon]\Big) && \text{by Theorem 3.5.3} \\
&\subseteq \overline{\mathrm{co}}(B + \varepsilon B_{X^*}) && \text{since } \bigcup_{n\in\mathbb{N}} B[x_n^*, \varepsilon] \subseteq B + \varepsilon B_{X^*} \\
&\subseteq \overline{\mathrm{co}}(\overline{\mathrm{co}}(B) + \varepsilon B_{X^*}) && \text{since } B \subseteq \overline{\mathrm{co}}(B) \\
&\subseteq \mathrm{co}(\overline{\mathrm{co}}(B) + \varepsilon B_{X^*}) + \varepsilon B_{X^*} && \\
&= (\overline{\mathrm{co}}(B) + \varepsilon B_{X^*}) + \varepsilon B_{X^*} && \text{since } \overline{\mathrm{co}}(B) + \varepsilon B_{X^*} \text{ is convex} \\
&= \overline{\mathrm{co}}(B) + 2\varepsilon B_{X^*} && \text{since } \varepsilon B_{X^*} \text{ is convex.}
\end{aligned}$$

This completes the proof. □

Theorem 3.5.6 ([217, Corollary 1.9]). *Every separable Banach space $(X, \|\cdot\|)$ whose duality mapping $D : S_X \to 2^{X^*}$ defined by $D(x) := \{x^* \in S_{X^*} : x^*(x) = 1\}$ for $x \in S_X$ is weakly upper semicontinuous on S_X is an Asplund space.*

Proof. By Corollary 3.3.40 we need to show that $(X^*, \|\cdot\|)$ is norm separable. In fact, to show that $(X^*, \|\cdot\|)$ is norm separable, it is sufficient to show that B_{X^*} is norm separable, and to show this, it is sufficient, because of Corollary 3.5.5, to show that B_{X^*} has a norm separable boundary. Let $\{x_n : n \in \mathbb{N}\}$ be a countable dense subset of S_X, and for each $n \in \mathbb{N}$, choose $x_n^* \in D(x_n) \subseteq B_{X^*}$. Let $B := \overline{\mathrm{co}}\{x_n^* : n \in \mathbb{N}\} \subseteq B_{X^*}$. We claim that B (which is clearly norm separable) is a boundary for B_{X^*}. To substantiate this claim, we need only establish that for each $x \in S_X$, there exists $b^* \in B$, such that

$$\widehat{x}(b^*) = \sup\{\widehat{x}(x^*) : x^* \in B_{X^*}\} = \|\widehat{x}\| = \|x\| = 1.$$

Since (i) D is weakly upper semicontinuous, (ii) D has nonempty images (see Example (xii) in Section 1.1.2), (iii) B is a weakly closed subset of X^*, as it is norm closed and convex, and (iv) $D(x)\cap B \neq \varnothing$ for each $x \in \{x_n : n \in \mathbb{N}\}$, we may deduce that $D(x)\cap B \neq \varnothing$ for all $x \in S_X$. Let $x \in S_X$ and $b^* \in D(x) \cap B$. Then

$$\widehat{x}(b^*) = b^*(x) = 1,$$

which completes the justification of the claim. □

An alternative way of seeing this result is appealing to Theorem 1.4.31 to conclude that D admits a Baire one selection $\sigma : S_X \to S_{X^*}$. Then use Exercise 1.4.26 to deduce that $\sigma(S_X)$ is norm separable. Finally, we only need to verify that $\sigma(S_X)$ is a boundary of B_{X^*} and, as before, apply Corollary 3.5.5.

Corollary 3.5.7 ([65, 144] and [217, Corollary 1.9]). *Every Banach space $(X, \|\cdot\|)$ whose duality mapping $D : S_X \to 2^{X^*}$ is weakly upper semicontinuous on S_X is an Asplund space.*

Proof. It follows from Theorem 3.3.45 that we need to show that every closed separable subspace $(Y, \|\cdot\|)$ of $(X, \|\cdot\|)$ is an Asplund space. So to accomplish this, let us consider an arbitrary closed separable subspace $(Y, \|\cdot\|)$ of $(X, \|\cdot\|)$. Let $D_Y : S_Y \to 2^{Y^*}$ denote the duality mapping on S_Y, and let $R : X^* \to Y^*$ denote the restriction mapping, that is, if $x^* \in X^*$, then $R(x^*) = x^*|_{Y^*}$. Then for each $y \in S_Y$, it follows from the Hahn–Banach extension theorem that $D_Y(y) = R(D(y)) = (R \circ D)(y)$. Since R is weak-to-weak continuous (as all norm continuous linear mappings are), we see that D_Y is weakly upper semicontinuous on S_Y. The result now follows from Theorem 3.5.6. □

We next present James' weak compactness theorem in the realm of separable Banach spaces. The general case (see [148]) requires some more technical manipulations. For a reasonably straightforward proof of the general case of James' theorem, see [228].

Our approach is based upon the following observation: For each $\mathscr{F} \in X^{***}$, there exists $x^* \in X^*$ such that $\mathscr{F}|_{\widehat{X}} = \widehat{x^*}|_{\widehat{X}}$. Indeed, if $\mathscr{F} \in X^{***}$ and we define $x^* : X \to \mathbb{R}$ by $x^*(x) := \mathscr{F}(\widehat{x})$ for $x \in X$, then x^* is linear, and $\|x^*\| \leq \|\mathscr{F}\|$. Therefore $x^* \in X^*$. Furthermore,

$$\widehat{x^*}(\widehat{x}) = \widehat{x}(x^*) = x^*(x) = \mathscr{F}(\widehat{x}) \quad \text{for all } \widehat{x} \in \widehat{X}.$$

Thus $\mathscr{F}|_{\widehat{X}} = \widehat{x^*}|_{\widehat{X}}$. In this way, we see that the relative weak topology on $\widehat{X}$ coincides with the relative weak* topology on $\widehat{X}$. In particular, each weak* compact subset of $\widehat{X}$ is weakly compact (and, of course, vice versa).

Theorem 3.5.8 (James' theorem [147, 219, 237]). *Let C be a closed bounded convex subset of a Banach space $(X, \|\cdot\|)$. If C is norm separable and every continuous linear functional on X attains its supremum over C, then C is weakly compact.*

Proof. Let $K := \overline{\widehat{C}}^{w^*}$. To show that C is weakly compact, it is sufficient to show $K \subseteq \widehat{X}$. Since $\widehat{C}$ is a norm separable boundary of K, we have by Corollary 3.5.5 that $K = \overline{\mathrm{co}}(\widehat{C}) \subseteq \widehat{X}$, which completes the proof. □

Our last result in this section provides an application of James' weak compactness theorem.

Theorem 3.5.9. *Every separable Banach space $(X, \|\cdot\|)$ whose (dual) duality mapping $D_{X^*} : S_{X^*} \to 2^{X^{**}}$ defined by $D_{X^*}(x^*) := \{x^{**} \in S_{X^{**}} : x^{**}(x^*) = 1\}$ for $x^* \in S_{X^*}$ is weakly upper semicontinuous on S_{X^*} is reflexive.*

Proof. From [75] we know that to prove that $(X, \|\cdot\|)$ is reflexive, we need to show that B_X is weakly compact. However, by Theorem 3.5.8 we see that to accomplish this, it is sufficient to show that every continuous linear functional on X attains its supremum over B_X or, equivalently, to show that $B_{\widehat{X}}$ is a boundary of $B_{X^{**}}$. In fact, to establish the result, we need only demonstrate that for every $x^* \in S_{X^*}$, there exists $\widehat{x} \in B_{\widehat{X}}$ such that

$$\widehat{x^*}(\widehat{x}) = \sup\{\widehat{x^*}(x^{**}) : x^{**} \in B_{X^{**}}\} = \|\widehat{x^*}\| = \|x^*\| = 1.$$

Since (i) D_{X^*} is weakly upper semicontinuous, (ii) D_{X^*} has nonempty images (see Example (xii) in Section 1.1.2), (iii) $B_{\widehat{X}}$ is a weakly closed subset of X^{**}, as it is norm closed and convex, and (iv) $\{x^* \in S_{X^*} : D_{X^*}(x^*) \cap B_{\widehat{X}} \neq \varnothing\}$ is dense in S_{X^*} by the Bishop–Phelps theorem [29], we may deduce that $D_{X^*}(x^*) \cap B_{\widehat{X}} \neq \varnothing$ for all $x^* \in S_{X^*}$. Let $x^* \in S_{X^*}$ and $\widehat{x} \in D_{X^*}(x^*) \cap B_{\widehat{X}}$. Then

$$\widehat{x^*}(\widehat{x}) = \widehat{x}(x^*) = 1,$$

which completes the justification of the proof. □

3.6 Differentiability of Lipschitz functions

In this section, we consider the differentiability of locally Lipschitz functions. Recall that a real-valued function $f : U \to \mathbb{R}$ defined on a nonempty open subset U of a normed linear space $(X, \|\cdot\|)$ is said to be *locally Lipschitz on U* if for every $x_0 \in U$, there exist $L > 0$ and $\delta > 0$ such that $B(x_0, \delta) \subseteq U$ and

$$|f(x) - f(y)| \leq L\|x - y\|$$

for all $x, y \in B(x_0, \delta)$.

For any such function f, we can define several types of directional derivatives. However, in this section, we restrict our attention to the following three types of derivatives: (i) the *upper Dini derivative at $x_0 \in U$ in the direction y*, given by

$$f^+(x; y) := \limsup_{\lambda \to 0^+} \frac{f(x_0 + \lambda y) - f(x_0)}{\lambda};$$

(ii) the *lower Dini derivative at $x_0 \in U$ in the direction y*, given by

$$f^-(x;y) := \liminf_{\lambda\to 0^+} \frac{f(x_0+\lambda y)-f(x_0)}{\lambda};$$

(iii) the *Clarke generalised derivative at $x_0 \in U$ in the direction y*, given by

$$f^0(x;y) := \limsup_{\substack{z\to x\\ \lambda\to 0^+}} \frac{f(z+\lambda y)-f(z)}{\lambda}.$$

It is immediate from these three definitions that for each $x \in U$ and each $y \in X$,

$$f^-(x;y) \le f^+(x;y) \le f^0(x;y).$$

Note that if f is a continuous convex function defined on a nonempty open convex subset U of a normed linear space $(X, \|\cdot\|)$, then $f'_+(x;y) = f^+(x;y) = f^-(x;y)$ for all $x \in U$ and $y \in X$; see the discussion just prior to Lemma 3.3.7. In fact, we have an even stronger result.

Theorem 3.6.1. *Let U be a nonempty open convex subset of a normed linear space $(X, \|\cdot\|)$, and let $\varphi : U \to \mathbb{R}$ be a continuous convex function. If $x_0 \in U$ and $y \in X$, then $\varphi'_+(x_0;y) = \varphi^0(x_0;y)$*

Proof. Let $x_0 \in U$ and $y \in X$. As observed already, $\varphi'_+(x_0;y) \le \varphi^0(x_0;y)$. So we need only show the reverse inequality. To this end, fix $\varepsilon > 0$. Choose $\delta' > 0$ such that
(i) $B[x_0, \delta'(\|y\|+1)] \subseteq U$,
(ii) φ is K-Lipschitz on $B[x_0, \delta'(\|y\|+1)]$ for some $0 < K$, and
(iii) $\frac{\varphi(x_0+\delta' y)-\varphi(x_0)}{\delta'} \le \varphi'_+(x_0;y) + \varepsilon/2$.

Let $\delta := \min\{\delta', (\varepsilon\delta')/(4K)\}$. Then $\delta > 0$, and

$$\varphi^0(x_0;y) \le \sup_{\substack{\|z-x_0\|\le\delta\\ 0<\lambda\le\delta}} \frac{\varphi(z+\lambda y)-\varphi(z)}{\lambda} \le \sup_{\|z-x_0\|\le\delta} \frac{\varphi(z+\delta' y)-\varphi(z)}{\delta'}$$

since $\lambda \to \frac{\varphi(z+\lambda y)-\varphi(z)}{\lambda}$ is increasing on $(0, \delta']$ for each $z \in B[x_0; \delta]$. Now let z be an arbitrary element of $B[x_0, \delta]$. Then since φ is K-Lipschitz on $B[x_0, \delta'(\|y\|+1)]$, we have that

$$\begin{aligned}
\frac{\varphi(z+\delta' y)-\varphi(z)}{\delta'} &\le \frac{\varphi(x_0+\delta' y)-\varphi(x_0)}{\delta'} + \frac{2K\|z-x_0\|}{\delta'}\\
&\le \frac{\varphi(x_0+\delta' y)-\varphi(x_0)}{\delta'} + \frac{2K\delta}{\delta'}\\
&\le \frac{\varphi(x_0+\delta' y)-\varphi(x_0)}{\delta'} + \varepsilon/2\\
&\le \varphi'_+(x_0;y) + \varepsilon \quad \text{by (iii) above.}
\end{aligned}$$

Since $z \in B[x_0, \delta]$ was arbitrary,

$$\varphi^0(x_0; y) \leq \sup_{\|z-x_0\|\leq\delta} \frac{\varphi(z + \delta' y) - \varphi(z)}{\delta'} \leq \varphi'_+(x_0; y) + \varepsilon.$$

Finally, since $\varepsilon > 0$ was arbitrary, $\varphi^0(x_0; y) \leq \varphi'_+(x_0; y)$. □

Next, we present some the most basic facts concerning the Clarke generalised directional derivative.

Lemma 3.6.2. *Let U be a nonempty open subset of a normed linear space $(X, \|\cdot\|)$, and let $f : U \to \mathbb{R}$ be a locally Lipschitz function. If $x_0 \in U$, then the mapping $y \mapsto f^0(x_0; y)$ from X into $\mathbb{R}$ is sublinear, and there exist positive real numbers δ and L such that $B(x_0, \delta) \subseteq U$ and $f^0(z; y) \leq L\|y\|$ for all $z \in B(x_0, \delta)$ and $y \in X$.*

Proof. Let $x_0 \in U$. We will first show that the mapping $y \mapsto f^0(x_0; y)$ is positively homogeneous. To this end, let $0 < \mu < \infty$ and $y \in X$. Then

$$\begin{aligned} f^0(x_0; \mu y) &= \limsup_{\substack{z\to x_0\\ \lambda\to 0^+}} \frac{f(z + \lambda(\mu y)) - f(z)}{\lambda} \\ &= \mu \limsup_{\substack{z\to x_0\\ \lambda\to 0^+}} \frac{f(z + (\lambda\mu)y) - f(z)}{\lambda\mu} \\ &= \mu \limsup_{\substack{z\to x_0\\ \lambda'\to 0^+}} \frac{f(z + \lambda' y) - f(z)}{\lambda'} \qquad (\text{where } \lambda' := \lambda\mu). \\ &= \mu f^0(x_0; y). \end{aligned}$$

Next, we show that the mapping $y \mapsto f^0(x_0; y)$ is subadditive. Let $x, y \in X$, and take arbitrary $\varepsilon > 0$. We will show that

$$f^0(x_0; x + y) \leq f^0(x_0; x) + f^0(x_0; y) + \varepsilon.$$

By the definitions of $f^0(x_0; x)$ and $f^0(x_0; y)$ there exist $\delta_1 > 0$ and $\delta_2 > 0$ such that

$$\sup_{\substack{\|z-x_0\|\leq\delta_1\\ 0<\lambda\leq\delta_1}} \frac{f(z + \lambda y) - f(z)}{\lambda} \leq f^0(x_0; y) + \varepsilon/2$$

and

$$\sup_{\substack{\|z-x_0\|\leq\delta_2\\ 0<\lambda\leq\delta_2}} \frac{f(z + \lambda x) - f(z)}{\lambda} \leq f^0(x_0; x) + \varepsilon/2.$$

Next, choose $0 < \delta < \min\{\delta_1/(1 + \|x\|), \delta_2\}$ such that $B[x_0, \delta(\|x\| + \|y\| + 1)] \subseteq U$. Let $z \in B(x_0, \delta)$ and $0 < \lambda < \delta$. Then

$$\|(z + \lambda x) - x_0\| \leq \|z - x_0\| + \delta\|x\| \leq \delta(1 + \|x\|) < \delta_1.$$

Therefore

$$\frac{f((z+\lambda x)+\lambda y)-f(z+\lambda x)}{\lambda}\le f^0(x_0;y)+\varepsilon/2.$$

Also, since, $\delta<\delta_2$,

$$\frac{f(z+\lambda x)-f(z)}{\lambda}\le f^0(x_0;x)+\varepsilon/2.$$

Thus

$$\begin{aligned}\frac{f(z+\lambda(x+y))-f(z)}{\lambda}&=\frac{f((z+\lambda x)+\lambda y)-f(z+\lambda x)}{\lambda}+\frac{f(z+\lambda x)-f(z)}{\lambda}\\&\le (f^0(x_0;y)+\varepsilon/2)+(f^0(x_0;x)+\varepsilon/2)\\&=f^0(x_0;x)+f^0(x_0;y)+\varepsilon.\end{aligned}$$

Since both $z\in B(x_0,\delta)$ and $0<\lambda<\delta$ were arbitrary,

$$f^0(x_0;x+y)\le \sup_{\substack{\|z-x_0\|<\delta\\0<\lambda<\delta}}\frac{f(z+\lambda(x+y))-f(z)}{\lambda}\le f^0(x_0;x)+f^0(x_0;y)+\varepsilon.$$

Furthermore, since $\varepsilon>0$ was arbitrary, $f^0(x_0;x+y)\le f^0(x_0;x)+f^0(x_0;y)$. This shows that the mapping $y\mapsto f^0(x_0;y)$ from X into $\mathbb{R}$ is sublinear.

We now verify the local boundedness condition. Since f is locally Lipschitz on U, there exist $\delta>0$ and $L>0$ such that $B(x_0,3\delta)\subseteq U$ and $|f(x)-f(y)|\le L\|x-y\|$ for all $x,y\in B(x_0,3\delta)$. Let $x\in B(x_0,\delta)$ and $y\in X\setminus\{0\}$. Then define $y':=y/\|y\|$. Note that

$$f^0(x;y')=\limsup_{\substack{z\to x\\\lambda\to0^+}}\frac{f(z+\lambda y')-f(z)}{\lambda}\le \sup_{\substack{\|z-x\|<\delta\\0<\lambda<\delta}}\frac{f(z+\lambda y')-f(z)}{\lambda}.$$

Fix $z\in B(x,\delta)$ and $0<\lambda<\delta$. Then

$$\begin{aligned}\|(z+\lambda y')-x_0\|&=\|z+\lambda y'-x+(x-x_0)\|\\&\le\|z+\lambda y'-x\|+\|x-x_0\|\\&\le\|z-x\|+\lambda+\|x-x_0\|<3\delta.\end{aligned}$$

Therefore $z+\lambda y'\in B(x_0,3\delta)$. Thus

$$\frac{f(z+\lambda y')-f(z)}{\lambda}\le\frac{L\|\lambda y'\|}{\lambda}=L\|y'\|=L.$$

Since both $z\in B(x,\delta)$ and $0<\lambda<\delta$ were arbitrary,

$$f^0(x;y')\le \sup_{\substack{\|z-x\|<\delta\\0<\lambda<\delta}}\frac{f(z+\lambda y')-f(z)}{\lambda}\le L.$$

Furthermore, from the positive homogeneity of $z \mapsto f^0(x;z)$ it follows that

$$f^0(x;y) \le L\|y\|.$$

This completes the proof. □

The following result follows immediately from the definition of the Clarke generalised directional derivative.

Proposition 3.6.3. *Let U be a nonempty open subset of a normed linear space $(X, \|\cdot\|)$, and let $f : U \to \mathbb{R}$ be a locally Lipschitz function. If $y \in X$, then the mapping $x \mapsto f^0(x;y)$ from U into $\mathbb{R}$ is upper semicontinuous on U.*

Exercise 3.6.4. Let U be a nonempty open subset of a normed linear space $(X, \|\cdot\|)$, and let $f : U \to \mathbb{R}$ be a locally Lipschitz function. Show that if $x_0 \in U$ and $y \in X$, then

$$f^0(x_0;-y) = (-f)^0(x_0;y).$$

Proposition 3.6.5. *Let U be a nonempty open subset of a normed linear space $(X, \|\cdot\|)$, and let $f : U \to \mathbb{R}$ be a locally Lipschitz function. If $x_0 \in U$ and $y \in X$, then*

$$-f^0(x_0;-y) = \liminf_{\substack{z \to x_0 \\ \lambda \to 0^+}} \frac{f(z+\lambda y) - f(z)}{\lambda} \le f^-(x_0;y) \le f^+(x_0;y).$$

Proof. From Exercise 3.6.4 we get that

$$\begin{aligned}
-f^0(x_0;-y) &= -[(-f)^0(x_0;y)] \\
&= -\limsup_{\substack{z \to x_0 \\ \lambda \to 0^+}} \frac{(-f)(z+\lambda y) - (-f)(z)}{\lambda} \\
&= \liminf_{\substack{z \to x_0 \\ \lambda \to 0^+}} \frac{f(z+\lambda y) - f(z)}{\lambda},
\end{aligned}$$

which completes the proof. □

Remarks 3.6.6. By replacing y with $-y$ in the above proposition we obtain that

$$-f^0(x_0;y) \le f^+(x_0;-y).$$

Associated with the Clarke directional derivative is the following set-valued mapping. Let U be a nonempty open subset of a normed linear space $(X, \|\cdot\|)$, and let $f : U \to \mathbb{R}$ be a locally Lipschitz function. Then the mapping $x \mapsto \partial_C f(x)$, called the *Clarke subdifferential mapping*, is defined by

$$\partial_C f(x) := \{x^* \in X^* : x^*(y) \le f^0(x;y) \text{ for all } y \in X\}.$$

It follows from Theorem 3.6.1 and Lemma 3.3.7 that if $\varphi : U \to \mathbb{R}$ is a continuous convex function defined on a nonempty open and convex subset U of a normed linear space $(X, \|\cdot\|)$, then $\partial\varphi(x) = \partial_C\varphi(x)$ for all $x \in U$. Note that $\partial_C\varphi(x)$ is defined since φ is locally Lipschitz on U by Proposition 3.3.6. In this way, we see that the Clarke subdifferential mapping, named after Frank Clarke [60, 61], is a legitimate generalisation of the classical convex subdifferential mapping.

Proposition 3.6.7. *Let U be a nonempty open subset of a normed linear space $(X, \|\cdot\|)$, and let $f : U \to \mathbb{R}$ be a locally Lipschitz function. If $x_0 \in U$ and $y \in X \setminus \{0\}$, then*

$$-f^0(x_0; -y) \le f^-(x_0; y) \le f^+(x_0; y) \le f^0(x_0; y)$$

and

$$\{x^*(y) \in \mathbb{R} : x^* \in \partial_C f(x_0)\} = [-f^0(x_0; -y), f^0(x_0; y)].$$

In particular, $\partial_C f(x_0) \neq \varnothing$.

Proof. The fact that $f^-(x_0; y) \le f^+(x_0; y) \le f^0(x_0; y)$ is obvious from their definitions. The fact that $-f^0(x_0; -y) \le f^-(x_0; y)$ follows from Proposition 3.6.5.

Let $x_0 \in U$ and $y \in X \setminus \{0\}$. We will first show that

$$\{x^*(y) \in \mathbb{R} : x^* \in \partial_C f(x_0)\} \subseteq [-f^0(x_0; -y), f^0(x_0; y)].$$

Indeed, let $x^* \in \partial_C f(x_0)$. Then $x^*(y) \le f^0(x_0; y)$ and $-x^*(y) = x^*(-y) \le f^0(x_0; -y)$ (or, equivalently, $-f^0(x_0; -y) \le x^*(y)$). Therefore $x^*(y) \in [-f^0(x_0; -y), f^0(x_0; y)]$. So now we consider the reverse set inclusion. By above we get that $-f^0(x_0; -y) \le f^0(x_0; y)$. Choose $\alpha \in [-f^0(x_0; -y), f^0(x_0; y)]$. Then define $g_\alpha : \operatorname{span}\{y\} \to \mathbb{R}$ by

$$g_\alpha(\lambda y) := \lambda\alpha \quad \text{for } \lambda \in \mathbb{R}.$$

Now fix $0 < \lambda < \infty$. Then by the positive homogeneity of $z \mapsto f^0(x_0; z)$ we have

$$-f^0(x_0; (-\lambda)y) \le \lambda\alpha \le f^0(x_0; \lambda y).$$

Therefore $g_\alpha(\lambda y) \le f^0(x_0; \lambda y)$ and $g_\alpha((-\lambda)y) \le f^0(x_0; (-\lambda)y)$. Hence we obtain

$$g_\alpha(\lambda y) \le f^0(x_0; \lambda y) \quad \text{for all } \lambda \in \mathbb{R}.$$

Thus by the Hahn–Banach theorem (see Theorem 3.3.5) there exists a linear functional $G_\alpha : X \to \mathbb{R}$ that extends g_α so that $G_\alpha(z) \le f^0(x_0; z)$ for all $z \in X$. Since $f^0(x_0; z) \le L\|z\|$ for some $L > 0$ and all $z \in X$, we have that $G_\alpha \in X^*$. Thus $G_\alpha \in \partial_C f(x_0)$. Furthermore, $\alpha = g_\alpha(y) = G_\alpha(y)$. Therefore $\alpha \in \{x^*(y) \in \mathbb{R} : x^* \in \partial_C f(x_0)\}$. □

Theorem 3.6.8 ([61]). *Let U be a nonempty open subset of a normed linear space $(X, \|\cdot\|)$, and let $f : U \to \mathbb{R}$ be a locally Lipschitz function. Then the Clarke subdifferential mapping $x \mapsto \partial_C f(x)$ is a weak*-cusco on U.*

Proof. Let $x_0 \in U$. We will show that $x \mapsto \partial_C f(x)$ is a weak*-cusco at x_0. Firstly, by Proposition 3.6.7, $\partial_C f(x_0)$ is nonempty. We next show that $\partial_C f(x_0)$ is a weak*-compact convex subset of X^*. For each $y \in X$, let

$$F_y := \{x^* \in X^* : x^*(y) \le f^0(x_0; y)\} = (\widehat{y})^{-1}(-\infty, f^0(x_0; y)].$$

Thus each F_y is weak*-closed and convex. Now $\partial_C f(x_0) = \bigcap_{y \in X} F_y$. Therefore $\partial_C f(x_0)$ is weak*-closed and convex. Let us now show that $x \mapsto \partial_C f(x)$ is locally bounded at x_0. (Note that then this will automatically show that $\partial_C f(x_0)$ is weak* compact by the Banach–Alaoglu theorem [1].) By Lemma 3.6.2 there exist positive numbers L and δ such that $B(x_0, \delta) \subseteq U$ and $f^0(x; y) \le L\|y\|$ for all $x \in B(x_0, \delta)$ and $y \in Y$. Let $x \in B(x_0, \delta)$ and $x^* \in \partial_C f(x)$. Then

$$x^*(y) \le f^0(x_0; y) \le L\|y\| \quad \text{for all } y \in X.$$

Therefore $\|x^*\| \le L$. Note that we used here the simple fact that if $x^*(y) \le L\|y\|$ for all $y \in X$, then $\|x^*\| \le L$. This shows that $x \mapsto \partial_C f(x)$ is locally bounded at x_0.

We now only need to show that $x \mapsto \partial_C f(x)$ is weak*-upper semicontinuous at x_0. So suppose, to obtain a contradiction, that $\partial_C f$ is not weak* upper semicontinuous at $x_0 \in U$. Then there exists a weak* open subset W of X^*, containing $\partial_C f(x_0)$, such that $\partial_C f(B(x_0, \delta)) \not\subseteq W$ for every $\delta > 0$. Therefore, in particular, there exist sequences $(x_n : n \in \mathbb{N})$ in U and $(x_n^* : n \in \mathbb{N})$ in X^* such that $x_0 = \lim_{n\to\infty} x_n$ and $x_n^* \in \partial_C f(x_n) \setminus W$ for all $n \in \mathbb{N}$. Furthermore, by above, we can assume that the sequence $(x_n^* : n \in \mathbb{N})$ is norm bounded in $(X^*, \|\cdot\|)$. Hence by the Banach–Alaoglu theorem [1] the sequence $(x_n^* : n \in \mathbb{N})$ has a weak* cluster-point x_∞^*, which must lie in $X^* \setminus W$. We will obtain our desired contradiction by showing that $x_\infty^* \in \partial_C f(x_0) \subseteq W$. To this end, fix $y \in X$ and $\varepsilon > 0$. Since $x \mapsto f^0(x; y)$ is upper semicontinuous at x_0, there exists $\delta > 0$ such that $B(x_0, \delta) \subseteq U$ and $f^0(x; y) < f^0(x_0; y) + \varepsilon$ for all $x \in B(x_0, \delta)$. Since $x_0 = \lim_{n\to\infty} x_n$, there exists $N \in \mathbb{N}$ such that $x_n \in B(x_0, \delta)$ for all $n > N$. Therefore as

$$x_n^*(y) \le f^0(x_n; y) < f^0(x_0; y) + \varepsilon \quad \text{for all } n > N,$$

we have that $x_\infty^*(y) \le f^0(x_0; y) + \varepsilon$. Since $y \in X$ and $\varepsilon > 0$ were both arbitrary,

$$x_\infty^*(y) \le f^0(x_0; y) \quad \text{for all } y \in X,$$

that is, $x_\infty^* \in \partial_C f(x_0)$, as desired. □

Motivated by the earlier results for continuous convex functions, we explore what it means when the Clarke subdifferential mapping is single-valued (single-valued and norm upper semicontinuous).

Our investigations will require, in addition to the notions of Gâteaux differentiability and Fréchet differentiability, the following stronger notions of differentiability.

Let $f : U \to \mathbb{R}$ be a locally Lipschitz function defined on a nonempty open subset U of a normed linear space $(X, \|\cdot\|)$. We say that f is *strictly differentiable* (*strictly Fréchet differentiable*) at $x_0 \in U$ if there exists a continuous linear functional $x^* \in X^*$ such that for all $\varepsilon > 0$ and $y \in S_X$, there exists $\delta > 0$ such that

$$\left|\frac{f(z+\lambda y) - f(z)}{\lambda} - x^*(y)\right| < \varepsilon$$

whenever $0 < \lambda < \delta$ and $\|z - x\| < \delta$ (uniformly over $y \in S_X$).

Clearly, strict differentiability implies Gâteaux differentiability, but the reverse implication does not hold; see Example 3.6.11.

Proposition 3.6.9. *Let U be a nonempty open subset of a normed linear space $(X, \|\cdot\|)$, and let $f : U \to \mathbb{R}$ be a locally Lipschitz function. If $x_0 \in U$, then f is strictly differentiable at x_0 if and only if $\partial_C f(x_0)$ is a singleton.*

Proof. Suppose that f is strictly differentiable at $x_0 \in U$. Then for each $y \in X$,

$$\begin{aligned} f^0(x_0; y) &= \limsup_{\substack{z\to x_0\\ \lambda\to 0^+}} \frac{f(z+\lambda y) - f(z)}{\lambda} \\ &= \lim_{\substack{z\to x_0\\ \lambda\to 0^+}} \frac{f(z+\lambda y) - f(z)}{\lambda} \\ &= \liminf_{\substack{z\to x_0\\ \lambda\to 0^+}} \frac{f(z+\lambda y) - f(z)}{\lambda} \\ &= -f^0(x_0; -y) \quad \text{by Proposition 3.6.5,} \end{aligned}$$

that is, $f^0(x_0; y) = -f^0(x_0; -y)$ for all $y \in Y$. It follows then from Proposition 3.6.7 that if $x^*, y^* \in \partial_C f(x_0)$. Then

$$f^0(x_0; y) = x^*(y) = -f^0(x_0; -y) \quad \text{and} \quad f^0(x_0; y) = y^*(y) = -f^0(x_0; -y) \quad \text{for all } y \in X,$$

that is, $x^*(y) = y^*(y)$ for all $y \in X$. Therefore $x^* = y^*$, which shows that $\partial_C f(x_0)$ is a singleton. On the other hand, if $x_0 \in U$ and $\partial_C f(x_0)$ is a singleton, then by the second part of Proposition 3.6.7 we have that $f^0(x_0; y) = -f^0(x_0; -y)$ for each $y \in X$. Let $y \in X$. Then

$$\begin{aligned} \limsup_{\substack{z\to x_0\\ \lambda\to 0^+}} \frac{f(z+\lambda y) - f(z)}{\lambda} &= f^0(x_0; y) \quad \text{by definition} \\ &= -f^0(x_0; -y) \\ &= \liminf_{\substack{z\to x_0\\ \lambda\to 0^+}} \frac{f(z+\lambda y) - f(z)}{\lambda} \quad \text{by Proposition 3.6.5,} \end{aligned}$$

that is, $\lim_{\substack{z\to x_0\\ \lambda\to 0^+}} \frac{f(z+\lambda y)-f(z)}{\lambda}$ exists for each $y \in X$. Furthermore, since $y \mapsto f^0(x_0; y)$ is sublinear and $f^0(x_0; -y) = -f^0(x_0; y)$ for all $y \in X$, we have by Exercise 3.3.4 that $y \mapsto$

$f^0(x_0; y)$ is linear. Moreover, by Lemma 3.6.2 there exists a positive number L such that $f^0(x_0; y) \le L\|y\|$ for all $y \in X$. Therefore $y \mapsto f^0(x_0; y)$ is a continuous linear functional on X, and

$$f^0(x_0; y) = \lim_{\substack{z \to x_0 \\ \lambda \to 0^+}} \frac{f(z + \lambda y) - f(z)}{\lambda} \quad \text{for all } y \in X.$$

Thus f is strictly differentiable at x_0. □

Corollary 3.6.10. *Let U be a nonempty open subset of a normed linear space $(X, \|\cdot\|)$, and let $f : U \to \mathbb{R}$ be a locally Lipschitz function. If (X^*, weak^*) possesses property (S) and $x \mapsto \partial_C f(x)$ is a minimal weak* cusco, then f is strictly differentiable at the points of a dense G_δ-subset of U.*

Proof. By Proposition 3.3.34, $x \mapsto \partial_C f(x)$ is single-valued at the points of a dense G_δ-subset of U. Then the result follows from Proposition 3.6.9. □

It might be tempting to think that the Clarke subdifferential mapping is always a minimal weak* cusco, as it is for continuous convex functions. However, there are many examples showing that this is false. Indeed, there is an example of a 1-Lipschitz function $f : \mathbb{R} \to \mathbb{R}$ such that $\partial_C f(x) = [-1, 1]$ for all $x \in \mathbb{R}$; see [273, p. 97] (also see [38, 39]). There are also examples of everywhere differentiable Lipschitz functions whose subdifferentials are not minimal weak* cuscos; see [37, Example 8.2].

For continuous convex functions, Gâteaux differentiability coincides with strict differentiability, as do Fréchet differentiability and strict Fréchet differentiability. However, in general, these notions are distinct.

Example 3.6.11. Let $f : \mathbb{R} \to \mathbb{R}$ be defined by

$$f(x) := \begin{cases} x^2 \sin(1/x) & \text{if } x \neq 0, \\ 0 & \text{if } x = 0. \end{cases}$$

Then f is differentiable everywhere on $\mathbb{R}$, but f is not strictly differentiable at $x = 0$. In fact, $f'(0) = 0$, whereas $\partial_C f(0) = [-1, 1]$.

Lemma 3.6.12 (Lebourg mean-value theorem [182]). *Let U be a nonempty open convex subset of a normed linear space $(X, \|\cdot\|)$, and let $f : U \to \mathbb{R}$ be a locally Lipschitz function. Suppose that $x_0 \in U$, $v \in X \setminus \{0\}$ and $0 < \lambda < \infty$. If $x_0 + \lambda v \in U$, then there exists a point $x' \in (x_0, x_0 + \lambda v)$ such that*

$$\frac{f(x_0 + \lambda v) - f(x_0)}{\lambda} = x^*(v) \quad \textit{for some } x^* \in \partial_C f(x').$$

Proof. We start by defining $g : [0, \lambda] \to \mathbb{R}$ by

$$g(t) := [f(x_0 + tv) - f(x_0)] - \left[\frac{f(x_0 + \lambda v) - f(x_0)}{\lambda}\right] \cdot t.$$

Note that $g(0) = g(\lambda) = 0$. We will show that there exists $0 < t_0 < \lambda$ such that either (i) $f^+(x'; v) \le \alpha$ and $f^+(x'; -v) \le -\alpha$ or (ii) $\alpha \le f^+(x'; v)$ and $-\alpha \le f^+(x'; -v)$, where $x' := x_0 + t_0 v$ and $\alpha := [f(x_0 + \lambda v) - f(x_0)]/\lambda$. If g is constant on $[0, \lambda]$ (i. e. $g(t) = 0$ for all $t \in [0, \lambda]$), then

$$f(x_0 + tv) = f(x_0) + \alpha t \quad \text{for all } t \in [0, \lambda].$$

Therefore $f^+(x_0 + tv; v) = \alpha$ and $f^+(x_0 + tv; -v) = -\alpha$ for all $t \in (0, \alpha)$. Hence we may suppose that g is not constant on $[0, \lambda]$. Thus either
(i) $0 = g(0) = g(\lambda) < \max\{g(t) : t \in [0, \lambda]\}$, or
(ii) $\min\{g(t) : t \in [0, \lambda]\} < 0 = g(0) = g(\lambda)$.

Case (i). Choose $t_0 \in (0, \lambda)$ such that $g(t_0) = \max\{g(t) : t \in [0, \lambda]\}$. Then

$$\frac{f(x' + \lambda' v) - f(x')}{\lambda'} - \alpha = \frac{g(t_0 + \lambda') - g(t_0)}{\lambda'} \le 0 \quad \text{for all } 0 < \lambda' < \lambda - t_0,$$

where $x' := x_0 + t_0 v$. Therefore $f^+(x'; v) \le \alpha$. Also,

$$\frac{f(x' + \lambda'(-v)) - f(x')}{\lambda'} + \alpha = \frac{g(t_0 - \lambda') - g(t_0)}{\lambda'} \le 0 \quad \text{for all } 0 < \lambda' < t_0,$$

where $x' := x_0 + t_0 v$. Therefore $f^+(x'; -v) \le -\alpha$.

Case (ii). Choose $t_0 \in (0, \lambda)$ such that $g(t_0) = \min\{g(t) : t \in [0, \lambda]\}$. Then

$$\frac{f(x' + \lambda' v) - f(x')}{\lambda'} - \alpha = \frac{g(t_0 + \lambda') - g(t_0)}{\lambda'} \ge 0 \quad \text{for all } 0 < \lambda' < \lambda - t_0,$$

where $x' := x_0 + t_0 v$. Therefore $f^+(x'; v) \ge \alpha$. Also,

$$\frac{f(x' + \lambda'(-v)) - f(x')}{\lambda'} + \alpha = \frac{g(t_0 - \lambda') - g(t_0)}{\lambda'} \ge 0 \quad \text{for all } 0 < \lambda' < t_0,$$

where $x' := x_0 + t_0 v$. Therefore $f^+(x'; -v) \ge -\alpha$.

It now follows from Proposition 3.6.5, Remark 3.6.6 and the first part of Proposition 3.6.7 that in both cases,

$$\alpha \in [-f^0(x'; -v), f^0(x'; v)].$$

Then the result follows from the second part of Proposition 3.6.7. □

Proposition 3.6.13. *Let U be a nonempty open subset of a normed linear space $(X, \|\cdot\|)$, and let $f : U \to \mathbb{R}$ be a locally Lipschitz function. If $x_0 \in U$, then f is strictly Fréchet differentiable at x_0 if and only if $x \mapsto \partial_C f(x)$ is single-valued and norm upper semicontinuous at x_0.*

Proof. Suppose that f is strictly Fréchet differentiable at $x_0 \in U$ with Fréchet derivative x_0^*. Let $\varepsilon > 0$. We will show that there exists $\delta > 0$ such that $\partial_C f(B(x_0, \delta)) \subseteq B[x_0^*, \varepsilon]$. Since f is strictly Fréchet differentiable at x_0, there exists $\delta > 0$ such that

$$\left|\frac{f(z+\lambda v) - f(z)}{\lambda} - x_0^*(v)\right| < \varepsilon \quad \text{for all } v \in S_X, \text{ all } \|z - x_0\| < \delta \text{ and all } 0 < \lambda < \delta.$$

Therefore

$$\frac{f(z+\lambda v) - f(z)}{\lambda} \leq x_0^*(v) + \varepsilon \quad \text{for all } v \in S_X, \text{ all } \|z - x_0\| < \delta \text{ and all } 0 < \lambda < \delta.$$

Thus $f^0(z; v) \leq x_0^*(v) + \varepsilon$ for all $v \in S_X$ and $z \in B(x_0, \delta)$. Now let x be any element of $B(x_0, \delta)$, x^* be any element of $\partial_C f(x)$, and v be any element of S_X. Then

$$(x^* - x_0^*)(v) = x^*(v) - x_0^*(v) \leq f^0(x; v) - x_0^*(v) \leq (x_0^*(v) + \varepsilon) - x_0^*(v) = \varepsilon.$$

Since $v \in S_X$ is arbitrary, $\|x^* - x_0^*\| \leq \varepsilon$. Thus $\partial_C f(B(x_0, \delta)) \subseteq B[x_0^*, \varepsilon]$. This shows that $x \mapsto \partial_C f(x)$ is single-valued and norm upper semicontinuous at x_0.

Conversely, suppose that, $x \mapsto \partial_C f(x)$, is single-valued and norm upper semicontinuous at $x_0 \in U$. Let $\{x_0^*\} := \partial_C f(x_0)$. We claim that x_0^* is the strict Fréchet derivative of f at x_0. To this end, consider $\varepsilon > 0$. Choose $\delta > 0$ such that $\partial_C f(B(x_0, 2\delta)) \subseteq B(x_0^*, \varepsilon)$. Let $v \in S_X$, $z \in B(x_0, \delta)$ and $0 < \lambda < \delta$. Then $z + \lambda v \in B(x_0, 2\delta)$, and by Lemma 3.6.12

$$\begin{aligned}\left|\frac{f(z+\lambda v) - f(z)}{\lambda} - x_0^*(v)\right| &= |x^*(v) - x_0^*(v)| \quad \text{for some } x^* \in \partial_C f(z'),\ z' \in (z, z+\lambda v) \\ &= |(x^* - x_0^*)(v)| \leq \|x^* - x_0^*\| < \varepsilon.\end{aligned}$$

This shows that x_0^* is the strict Fréchet derivative of f at x_0. □

Corollary 3.6.14. *Let U be a nonempty open subset of a normed linear space $(X, \|\cdot\|)$, and let $f : U \to \mathbb{R}$ be a locally Lipschitz function. If $(X, \|\cdot\|)$ is an Asplund space and $x \mapsto \partial_C f(x)$ is a minimal weak* cusco, then f is strictly Fréchet differentiable at the points of a dense G_δ-subset of U.*

Proof. Following the proof of (iii)⇒(iv) in Theorem 3.3.37, we see that $x \mapsto \partial_C f(x)$ is single-valued and norm upper semicontinuous at the points of a dense G_δ-subset of U. Then the result follows from Proposition 3.6.13. □

We now turn to the classical results on D-representability.

Proposition 3.6.15. *Let Ω be a densely defined set-valued mapping from a topological space (A, τ') into subsets of a Hausdorff topological space (X, τ) (Hausdorff locally convex space $(X, +, \cdot, \tau)$). If the graph of Ω is contained in the graph of an usco (cusco) mapping $\Phi : A \to 2^X$, then there exists a unique smallest usco (cusco) containing Ω, denoted* USC(Ω) (CSC(Ω))*, given by*

$$\mathrm{USC}(\Omega)(x) := \bigcap\{\overline{\Omega(U)} : U \text{ is an open neighbourhood of } x\},$$
$$\mathrm{CSC}(\Omega)(x) := \bigcap\{\overline{\mathrm{co}}\, \Omega(U) : U \text{ is an open neighbourhood of } x\}.$$

Proof. We only prove that CSC(Ω) is the smallest cusco containing Ω, as the proof that USC(Ω) is the smallest usco containing Ω, is identical to this. We begin with the following four observations:
(i) for each $x \in \mathrm{Dom}(\Omega)$, $\Omega(x) \subseteq \mathrm{CSC}(\Omega)(x)$;
(ii) for any set-valued mapping $\Psi : A \to 2^X$, CSC(Ψ) possesses a closed graph;
(iii) if $\Psi : A \to 2^X$ is a cusco, then $\Psi = \mathrm{CSC}(\Psi)$;
(iv) for any set-valued mapping $\Psi : X \to 2^X$, CSC(Ψ) has convex images.

We now show that CSC(Ω) is a cusco mapping on (A, τ'). From (iii) and the definition of CSC(Ω) it follows that

$$\mathrm{Gr}(\mathrm{CSC}(\Omega)) \subseteq \mathrm{Gr}(\mathrm{CSC}(\Phi)) = \mathrm{Gr}(\Phi).$$

Furthermore, by (ii) we have that the graph of CSC(Ω) is closed, so by Corollary 1.1.14 CSC(Ω) is an usco. Therefore by (iv) CSC(Ω) is a cusco. To see that CSC(Ω) is the smallest cusco containing Ω, it suffices to observe that for any cusco $\Psi : A \to 2^X$ containing Ω,

$$\mathrm{Gr}(\mathrm{CSC}(\Omega)) \subseteq \mathrm{Gr}(\mathrm{CSC}(\Psi)) = \mathrm{Gr}(\Psi).$$

This completes the proof. □

Note that in the above proposition the set-valued mapping CSC(Ω) is called the *cusco generated by* Ω and USC(Ω) is called the *usco generated by* Ω .

Remarks 3.6.16. It is possible to strengthen the previous proposition so as to only require that for each $x \in A$, there exist an open neighbourhood U_x of x and an usco (cusco) Φ_x defined on U_x such that $\Omega(y) \subseteq \Phi_x(y)$ for each $y \in U_x \cap \mathrm{Dom}(\Omega)$. In this way, we see that the graph of any densely defined locally bounded mapping into the dual of a normed linear space is contained in the graph of a weak* cusco.

Let U be a nonempty open subset of a normed linear space $(X, \|\cdot\|)$, and let $f : U \to \mathbb{R}$ be a locally Lipschitz function. Then we say that f is *D-representable on* U if:
(i) $D := \{x \in U : f \text{ is Gâteaux differentiable at } x\}$ is dense in U and
(ii) for each dense subset D^* of D, we have that $\partial_C f = \mathrm{CSC}(\Omega_{D^*})$, where $\Omega_{D^*} : D^* \to 2^{X^*}$ is defined by $\Omega_{D^*}(x) := \{\nabla f(x)\}$ for $x \in D^*$.

Here $\nabla f(x)$ denotes the Gâteaux derivative of f at x. Note that, in particular, we have that $\partial_C f = \mathrm{CSC}(\Omega_D)$.

A set-valued mapping $\Phi : A \to 2^X$ from a topological space (A, τ') into subsets of a linear topological space $(X, +, \cdot, \tau)$ is *hyperplane minimal* if for any open half-space W in $(X, +, \cdot, \tau)$ and open set U in (A, τ') with $\Phi(U) \cap W \neq \varnothing$, there exists a nonempty τ'-open subset V of U such that $\Phi(V) \subseteq W$. It follows from Proposition 3.3.22 that a cusco $\Phi : A \to 2^X$ from a topological space (A, τ') into subsets of a Hausdorff locally topological space $(X, +, \cdot, \tau)$ is a minimal cusco if and only if it is hyperplane minimal.

Note also that if $\Phi : A \to 2^X$ is a minimal cusco and $\Psi : A \to 2^X$ has $\mathrm{Gr}(\Psi) \subseteq \mathrm{Gr}(\Phi)$, then Ψ will also be hyperplane minimal. If $f : A \to X$, then we say that f is *hyperplane minimal* if the set-valued mapping $\Phi : A \to 2^X$ defined by $\Phi(a) := \{f(a)\}$ for $a \in A$ is hyperplane minimal. Clearly, if $f : A \to X$ is a quasicontinuous function, then f is hyperplane minimal. However, the following simple example shows that in general a hyperplane minimal function need not be quasicontinuous.

Example 3.6.17. Let $X := Y := \mathbb{R}$ with the usual topology. Define $f : X \to 2^Y$ by

$$f(x) := \begin{cases} -1 & \text{if } x < 0, \\ 0 & \text{if } x = 0, \\ 1 & \text{if } 0 < x. \end{cases}$$

Then f is hyperplane minimal but not quasicontinuous.

To see the relationship between minimal cuscos and D-representability, we need the following cusco version of Theorem 1.3.10.

Theorem 3.6.18 ([36, 37]). *Let $\Phi : X \to 2^Y$ be a cusco mapping acting from a topological space (X, τ) into subsets of a Hausdorff locally convex space $(Y, +, \cdot, \tau')$. Then the following properties are equivalent:*

(i) *Φ is a minimal cusco on (X, τ);*

(ii) *for every densely defined selection σ of Φ, $CSC(\sigma) = \Phi$;*

(iii) *there exists a densely defined selection σ of Φ such that $CSC(\sigma|_{D^*}) = \Phi$ for every dense subset D^* of $Dom(\sigma)$;*

(iv) *there is a densely defined hyperplane minimal selection σ of Φ such that $CSC(\sigma) = \Phi$.*

Proof. (i)⇒(ii). Suppose that Φ is a minimal cusco on (X, τ). Let $\sigma : D \to Y$ be a densely defined selection of Φ. By Proposition 3.6.15 $\mathrm{CSC}(\sigma)$ is a cusco on (X, τ) with $\mathrm{Gr}(\mathrm{CSC}(\sigma)) \subseteq \mathrm{Gr}(\Phi)$. Since Φ is a minimal usco on (X, τ), we have that $\mathrm{CSC}(\sigma) = \Phi$.
(ii)⇒(iii). This implication is obvious.

(iii)⇒(iv). Let $\sigma : D \to Y$ be a densely defined selection of Φ such that $\mathrm{CSC}(\sigma|_{D^*}) = \Phi$ for every dense subset D^* of D. We claim that σ is hyperplane minimal on D. We will verify this indirectly. So suppose, to obtain a contradiction, that σ is not hyperplane minimal on D. Then there exist an open subset U of (X, τ) and an open half-space W of $(Y, +, \cdot, \tau')$ with $\sigma(U) \cap W \neq \varnothing$ such that $\sigma(V \cap D) \not\subseteq W$ for any nonempty τ-open subset V of U. This means that $D' := \{x \in U \cap D : \sigma(x) \notin W\}$ is dense in $U \cap D$. Let $D^* := (D \setminus U) \cup D'$. Then D^* is dense in D. Therefore

$$\Phi(U) = \mathrm{CSC}(\sigma|_{D^*})(U) \subseteq \overline{\mathrm{co}}(\sigma|_{D^*}(U \cap D^*)) = \overline{\mathrm{co}}(\sigma|_{D^*}(D')) \subseteq Y \setminus W,$$

and so $\sigma(U) \subseteq Y \setminus W$, that is, $\sigma(U) \cap W = \varnothing$. However, this contradicts the fact that $\sigma(U) \cap W \neq \varnothing$. Hence σ must be hyperplane minimal on D. Finally, by assumption, $\Phi = \mathrm{CSC}(\sigma)$.

(iv)⇒(i). Suppose that $\sigma : D \to Y$ is a densely defined hyperplane minimal selection of Φ such that $\Phi = \mathrm{CSC}(\sigma)$. We will proceed via Proposition 3.3.22. To this end, suppose that U is an open subset of (X, τ) and K is a closed and convex subset of $(Y, +, \cdot, \tau')$ such that $\Phi(U) \not\subseteq K$. Since $\Phi = \mathrm{CSC}(\sigma)$, it follows from the definition of $\mathrm{CSC}(\sigma)$ that $\sigma(U \cap D) \not\subseteq K$. Let $y_0 \in \sigma(U \cap D) \setminus K$. Then there exist $x^* \in Y^*$ and $\varepsilon > 0$ such that

$$s := \sup\{x^*(k) : k \in K\} < x^*(y_0) - \varepsilon.$$

Let $W := \{y \in Y : (s + \varepsilon/2) < x^*(y)\}$. Then W is an open half-space of $(Y, +, \cdot, \tau')$, containing y_0, such that $K \cap \overline{W} = \varnothing$. In particular, $\sigma(U \cap D) \cap W \neq \varnothing$. Since σ is hyperplane minimal on D, there exists a nonempty τ-open subset V of U such that $\sigma(V \cap D) \subseteq W$. Now by the definition of $\mathrm{CSC}(\sigma)$ we have that

$$\Phi(V) = \mathrm{CSC}(\sigma)(V) \subseteq \overline{\mathrm{co}}(\sigma(V \cap D)) \subseteq \overline{\mathrm{co}}(W) \subseteq Y \setminus K,$$

that is, $\Phi(V) \cap K = \varnothing$. □

Corollary 3.6.19. *Let U be a nonempty open subset of a normed linear space $(X, \|\cdot\|)$, and let $f : U \to \mathbb{R}$ be densely Gâteaux differentiable locally Lipschitz function on U. Then f is D-representable on U if and only if $x \mapsto \partial_C f(x)$ is a minimal weak* cusco on U.*

Proof. The result follows from parts (i) and (iii) of Theorem 3.6.18. □

So we see that by a desire to consider locally Lipschitz functions that are D-representable we are inextricably lead to consider locally Lipschitz functions whose Clarke sudifferential mappings are minimal weak* cuscos.

In the next part of this section, we characterise the minimality of the Clarke subdifferential mapping.

Proposition 3.6.20. *Let $\Phi : X \to 2^Y$ be a cusco acting from a topological space (X, τ) into subsets of a Hausdorff locally convex space $(Y, +, \cdot, \tau')$. Then Φ is a minimal cusco on X if and only if for each $y^* \in Y^*$, $(y^* \circ \Phi) : X \to 2^{\mathbb{R}}$ is a minimal cusco.*

Proof. The proof is left to the reader. It closely follows the proof of Theorem 1.3.7. □

Proposition 3.6.21. *Let $\Phi : X \to 2^{\mathbb{R}}$ be a minimal cusco defined on a topological space (X, τ), and let $\sigma : D \to \mathbb{R}$ be a densely defined selection of Φ. Then* $\mathrm{CSC}(\sigma) = \Phi$ *if and only if for each $x \in X$,*

$$\limsup_{\substack{z \to x \\ z \in D}} \sigma(z) = \max\{y : y \in \Phi(x)\} \quad \textit{and} \quad \liminf_{\substack{z \to x \\ z \in D}} \sigma(z) = \min\{y : y \in \Phi(x)\}.$$

Proof. This result follows directly from the definition of $\mathrm{CSC}(\sigma)$. □

We now need to recall some classical results concerning the differentiability of real-valued Lipschitz functions defined on open subsets of $\mathbb{R}$.

Theorem 3.6.22 (Lebesgue differentiation theorem [296, p. 320]). *Let $f : U \to \mathbb{R}$ be a locally Lipschitz function defined on a nonempty open subset U of $\mathbb{R}$. Then: (i) $D := \{t \in U : f'(t) \text{ exists}\}$ is a Borel subset of $\mathbb{R}$; (ii) $\mu(U \setminus D) = 0$; and (iii) if $[a, b]$ is a non-degenerate subinterval of U, then*

$$\int_a^b f'(t)\,\mathrm{d}\mu(t) = f(b) - f(a)$$

(Here and latter, μ denotes the Lebesgue measure on $\mathbb{R}$.)

From this theorem we can easily deduce the following useful mean-value theorem.

Corollary 3.6.23 (Lebesgue mean-value theorem). *Let $f : U \to \mathbb{R}$ be a locally Lipschitz function defined on a nonempty open subset U of $\mathbb{R}$. If $[a, b]$ is a non-degenerate subinterval of U, then there exists a Borel subset M of $[a, b]$ of positive Lebesgue measure such that for each $t \in M$, $f'(t)$ exists, and*

$$\frac{f(b) - f(a)}{b - a} \leq f'(t).$$

To expedite the latter results of this section, we introduce the following definitions. Let A be a nonempty Borel subset of a normed linear space $(X, \|\cdot\|)$, and let $y \in X \setminus \{0\}$. Then a Borel subset S of A is $1 - D$ *almost everywhere in A in the direction y* if for each $x \in A$,

$$\mu(\{t \in \mathbb{R} : x + tu \in A \text{ and } x + ty \notin S\}) = 0.$$

If U is a nonempty open subset of a normed linear space $(X, \|\cdot\|)$ and $f : U \to \mathbb{R}$ is a locally Lipschitz function, then for $y \in X \setminus \{0\}$, we define $T_y^f : U \to 2^{\mathbb{R}}$ by

$$T_y^f(x) := (\widehat{y} \circ \partial_C f)(x) = \{x^*(y) : x^* \in \partial_C f(x)\} = [-f^0(x; -y), f^0(x; y)].$$

For us, the most important example of a $1 - D$ almost everywhere set is the following.

Proposition 3.6.24. *Let $f : U \to \mathbb{R}$ be a locally Lipschitz function defined on a nonempty open subset U of a normed linear space $(X, \|\cdot\|)$. Then for each $y \in X \setminus \{0\}$,*

$$D_y := \{x \in U : f'(x; y) \text{ exists}\}$$

is $1 - D$ almost everywhere in U in the direction y.

Proof. This follows directly from the Lebesgue differentiation theorem. □

Lemma 3.6.25. *Let U be a nonempty open subset of a normed linear space $(X, \|\cdot\|)$, and let $f : U \to \mathbb{R}$ be a locally Lipschitz function. If $y \in X \setminus \{0\}$ and S is a Borel subset of U that is $1 - D$ almost everywhere in U in the direction y, then for each $x \in U$,*

$$f^0(x; y) = \limsup_{\substack{z \to x \\ z \in S}} f^+(z; y) \quad \text{and} \quad f^0(x; -y) = \limsup_{\substack{z \to x \\ z \in S}} f^+(z; -y).$$

Proof. Since both cases are identical, we consider only the first case. We begin by observing that for any $x \in U$, $\limsup_{z\to x} f^+(z;y) \le f^0(x;y)$, and so $\limsup_{\substack{z\to x\\ z\in S}} f^+(z;y) \le f^0(x;y)$. Therefore it is sufficient to show that for each $x \in U$,

$$f^0(x;y) \le \limsup_{\substack{z\to x\\ z\in S}} f^+(z;y).$$

Indeed, to accomplish this, it is sufficient to show the for each $\varepsilon > 0$ and each $\delta > 0$,

$$f^0(x;y) - \varepsilon \le \sup\{f^+(z;y) : \|z - x\| < \delta \text{ and } z \in S\}.$$

So suppose $\varepsilon > 0$ and $\delta > 0$ are given. By the definition of $f^0(x;y)$ there exist $z \in U$ and $\lambda \in (0,1)$ such that $\|(z + \lambda y) - x\| < \delta$, $\|z - x\| < \delta$ and

$$f^0(x;y) - \varepsilon < \frac{f(z + \lambda y) - f(z)}{\lambda}.$$

Let $g : [0,\lambda] \to \mathbb{R}$ be defined by $g(t) := f(z + ty)$. Then g is Lipschitz, and so by the Lebesgue mean-value theorem there exists $t_0 \in (0,\lambda)$ such that $x' := z + t_0 y \in S$ and

$$\frac{f(z + \lambda y) - f(z)}{\lambda} = \frac{g(\lambda) - g(0)}{\lambda} \le g'(t_0) = f'(x';y).$$

Now $\|x' - x\| < \delta$, and so

$$f^0(x;y) - \varepsilon < \frac{f(z + \lambda y) - f(z)}{\lambda} \le f'(x';y) \le \sup\{f^+(z;y) : \|z - x\| < \delta \text{ and } z \in S\},$$

which completes the proof. □

Theorem 3.6.26. *Let U be a nonempty open subset of a normed linear space $(X, \|\cdot\|)$, and let $f : U \to \mathbb{R}$ be a locally Lipschitz function. Then $x \mapsto \partial_C f(x)$ is a minimal weak* cusco on U if and only if for each $y \in S_X$, one of the following conditions holds:*
(i) *the mapping T_y^f is a minimal cusco on U;*
(ii) *the function $D_y : U \to \mathbb{R}$ defined by $D_y(x) := f^+(x;y)$ is hyperplane minimal on U;*
(iii) *the restriction of D_y to a Borel subset S_y of U, which is $1 - D$ almost everywhere in U in the direction y, is hyperplane minimal on S_y.*

Proof. The fact that $x \mapsto \partial_C f(x)$ is a minimal weak* cusco on U if and only if condition (i) holds for all $y \in S_X$ follows from Proposition 3.6.20. We now show that for every $y \in S_X$, conditions (i), (ii) and (iii) are equivalent. So suppose that $y \in S_X$.

(i)⇒(ii). Suppose that T_y^f is a minimal cusco on U. By Proposition 3.6.7 D_y is a selection of T_y^f. Hence by the discussion just after the definition of hyperplane minimal we see that D_y is hyperplane minimal on U.

(ii)⇒(iii). This implication follows from the general fact that if $\Phi : A \to 2^Z$ is a hyperplane minimal mapping from a topological space (A, τ') into subsets of a Hausdorff locally convex space $(Z, +, \cdot, \tau)$ and D is a dense subset of A, then $\Phi|_D$ is a hyperplane minimal mapping on D.

(iii)⇒(i). Suppose that the restriction of D_y to a Borel subset S_y of U, which is $1-D$ almost everywhere in U in the direction y, is hyperplane minimal on S_y. To show that T_y^f is a minimal cusco on U, it is sufficient, because of Theorem 3.6.18 part (iv), to show that $\mathrm{CSC}(\sigma) = T_y^f$, where $\sigma := (D_y)|_{S_y}$. In fact, to show this, it is sufficient to show, because of Proposition 3.6.21, that for every $x \in U$,

$$\limsup_{\substack{z\to x\\ z\in S_y}} f^+(z;y) = f^0(x;y) \quad \text{and} \quad \liminf_{\substack{z\to x\\ z\in S_y}} f^+(z;y) = -f^0(x;-y).$$

To this end, fix $x_0 \in U$ and note that, by possibly making S_y smaller but still $1-D$ almost everywhere in U in the direction y, we may assume that $f'(x;y)$ exists for each $x \in S_y$. Then by Lemma 3.6.25 we have that

$$\limsup_{\substack{z\to x_0\\ z\in S_y}} f^+(z;y) = f^0(x_0;y) \quad \text{and} \quad \limsup_{\substack{z\to x_0\\ z\in S_y}} f^+(z;-y) = f^0(x_0;-y).$$

Furthermore,

$$\limsup_{\substack{z\to x_0\\ z\in S_y}} f^+(z;-y) = \limsup_{\substack{z\to x_0\\ z\in S_y}} f'(z;-y) = \limsup_{\substack{z\to x_0\\ z\in S_y}} -f'(z;y) = -\liminf_{\substack{z\to x_0\\ z\in S_y}} f^+(z;y).$$

Therefore $-f^0(x_0;-y) = \liminf_{\substack{z\to x_0\\ z\in S_y}} f^+(z;y)$. Thus $\mathrm{CSC}(\sigma) = T_y^f$. □

Motivated by Theorem 3.6.26, we consider the following class of locally Lipschitz functions.

We say that a locally Lipschitz function $f : U \to \mathbb{R}$ defined on a nonempty open subset U of a normed linear space $(X, \|\cdot\|)$ is *directionally essentially smooth* if for each $y \in S_X$,

$$S_y^f := \{x \in U : T_y^f \text{ is single-valued}\}$$

is $1-D$ almost everywhere in U in the direction y. Note that the set S_y^f is always a Borel set (in fact, a G_δ-set). Furthermore, the mapping, $x \mapsto f^+(x;y)$ restricted to S_y^f is continuous. Thus by Theorem 3.6.26 every directionally essentially smooth locally Lipschitz function has a Clarke subdifferential mapping that is a minimal weak* cusco.

Exercise 3.6.27. Let U be a nonempty open subset of a normed linear space $(X, \|\cdot\|)$, and let $f : U \to \mathbb{R}$ be a locally Lipschitz function. Show that if $y \in X \setminus \{0\}$ and

$$S_y^f := \{x \in U : T_y^f \text{ is single-valued}\},$$

then $f'(x;y)$ exists for each $x \in S_y^f$, and $z \mapsto f'(z;y)$ is continuous on S_y^f.

For a nonempty open subset U of a normed linear space $(X, \|\cdot\|)$, we denote by $DS_e(U)$ the set of all real-valued directionally essentially smooth locally Lipschitz functions on U.

Proposition 3.6.28. *Let U be a nonempty open subset of a normed linear space $(X, \|\cdot\|)$. Then $DS_e(U)$ is a vector space (in fact, an algebra) that contains all the continuous convex functions on U (in the case where U is convex). Furthermore, each element of $DS_e(U)$ has a Clarke subdifferential mapping that is a minimal weak* cusco.*

Proof. The fact that $DS_e(U)$ is a vector space follows from Proposition 3.6.29 and the fact that for any locally Lipschitz function $f : U \to \mathbb{R}$ and any $0 \leq \lambda < \infty$,
(i) $\lambda \partial_C f(x) = \partial_C(\lambda f)(x)$, and
(ii) $-\partial f_C(x) = \partial_C(-f)(x)$ (see Exercise 3.6.4)

for all $x \in U$. The fact that $DS_e(U)$ contains all the continuous convex functions (when U is convex) follows from Theorem 3.6.1 and Proposition 3.6.24. □

Next, we show that all the elements of $DS_e(U)$ are, up to an additive constant, uniquely determined by their Clarke subdifferentials.

Let U be a nonempty open subset of a normed linear space $(X, \|\cdot\|)$, and let $f : U \to \mathbb{R}$ be a locally Lipschitz function. Then we say that f is *integrable* on U if $\partial_C(f - g)(x) = \{0\}$ for all $x \in U$ for each locally Lipschitz function $g : U \to \mathbb{R}$ with $\partial_C g(x) \subseteq \partial_C f(x)$ for all $x \in U$.

To study integrability further, we need the following "sum rule" for the Clarke generalised subdifferential mapping.

Proposition 3.6.29 (Sum rule [61, Proposition 2.3.3]). *Let U be a nonempty open subset of a normed linear space $(X, \|\cdot\|)$. If $f : U \to \mathbb{R}$ and $g : U \to \mathbb{R}$ are locally Lipschitz functions on U and $x_0 \in U$, then $\partial_C(f + g)(x_0) \subseteq \partial_C f(x_0) + \partial_C g(x_0)$.*

Proof. Let $x_0 \in U$ and $y \in X \setminus \{0\}$. We will first show that

$$(f + g)^0(x_0; y) \leq f^0(x_0; y) + g^0(x_0; y). \qquad (*)$$

To this end, let $\varepsilon > 0$ be arbitrary. Then choose $\delta_1 > 0$ and $\delta_2 > 0$ such that

$$\sup_{\substack{\|z - x_0\| < \delta_1 \\ 0 < \lambda < \delta_1}} \frac{f(z + \lambda y) - f(z)}{\lambda} < f^0(x_0; y) + \frac{\varepsilon}{2} \quad \text{and} \quad \sup_{\substack{\|z - x_0\| < \delta_2 \\ 0 < \lambda < \delta_2}} \frac{g(z + \lambda y) - g(z)}{\lambda} < g^0(x_0; y) + \frac{\varepsilon}{2}.$$

Let $\delta := \min\{\delta_1, \delta_2\} > 0$. Then for any $0 < \lambda < \delta$ and $\|z - x_0\| < \delta$, we have

$$\begin{aligned}
\frac{(f + g)(z + \lambda y) - (f + g)(z)}{\lambda} &= \frac{f(z + \lambda y) - f(z)}{\lambda} + \frac{g(z + \lambda y) - g(z)}{\lambda} \\
&< \left(f^0(x_0; y) + \frac{\varepsilon}{2}\right) + \left(g^0(x_0; y) + \frac{\varepsilon}{2}\right) \\
&= f^0(x_0; y) + g^0(x_0; y) + \varepsilon.
\end{aligned}$$

Therefore $(f+g)^0(x_0;y) \le f^0(x_0;y) + g^0(x_0;y) + \varepsilon$. Since $\varepsilon > 0$ was arbitrary, $(f+g)^0(x_0;y) \le f^0(x_0;y) + g^0(x_0;y)$. Next, we show that

$$\partial_C(f+g)(x_0) \subseteq \partial_C f(x_0) + \partial_C g(x_0).$$

So suppose, to obtain a contradiction, that $\partial_C(f+g)(x_0) \nsubseteq \partial_C f(x_0)+\partial_C g(x_0)$. Then there exists $x_0^* \in \partial_C(f+g)(x_0) \setminus (\partial_C f(x_0) + \partial_C g(x_0))$. By the separation theorem [75, p. 418] in (X^*, weak^*) there exists $y \in X$ such that

$$\begin{aligned}\sup\{x^*(y) : x^* \in \partial_C f(x_0) + \partial_C g(x_0)\} &= \sup\{\hat{y}(x^*) : x^* \in \partial_C f(x_0) + \partial_C g(x_0)\} \\ &< \hat{y}(x_0^*) = x_0^*(y) \le \sup\{x^*(y) : x^* \in \partial_C(f+g)(x_0)\}.\end{aligned}$$

Now

$$\begin{aligned}&\sup\{x^*(y) : x^* \in \partial_C f(x_0) + \partial_C g(x_0)\} \\ &\quad = \sup\{x^*(y) : x^* \in \partial_C f(x_0)\} + \sup\{x^*(y) : x^* \in \partial_C g(x_0)\}.\end{aligned}$$

Therefore

$$\sup\{x^*(y) : x^* \in \partial_C f(x_0)\} + \sup\{x^*(y) : x^* \in \partial_C g(x_0)\} < \sup\{x^*(y) : x^* \in \partial_C(f+g)(x_0)\}.$$

Thus applying the second part of Proposition 3.6.7, we obtain that

$$f^0(x_0;y) + g^0(x_0;y) < (f+g)^0(x_0;y),$$

which contradicts inequality $(*)$. Hence $\partial_C(f+g)(x_0) \subseteq \partial_C f(x_0) + \partial_C g(x_0)$. □

Our main result on integrability relies upon the following result from convex analysis. We hope that this result might be of some independent interest to the reader.

Proposition 3.6.30 (Convex set cancellation law). *Let B, C and D be nonempty subsets of a normed linear space $(X, \|\cdot\|)$. If (i) $B + D \subseteq C + D$, (ii) C is closed and convex, and (iii) D is bounded, then $B \subseteq C$. Furthermore, if B, C and D are all nonempty closed and bounded convex subsets of X and $B + D = C + D$, then $B = C$ (i. e. we have a cancellation law).*

Proof. Suppose, to obtain a contradiction, that there exists $b_0 \in B \setminus C$. Then by the separation theorem [75, p. 418] there exists $x^* \in X^*$ such that $\sup\{x^*(c) : c \in C\} < x^*(b_0)$. Therefore

$$\begin{aligned}\sup\{x^*(x) : x \in C + D\} &= \sup\{x^*(c) : c \in C\} + \sup\{x^*(d) : d \in D\} \\ &< x^*(b_0) + \sup\{x^*(d) : d \in D\} \\ &\le \sup\{x^*(b) : b \in B\} + \sup\{x^*(d) : d \in D\} \\ &= \sup\{x^*(x) : x \in B + D\}.\end{aligned}$$

However, this contradicts the assumption that $B + D \subseteq C + D$. Therefore $B \subseteq C$. For the second conclusion, suppose that B, C and D are all nonempty closed bounded convex subsets of X and $B + D = C + D$. Then $B + D \subseteq C + D$, and so, by above, $B \subseteq C$. Again, repeating this argument but with the roles of B and C reversed, we get that $C \subseteq B$, that is, $B = C$. □

Proposition 3.6.31. *Let U be a nonempty open subset of a normed linear space $(X, \|\cdot\|)$, and let $f : U \to \mathbb{R}$ be a locally Lipschitz function. If $g : U \to \mathbb{R}$ is a locally Lipschitz function on U such that $\partial_C g(x) \subseteq \partial_C f(x)$ for all $x \in U$, then $\partial_C(f-g)(x) = \{0\}$ for all $x \in U$ if and only if $x \mapsto \partial_C(f-g)(x)$ is a minimal weak* cusco on U.*

Proof. Since $f = (f-g) + g$, we have by the sum rule (see Proposition 3.6.29) that

$$\partial_C f(x) \subseteq \partial_C(f-g)(x) + \partial_C g(x) \subseteq \partial_C(f-g)(x) + \partial_C f(x).$$

Therefore by the cancellation law for closed bounded convex sets (see Proposition 3.6.30) $0 \in \partial_C(f-g)(x)$ for all $x \in U$. So if $x \mapsto \partial_C(f-g)(x)$ is a minimal weak* cusco on U, then $\partial_C(f-g)(x) = \{0\}$ for all $x \in U$. The converse is obvious. □

Corollary 3.6.32. *Let U be a nonempty open connected subset of a normed linear space $(X, \|\cdot\|)$, and let $f : U \to \mathbb{R}$ be a locally Lipschitz function. If $g : U \to \mathbb{R}$ is a locally Lipschitz function on U such that $\partial_C g(x) \subseteq \partial_C f(x)$ for all $x \in U$, then $f - g$ is constant on U if and only if $x \mapsto \partial_C(f-g)(x)$ is a minimal weak* cusco on U.*

Proof. Suppose that $x \mapsto \partial_C(f-g)(x)$ is a minimal weak* cusco on U. Let $x_0 \in U$, and let $U_0 := \{x \in U : (f-g)(x) = (f-g)(x_0)\}$. It follows from Proposition 3.6.31 and the Lebourg mean-value theorem (Lemma 3.6.12) that both U_0 and $U \setminus U_0$ are open subsets of U. Now $U_0 \neq \varnothing$ as $x_0 \in U_0$. Therefore, as U is connected, we have that $U_0 = U$, which shows that $f - g$ is constant on U. The converse is again obvious. □

We saw earlier that to investigate D-representability, we were forced to consider locally Lipschitz functions whose Clarke subdifferential mappings are minimal weak* cuscos. Here we again see that to consider locally Lipschitz functions that are integrable, we are compelled to examine locally Lipschitz functions whose Clarke subdifferential mappings are minimal weak* cuscos.

Proposition 3.6.33. *Let U be a nonempty open subset of a normed linear space $(X, \|\cdot\|)$. Then each member of $DS_e(U)$ is integrable.*

Proof. Suppose that $f \in DS_e(U)$ and $g : U \to \mathbb{R}$ is a locally Lipschitz function on U such that $\partial_C g(x) \subseteq \partial_C f(x)$ for all $x \in U$. Clearly, then $g \in DS_e(U)$. Moreover, since $DS_e(U)$ is a vector space (see Proposition 3.6.28), $f - g \in DS_e(U)$, and so the subdifferential mapping $x \mapsto \partial_C(f-g)(x)$ is a minimal weak* cusco on U (see Proposition 3.6.28). The result now follows from Proposition 3.6.31. □

Corollary 3.6.34. *Let $\varphi : U \to \mathbb{R}$ be a continuous convex function defined on a nonempty open convex subset U of a normed linear space $(X, \|\cdot\|)$. Then φ is integrable on U.*

Proof. From Proposition 3.6.28 it follows that $\varphi \in DS_e(U)$. Thus, by Proposition 3.6.33, φ is integrable on U. □

Theorem 3.6.35 (Identity theorem). *Let $f : U \to \mathbb{R}$ and $g : U \to \mathbb{R}$ be locally Lipschitz functions defined on a nonempty open connected subset U of a normed linear space $(X, \|\cdot\|)$. If $f \in DS_e(U)$, $x \mapsto \partial_C g(x)$ is a minimal weak* cusco on U, and $\partial_C f(x) \cap \partial_C g(x) \neq \varnothing$ for each x in a dense subset of U, then $f - g$ is a constant function on U.*

Proof. Consider the set-valued mapping $T : U \to 2^{X^*}$ defined by $T(x) := \partial_C f(x) \cap \partial_C g(x)$ for $x \in U$. By Theorem 1.1.11, T is a weak* usco on U. Furthermore, since both $\partial_C f$ and $\partial_C g$ have convex images, T has convex images too. Thus T is in fact a weak* cusco on U. Now since $T(x) \subseteq \partial_C f(x)$ and $T(x) \subseteq \partial_C g(x)$ for all $x \in U$, we have by the minimality of both $\partial_C f$ and $\partial_C g$ that $\partial_C f = T = \partial_C g$. The result now follows from Corollary 3.6.32. □

In conclusion, we see that the elements of $DS_e(U)$ are D-representable and integrable and possess differentiability properties similar to those enjoyed by continuous convex functions.

3.7 Exercises and commentary

In Section 3.1, we reviewed several known results from general topology from the perspective of set-valued and usco mappings. This brief section is intended to give just a small taste of how usco mappings may be utilised in the study of general topology. In addition to some extension theorems, some results on the class of Lindelöf Σ-spaces are given.

Section 3.2 concerns the application of usco mappings to the theory of abstract approximation in Banach spaces. The main two topics considered here are the study of nearest points and the study of farthest points. On these topics, we barely scratched the surface. However, we did illustrate the central role played by set-valued mappings. Further information on these topics can be found in [14, 68, 83, 283, 306].

In Section 3.3, we examined the differentiability theory of continuous convex functions defined on Banach spaces from a set-valued analysis perspective. This area of research is vast, and we have not even begun to describe all the various avenues of research on this topic. However, there are several excellent monographs on this topic, namely, [81, 91, 257]. The book [44] gives an excellent account of what was known about the Radon–Nikodým property in 1983. For the latest state of play on weak Asplund spaces, see [218, 231, 234].

In Section 3.4, we investigated the role of uscos in the study of variational analysis. The father of all variational theorems is the "Ekeland variational principle', [76].

This was later refined by the "Borwein–Preiss smooth variational principle", [40] and then further refined in [69]. The paper [69] is important from our perspective as it relies upon Baire category arguments. For further information on "generic variational principles", see [62, 67, 171]. The "Stegall variational principle" is also a notable theorem in this section.

Section 3.5 considers the implications of weak upper semicontinuity of the duality mapping on a Banach space. It also gives a short proof of James' weak compactness theorem in the setting of separable Banach spaces.

In the penultimate section of this chapter, Section 3.6, we investigated the use of cusco mappings in the study of the differentiability theory of locally Lipschitz functions defined on Banach spaces. For further information on this topic, see [39, 60, 61]. In particular, the authoritative book [186] is the Bible in this area of research.

Exercises

1. Construction of the Stone–Čech compactification. Let (X, τ) be a completely regular topological space, and let $C := \{f \in [-1,1]^X : f \text{ is continuous}\}$. Then $Y := [-1,1]^C$, endowed with the topology of pointwise convergence on C, is compact by Tychonoff's theorem.
   (a) Show that the mapping $e_0 : X \to Y$ defined by $e_0(x)(f) := f(x)$ for $f \in C$ is well-defined, one-to-one and continuous.
   (b) Let $\beta X := \overline{e_0(X)}^{\tau_p(C)}$, and let τ_β denote the relative $\tau_p(C)$-topology on βX. Show that $(\beta X, \tau_\beta)$ is a compact Hausdorff space.
   (c) Show that the mapping $e : X \to \beta X$ defined by $e(x)(f) := f(x)$ for $f \in C$ is a homeomorphic embedding of X into a dense subset of βX. (Note that the only difference between the mappings e_0 and e is that their codomains are different.)
   (d) Show that if $g : X \to [-1,1]$ is a continuous function, then there exists a continuous function $\widehat{g} : \beta X \to [-1,1]$ such that $\widehat{g}(e(x)) = g(x)$ for all $x \in X$. *Hint*: $g \in C$, and the mapping $\delta_g : [-1,1]^C \to [-1,1]$ defined by $\delta_g(y) := y(g)$ for $y \in [-1,1]^C$ is $\tau_p(C)$-continuous.
2. (Lavrentiev theorem [179]) Suppose that (X, τ) and (Y, τ') are Čech-complete spaces and D is a dense subset of (X, τ). If both (X, τ) and (Y, τ') have G_δ-diagonals and $f : D \to Y$ is a homeomorphic embedding into a dense subset of (Y, τ'), then there exist G_δ-subsets G_1 of (X, τ) containing D and G_2 of (Y, τ') containing $f(D)$ and a homeomorphism $\tilde{f} : G_1 \to G_2$ such that $\tilde{f}|_D = f$.
   Hint: First, let $g := f^{-1}$.
   (a) Show that there exists a G_δ-subset G of (Y, τ') containing $f(D)$ and continuous function $\tilde{g} : G \to X$ such that $\tilde{g}|_{f(D)} = g$.
   (b) Show that a dense G_δ-subset of a Čech-complete space is again Čech-complete (hence, in particular, G is Čech-complete).

(c) Show that there exists a G_δ-subset G_1 of (X,τ) containing D and a continuous function $\tilde{f} : G_1 \to G$ such that $\tilde{f}|_{G_1} = f$.

(d) Let $I_X : G_1 \to X$ be the inclusion mapping, that is, $I_X(x) := x$ for all $x \in G_1$. Note that $(\tilde{g} \circ \tilde{f}) : G_1 \to X$ and $(\tilde{g} \circ \tilde{f})(x) = I_X(x)$ for all $x \in D$. Hence show that, by continuity, $(\tilde{g} \circ \tilde{f}) = I_X$. In particular, $\tilde{f}$ is one-to-one, and $\tilde{g}(\tilde{f}(G_1)) = G_1$, that is, $\tilde{f}(G_1)$ is homeomorphic to G_1 and is hence Čech-complete.

(e) Show that if $G_2 := \tilde{f}(G_1)$, then G_2 is a G_δ-subset of (Y,τ'); see Remark 3.1.2.

3. Suppose that (X,τ) and (Y,τ') are topological spaces and $\Phi : X \to 2^Y$ is a surjective usco from (X,τ) onto (Y,τ'). Show that if (X,τ) is a Lindelöf Σ-space, then so is (Y,τ'). *Hint*: use Construction (v) in Section 1.1.3, and Proposition 3.1.11.

4. Let K be a subset of a normed linear space $(X, \|\cdot\|)$. We say that K is *boundedly weakly compact* if for any $r > 0$, $rB_X \cap K$ is weakly compact. Suppose that K is a nonempty boundedly weakly compact subset of a rotund normed linear space $(X, \|\cdot\|)$.

   (a) Show that the metric projection mapping $y \mapsto P_K(y)$ is a minimal (weak)-usco on $(X, \|\cdot\|)$.

   (b) Hence show that $\{x \in X : P_K(x) \text{ is a singleton}\}$ is residual in $(X, \|\cdot\|)$.

5. Let $(X, \|\cdot\|)$ be a finite-dimensional rotund normed linear space, and let C be a nonempty closed and bounded subset of $(X, \|\cdot\|)$.

   (a) Show that the set-valued mapping $F_C : X \to 2^C$ defined by

$$F_C(x) := \left\{y \in C : \|x-y\| = \sup_{c\in C} \|x-c\|\right\} \quad \text{for } x \in X$$

   is an minimal usco on $(X, \|\cdot\|)$ (see Example (viii) in Section 1.1.2).
   Hint: Apply Lemma 3.2.11 to show that if $x \in X$, $z \in F_C(x)$ and $x \in (z,y)$ for some $y \in X$, then $F_C(y) = \{z\}$.

   (b) Hence show that $\{x \in X : F_C(x) \text{ is a singleton}\}$ is residual in $(X, \|\cdot\|)$.

6. Use Proposition 3.2.24 and Corollary 3.2.28 to show that every uniquely remotal subset of a finite-dimensional rotund Banach space is a singleton.

7. Let U be a nonempty open convex subset of a normed linear space $(X, \|\cdot\|)$, and let $\varphi_1 : U \to \mathbb{R}$ and $\varphi_2 : U \to \mathbb{R}$ be continuous convex functions defined on U. Show that

$$\partial(\varphi_1 + \varphi_2)(x) = \partial\varphi_1(x) + \partial\varphi_2(x) \quad \text{for all } x \in U.$$

   Hint: First, show that $\partial\varphi_1(x) + \partial\varphi_2(x) \subseteq \partial(\varphi_1 + \varphi_2)(x)$ for all $x \in U$ and then use the fact that $x \mapsto \partial(\varphi_1 + \varphi_2)(x)$ is a minimal weak*-cusco on U.

8. Let U be a nonempty open convex subset of a finite-dimensional normed linear space $(X, \|\cdot\|)$, and let $\varphi : U \to \mathbb{R}$ be a continuous convex function on U. Show that φ is Gâteaux differentiable at a point $x_0 \in U$ if and only if φ is Fréchet differentiable at $x_0 \in U$. *Hint*: Appeal to Theorems 3.3.20 and 3.3.35; then use the fact that in finite-dimensional spaces the weak* topology coincides with the norm topology.

9. Show that if $\Phi : A \to 2^{X^*}$ is a minimal weak*-cusco acting from a Baire space (A, τ) into subsets of the dual of a normed linear space $(X, \|\cdot\|)$, then there exists a dense open subset B of (A, τ) such that Φ is locally bounded at each point of B. *Hint*: Consider the sets $B_n := \{x \in A : \Phi(x) \cap nB_{X^*} \neq \varnothing\}$. Show that the sets $\{B_n : n \in \mathbb{N}\}$ form closed cover of (A, τ). Then apply Proposition 3.3.22.
10. Let (X, τ) be a regular Hausdorff space. Show that (X, τ) has property (S) if and only if every minimal τ-usco $\Phi : A \to 2^X$ acting from a Baire space (A, τ') admits a selection that is continuous at the points of a dense G_δ-subset of (A, τ').
11. Let (X, τ) and (Y, τ') be Hausdorff topological spaces, and let $f : X \to Y$ be a perfect function. Show that if (X, τ) has property (S), then so does (Y, τ'). *Hint*: see Exercise 16 in Section 1.6.
12. Let $\{(X_n, \tau_n) : n \in \mathbb{N}\}$ be Hausdorff topological spaces. Show that if each (X_n, τ_n) has property (S), then so does $\prod_{n\in\mathbb{N}} X_n$.
13. Let $(X, \|\cdot\|)$ be a Banach space. Show that $(X, \|\cdot\|)$ is an Asplund space if and only if every locally bounded minimal weak*-cusco acting from a Baire space (A, τ) into subsets of X^* is single-valued and norm upper semicontinuous at the points of a dense G_δ-subset of (A, τ).
14. Use the Baire–Osgood theorem and Theorem 1.4.31 to show that every reflexive Banach space is an Asplund space.
15. Let $(X, \|\cdot\|)$ be a Banach space, and let Y be a closed subspace of $(X, \|\cdot\|)$. Use Michael's selection theorem to show that if both $(Y, \|\cdot\|)$ and $(X/Y, \|\cdot\|)$ are Asplund spaces, then so is $(X, \|\cdot\|)$. (Note that this question requires some knowledge of Banach space theory.)
16. Use Stegall's variational theorem to show that every nonempty closed and bounded convex subset of a Banach space $(X, \|\cdot\|)$ with the Radon–Nikodým property is the closed convex hull of its strongly exposed points. *Hint*: see Theorem 3.4.6.
17. Let $f : X \to (-\infty, \infty]$ be a proper lower semicontinuous function on a Banach space $(X, \|\cdot\|)$ with the Radon–Nikodým property. Show that if there exist real numbers a and b such that (i) $a > 0$ and (ii) $b + a\|x\| \leq f(x)$ for all $x \in X$, then for every $\varepsilon > 0$, there exists a continuous linear functional $x^* \in X^*$ such that (i) $\|x^*\| < \varepsilon$ and (ii) $f + x^*$ attains a strong minimum on $(X, \|\cdot\|)$.
18. Use the abstract variational theorem to show that every nonempty weakly compact convex subset of a Banach space $(X, \|\cdot\|)$ is the closed convex hull of its strongly exposed points. *Hint*: see Theorem 3.4.6.
19. Let U be a nonempty open convex subset of a normed linear space $(X, \|\cdot\|)$, and let $\varphi : U \to \mathbb{R}$ be a continuous convex function on U. Show that φ is Gâteaux differentiable (Fréchet differentiable) at a point $x_0 \in U$ if and only if φ is strictly differentiable (strictly Fréchet differentiable) at $x_0 \in U$. *Hint*: Consider Theorems 3.3.20 and 3.3.35 and Propositions 3.6.9 and 3.6.13.
20. Let U be a nonempty open subset of a normed linear space $(X, \|\cdot\|)$, and let $f : U \to \mathbb{R}$ be a locally Lipschitz function on U. Let $x_0 \in U$ and $y \in X \setminus \{0\}$. Then we say that f is *pseudoregular at* $x_0 \in U$ *in the direction* y if $f^0(x_0; y) = f^+(x_0; y)$.

Show that f is pseudoregular at $x_0 \in U$ in the direction y if and only if the mapping $x \mapsto f^+(x; y)$ is upper semicontinuous at x_0.

21. Let $f : (a, b) \to \mathbb{R}$ be a locally Lipschitz function. Show that if f is integrable, then $x \mapsto \partial_C f(x)$, is a minimal cusco on (a, b). *Hint*: First, show that if $\Phi : (a, b) \to 2^{\mathbb{R}}$ is a minimal cusco, then there exists a locally Lipschitz function $g : (a, b) \to \mathbb{R}$ such that $\partial_C g(x) = \Phi(x)$ for all $x \in (a, b)$. Then use the fact that the graph of every cusco contains the graph of a minimal cusco (simply modify the proof of Proposition 1.3.3). Note that this result does not generalise beyond one-dimensional spaces; see [37, Example 7.1].

4 Topological properties of the space of usco mappings

In this chapter, we consider some topologies on the space of usco mappings. We are mainly interested in the topology of uniform convergence on compacta and the Vietoris topology on the space of usco, minimal usco and minimal cusco mappings.

4.1 Minimal usco mappings

In this part, we give further characterisations of minimal usco mappings. Theorem 1.3.10 gives characterisations of minimal usco mappings in the class of usco mappings. Minimal usco mappings were also studied by Drewnowski and Labuda [73], who also gave an interesting characterisation of minimal usco mappings. We also extend some results of [73].

We recall the definition of subcontinuity. Let (X,τ) and (Y,τ') be topological spaces. A function $f : X \to Y$ is *subcontinuous at* $x \in X$ (see [87]) if for every net $(x_\sigma : \sigma \in \Sigma)$ in (X,τ) converging to x, there is a convergent subnet of $(f(x_\sigma) : \sigma \in \Sigma)$. A function f is *subcontinuous on* (X,τ) if it is subcontinuous at every point of X.

The following definition is the local version of subcontinuity for set-valued mappings introduced by Hrycay (see [142, Definition 3.4]). Let (X,τ) and (Y,τ') be topological spaces, and let $F : X \to 2^Y$ be a set-valued mapping. Then F is *subcontinuous at* $x \in X$ if for every net $(x_\sigma : \sigma \in \Sigma)$ in (X,τ) converging to x and for every choice of $y_\sigma \in F(x_\sigma)$, the net $(y_\sigma : \sigma \in \Sigma)$ has a cluster point in (Y,τ'). We say that F is *subcontinuous on* (X,τ) if it is subcontinuous at every point of X.

For $x \in X$ and $y \in Y$, we use $\mathcal{U}(x)$ and $\mathcal{V}(y)$ to denote bases of open neighbourhoods of x in X and $y \in Y$, respectively.

Proposition 4.1.1 ([118, 183]). *Let (X,τ) and (Y,τ') be topological spaces. If $F : X \to 2^Y$ is an usco, then F is subcontinuous, and every selection f of F is subcontinuous too.*

Proof. Suppose that there is a selection f of F that is not subcontinuous. Then there is a net $(x_\sigma : \sigma \in \Sigma)$ in (X,τ) convergent to a point x such that $(f(x_\sigma) : \sigma \in \Sigma)$ has no cluster points in (Y,τ'). The compactness of $F(x)$ implies that there are an open set $F(x) \subseteq O$ and $\sigma_0 \in \Sigma$ with $f(x_\sigma) \notin O$ for every $\sigma_0 \leq \sigma$. The upper semicontinuity of F at x implies that there is $V \in \mathcal{U}(x)$ with $F(z) \subseteq O$ for every $z \in V$. There is $\sigma_0 \leq \sigma_1$ such that $x_\sigma \in V$ for every $\sigma_1 \leq \sigma$, that is, $f(x_\sigma) \in F(x_\sigma) \subseteq O$ for every $\sigma_1 \leq \sigma$, a contradiction. The proof of the subcontinuity of F is similar. □

Note that the notion of subcontinuity can be extended to densely defined functions. Let A be a dense subset of a topological space (X,τ), and let (Y,τ') be a topological space. If $f : A \to Y$ is a function, then we say that $f : A \to Y$ is *subcontinuous at* $x \in X$ [184] if for every net $(x_\sigma : \sigma \in \Sigma)$ in A convergent to $x \in X$, $(f(x_\sigma) : \sigma \in \Sigma)$ has a

https://doi.org/10.1515/9783110750188-004

convergent subnet. It is easy to verify that $f : A \to Y$ is subcontinuous at $x \in X$ if and only if

$$(*) \quad \text{for every open cover } \mathcal{H} \text{ of } Y, \text{ there are a finite subset } \mathcal{F} \text{ of } \mathcal{H} \text{ and } U \in \mathcal{U}(x) \text{ such that } f(U \cap A) \subseteq \bigcup \mathcal{F}.$$

(This is a slight modification of [254, Theorem 2.1].) A function $f : A \to Y$ is *subcontinuous on* (X, τ) if it is subcontinuous at every $x \in X$. As before, we will identify a function $f : A \to Y$ defined on a subset A of X with the set-valued mapping defined on X that maps x to $\{f(x)\}$ if $x \in A$ and to $\emptyset$ if $x \notin A$. We will denote this set-valued mapping by f as well (i. e. the same notation as the original function $f : A \to Y$).

Theorem 4.1.2 ([122]). *Let (X, τ) and (Y, τ') be topological spaces and suppose that (Y, τ') is a regular T_1 space. If A is a dense subset of (X, τ) and $f : A \to Y$ is a subcontinuous function, then* USC(f) *is an usco on (X, τ).*

Proof. Let A be a dense subset of (X, τ), and let $f : A \to Y$ be a subcontinuous function. We show that the subcontinuity of f implies that USC(f)(x) is a nonempty compact set for every $x \in X$. Let $x \in X$. Of course, $\mathrm{USC}(f)(x) \neq \emptyset$. Let $\mathcal{H}$ be an open cover of USC(f)(x). Let $\mathcal{H}'$ be a refinement of $\mathcal{H}$ such that for every $H' \in \mathcal{H}'$, there is $H \in \mathcal{H}$ with $\overline{H'} \subseteq H$ and $\mathrm{USC}(f)(x) \subseteq \bigcup \mathcal{H}'$. For every $y \in Y \setminus \mathrm{USC}(f)(x)$, let O_y be an open neighbourhood of y such that $\overline{O_y} \cap \mathrm{USC}(f)(x) = \emptyset$. Then the family $\mathcal{H}' \cup \{O_y : y \in Y \setminus \mathrm{USC}(f)(x)\}$ is an open cover of Y. By $(*)$ there are $U \in \mathcal{U}(x)$, $H'_1, H'_2, \ldots, H'_n \in \mathcal{H}'$ and a finite indexed set I such that $f(U \cap A) \subseteq \bigcup\{H'_i : i = 1, 2, \ldots, n\} \cup (\bigcup\{O_{y_i} : i \in I\})$. Thus

$$\mathrm{USC}(f)(x) \subseteq \overline{f(U \cap A)} \subseteq (\overline{H'_1} \cup \overline{H'_2} \cup \cdots \cup \overline{H'_n}) \cup \Big(\bigcup\{\overline{O_{y_i}} : i \in I\}\Big).$$

Hence $\mathrm{USC}(f)(x) \subseteq H_1 \cup H_2 \cup \cdots \cup H_n$, where $H_i \in \mathcal{H}$ for $i = 1, 2, \ldots, n$.

We will now show that USC(f) is upper semicontinuous. Suppose there is $x \in X$ such that USC(f) is not upper semicontinuous at x. Let V be an open set in (Y, τ') with $\mathrm{USC}(f)(x) \subseteq V$ such that for every $U \in \mathcal{U}(x)$, there are $x_U \in U$ and $y_U \in \mathrm{USC}(f)(x_U) \setminus V$. The regularity of (Y, τ') implies that there is an open set G in (Y, τ') such that $\mathrm{USC}(f)(x) \subseteq G \subseteq \overline{G} \subseteq V$. Thus for every $U \in \mathcal{U}(x)$, we have $x_U \in U$ and $y_U \in \mathrm{USC}(f)(x_U) \cap (Y \setminus \overline{G})$. For every $U \in \mathcal{U}(x)$, there is $a_U \in A \cap U$ such that $f(a_U) \in Y \setminus \overline{G}$. Since the net $(a_U : U \in \mathcal{U}(x))$ converges to x, the subcontinuity of f at x implies that there is a cluster point $y \in Y \setminus G$ of the net $(f(a_U) : U \in \mathcal{U}(x))$, a contradiction, since $y \in \mathrm{USC}(f)(x) \subseteq G$. □

The following theorem gives a characterisation of minimal usco mappings in the class of all mappings from a topological space (X, τ) to subsets of a topological space (Y, τ'). By a *densely defined quasicontinuous function* we mean a function defined on a dense subset A of (X, τ) that is quasicontinuous with respect to the induced topology on A.

Theorem 4.1.3 ([118, 122]). *Let (X, τ) and (Y, τ') be topological spaces and suppose that (Y, τ') is a regular T_1 space. If $F : X \to 2^Y$ is a set-valued mapping, then the following are equivalent:*
(i) *F is a minimal usco mapping;*
(ii) *There is a quasicontinuous subcontinuous selection f of F such that* $\mathrm{USC}(f) = F$;
(iii) *Every selection f of F is quasicontinuous and subcontinuous, and* $\mathrm{USC}(f) = F$;
(iv) *There is a densely defined quasicontinuous subcontinuous selection f of F such that* $\mathrm{USC}(f) = F$.

Proof. (i) ⇒ (iii). Suppose that F is a minimal usco. By Proposition 4.1.1 every selection f of F is subcontinuous. By Theorem 1.3.10 (i)⇒(ii), for every selection f of F, we have $\mathrm{USC}(f) = F$. Using Proposition 1.3.5 and Exercise 1.3.9, every selection f of F is quasicontinuous. (iii) ⇒ (ii) and (ii)⇒ (iv) are clear. To prove (iv)⇒(i), let f be a densely defined quasicontinuous subcontinuous selection of F such that $\mathrm{USC}(f) = F$. By Theorem 4.1.2, F is an usco. To prove that F is a minimal usco, we use Theorem 1.3.10 (iv)⇒(i). □

Now we mention some applications of the above characterisations. Conditions (ii) and (iv) in Theorem 4.1.3 give a useful method for constructing minimal usco mappings.

The following remark shows that condition (iv) in Theorem 4.1.3 is also useful for constructing a minimal usco mapping inside a given usco mapping.

Remarks 4.1.4. Let (X, τ) be a Baire space, and let $F : X \to 2^{\mathbb{R}}$ be an usco. By Lemma 3.3.27 the function $f(x) := \inf\{t \in \mathbb{R} : t \in F(x)\}$ for $x \in X$ is lower semicontinuous. By Lemma 3.3.28 the set $C(f)$ is a dense G_δ-set in (X, τ). Since f is also subcontinuous by Theorem 4.1.3, the mapping $G := \mathrm{USC}(f|_{C(f)})$ is a minimal usco mapping from (X, τ) to $2^{\mathbb{R}}$, and $\mathrm{Gr}(G) \subseteq \mathrm{Gr}(F)$. Note that [43, Proposition 5.1.24] also gives a construction of a minimal usco mapping contained inside a given usco mapping from a general topological space with values in $\mathbb{R}$.

Note that the minimal usco mapping from Remark 4.1.4 is single-valued at the points of a dense G_δ-set. By Corollary 1.3.12 every minimal usco mapping $F : X \to 2^Y$ acting from a topological space (X, τ) into subsets of a Hausdorff topological space (Y, τ') is single-valued at the points of continuity of a densely defined selection f of F; in fact, $F(x_0) = \{f(x_0)\}$ if f is continuous at x_0.

Let (X, τ) and (Y, τ') be topological spaces. For a set-valued mapping $F : X \to 2^Y$, we denote

$$S(F) := \{x \in X : F(x) \text{ is a singleton}\},$$

and for a function $f : X \to Y$, we denote

$$C(f) := \{x \in X : f \text{ is continuous at } x\}.$$

Lemma 4.1.5. *Let (X, τ) be a topological space, and let (Y, τ') be a Hausdorff topological space. If $F : X \to 2^Y$ is a minimal usco and $f : X \to Y$ is a selection of F, then $S(F) = C(f)$.*

Proof. Let $x \in C(f)$. By Corollary 1.3.12, $x \in S(F)$. Now suppose that $x \in S(F)$. Since F is upper semicontinuous at x, $x \in C(f)$. □

By Proposition 1.3.5 every selection of a minimal usco mapping $F : X \to 2^Y$ from a topological space (X, τ) into subsets of Hausdorff topological space (Y, τ') is quasicontinuous. Thus the study of the sets of continuity points of quasicontinuous functions will help us to study the sets of points of single-valuedness of minimal usco mappings.

Let us now recall that a topological space (Y, τ') is called a *Piotrowski space* if every quasicontinuous function $f : X \to Y$ from a Baire space (X, τ) has a point of continuity. Let us also recall that a topological space (Y, τ') is called a *Stegall space* if for every Baire space (X, τ) and every minimal usco mapping $F : X \to 2^Y$, F is single-valued at some point of (X, τ).

By Proposition 2.3.15 every Hausdorff Piotrowski space is a Stegall space.

In Proposition 2.3.17, it is shown that if (X, τ) is a Baire space and (Y, τ') is a Hausdorff Stegall space, then for every minimal usco mapping $F : X \to 2^Y$, the set $S(F)$ is residual (comeager), that is, the set $X \setminus S(F)$ is of the first Baire category. By contrast, Example 1.3.1 gives a minimal usco mapping $F : \mathbb{R} \to 2^{\mathbb{R}}$ for which the set $X \setminus S(F)$ has positive Lebesgue measure.

Proposition 2.3.16 states that if (X, τ) is a Baire space and (Y, τ') is fragmentable, then every minimal usco mapping $F : X \to 2^Y$ is single-valued at the points of a dense G_δ-subset of (X, τ). This generalises Corollary 1.5.18, as every metric space (X, d) is fragmentable (by its metric d).

Note that by Example 2.3.8 the space $Y = \mathbb{R}$ endowed with the Sorgenfrey topology τ_S is not a Piotrowski space. However, since (Y, τ_S) is fragmentable, it is a Stegall space.

Example 4.1.6 ([164, 218, 230, 231]). The"double arrow" space L is the set of points

$$((0,1] \times \{0\}) \cup ([0,1) \times \{1\}) \subseteq \mathbb{R}^2$$

endowed with the order topology from the lexicographical order, that is,

$$(s,i) \leq (t,j) \text{ if either } s < t \text{ or } s = t \text{ and } i < j.$$

This space is compact and Hausdorff. The mapping $\Phi : (0,1) \to 2^L$ defined by

$$\Phi(t) := \{(t,0),(t,1)\} \quad \text{for } t \in (0,1)$$

is clearly an uso mapping on $(0,1)$ endowed with the usual topology. It also easily follows from Proposition 1.3.5 that Φ is a minimal usco on $(0,1)$. Hence Φ is an everywhere two-valued minimal usco. Drewnowski and Labuda [73] also gave an example of a minimal usco mapping that is two-valued at every point.

In [157], some consistent (with ZFC) examples of nonfragmentable compact Hausdorff spaces that belong to the class of Stegall spaces were given.

Example 4.1.7 ([115]). Let X be the space of rational numbers with the usual topology. Enumerate X as $\{q_n : n \in \mathbb{Z}^+\}$ and define the set-valued mapping $F : X \to 2^{\mathbb{R}}$ by

$$F(x) := \left\{ \sum_{n:q_n<x} \frac{1}{2^n}, \sum_{n:q_n\leq x} \frac{1}{2^n} \right\} \quad \text{for } x \in X.$$

Then F is a minimal usco mapping on X, which is nowhere single-valued. To prove that F is upper semicontinuous, let $x \in X$, and let $\varepsilon > 0$. There is $n_0 \in \mathbb{N}$ such that $\sum_{n\geq n_0} \frac{1}{2^n} < \varepsilon$. Put $\delta_x := \min\{|q_i - x| : i \leq n_0, q_i \neq x\}$. For every $z \in B(x, \delta_x/2)$, we have $F(z) \subseteq B(F(x), \varepsilon)$, where $B(F(x), \varepsilon) := \{s \in \mathbb{R} : d(s, F(x)) < \varepsilon\}$. (If $z \in B(x, \delta_x/2)$ and $x < z$, then $\sum_{n:q_n\leq z} \frac{1}{2^n} - \sum_{n:q_n\leq x} \frac{1}{2^n} = \sum_{x<q_n\leq z} \frac{1}{2^n} < \varepsilon$.)

To prove that F is minimal, suppose there is an usco mapping G such that $\mathrm{Gr}(G) \subseteq \mathrm{Gr}(F)$ and there is q_n such that $F(q_n) \neq G(q_n)$. Suppose that $G(q_n) = \{\sum_{i:q_i<q_n} \frac{1}{2^i}\}$ (the other case is similar). The upper semicontinuity of G implies that there is a neighbourhood O of q_n such that $G(z) \subseteq B(G(q_n), 1/2^{q_n})$ for every $z \in O$, a contradiction, since for every $z \in O$ with $q_n < z$, we have $G(z) \subseteq (\sum_{i:q_i\leq q_n} \frac{1}{2^i}, \infty)$.

Exercises

1. Let (X, τ) be a Baire space, and let $F : X \to 2^{\mathbb{R}}$ be an usco on (X, τ). If $h : X \to \mathbb{R}$ is defined by $h(x) := \sup\{t \in \mathbb{R} : t \in F(x)\}$ for $x \in X$, then h is upper semicontinuous, and by Lemma 3.3.28 the set $C(h)$ of the points of continuity of h is a dense G_δ-set in (X, τ). Show that the mapping $H := \mathrm{USC}(h|_{C(h)})$ is a minimal usco from (X, τ) into $2^{\mathbb{R}}$ and $\mathrm{Gr}(H) \subseteq \mathrm{Gr}(F)$.
2. Let (X, τ) and (Y, τ') be topological spaces and suppose that (Y, τ') is Hausdorff. Let $f : X \to Y$ be a function. Put $F := \mathrm{USC}(f)$ and let $x \in X$. Show that if F is upper semicontinuous at x, then the following are equivalent:
   (i) The function f is continuous at x;
   (ii) $F(x) = \{f(x)\}$.
3. Let (X, τ) and (Y, τ') be topological spaces, and let A be a dense subset of (X, τ). Prove that $f : A \to Y$ is subcontinuous at $x \in X$ if and only if

   for every open cover $\mathcal{H}$ of Y, there is a finite subset $\mathcal{F}$ of $\mathcal{H}$
   and $U \in \mathcal{U}(x)$ such that $f(U \cap A) \subseteq \bigcup \mathcal{F}$.

4. Let (X, τ) be a topological space, and let $f : X \to \mathbb{R}$ be a function. Prove that f is subcontinuous if and only if f is locally bounded.
5. Let (X, τ) be a topological space, let (Y, d) be a boundedly compact metric space, and let $f : X \to Y$ be a function. Prove that f is subcontinuous if and only if f is locally bounded.

6. Show that if f is a subcontinuous function from a topological space (X,τ) into a regular T_1 space (Y,τ'), then for every compact set K in (X,τ), $\overline{f(K)}$ is a compact set in (Y,τ') (see [134, Proposition 3.13]).

4.2 Densely continuous forms

In this section, we will mention some results concerning so-called densely continuous forms and their connection to minimal usco mappings. Densely continuous forms were introduced by McCoy and Hammer [102] and then studied by Holá, McCoy, Holý, and Vadovič [114, 130, 140, 141].

Densely continuous forms can be considered as set-valued mappings that have a kind of minimality property found in the theory of minimal usco mappings. In particular, if (X,τ) is a Baire space and (Y,τ') is a Hausdorff fragmentable space, then every minimal usco mapping from (X,τ) into 2^Y is a densely continuous form (see Corollary 4.2.5).

To define a densely continuous form from a topological space (X,τ) to a topological space (Y,τ') [102], let $DC(X,Y)$ be the set of all functions f from (X,τ) to (Y,τ') such that the set $C(f)$ of the continuity points of f is dense in (X,τ). We call such functions densely continuous. Of course, $DC(X,Y)$ contains the set $C(X,Y)$ of all continuous functions from (X,τ) to (Y,τ'). There are many other interesting subsets in $DC(X,Y)$. For example, if (Y,τ') is a locally compact second countable space and (X,τ) is a Baire space, then $DC(X,Y)$ contains all functions from (X,τ) to (Y,τ') with closed graphs [114]. Also, by Theorem 2.3.18, if (X,τ) is a Baire space and (Y,τ') is a topological space fragmentable by a metric that majorises the topology τ' of Y, then $DC(X,Y)$ contains all quasicontinuous functions from (X,τ) to (Y,τ').

If Y is the set $\mathbb{R}$ of all real numbers and (X,τ) is a Baire space, then all upper and lower semicontinuous functions on (X,τ) belong to $DC(X,Y)$ as well (see Lemma 3.3.28).

Let (X,τ) and (Y,τ') be topological spaces. We define the set $D(X,Y)$ of *densely continuous forms* from (X,τ) to (Y,τ') by

$$D(X,Y) := \{\mathrm{USC}(f|_{C(f)}) : f \in DC(X,Y)\}.$$

The densely continuous forms from (X,τ) to (Y,τ') may be considered as set-valued mappings. For $x \in X$ and $\Phi \in D(X,Y)$, define $\Phi(x) := \{y \in Y : (x,y) \in \mathrm{Gr}(\Phi)\}$.

Note that by Exercise 1.1.16, if (Y,τ') is a Hausdorff topological space, then every densely continuous form from (X,τ) to (Y,τ') is single-valued at the points of a dense subset of (X,τ). In fact, if $F \in D(X,Y)$, then there is $f \in DC(X,Y)$ such that $F = \mathrm{USC}(f|_{C(f)})$ and $F(x) = \{f(x)\}$ for every $x \in C(f)$.

Denote by $A(X, Y)$ [118] the set of functions

$$A(X,Y) := \{f : X \to Y : \text{ for every } x \in X \text{ and for every neighbourhood } U \text{ of } (x, f(x)), \text{ there exists } y \in C(f) \text{ such that } (y, f(y)) \in U\}.$$

Of course, $A(X,Y) \subseteq DC(X,Y)$. The inclusion is proper. The function $f : \mathbb{R} \to \mathbb{R}$ defined by $f(0) := 0$ and $f(x) := 1/x$ otherwise belongs to $DC(X,Y)$ but not to $A(X,Y)$. For the densely continuous form $\Phi := \mathrm{USC}(f|_{C(f)})$, we have $\Phi(0) = \emptyset$.

It is very easy to verify that $A(X,Y) \subseteq Q(X,Y)$; however, the opposite is not true in general. Example 2.3.8 shows a quasicontinuous function f with $C(f) = \emptyset$.

If (X,τ) is a Baire space and (Y,τ') is a topological space fragmentable by a metric that majorises the topology τ', then every quasicontinuous function $f : X \to Y$ has a residual set $C(f)$ of the points of continuity (see Theorem 2.3.18), that is, $A(X,Y) = Q(X,Y)$.

Clearly, if $f : X \to Y$, then $f \in A(X,Y)$ if and only if $\mathrm{USC}(f) = \mathrm{USC}(f|_{C(f)})$.

We have the following characterisation of elements of $D(X,Y)$.

Proposition 4.2.1 ([118]). *Let (X,τ) and (Y,τ') be topological spaces, let (Y,τ') be a regular T_1 space, and let $F : X \to 2^Y$. Then the following are equivalent:*
(i) *$F \in D(X,Y)$ and $F(x) \neq \emptyset$ for every $x \in X$;*
(ii) *There is a function $f \in A(X,Y)$ such that* $\mathrm{USC}(f) = F$;
(iii) *Every selection f of F belongs to $A(X,Y)$, and* $\mathrm{USC}(f) = F$.

Proof. (i)⇒(iii). Let f be a selection of F. There is $g \in DC(X,Y)$ such that $F = \mathrm{USC}(g|_{C(g)})$. Of course, $F(x) = \{g(x)\}$ for every $x \in C(g)$, that is, $f(x) = g(x)$ for every $x \in C(g)$. It is easy to verify that $C(g) \subseteq C(f)$. (Let $x \in C(g)$. Suppose $x \notin C(f)$. There is $V \in \mathcal{V}(f(x))$ such that for every $U \in \mathcal{U}(x)$, there is $x_U \in U$ with $f(x_U) \notin V$. Let $H \in \mathcal{V}(f(x))$ be such that $\overline{H} \subseteq V$. The continuity of g at x implies that there is $O \in \mathcal{U}(x)$ such that $g(O) \subseteq H$. Then $O \times (Y \setminus \overline{H})$ is a neighbourhood of $(x_O, f(x_O))$ that has an empty intersection with the graph of g, a contradiction, since $\mathrm{Gr}(f) \subseteq \mathrm{Gr}(F)$ and $F = \mathrm{USC}(g|_{C(g)})$.) Thus the set $C(f)$ of the points of continuity of f is dense in X, that is, $f \in DC(X,Y)$. Since $\mathrm{Gr}(f) \subseteq \mathrm{Gr}(F)$ and $F = \mathrm{USC}(f|_{C(f)})$, we have that $F = \mathrm{USC}(f)$ and $f \in A(X,Y)$.

(iii)⇒(ii) is trivial. (ii)⇒(i) is also trivial since if $f \in A(X,Y)$, then, of course, $f \in DC(X,Y)$, and by the above $\mathrm{USC}(f) = \mathrm{USC}(f|_{C(f)})$. □

Corollary 4.2.2. *Let (X,τ) be a Baire space, and let (Y,τ') be a regular T_1 topological space that is fragmentable by a metric that majorises the topology on (Y,τ'). Let $F : X \to 2^Y$. Then the following are equivalent:*
(i) *$F \in D(X,Y)$ and $F(x) \neq \emptyset$ for every $x \in X$;*
(ii) *There is a quasicontinuous function $f : X \to Y$ such that* $\mathrm{USC}(f) = F$;
(iii) *Every selection f of F is quasicontinuous, and* $\mathrm{USC}(f) = F$.

Let (X,τ) and (Y,τ') be topological spaces. Denote by $MU(X,Y)$ the space of all minimal usco mappings from (X,τ) to 2^Y. A natural question arises: Under what conditions does $F \in D(X,Y)$ belong to $MU(X,Y)$? We have the following answer.

Proposition 4.2.3 ([123]). *Let (X,τ) and (Y,τ') be topological spaces and suppose that (Y,τ') is a regular T_1 topological space and $F \in D(X,Y)$. Then the following are equivalent:*
(i) $F \in MU(X,Y)$;
(ii) $F(x) \neq \varnothing$ *for every* $x \in X$, *and every selection* f *of* F *is subcontinuous.*

It is also interesting to know when $F \in MU(X,Y)$ belongs to $D(X,Y)$.

Proposition 4.2.4 ([123]). *Let (X,τ) and (Y,τ') be topological spaces and suppose that (Y,τ') is a Hausdorff topological space and $F \in MU(X,Y)$. Then the following are equivalent:*
(i) $F \in D(X,Y)$;
(ii) $S(F) = \{x \in X : F(x)$ *is a singleton*$\}$ *is dense in* (X,τ).

Proof. (i)⇒(ii) is clear. To prove (ii)⇒(i), let f be a selection of F. Then $S(F) \subseteq C(f)$. Thus $f \in DC(X,Y)$, and $\mathrm{USC}(f|_{C(f)}) \in D(X,Y)$. It is sufficient to show that $\mathrm{USC}(f|_{C(f)}) = F$. By Proposition 1.1.8, $\mathrm{Gr}(F)$ is a closed set in $X \times Y$. By Exercise 1.1.16, $\mathrm{Gr}(\mathrm{USC}(f|_{C(f)}))$ is a closed set in $X \times Y$, and $\mathrm{Gr}(\mathrm{USC}(f|_{C(f)})) \subseteq \mathrm{Gr}(F)$. By Corollary 1.1.14, $\mathrm{USC}(f|_{C(f)})$ is an usco, and thus $\mathrm{USC}(f|_{C(f)}) = F$. ☐

Corollary 4.2.5. *Let (X,τ) be a Baire space, and let (Y,τ') be a Hausdorff fragmentable topological space. Then $MU(X,Y) \subseteq D(X,Y)$.*

Proof. Let $F \in MU(X,Y)$. By Proposition 2.3.16, $S(F) = \{x \in X : F(x)$ is a singleton$\}$ contains a dense G_δ-set. Thus, by Proposition 4.2.4, $F \in D(X,Y)$. ☐

McCoy and Hammer [102, 199] also studied the space $D^*(X,\mathbb{R})$ of locally bounded densely continuous forms from (X,τ) to $\mathbb{R}$. If $\Phi : X \to 2^{\mathbb{R}}$ and $A \subseteq X$, then we say that Φ is *bounded on A* if the set $\Phi(A) := \bigcup\{\Phi(x) : x \in A\}$ is a bounded subset of $\mathbb{R}$. We say that Φ is *locally bounded* if each point of X has a neighbourhood on which Φ is bounded. Then we define $D^*(X,\mathbb{R})$ to be the set of all elements of $D(X,\mathbb{R})$ that are locally bounded.

Remarks 4.2.6. Let $U(X,\mathbb{R})$ be the set of all usco mappings from (X,τ) to $2^{\mathbb{R}}$. Then $D^*(X,\mathbb{R}) \subseteq U(X,\mathbb{R})$. In fact, if $\Phi \in D^*(X,\mathbb{R})$, then for all $x \in X$, $\Phi(x)$ is a nonempty compact set. By Corollary 1.1.15 any set-valued mapping with a closed graph that has a compact range is upper semicontinuous. Since upper semicontinuity is a local property, every $\Phi \in D^*(X,\mathbb{R})$ belongs to $U(X,\mathbb{R})$. By Proposition 4.1.1 every selection of $\Phi \in U(X,\mathbb{R})$ is subcontinuous, and thus by Proposition 4.2.3 Φ is a minimal usco, and $D^*(X,\mathbb{R}) \subseteq MU(X,\mathbb{R})$.

If (X,τ) is a Baire space, then $MU(X,\mathbb{R}) \subseteq D^\star(X,\mathbb{R})$. In fact, by Corollary 4.2.5, $MU(X,\mathbb{R}) \subseteq D(X,\mathbb{R})$. It is easy to show that every usco mapping from (X,τ) to $2^{\mathbb{R}}$ is locally bounded. Thus $MU(X,\mathbb{R}) \subseteq D^\star(X,\mathbb{R})$. Therefore if (X,τ) is a Baire space, then $MU(X,\mathbb{R}) = D^\star(X,\mathbb{R})$.

Exercises

1. Let (X,τ) be a Baire space, and let (Y,d) be a boundedly compact metric space, thats is, every closed bounded set in (Y,d) is compact. Show that $MU(X,Y) = D^\star(X,Y)$.
2. Let (X,τ) be a topological space, and let (Y,τ') be a regular T_1 topological space. Show that if $f,g \in DC(X,Y)$ and $\mathrm{USC}(f|_{C(f)}) = \mathrm{USC}(g|_{C(g)})$, then $\{f(x)\} = \{g(x)\}$ for every $x \in C(f) \cap C(g)$ [199].
3. Let (X,τ) be a topological space, and let (Y,τ') be a regular T_1 topological space. If $f,g \in DC(X,Y)$ and $f(x) = g(x)$ for every x in some dense set in (X,τ), then $\mathrm{USC}(f|_{C(f)}) = \mathrm{USC}(g|_{C(g)})$ [199].
4. Prove that if (X,τ) is a Baire space, then $D(X,\mathbb{R})$ has a natural vector space structure defined by
$$\mathrm{USC}(f|_{C(f)}) + \mathrm{USC}(g|_{C(g)}) := \mathrm{USC}((f+g)|_{(C(f)\cap C(g))}) \quad \text{for all } f,g \in D(X,\mathbb{R})$$
and
$$a\mathrm{USC}(f|_{C(f)}) := \mathrm{USC}(af|_{C(f)}) \quad \text{for all } f \in D(X,\mathbb{R}) \text{ and } a \in \mathbb{R}.$$
5. Let (X,τ) and (Y,τ') be topological spaces. If (Y,τ') is Hausdorff, then prove that every densely continuous form from (X,τ) to (Y,τ') is single-valued at the points of a dense subset of (X,τ).
6. Let (X,τ), (Y,τ') be topological spaces and let $f : X \to Y$. Show that $f \in A(X,Y)$ if and only if $\mathrm{USC}(f) = \mathrm{USC}(f|_{C(f)})$.

4.3 Minimal cusco mappings

In this section, we revisit the class of minimal cuscos mappings.

Let (X,τ) be a topological space, let $(Y,+,\cdot,\tau')$ be a linear topological space, and let $\Phi : X \to 2^Y$ be a cusco on (X,τ). Then we say that Φ is a *minimal cusco* if the graph of Φ does not contain as a proper subset the graph of any other cusco defined on (X,τ).

Proposition 4.3.1 ([41, 43, 73]). *Let (X,τ) be a topological space, let $(Y,+,\cdot,\tau')$ be a linear topological space, and let $\Phi : X \to 2^Y$ be a cusco on (X,τ). If $(Y,+,\cdot,\tau')$ is Hausdorff, then there exists a minimal cusco mapping $\Psi : X \to 2^Y$ such that* $\mathrm{Gr}(\Psi) \subseteq \mathrm{Gr}(\Phi)$ *(i. e. every cusco contains a minimal cusco).*

Proof. The proof of this proposition is identical to that of Proposition 1.3.3. So the interested reader is directed to the proof of Proposition 1.3.3. ☐

Proposition 3.3.22 gives a very useful characterisation of when a cusco mapping is a minimal cusco.

Lemma 4.3.2. *Let (X,τ) be a topological space, and let $(Y,+,\cdot,\tau')$ be a Hausdorff locally convex topological space. If A is a dense subset of (X,τ) and $f : A \to Y$ is a subcontinuous function such that $\overline{\text{co}}(\text{USC}(f)(x))$ is compact for every $x \in X$, then $\text{CSC}(f)(x) = \overline{\text{co}}(\text{USC}(f)(x))$ for every $x \in X$.*

Proof. Recall that for every $x \in X$, $\text{USC}(f)(x) = \bigcap\{\overline{f(V)} : V \in \mathcal{U}(x)\}$ and for every $x \in X$, $\text{CSC}(f)(x) = \bigcap\{\overline{\text{co}}(f(V)) : V \in \mathcal{U}(x)\}$. Thus the inclusion

$$\overline{\text{co}}(\text{USC}(f)(x)) \subseteq \text{CSC}(f)(x) \quad \text{for all } x \in X$$

is obvious. So it is sufficient to show that $\text{CSC}(f)(x) \subseteq \overline{\text{co}}(\text{USC}(f)(x))$ for every $x \in X$. To this end, let us suppose, to obtain a contradiction, that there exist $x_0 \in X$ and $y_0 \in Y$ such that

$$y_0 \in \text{CSC}(f)(x_0) \setminus \overline{\text{co}}(\text{USC}(f)(x_0)).$$

Since $\overline{\text{co}}(\text{USC}(f)(x_0))$ is a closed convex set and $y_0 \notin \overline{\text{co}}(\text{USC}(f)(x_0))$, there is a nonzero continuous linear functional strongly separating $\overline{\text{co}}(\text{USC}(f)(x_0))$ and y_0. So let $h : Y \to \mathbb{R}$ be a continuous linear functional, and let $\lambda \in \mathbb{R}$ be such that

$$\overline{\text{co}}(\text{USC}(f)(x_0)) \subseteq \{y \in Y : h(y) < \lambda\} \quad \text{and} \quad \lambda < h(y_0).$$

By Theorem 4.1.2, $\text{USC}(f)$ is an usco. By Proposition 3.3.31 the mapping $L : X \to 2^Y$ defined by

$$L(x) := \overline{\text{co}}(\text{USC}(f)(x)) \quad \text{for } x \in X$$

is a cusco on (X,τ). Since L is upper semicontinuous, there is $U \in \mathcal{U}(x_0)$ such that $L(U) \subseteq \{y \in Y : h(y) < \lambda\}$. Thus $f(U) \subseteq \{y \in Y : h(y) < \lambda\}$. Since $\{y \in Y : h(y) \leq \lambda\}$ is a closed convex set

$$\overline{\text{co}}(f(U)) \subseteq \{y \in Y : h(y) \leq \lambda\},$$

a contradiction, since $y_0 \in \overline{\text{co}}(f(U))$. ☐

Note that the condition of subcontinuity of f is essential in the previous lemma as the following example shows.

Example 4.3.3. Let $X := Y := \mathbb{R}$ with the usual topology. Let $f : X \to Y$ be defined as follows: $f(x) := 0$ for $x \leq 0$ and $f(x) := 1/x$ for $x > 0$. Then $\overline{\text{co}}(\text{USC}(f)(x)) = \{f(x)\}$ for every $x \in X$, whereas $\text{CSC}(f)(0) = [0,\infty)$. Of course, f is not subcontinuous at 0.

Theorem 3.6.18 gives some characterisations of minimal cusco mappings in the class of cusco mappings (see also [41, 94]). So the following characterisation of minimal cusco mappings in the class of all set-valued mappings can be of some interest.

Theorem 4.3.4 ([122]). *Let (X,τ) be a topological space, and let $(Y,+,\cdot,\tau')$ be a Hausdorff locally convex topological space. If $F : X \to 2^Y$, then the following are equivalent:*
(i) *F is minimal cusco;*
(ii) *F is nonempty compact valued, and there is a quasicontinuous and subcontinuous selection f of F such that $\overline{\text{co}}(\text{USC}(f)(x)) = F(x)$ for every $x \in X$;*
(iii) *F is nonempty compact valued, and there is a densely defined quasicontinuous and subcontinuous selection f of F such that $\overline{\text{co}}(\text{USC}(f)(x)) = F(x)$ for every $x \in X$;*
(iv) *F is nonempty compact valued, and there is a densely defined hyperplane minimal subcontinuous selection f of F such that $\overline{\text{co}}(\text{USC}(f)(x)) = F(x)$ for every $x \in X$;*
(v) *F is nonempty compact-valued, every selection f of F is hyperplane minimal subcontinuous, and $\overline{\text{co}}(\text{USC}(f)(x)) = F(x)$ for every $x \in X$.*

Proof. (i)⇒(ii). Let $G : X \to 2^Y$ be a minimal usco mapping such that $\text{Gr}(G) \subseteq \text{Gr}(F)$. Let f be a selection of G. By Theorem 4.1.3, f is a quasicontinuous and subcontinuous selection of G such that $\text{USC}(f) = G$. So f is also a selection of F. By Proposition 3.3.31 we have $\overline{\text{co}}(\text{USC}(f)(x)) = F(x)$ for every $x \in X$.

(ii)⇒(iii) is trivial. (iii)⇒(iv) is also trivial since every quasicontinuous function from (X,τ) to $(Y,+,\cdot,\tau')$ is hyperplane minimal.

(iv)⇒(i). Let f be a densely defined hyperplane minimal subcontinuous selection of F such that $\overline{\text{co}}(\text{USC}(f)(x)) = F(x)$ for every $x \in X$ and $F(x)$ is compact for every $x \in X$. Since f is subcontinuous, $\text{USC}(f)$ is an usco by Theorem 4.1.2. Since $\overline{\text{co}}(\text{USC}(f)(x)) = F(x)$ for every $x \in X$ and $F(x)$ is compact for every $x \in X$, F is a cusco by Proposition 3.3.31. Thus it is sufficient to show that F is a minimal cusco. By Lemma 4.3.2, $\text{CSC}(f)(x) = \overline{\text{co}}(\text{USC}(f)(x))$ for every $x \in X$. By Theorem 3.6.18 ((iv)⇒(i)), $F = \text{CSC}(f)$ is a minimal cusco on (X,τ).

Since (v)⇒(iv) is trivial, it is sufficient to prove that (i)⇒(v). Let f be a selection of F. Since every selection of an usco mapping is subcontinuous, f must be subcontinuous. $\text{USC}(f)$ is an usco, and $\text{Gr}(\text{USC}(f)) \subseteq \text{Gr}(F)$ implies that $\overline{\text{co}}(\text{USC}(f)(x))$ is compact for every $x \in X$. By Proposition 3.3.31 the mapping G defined by $G(x) := \overline{\text{co}}(\text{USC}(f)(x))$ for every $x \in X$ is a cusco on (X,τ). Since $\text{Gr}(G) \subseteq \text{Gr}(F)$ and F is a minimal cusco, we have $\overline{\text{co}}(\text{USC}(f)(x)) = F(x)$ for every $x \in X$. It follows from Proposition 3.3.22 that F is hyperplane minimal. It is easy to verify that every selection of a hyperplane minimal mapping is also hyperplane minimal. □

We have the following variant of Theorem 4.3.4.

Theorem 4.3.5 ([122]). *Let (X,τ) be a topological space, and let $(Y,+,\cdot,\tau')$ be a Hausdorff locally convex topological space in which the closed convex hull of every compact set is compact. If $F : X \to 2^Y$, then the following are equivalent:*

(i) *F is minimal cusco;*
(ii) *There is a quasicontinuous subcontinuous function $f : X \to Y$ such that for every $x \in X$, $\overline{\text{co}}(\text{USC}(f)(x)) = F(x)$;*
(iii) *There is a densely defined quasicontinuous subcontinuous selection of F such that $\overline{\text{co}}(\text{USC}(f)(x)) = F(x)$ for every $x \in X$;*
(iv) *There is a densely defined hyperplane minimal subcontinuous selection of F such that $\overline{\text{co}}(\text{USC}(f)(x)) = F(x)$ for every $x \in X$;*
(v) *Every selection f of F is hyperplane minimal and subcontinuous, and for every $x \in X$, $\overline{\text{co}}(\text{USC}(f)(x)) = F(x)$.*

As was also mentioned in Chapter 3, there are some important cases where the closed convex hull of a compact set is compact. The first is where the compact set is a finite union of compact convex sets. The second is where the space is completely metrisable and locally convex. This includes the case of all Banach spaces with their norm topologies. The third case is a compact set in the weak topology on a Banach space (see [3]).

Note that Theorem 4.3.5 provides a method of constructing a minimal cusco with values in a Hausdorff locally convex topological space in which the closed convex hull of every compact set is compact.

Remarks 4.3.6. Let (X, τ) be a Baire space, and let $F : X \to 2^{\mathbb{R}}$ be a cusco. Let

$$f(x) := \min\{t \in \mathbb{R} : t \in F(x)\} \quad \text{for } x \in X.$$

By Remark 4.1.4 the mapping $G := \text{USC}(f|_{C(f)})$ is a minimal usco from (X, τ) to $2^{\mathbb{R}}$, and $\text{Gr}(G) \subseteq \text{Gr}(F)$. By Corollary 3.3.32 the mapping $x \to \overline{\text{co}}(\text{USC}(f|_{C(f)})(x))$ is a minimal cusco, the graph of which is contained in the graph of F.

Similarly, if

$$h(x) := \max\{t \in \mathbb{R} : t \in F(x)\} \quad \text{for } x \in X,$$

then the mapping $x \to \overline{\text{co}}(\text{USC}(h|_{C(h)})(x))$ is a minimal cusco, the graph of which, is contained in the graph of F.

Exercises

1. Let (T, d) be a metric space, and let $(G_n : n \in \mathbb{N})$ be a decreasing sequence of open subsets of T with dense intersection. Let

$$f(t) := \sum_{n=1}^{\infty} 10^{-n} \sin(1/d(t, T \setminus G_n)) \quad \text{for } t \in G := \bigcap\{G_n : n \in \mathbb{N}\}.$$

Show that f is continuous on G and no extension of f to T is continuous at any point of $T \setminus G$ [41, Lemma 2.13].

2. For f in Exercise 1, define

$$\Phi(t) := \left[\liminf_{\substack{s\to t\\ s\in G}} f(s), \limsup_{\substack{s\to t\\ s\in G}} f(s)\right] \quad \text{for } t \in T.$$

Show that Φ is a minimal cusco that is single valued at t if and only if $t \in G$ [41, Theorem 2.14]. (The function $f : G \to \mathbb{R}$ is defined on a dense G_δ-set G of T. It is easy to verify that f is subcontinuous. Since f is continuous on G, by Theorem 4.3.4 the mapping $x \to \overline{\text{co}}(\text{USC}(f)(x))$ for $x \in T$ is a minimal cusco. Of course, we have $\Phi(x) = \overline{\text{co}}(\text{USC}(f)(x))$ for every $x \in T$.)

4.4 Minimal cusco mappings and extreme functions

Let B be a nonempty subset of a linear topological space $(V, +, \cdot)$. We say that a point $x \in B$ is an *extreme point* of B if for $y, z \in B$, if $x = \frac{1}{2}(y + z)$, then $x = y = z$. By $\mathcal{E}(B)$ we denote the set of all extreme points of B.

Let (X, τ) be a topological space, and let $(Y, +, \cdot, \tau')$ be a Hausdorff locally convex topological space. If $F : X \to 2^Y$ is a mapping with nonempty compact values, then a selection f of F such that $f(x) \in \mathcal{E}(F(x))$ for every $x \in X$ is called an *extreme function of* F. (By [3, Corollary 7.66] every nonempty compact subset of a Hausdorff locally convex topological space has an extreme point. This is called the *Krein–Milman theorem*. The hypothesis of local convexity cannot be dispensed [3, p. 298].)

Lemma 4.4.1. *Let (X, τ) be a topological space, and let $(Y, +, \cdot, \tau')$ be a Hausdorff locally convex topological space. If $F : X \to 2^Y$ is a minimal cusco and $G : X \to 2^Y$ is an usco such that* $\text{Gr}(G) \subseteq \text{Gr}(F)$, *then $\mathcal{E}(F(x)) \subseteq G(x)$ for every $x \in X$.*

Proof. Let $x \in X$. By Proposition 3.3.31 we have that $F(x) = \overline{\text{co}}(G(x))$ for every $x \in X$. By [203, Theorem 2.10.15], which was proved by Milman [212], every extreme point of $\overline{\text{co}}(G(x))$ is contained in $G(x)$. Thus $\mathcal{E}(F(x)) \subseteq G(x)$ for every $x \in X$. □

Theorem 4.4.2 ([122]). *Let (X, τ) be a topological space, and let $(Y, +, \cdot, \tau')$ be a Hausdorff locally convex topological space. If $F : X \to 2^Y$, then the following are equivalent:*
(i) *F is minimal cusco;*
(ii) *F is nonempty compact and convex valued and has a closed graph, every extreme function of F is quasicontinuous and subcontinuous, and any two extreme functions of F have the same closures of their graphs;*
(iii) *F is nonempty compact valued, every extreme function f of F is quasicontinuous and subcontinuous, and $F(x) = \overline{\text{co}}(\text{USC}(f)(x))$ for every $x \in X$.*

Proof. (i)⇒(ii). Of course, F must have nonempty compact convex values and a closed graph. Let f be an extreme function of F. By Proposition 4.1.1 every selection of usco mapping is subcontinuous. We will show that f is quasicontinuous. Let G be a minimal usco mapping contained in F. By Lemma 4.4.1 we have $\mathcal{E}(F(x)) \subseteq G(x)$ for every $x \in X$. Since $f(x) \in \mathcal{E}(F(x))$ for every $x \in X$, f is a selection of G. By Theorem 4.1.3, f must be quasicontinuous, and $\mathrm{USC}(f) = G$. Thus every two extreme functions have the same closures of their graphs.

(ii)⇒(iii). Let f be an extreme function of F. (Such a function exists since $F(x)$ is a nonempty compact set for every $x \in X$.) Since f is quasicontinuous and subcontinuous, $\mathrm{USC}(f)$ is a minimal usco mapping by Theorem 4.1.3, and $\mathrm{Gr}(\mathrm{USC}(f)) \subseteq \mathrm{Gr}(F)$. We claim that $F(x) = \overline{\mathrm{co}}(\mathrm{USC}(f)(x))$ for every $x \in X$. It is sufficient to prove that $\mathcal{E}(F(x)) \subseteq \mathrm{USC}(f)(x)$ for every $x \in X$ since by the Krein–Milman theorem every compact convex set is the closed convex hull of its extreme points.

Suppose there is $(x,y) \in X \times Y$ such that $y \in \mathcal{E}(F(x)) \setminus \mathrm{USC}(f)(x)$. There are two open and disjoint sets O_1, O_2 in Y such that

$$\mathrm{USC}(f)(x) \subseteq O_1 \quad \text{and} \quad y \in O_2.$$

Let $U \in \mathcal{U}(x)$ be such that $\mathrm{USC}(f)(U) \subseteq O_1$. Let g be an extreme function of F such that $g(x) = y$, a contradiction with the fact that every two extreme functions of F have the same closures of their graphs.

(iii)⇒(i). To prove that F is a minimal cusco, let f be an extreme function of F. Since f is quasicontinuous and subcontinuous, by Theorem 4.1.3, $\mathrm{USC}(f)$ is a minimal usco, and by Corollary 3.3.32 the mapping $x \mapsto \overline{\mathrm{co}}(\mathrm{USC}(f)(x))$ is minimal cusco. Since $F(x) = \overline{\mathrm{co}}(\mathrm{USC}(f)(x))$ for every $x \in X$, we are done. □

Let (X, τ) be a topological space, and let $\mathbb{R}$ be equipped with the usual Euclidean topology. Let $F : X \to 2^{\mathbb{R}}$ be such that $F(x)$ is a nonempty bounded set for every $x \in X$. We will define $\sup F$ and $\inf F$ by

$$\sup F(x) := \sup\{t \in \mathbb{R} : t \in F(x)\} \quad \text{for } x \in X$$

and

$$\inf F(x) := \inf\{t \in \mathbb{R} : t \in F(x)\} \quad \text{for } x \in X.$$

Theorem 4.4.3 ([122]). *Let (X, τ) be a topological space, and let $F : X \to 2^{\mathbb{R}}$. Then the following are equivalent:*

(i) *F is minimal cusco;*

(ii) *F is nonempty compact convex valued and has a closed graph, $\sup F$ and $\inf F$ are quasicontinuous and subcontinuous functions, and $\mathrm{USC}(\sup F) = \mathrm{USC}(\inf F)$;*

(iii) *F is nonempty compact valued, $\sup F$ and $\inf F$ are quasicontinuous and subcontinuous functions, and $F(x) = \overline{\mathrm{co}}(\mathrm{USC}(\sup F)(x)) = \overline{\mathrm{co}}(\mathrm{USC}(\inf F)(x))$.*

Proof. (i)⇒(ii) is clear from Theorem 4.4.2. (ii)⇒(iii). We will prove that $F(x) = \overline{\mathrm{co}}(\mathrm{USC}(\sup F)(x))$ for every $x \in X$. Suppose there is $(x,y) \in X \times \mathbb{R}$ such that $y \in$

$F(x) \setminus \overline{\text{co}}(\text{USC}(\sup F)(x))$. Let $\varepsilon > 0$ be such that

$$(y - 2\varepsilon, y + 2\varepsilon) \cap \overline{\text{co}}(\text{USC}(\sup F)(x)) = \emptyset.$$

The upper semicontinuity of $z \mapsto \overline{\text{co}}(\text{USC}(\sup F)(z))$ at $x \in X$ implies that there is $U \in \mathcal{U}(x)$ such that $\overline{\text{co}}(\text{USC}(\sup F)(z)) \subseteq (y + \varepsilon, \infty)$ for every $z \in U$. Since $\text{USC}(\sup F) = \text{USC}(\inf F)$ and $\inf F(x) \leq y < y + \varepsilon$, we have a contradiction.

(iii)⇒(i). By Theorem 4.1.3, $\text{USC}(\sup F)$ is a minimal usco. By Corollary 3.3.32 the mapping $x \mapsto \overline{\text{co}}(\text{USC}(\sup F)(x))$ is a minimal cusco, so we are done. □

4.5 Minimal usco and minimal cusco mappings

In this section, we give a characterisation of those mappings that are simultaneously minimal usco and minimal cusco. We also show that there is a bijection between the space of minimal usco mappings and the space of minimal cusco mappings.

Let (X, τ) be a topological space, and let $(Y, +, \cdot, \tau')$ be a Hausdorff locally convex topological space. As in Section 4.2 of this chapter, we denote by $MU(X, Y)$ the set of all minimal usco mappings from X to 2^Y and by $MC(X, Y)$ the set of all minimal cusco mappings from (X, τ) to 2^Y. Of course, $MU(X, Y) \cap MC(X, Y) \neq \emptyset$. It follows from the next example that $MU(X, Y) \setminus MC(X, Y) \neq \emptyset$ and also $MC(X, Y) \setminus MU(X, Y) \neq \emptyset$.

Example 4.5.1. Consider $X := [-1, 1]$ with the usual Euclidean topology and consider the mappings $F : X \to 2^{\mathbb{R}}$ and $G : X \to 2^{\mathbb{R}}$ defined by

$$F(x) := \begin{cases} \{1\} & \text{if } x \in [-1, 0), \\ \{-1, 1\} & \text{if } x = 0, \\ \{-1\} & \text{if } x \in (0, 1], \end{cases} \quad \text{and} \quad G(x) := \begin{cases} \{1\} & \text{if } x \in [-1, 0), \\ [-1, 1] & \text{if } x = 0, \\ \{-1\} & \text{if } x \in (0, 1]. \end{cases}$$

Let (X, τ) be a topological space, and let $(Y, +, \cdot, \tau')$ be a Hausdorff locally convex topological space. We say that f is $*$-*quasicontinuous at* $x \in X$ if for every $y \in \overline{\text{co}}(\text{USC}(f)(x))$, every $V \in \mathcal{U}(y)$ and every $U \in \mathcal{U}(x)$, there is a nonempty open set $W \subseteq U$ such that $f(W) \subseteq V$. If f is $*$-quasicontinuous at every point of X, we say that f is $*$-*quasicontinuous on* (X, τ).

It is easy to verify that every $*$-quasicontinuous function is quasicontinuous.

Example 4.5.2. Consider the function $f : \mathbb{R} \to \mathbb{R}$ defined by

$$f(x) := \begin{cases} \sin \frac{1}{x} & \text{if } x \neq 0, \\ 0 & \text{if } x = 0. \end{cases}$$

The function f is not continuous at $x = 0$, but it is $*$-quasicontinuous at 0.

Theorem 4.5.3 ([120]). *Let (X, τ) be a topological space, and let $(Y, +, \cdot, \tau')$ be a Hausdorff locally convex topological space. If $f : X \to Y$, then f is $*$-quasicontinuous at x if and only if every selection of* $\text{USC}(f)$ *is quasicontinuous at x and* $\text{USC}(f)(x) = \overline{\text{co}}(\text{USC}(f)(x))$.

Theorem 4.5.4 ([120]). *Let (X,τ) be a topological space, and let $(Y,+,\cdot,\tau')$ be a Hausdorff locally convex topological space. If $F: X \to 2^Y$, then the following are equivalent:*
(i) $F \in MU(X,Y) \cap MC(X,Y)$;
(ii) *There exists a $*$-quasicontinuous and subcontinuous function f from X to Y such that* $\mathrm{USC}(f) = F$;
(iii) *Every selection f of F is $*$-quasicontinuous and subcontinuous and* $\mathrm{USC}(f) = F$.

Proof. (i)⇒(iii). Let f be a selection of F. Since $F \in MU(X,Y)$, by Theorem 4.1.3, f is quasicontinuous and subcontinuous and $\mathrm{USC}(f) = F$. Since $F \in MC(X,Y)$, $\overline{\mathrm{co}}(\mathrm{USC}(f)(x)) = F(x)$. So, by Theorem 4.5.3, f is $*$-quasicontinuous.

(iii)⇒(ii) is trivial.

(ii)⇒(i). Let f be a $*$-quasicontinuous and subcontinuous function from X to Y such that $\mathrm{USC}(f) = F$. Thus f is quasicontinuous, and by Theorem 4.5.3

$$F(x) = \mathrm{USC}(f)(x) = \overline{\mathrm{co}}(\mathrm{USC}(f)(x)).$$

By Theorem 4.1.3, F is a minimal usco mapping, and by Corollary 3.3.32, F is a minimal cusco mapping. □

Let (X,τ) be a topological space, and let $(Y,+,\cdot,\tau')$ be a Hausdorff locally convex topological space. By $F(X,2^Y)$ we mean the set of all mappings $F: X \to 2^Y$. Define the function $\varphi : MU(X,Y) \to F(X,2^Y)$ as follows: $\varphi(F)(x) := \overline{\mathrm{co}}(F(x))$ for all $x \in X$ and $F \in MU(X,Y)$.

To study properties of the mapping φ, we start with the following result.

Theorem 4.5.5 ([120]). *Let (X,τ) be a topological space, and let $(Y,+,\cdot,\tau')$ be a Hausdorff locally convex topological space. If $F: X \to 2^Y$ is a minimal cusco, then there is a unique minimal usco mapping contained in F.*

Proof. The existence follows from Proposition 1.3.3. So we only need to prove the uniqueness. Let $G: X \to 2^Y$ and $H: X \to 2^Y$ be minimal usco mappings such that $\mathrm{Gr}(G) \subseteq \mathrm{Gr}(F)$ and $\mathrm{Gr}(H) \subseteq \mathrm{Gr}(F)$. It is sufficient to prove that $G(x) \cap H(x) \neq \emptyset$ for every $x \in X$. Indeed, if $G(x) \cap H(x) \neq \emptyset$ for every $x \in X$, then the mapping $L: X \to 2^Y$ defined by $L(x) := G(x) \cap H(x)$ for $x \in X$ is an usco on (X,τ) (see Proposition 1.1.8 and Theorem 1.1.11), and $\mathrm{Gr}(L) \subseteq \mathrm{Gr}(G)$, $\mathrm{Gr}(L) \subseteq \mathrm{Gr}(H)$. Thus $G = L = H$. Now, by Proposition 3.3.31 and the minimality of F, $\varphi(G) = \varphi(H) = F$. Thus, by the Krein–Milman theorem and Lemma 4.4.1 we have that

$$\emptyset \neq \mathcal{E}(F(x)) \subseteq G(x) \cap H(x) \quad \text{for all } x \in X.$$

This completes the proof. □

Theorem 4.5.6 ([120]). *Let (X,τ) be a topological space, and let $(Y,+,\cdot,\tau')$ be a Hausdorff locally convex linear topological space in which the closed convex hull of every compact set is compact. Then the mapping φ is a bijection from $MU(X,Y)$ to $MC(X,Y)$.*

Proof. Let $F \in MU(X,Y)$. The fact that $\varphi(F) \in MC(X,Y)$ follows directly from Corollary 3.3.32. Hence the mapping φ is well-defined. We now show that φ is surjective. To this end, let $G \in MC(X,Y)$. By Theorem 4.5.5 there exists a unique $F \in MU(X,Y)$ such that $\mathrm{Gr}(F) \subseteq \mathrm{Gr}(G)$. By the definition of $\varphi(F)$ we have that $\mathrm{Gr}(\varphi(F)) \subseteq \mathrm{Gr}(G)$. Since, by the above, $\varphi(F) \in MC(X,Y)$, we have from the minimality of G that $\varphi(F) = G$. This shows that φ is surjective. To show that φ is injective, consider the following. Suppose that $F, H \in MU(X,Y)$ and $G := \varphi(F) = \varphi(H)$. Then, by the definition of φ, $\mathrm{Gr}(F) \subseteq \mathrm{Gr}(G)$ and $\mathrm{Gr}(H) \subseteq \mathrm{Gr}(G)$. Hence, by Theorem 4.5.5, $F = H$, which shows that φ is injective. □

4.6 Topological properties

In this part, we consider topologies on the space of minimal usco and minimal cusco mappings, and we study topological properties of the mapping φ defined in the end of the previous section and also important mappings defined on the space of minimal uscos (see [118, 120]).

Following [102], we define the topologies of pointwise convergence, uniform convergence on compacta and uniform convergence.

Let (Z,τ) be a topological space. Denote by $CL(Z)$ the space of all nonempty closed subsets of (Z,τ), by $S(Z)$ the space of all closed subsets of (Z,τ) (including the empty set) and by $K(Z)$ the space of all nonempty compact subsets of (Z,τ).

Recall that if (Y,d) is a metric space, then the distance functional $d(\cdot,A) : Y \to [0,\infty)$ is defined by the familiar formula

$$d(y,A) := \inf\{d(y,a) : a \in A\} \quad \text{for } y \in Y.$$

Let A and C be nonempty subsets of (Y,d). The *excess of A over C with respect to d* is defined as

$$e_d(A,C) := \sup\{d(a,C) : a \in A\}.$$

The *Hausdorff (extended-valued) metric H_d* on $S(Y)$ [24] is defined as

$$H_d(A,C) := \max\{\sup\{d(a,C) : a \in A\}, \sup\{d(c,A) : c \in C\}\}$$

if A and C are nonempty; if $A \neq \varnothing$, then $H_d(A,\varnothing) := H_d(\varnothing,A) := \infty$.

We will often use the following equalities on $CL(Y)$ [24]:

$$H_d(A,C) = \inf\{\varepsilon > 0 : A \subseteq B(C,\varepsilon) \text{ and } C \subseteq B(A,\varepsilon)\} \quad \text{and}$$
$$H_d(A,C) = \max\{e_d(A,C), e_d(C,A)\}.$$

The topology generated by H_d is called the *Hausdorff metric topology*.

We will often use the well-known result that if (Y,d) is a complete metric space, then $(CL(Y),H_d)$ is a complete metric space too, and if (Y,d) is a compact metric space, then $(K(Y),H_d)$ is also a compact metric space [24].

Let (X,τ) be a Hausdorff topological space, and let (Y,d) be a non-trivial metric space. By $F(X,S(Y))$ we mean the space of all mappings from X to $S(Y)$. The topology τ_p of *pointwise convergence on* $F(X,S(Y))$ is induced by the uniformity $\mathfrak{U}_p$ of pointwise convergence, which has a base consisting of sets of the form

$$W(A,\varepsilon) := \{(\Phi,\Psi) : H_d(\Phi(x),\Psi(x)) < \varepsilon \text{ for all } x \in A\},$$

where A is a finite set in X, and $\varepsilon > 0$. A general τ_p-basic neighbourhood of $\Phi \in F(X,S(Y))$ is denoted by $W(\Phi,A,\varepsilon)$, that is,

$$W(\Phi,A,\varepsilon) := W(A,\varepsilon)[\Phi] = \{\Psi : H_d(\Phi(x),\Psi(x)) < \varepsilon \text{ for every } x \in A\}.$$

If $A := \{a\}$, then we will write $W(\Phi,a,\varepsilon)$ instead of $W(\Phi,\{a\},\varepsilon)$. We define the topology τ_{UC} of *uniform convergence on compact sets on* $F(X,S(Y))$. This topology is induced by the uniformity $\mathfrak{U}_{UC}$, which has a base consisting of sets of the form

$$W(K,\varepsilon) := \{(\Phi,\Psi) : H_d(\Phi(x),\Psi(x)) < \varepsilon \text{ for all } x \in K\},$$

where $K \in K(X)$, and $\varepsilon > 0$. The general τ_{UC}-basic neighbourhood of $\Phi \in F(X,S(Y))$ is denoted by $W(\Phi,K,\varepsilon)$, that is, $W(\Phi,K,\varepsilon) := W(K,\varepsilon)[\Phi]$. Finally, we define the topology τ_U of *uniform convergence on* $F(X,S(Y))$ [102]. Let ϱ be the (extended-valued) metric on $F(X,S(Y))$ defined by

$$\varrho(\Phi,\Psi) := \sup\{H_d(\Phi(x),\Psi(x)) : x \in X\}$$

for $\Phi,\Psi \in F(X,S(Y))$. Then the *topology of uniform convergence* for the space $F(X,S(Y))$ is the topology generated by the metric ϱ.

We need the following lemma.

Lemma 4.6.1. *Let $(Y,\|\cdot\|)$ be a normed linear space. Let A,B be nonempty closed subsets of $(Y,\|\cdot\|)$. Then $H_d(\overline{\text{co}}(A),\overline{\text{co}}(B)) \le H_d(A,B)$, where d is the metric induced by the norm on $(Y,\|\cdot\|)$.*

Proof. First, we show that $e_d(\overline{\text{co}}(A),\overline{\text{co}}(B)) \le e_d(A,B)$. It is known (see [24] Exercise 1.5.3, part b), that if C is convex, then $e_d(\overline{\text{co}}(A),C) = e_d(A,C)$. So $e_d(\overline{\text{co}}(A),\overline{\text{co}}(B)) = e_d(A,\overline{\text{co}}(B))$. Since $B \subseteq \overline{\text{co}}(B)$, we have that

$$e_d(\overline{\text{co}}(A),\overline{\text{co}}(B)) = e_d(A,\overline{\text{co}}(B)) \le e_d(A,B).$$

Similarly, we can show that

$$e_d(\overline{\text{co}}(B),\overline{\text{co}}(A)) = e_d(B,\overline{\text{co}}(A)) \le e_d(B,A).$$

Since

$$H_d(C,D) = \max\{e_d(C,D), e_d(D,C)\}$$

for all $C, D \in CL(Y)$, we are done. □

Theorem 4.6.2 ([120]). *Let (X,τ) be a topological space, and let $(Y, \|\cdot\|)$ be a Banach space. The mapping φ from $(MU(X,Y),\tau)$ onto $(MC(X,Y),\tau)$ defined by $\varphi(F)(x) := \overline{\text{co}}(F(x))$ for $x \in X$ is continuous if τ is one of the following topologies $\tau_p, \tau_{UC}, \tau_U$.*

Proof. The proof follows from Lemma 4.6.1. □

The following example shows that the mapping φ^{-1} from $(MC([-1,1],\mathbb{R}),\tau_p)$ onto $(MU([-1,1],\mathbb{R}),\tau_p)$ is not continuous.

Example 4.6.3. Consider $X := [-1,1]$ with the usual Euclidean topology. Let F and G be the mappings from Example 4.5.1. Then $F = \varphi^{-1}(G)$. We claim that φ^{-1} is not continuous at G. For $n \in \mathbb{N}$, let P_n be the mapping from $[-1,1]$ to $\mathbb{R}$ defined by

$$P_n(x) := \begin{cases} \{1\} & \text{if } x \in [-1,0), \\ [-1,1] & \text{if } x = 0, \\ \{\sin \frac{1}{x}\} & \text{if } x \in (0, \frac{2}{(4n-1)\pi}], \\ \{-1\} & \text{if } x \in [\frac{2}{(4n-1)\pi}, 1]. \end{cases}$$

It is easy to see that for every finite subset A of X and every $\varepsilon > 0$, there exists $n_0 \in \mathbb{N}$ such that $P_n \in W(G,A,\varepsilon)$ for every $n \geq n_0$. For every $n \in \mathbb{N}$, we have that $H_d(F(0), \varphi^{-1}(P_n)(0)) = 1$. Then for every $n \in \mathbb{N}$, $P_n \notin W(F, 0, \frac{1}{2})$, and so the mapping φ^{-1} is not continuous at G.

Lemma 4.6.4. *Let (X,τ) be a topological space, and let (Y,d) be a metric space. Let $\Phi, \Psi \in MU(X,Y)$. If $\varepsilon > 0$ and $H_d(\Phi(x),\Psi(x)) \leq \varepsilon$ for all x in a dense subset of (X,τ), then $H_d(\Phi(x),\Psi(x)) \leq \varepsilon$ for all $x \in X$.*

Proof. Let $\varepsilon > 0$ and consider the mapping $F_\varepsilon : X \to CL(Y)$ defined by

$$F_\varepsilon(x) := \{y \in Y : d(y,\Phi(x)) \leq \varepsilon\} \quad \text{for } x \in X.$$

Then F_ε has a closed graph, and $\Psi(x) \cap F_\varepsilon(x) \neq \varnothing$ for all x in a dense subset of (X,τ). Therefore by Theorem 1.1.11 and the minimality of Ψ it follows that $\Psi(x) \subseteq F_\varepsilon(x)$ for all $x \in X$. Similarly, we can consider the mapping $G_\varepsilon : X \to CL(Y)$ defined by

$$G_\varepsilon(x) := \{y \in Y : d(y,\Psi(x)) \leq \varepsilon\}.$$

Then G_ε has a closed graph and $\Phi(x) \cap G_\varepsilon(x) \neq \varnothing$ for all x in a dense subset of (X,τ). Therefore by Theorem 1.1.11 and the minimality of Φ it follows that $\Phi(x) \subseteq G_\varepsilon(x)$ for all $x \in X$. Thus $H_d(\Phi(x),\Psi(x)) \leq \varepsilon$ for all $x \in X$. □

Theorem 4.6.5 ([120]). *Let (X, τ) be a locally compact space, and let $(Y, \|\cdot\|)$ be a Banach space. The mapping φ from $(MU(X, Y), \tau_{UC})$ onto $(MC(X, Y), \tau_{UC})$ is homeomorphism.*

Proof. By Theorem 4.5.6, φ is a bijection. By Theorem 4.6.2 it is sufficient to prove that φ^{-1} is continuous. Let $G \in MC(X, Y)$ and $F = \varphi^{-1}(G)$. Let $K \in K(X)$ and $\varepsilon > 0$. We show that there exist $K_1 \in K(X)$ and $\varepsilon_1 > 0$ such that $\varphi^{-1}(W(G, K_1, \varepsilon_1)) \subseteq W(F, K, \varepsilon)$. Let $K_1 \in K(X)$ be such that $K \subseteq \operatorname{int}(K_1)$. Put $\varepsilon_1 = \frac{\varepsilon}{2}$. Let $H \in W(G, K_1, \varepsilon_1)$. Let d be the metric induced by the norm $\|\cdot\|$ on Y. We show that $H_d(F(x), \varphi^{-1}(H)(x)) < \varepsilon$ for every $x \in K$. By Corollary 1.5.18 there is a residual set E in (X, τ) such that $F(x)$ and $\varphi^{-1}(H)(x)$ are singletons for every $x \in E$. Of course, $F(x) = G(x)$ and $H(x) = \varphi^{-1}(H)(x)$ for every $x \in E$. Since $H \in W(G, K_1, \varepsilon_1)$, we have $H_d(F(x), \varphi^{-1}(H)(x)) = H_d(G(x), H(x)) < \varepsilon_1$ for every $x \in \operatorname{int}(K_1) \cap E$. By Lemma 4.6.4 we have $H_d(F(x), \varphi^{-1}(H)(x)) \le \varepsilon_1 < \varepsilon$ for every $x \in \operatorname{int}(K_1)$. Since $K \subseteq \operatorname{int}(K_1)$, we are done. □

The following example shows that the condition of local compactness in Theorem 4.6.5 is essential.

Example 4.6.6. Let $X := [-1, 1]$ with the topology, where the open sets in X are all subsets of X not containing 0 and all subsets of X containing 0 that have countable complement. Every compact set in X is finite. Thus the topology τ_{UC} on $MU(X, \mathbb{R})$ and $MC(X, \mathbb{R})$ reduces to the topology τ_p. So we can use Example 4.6.3.

Theorem 4.6.7 ([120]). *Let (X, τ) be a Baire space, and let $(Y, \|\cdot\|)$ be a Banach space. The mapping φ from $(MU(X, Y), \tau_U)$ onto $(MC(X, Y), \tau_U)$ is homeomorphism.*

Proof. The proof is similar to that of Theorem 4.6.5. □

Let (X, τ) be a topological space. We will also consider the *Vietoris topology* τ_V on $MU(X, \mathbb{R})$ and $MC(X, \mathbb{R})$. First, we consider the Vietoris topology τ_V on the space $CL(X \times \mathbb{R})$ of nonempty closed subsets of $X \times \mathbb{R}$. The basic open subsets of $CL(X \times \mathbb{R})$ in τ_V are the subsets of the form

$$W^+ \cap W_1^- \cap \cdots \cap W_n^-,$$

where $W, W_1, \ldots, W_n$ are open subsets of $X \times \mathbb{R}$, $W^+ := \{F \in CL(X \times \mathbb{R}) : F \subseteq W\}$, and each $W_i^- := \{F \in CL(X \times \mathbb{R}) : F \cap W_i \ne \emptyset\}$.

Under the identification of every element of $MU(X, \mathbb{R})$ and $MC(X, \mathbb{R})$ with its graph, we can consider $MU(X, \mathbb{R})$ and $MC(X, \mathbb{R})$ as subsets of $CL(X \times \mathbb{R})$. We will consider the induced Vietoris topology τ_V on $MU(X, \mathbb{R})$ and $MC(X, \mathbb{R})$.

Theorem 4.6.8 ([120]). *Let (X, τ) be a locally connected space. Then the mapping φ from $(MU(X, \mathbb{R}), \tau_V)$ onto $(MC(X, \mathbb{R}), \tau_V)$ is continuous.*

Proof. Let $F \in MU(X, \mathbb{R})$, and let $W^+ \cap W_1^- \cap \cdots \cap W_n^-$ be a basic open set in $(MC(X, \mathbb{R}), \tau_V)$ such that $\operatorname{Gr}(\varphi(F)) \in W^+ \cap W_1^- \cap \cdots \cap W_n^-$. Let $G := \varphi(F)$. By [126, Lemma 4.1] there is an open set $H \subseteq X \times \mathbb{R}$ such that $\operatorname{Gr}(G) \subseteq H \subseteq W$, and $H(x)$ is connected for every $x \in X$.

Without loss of generality, we can also suppose that for every $i = 1, 2, \dots, n$, $W_i \subseteq H$ and $W_i = U_i \times V_i$ with U_i open in X and V_i an open interval in $\mathbb{R}$.

For every $i \in \{1, 2, \dots, n\}$, we define an open set $\mathcal{H}_i$ as follows. Let $i \in \{1, 2, \dots, n\}$. Let $(x_i, y_i) \in \mathrm{Gr}(G) \cap W_i$. If $y_i = \inf F(x_i)$ or $y_i = \sup F(x_i)$, then we put $\mathcal{H}_i := W_i^-$. Otherwise, let C_i be a connected set in (X, τ) such that $x_i \in \mathrm{int}(C_i) \subseteq C_i \subseteq U_i$, and let $\varepsilon > 0$ be such that $\inf F(x_i) + \varepsilon < y_i < \sup F(x_i) - \varepsilon$. Put

$$\mathcal{H}_i := (\mathrm{int}(C_i) \times (\inf F(x_i) - \varepsilon, \inf F(x_i) + \varepsilon))^- \cap (\mathrm{int}(C_i) \times (\sup F(x_i) - \varepsilon, \sup F(x_i) + \varepsilon))^-.$$

It is easy to verify that $L \in MU(X, \mathbb{R})$ and $\mathrm{Gr}(L) \in \mathcal{H}_i$ imply that $\mathrm{Gr}(\varphi(L)) \in \mathcal{H}_i$. Since $\varphi(L)$ is an upper semicontinuous set-valued mapping with connected values, $\varphi(L)(\mathrm{int}(C_i))$ must be a connected set (see Exercise 5 in Section 1.6 of Chapter 1 or [24, Proposition 6.2.12]), that is, $y_i \in \varphi(L)(\mathrm{int}(C_i))$. Thus $\mathrm{Gr}(\varphi(L)) \in \cap W_i^-$.

Put $\mathcal{G} := H^+ \cap \mathcal{H}_1 \cap \cdots \cap \mathcal{H}_n$. Then $\mathrm{Gr}(F) \in \mathcal{G}$ and $\mathrm{Gr}(\varphi(S)) \in W^+ \cap W_1^- \cap \cdots \cap W_n^-$ for every $S \in MU(X, \mathbb{R})$ with $\mathrm{Gr}(S) \in \mathcal{G}$. □

The following example shows that the condition of local connectedness in Theorem 4.6.8 is essential.

Example 4.6.9. Let $X := [-1, 1] \setminus \{\frac{1}{n} : n \in \mathbb{N}\}$ with the usual Euclidean topology. Consider the mappings F and G from X to $\mathbb{R}$ defined by

$$F(x) := \begin{cases} \{1\} & \text{if } x \in X \cap [-1, 0), \\ \{-1, 1\} & \text{if } x = 0, \\ \{-1\} & \text{if } x \in X \cap (0, 1], \end{cases} \quad \text{and} \quad G(x) := \begin{cases} \{1\} & \text{if } x \in X \cap [-1, 0), \\ [-1, 1] & \text{if } x = 0, \\ \{-1\} & \text{if } x \in X \cap (0, 1]. \end{cases}$$

Then $G = \varphi(F)$, and we claim that φ is not continuous at F. For every $n \in \mathbb{N}$, let f_n be the function from X to $\mathbb{R}$ defined by

$$f_n(x) := \begin{cases} 1 & \text{if } x \in X \cap [-1, \frac{1}{n}), \\ -1 & \text{if } x \in X \cap (\frac{1}{n}, 1]. \end{cases}$$

We have that $f_n = \varphi(f_n)$ for every $n \in \mathbb{N}$. The sequence $(f_n : n \in \mathbb{N})$ converges in $(MU(X, \mathbb{R}), \tau_V)$ to F, but $(f_n : n \in \mathbb{N})$ does not converge to G in $(MC(X, \mathbb{R}), \tau_V)$.

The following example shows that $\varphi^{-1} : (MC([-1, 1], \mathbb{R}), \tau_V) \to (MU([-1, 1], \mathbb{R}), \tau_V)$ is not continuous.

Example 4.6.10. Let $X := [-1, 1]$ with the usual Euclidean topology. Let F, G be mappings from Example 4.5.1. Then $F = \varphi^{-1}(G)$, and we claim that φ^{-1} is not continuous at G. For every $n \in \mathbb{N}$, let g_n be the function from $[-1, 1]$ to $\mathbb{R}$ defined by

$$g_n(x) := \begin{cases} 1 & \text{if } x \in [-1, 0], \\ 1 - 2nx & \text{if } x \in (0, \frac{1}{n}), \\ -1 & \text{if } x \in [\frac{1}{n}, 1]. \end{cases}$$

Evidently, $g_n = \varphi^{-1}(g_n)$ for every $n \in \mathbb{N}$. It is easy to see that the sequence $(g_n : n \in \mathbb{N})$ converges in $(MC([-1,1],\mathbb{R}),\tau_V)$ to G, but $(g_n : n \in \mathbb{N})$ does not converge to F in $(MU([-1,1],\mathbb{R}),\tau_V)$.

In the last part, we will prove some further results concerning interesting bijections and homeomorphisms from the space $MU(X,\mathbb{R})$.

Let $F : X \to 2^{\mathbb{R}}$ be an usco mapping from a topological space (X,τ) into subsets of $\mathbb{R}$. Recall that the function $\sup F$ defined by

$$\sup F(x) := \sup\{t : t \in F(x)\} \quad \text{for } x \in X$$

is a selection of F, and $\sup F$ is upper semicontinuous and locally bounded. If $F : X \to 2^{\mathbb{R}}$ is a minimal usco, then by Theorem 4.1.3, $\sup F$ is also quasicontinuous.

In what follows, denote by $Q^\star(X,\mathbb{R})(A^\star(X,\mathbb{R}))$ the set of all locally bounded elements of the space $Q(X,\mathbb{R})(A(X,\mathbb{R}))$, where $Q(X,\mathbb{R})$ is the space of all quasicontinuous functions from X to $\mathbb{R}$, and $A(X,\mathbb{R})$ is defined in Section 4.2 of this chapter. By $UC(X,\mathbb{R})$ we denote the set of all upper semicontinuous functions.

Define the mapping $\Omega : MU(X,\mathbb{R}) \to Q^\star(X,\mathbb{R}) \cap UC(X,\mathbb{R})$ by

$$\Omega(F) := \sup F \quad \text{for } F \in MU(X,\mathbb{R}).$$

Proposition 4.6.11 ([118]). *The mapping* $\Omega : MU(X,\mathbb{R}) \to Q^\star(X,\mathbb{R}) \cap UC(X,\mathbb{R})$ *is a bijection, and* $\Omega(D^\star(X,\mathbb{R})) = A^\star(X,\mathbb{R}) \cap UC(X,\mathbb{R})$.

Proof. To show that Ω is one-to-one, let $F, G \in MU(X,\mathbb{R})$ be such that $F \neq G$. By the above, $\sup F$ is upper semicontinuous and locally bounded selection of F. By Theorem 4.1.3, $\sup F$ is also quasicontinuous, and $F = \mathrm{USC}(\sup F)$. Also, $\sup G$ is upper semicontinuous locally bounded and quasicontinuous selection of G, and $\mathrm{USC}(\sup G) = G$. Thus $\sup F \neq \sup G$; otherwise, $F = \mathrm{USC}(\sup F) = \mathrm{USC}(\sup G) = G$.

To show that the mapping Ω is onto, let $f \in Q^\star(X,\mathbb{R}) \cap UC(X,\mathbb{R})$. By Theorem 4.1.3, $\mathrm{USC}(f)$ is a minimal usco mapping. The upper semicontinuity of f guarantees the equality $f(x) = \sup\{y : y \in \mathrm{USC}(f)(x)\}$ for every $x \in X$, that is, $\Omega(\mathrm{USC}(f)) = f$. Thus Ω is onto.

If $F \in D^\star(X,\mathbb{R})$, then by Proposition 4.2.1, $\sup F \in A(X,\mathbb{R})$. Of course, $\sup F$ is upper semicontinuous and locally bounded, that is, $\Omega(D^\star(X,\mathbb{R})) \subseteq A^\star(X,\mathbb{R}) \cap UC(X,\mathbb{R})$. Now we prove the equality. Let $f \in A^\star(X,\mathbb{R}) \cap UC(X,\mathbb{R})$. By Proposition 4.2.1, $\mathrm{USC}(f) \in D(X,\mathbb{R})$, and since f is locally bounded, $\mathrm{USC}(f) \in D^\star(X,\mathbb{R})$. The upper semicontinuity of f guarantees the equality $f(x) = \sup\{y : y \in \mathrm{USC}(f)(x)\}$ for every $x \in X$. Thus $\Omega(\mathrm{USC}(f)) = f$. □

As before, for an usco mapping $F : X \to 2^{\mathbb{R}}$, the function $\inf F$ defined by

$$\inf F(x) := \inf\{t : t \in F(x)\} \quad \text{for } x \in X$$

is also a selection of F, which is lower semicontinuous and locally bounded. We can give a similar result for lower semicontinuous functions as we gave above for upper semicontinuous functions. The result for lower semicontinuous functions is dual.

Denote by $LC(X,\mathbb{R})$ the set of all lower semicontinuous functions and define the mapping $\mathcal{S} : MU(X,\mathbb{R}) \to Q^\star(X,\mathbb{R}) \cap LC(X,\mathbb{R})$ by $\mathcal{S}(F) := \inf F$ for $F \in MU(X,\mathbb{R})$.

Proposition 4.6.12 ([118]). *The mapping $\mathcal{S} : MU(X,\mathbb{R}) \to Q^\star(X,\mathbb{R}) \cap LC(X,\mathbb{R})$ is a bijection, and $\mathcal{S}(D^\star(X,\mathbb{R})) = A^\star(X,\mathbb{R}) \cap LC(X,\mathbb{R})$.*

Remarks 4.6.13. It is easy to see that if A and B are nonempty compact subsets of $\mathbb{R}$, then $d(\sup A, \sup B) \leqq H_d(A,B)$.

Proposition 4.6.14 ([118]). *Let (X,τ) be a topological space. Then the mapping Ω from $(MU(X,\mathbb{R}),\mathfrak{U}_p)$ onto $(Q^\star(X,\mathbb{R}) \cap UC(X,\mathbb{R}),\mathfrak{U}_p)$ is uniformly continuous.*

Proof. The proof follows from Remark 4.6.13. □

The following example shows that even τ_{UC}-convergence in $Q^\star(X,\mathbb{R}) \cap UC(X,\mathbb{R})$ does not imply the convergence in $(MU(X,\mathbb{R}),\tau_p)$.

Example 4.6.15. Let W be the set of all ordinal numbers less than or equal to the first uncountable ordinal number ω_1, with the usual order topology. Let L be the set of all limit ordinal numbers different from ω_1. Put $X := W \setminus L$ and equip X with the induced topology from W. If λ is a non-limit ordinal, there are a unique integer $I(\lambda) \in \mathbb{N}$ and a limit ordinal β such that $\lambda = \beta + I(\lambda)$.

For every $n \in \mathbb{N}$, put $C_n := \{\lambda \in X \setminus \omega_1 : I(\lambda) = n\}$. Then $\omega_1 \in \overline{C_n}$ for every $n \in \mathbb{N}$. Further, for every $n \in \mathbb{N}$, let $f_n \in Q^\star(X,\mathbb{R}) \cap UC(X,\mathbb{R})$ be defined as follows: $f_n(x) := 0$ if $x \in C_n$ and $f_n(x) := 1$ otherwise. It is easy to verify that $(f_n : n \in \mathbb{N})$ τ_{UC}-converges to the function f identically equal to 1. However, the sequence $(\Omega^{-1}(f_n) : n \in \mathbb{N})$ fails to converge to $\Omega^{-1}(f)$ in $(MU(X,\mathbb{R}),\tau_p)$ since $\Omega^{-1}(f_n) = \mathrm{USC}(f_n)$ takes the value $\{0,1\}$ at ω_1 for every $n \in \mathbb{N}$ and $\Omega^{-1}(f)(\omega_1) = \{1\}$.

Theorem 4.6.16 ([118]). *Let (X,τ) be a topological space. Then the spaces $(MU(X,\mathbb{R}),\varrho)$ and $(Q^\star(X,\mathbb{R}) \cap UC(X,\mathbb{R}),\varrho)$ are uniformly isomorphic. Also, the spaces $(D^\star(X,\mathbb{R}),\varrho)$ and $(A^\star(X,\mathbb{R}) \cap UC(X,\mathbb{R}),\varrho)$ are uniformly isomorphic.*

Proof. As we proved above, the mapping Ω from $MU(X,\mathbb{R})$ to $Q^\star(X,\mathbb{R}) \cap UC(X,\mathbb{R})$ is a bijection. By Remark 4.6.13 we have that $\Omega : (MU(X,\mathbb{R}),\varrho) \to (Q^\star(X,\mathbb{R}) \cap UC(X,\mathbb{R}),\varrho)$ is uniformly continuous. To prove that Ω^{-1} is also uniformly continuous, it is sufficient to show that if for $f, g \in Q^\star(X,\mathbb{R}) \cap UC(X,\mathbb{R})$, $d(f(x),g(x)) < \varepsilon$ for every $x \in X$, then $H_d(\mathrm{USC}(f)(x),\mathrm{USC}(g)(x)) \leq \varepsilon$ for every $x \in X$. Suppose that this is not true. Then there exists $x_0 \in X$ such that

$$\varepsilon < H_d(\mathrm{USC}(f)(x_0),\mathrm{USC}(g)(x_0)).$$

There is $r \in \mathrm{USC}(f)(x_0)$ such that $\varepsilon < d(r,\mathrm{USC}(g)(x_0))$, or there is $s \in \mathrm{USC}(g)(x_0)$ such that $\varepsilon < d(s,\mathrm{USC}(f)(x_0))$. Suppose the first case occurs; the proof of the second one is analogous. Put $\beta := d(r,\mathrm{USC}(g)(x_0)) - \varepsilon$. Let $(x_\sigma : \sigma \in \Sigma)$ be a net in (X,τ) converging to x_0 such that the net $(f(x_\sigma) : \sigma \in \Sigma)$ converges to r. Then for $\frac{\beta}{4}$, there is $\sigma_0 \in \Sigma$ such that $f(x_\sigma) \in B(r, \frac{\beta}{4})$ for all $\sigma > \sigma_0$. The upper semicontinuity of USC(g) at x_0 implies that there is $U \in \mathcal{U}(x_0)$ such that $\mathrm{USC}(g)(x) \subseteq B(\mathrm{USC}(g)(x_0), \frac{\beta}{4})$ for all $x \in U$. Let $\sigma \in \Sigma$ be such that $\sigma > \sigma_0$ and $x_\sigma \in U$. Then, of course, $\varepsilon < d(f(x_\sigma), g(x_\sigma))$, a contradiction.

Concerning the proof of the second statement of the theorem, by Proposition 4.6.11 we have that $\Omega(D^\star(X, \mathbb{R})) = A^\star(X, \mathbb{R}) \cap UC(X, \mathbb{R})$, and by the above we know that $\Omega : (MU(X, \mathbb{R}), \varrho) \to (Q^\star(X, \mathbb{R}) \cap UC(X, \mathbb{R}), \varrho)$ is uniformly isomorphic. Thus the restriction of Ω on $D^\star(X, \mathbb{R})$ to $A^\star(X, \mathbb{R}) \cap UC(X, \mathbb{R})$ is also uniformly isomorphic. □

Theorem 4.6.17 ([118]). *Let (X, τ) be a locally compact topological space. The spaces $(MU(X, \mathbb{R}), \mathfrak{U}_{UC})$ and $(Q^\star(X, \mathbb{R}) \cap UC(X, \mathbb{R}), \mathfrak{U}_{UC})$ are uniformly isomorphic. Also, the spaces $(D^\star(X, \mathbb{R}), \mathfrak{U}_{UC})$ and $(A^\star(X, \mathbb{R}) \cap UC(X, \mathbb{R}), \mathfrak{U}_{UC})$ are uniformly isomorphic.*

Proof. As we proved above, the mapping Ω from $MU(X, \mathbb{R})$ to $Q^\star(X, \mathbb{R}) \cap UC(X, \mathbb{R})$ is a bijection. By Remark 4.6.13 we have that $\Omega : (MU(X, \mathbb{R}), \mathfrak{U}_{UC}) \to (O^\star(X, \mathbb{R}) \cap UC(X, \mathbb{R}), \mathfrak{U}_{UC})$ is uniformly continuous. To prove that also Ω^{-1} is uniformly continuous let $K \in K(X)$ and $\varepsilon > 0$. The local compactness of (X, τ) implies that there is an open set G in (X, τ) such that $K \subseteq G$ and $\overline{G}$ is compact. Let $f, g \in Q^\star(X, \mathbb{R}) \cap UC(X, \mathbb{R})$ be such that $d(f(x), g(x)) < \varepsilon$ for every $x \in \overline{G}$. To prove that $H_d(\mathrm{USC}(f), \mathrm{USC}(g)) \le \varepsilon$ for every $x \in K$, we can use a similar idea as in the proof of Theorem 4.6.16. □

Topological properties and the cardinal function properties of character, pseudo character, density, weight, net weight and cellularity on $(D^\star(X, \mathbb{R}), \tau_p)$, $(D^\star(X, \mathbb{R}), \tau_{UC})$ and $(D^\star(X, \mathbb{R}), \tau_U)$ were studied in [114, 115, 118, 130, 138, 141].

4.7 Metrisation of τ_{UC}

We start this section with a result concerning metrisability and complete metrisability of $(F(X, S(Y)), \mathfrak{U}_{UC})$, which will be useful later. Throughout this part, all spaces are assumed to be Hausdorff and nontrivial, that is, all spaces contain at least two distinct points.

Recall that a topological space (X, τ) is *hemicompact* [78] if in the family of all compact subspaces of (X, τ) ordered by set-inclusion, there exists a countable cofinal subfamily.

Theorem 4.7.1 ([114]). *Let (X, τ) be a topological space, and let (Y, d) be a metric space. Then the following are equivalent:*
(i) *The uniformity $\mathfrak{U}_{UC}$ on $F(X, S(Y))$ is induced by a metric;*
(ii) *$(F(X, S(Y)), \tau_{UC})$ is metrisable;*
(iii) *$(F(X, S(Y)), \tau_{UC})$ is first countable;*
(iv) *(X, τ) is hemicompact.*

Proof. (iv)⇒(i). Let $\{K_n : n \in \mathbb{N}\}$ be a countable cofinal subfamily in $K(X)$ with respect to the inclusion. The family

$$\left\{W\left(K_n, \frac{1}{m}\right) : n, m \in \mathbb{N}\right\}$$

is a countable base of $\mathfrak{U}_{UC}$. By the metrisation theorem in [159], $(F(X, S(Y)), \mathfrak{U}_{UC})$ is metrisable.

(i)⇒(ii) and (ii)⇒(iii) are obvious.

(iii)⇒(iv). Let $y_0 \in Y$, and let f be the constant function on X mapping each point to y_0. By assumption f has a countable base $\{W(f, K_n, \varepsilon_n) : \varepsilon_n < 1/n \text{ and } n \in \mathbb{N}\}$. We claim that $\{K_n : n \in \mathbb{N}\}$ is a countable cofinal family in $K(X)$ with respect to the inclusion. Suppose that this is not true. Then there is $K \in K(X)$ such that for each $n \in \mathbb{N}$, there is $k_n \in K \setminus K_n$. For every $n \in \mathbb{N}$, there is an open neighbourhood U_n of k_n such that $\overline{U_n} \cap K_n = \emptyset$. Let $y_1 \in Y$ be a point different from y_0. For every $n \in \mathbb{N}$, let $f_n : X \to Y$ be the function defined by

$$f_n(x) := \begin{cases} y_1 & \text{if } x \in \overline{U_n}, \\ y_0 & \text{otherwise.} \end{cases}$$

Without loss of generality, we may suppose that $\mathrm{USC}(f_n) \notin W(f, K, d(y_0, y_1))$ for every $n \in \mathbb{N}$. Thus the family $\{W(f, K_n, \varepsilon_n) : n \in \mathbb{N}\}$ fails to be a base of neighbourhoods of f (otherwise, there is $n \in \mathbb{N}$ with $\mathrm{USC}(f_n) \in W(f, K_n, \varepsilon_n) \subseteq W(f, K, d(y_0, y_1))$, a contradiction). □

We have the following result concerning the complete metrisability of $F(X, S(Y))$.

Theorem 4.7.2 ([114]). *Let (X, τ) be a topological space, and let (Y, d) be a complete metric space. Then the following are equivalent:*
(i) *The uniformity $\mathfrak{U}_{UC}$ on $F(X, S(Y))$ is induced by a complete metric;*
(ii) *$(F(X, S(Y)), \tau_{UC})$ is completely metrisable;*
(iii) *(X, τ) is hemicompact.*

Proof. Of course, it is sufficient to prove that (iii)⇒(i). It is known [24] that if d is a complete metric on Y, then H_d is also a complete metric on $S(Y)$. Thus by [159, Theorem 7.10] the uniformity $\mathfrak{U}_{UC}$ on $F(X, S(Y))$ is complete. By Theorem 4.7.1, $\mathfrak{U}_{UC}$ is also induced by a metric, and thus $\mathfrak{U}_{UC}$ on $F(X, S(Y))$ is induced by a complete metric. □

In the next part of this section, we are interested mainly in metrisability and complete metrisability of $D(X, Y)$, the space of densely continuous forms from a topological space (X, τ) into a metric space (Y, d), equipped with the uniformity $\mathfrak{U}_{UC}$ induced from $F(X, S(Y))$ on $D(X, Y)$ and the corresponding topology of uniform convergence on compact sets on $D(X, Y)$. Our results improve those from [102, 199].

For our analysis, the notion of set-valued mappings with closed graphs will also be important. Denote by $G(X, S(Y))$ the set of all set-valued mappings with closed graphs, that is, if $\Phi \in G(X, S(Y))$, then the set $\{(x, y) : y \in \Phi(x)\}$ is a closed set in $X \times Y$. Note that set-valued mappings (as well as single-valued functions) with closed graphs were intensively studied in the literature; see, for example, [24, 87, 89, 117, 153]. There is also a connection between c-upper semicontinuous set-valued mappings [117] and set-valued mappings with closed graphs with values in locally compact spaces.

Of course, $D(X, Y)$ is a subset of $G(X, S(Y))$. We equip both these spaces with the uniformity $\mathfrak{U}_{UC}$ induced from $F(X, S(Y))$ and the corresponding topology of uniform convergence on compact sets. We have the following result concerning metrisability and first countability of $(D(X, Y), \tau_{UC})$ and $(G(X, S(Y)), \tau_{UC})$, which improves Theorem 4.3 in [102].

Theorem 4.7.3 ([114]). *Let (X, τ) be a topological space, and let (Y, d) be a metric space. Then the following are equivalent:*
(i) *$(G(X, S(Y)), \tau_{UC})$ is metrisable;*
(ii) *$(G(X, S(Y)), \tau_{UC})$ is first countable;*
(iii) *$(D(X, Y), \tau_{UC})$ is metrisable;*
(iv) *$(D(X, Y), \tau_{UC})$ is first countable;*
(v) *(X, τ) is hemicompact.*

Proof. (i)⇒(ii), (ii)⇒(iv), (i)⇒(iii), (iii)⇒(iv) are trivial. (v)⇒(i). From Theorem 4.7.1 we know that the hemicompactness of (X, τ) implies that $(F(X, S(Y)), \mathfrak{U}_{UC})$ is metrisable, so also $(G(X, S(Y)), \mathfrak{U}_{UC})$ is metrisable. Thus also $(G(X, S(Y)), \tau_{UC})$.

To prove (iv)⇒(v), we can argue in the same way as in (iii)⇒(iv) in Theorem 4.7.1. In fact, realise that all USC(f_n) are densely continuous forms ($C(f_n)$ contains $U_n \cup (X \setminus \overline{U_n})$, which is dense in (X, τ)). □

Remarks 4.7.4. Of course, the conditions of Theorem 4.7.3 are also equivalent to the metrisability of $(G(X, S(Y)), \mathfrak{U}_{UC})$ and $(D(X, Y), \mathfrak{U}_{UC})$.

Every hemicompact space is σ-compact, but not vice versa. The space of rationals with the usual topology is a σ-compact space that is not hemicompact. A locally compact σ-compact space is hemicompact. There is a hemicompact space that is not a k-space [78, 3.4.E], thus our Theorem 4.7.3 is a generalisation of a part of [102, Theorem 4.3].

The following result lists some properties of the space $(D^*(X, \mathbb{R}), \tau_{UC})$ of locally bounded densely continuous forms from X to $\mathbb{R}$, which are equivalent to metrisability. One of these properties is that of being a space of *pointwise countable type* [78], that is, each point is contained in a compact of countable character, and another one is the more general property of being *q-space*. Recall that this is a space such that for each point, there exists a sequence $(U_n : n \in \mathbb{N})$ of neighbourhoods of that point, so that if $x_n \in U_n$ for each $n \in \mathbb{N}$, then $(x_n : n \in \mathbb{N})$ has a cluster point [201]. Moreover, the following result gives a positive answer to Question 4.1 in [199].

Theorem 4.7.5 ([114]). *Let (X,τ) be a topological space. Then the following are equivalent:*
(i) *$(D^*(X,\mathbb{R}),\tau_{UC})$ is metrisable;*
(ii) *$(D^*(X,\mathbb{R}),\tau_{UC})$ is first countable;*
(iii) *$(D^*(X,\mathbb{R}),\tau_{UC})$ is of pointwise countable type;*
(iv) *$(D^*(X,\mathbb{R}),\tau_{UC})$ is a q-space;*
(v) *(X,τ) is hemicompact.*

Proof. (i) $\Rightarrow$ (ii) $\Rightarrow$ (iii) $\Rightarrow$ (iv) are all immediate.

(iv) $\Rightarrow$ (v). Suppose that $(D^*(X,\mathbb{R}),\tau_{UC})$ is a q-space. Let f be the zero function on X. Put $\Phi := \mathrm{USC}(f|_X) = f$. By the assumption there is a sequence $(W(\Phi,K_n,\varepsilon_n) : n \in \mathbb{N})$ such that if $\Phi_n \in W(\Phi,K_n,\varepsilon_n)$ for each $n \in \mathbb{N}$, then $(\Phi_n : n \in \mathbb{N})$ has a cluster point in $(D^*(X,R),\tau_{UC})$. We claim that $\{K_n : n \in \mathbb{N}\}$ is a cofinal family in $K(X)$ with respect to set-inclusion. Suppose that this is not true. Then there is $K \in K(X)$ such that for each $n \in \mathbb{N}$, there is $k_n \in K \setminus K_n$. For every $n \in \mathbb{N}$, let U_n be an open neighbourhood of k_n such that $\overline{U_n} \cap K_n = \varnothing$. Let $f_n \in D(X,\mathbb{R})$ be defined as follows:

$$f_n(z) := \begin{cases} n & \text{if } z \in \overline{U_n}, \\ 0 & \text{otherwise.} \end{cases}$$

Put $\Phi_n := \mathrm{USC}(f_n|_{C(f_n)})$. Of course, $\Phi_n \in D^*(X,\mathbb{R})$ and $\Phi_n \in W(\Phi,K_n,\varepsilon_n)$ for every $n \in \mathbb{N}$. Since every $\Psi \in D^*(X,\mathbb{R})$ is bounded on K, $(\Phi_n : n \in \mathbb{N})$ cannot have a cluster point in $(D^*(X,\mathbb{R}),\tau_{UC})$, a contradiction. □

Remarks 4.7.6. For a Tychonoff space (X,τ), the proof of (iv) $\Rightarrow$ (v) in Theorem 4.7.5 can be done in a simpler way. Since, by [199], $C(X,\mathbb{R})$ is a closed subspace of $(D^*(X,\mathbb{R}),\tau_{UC})$, closed subspaces of a q-space are also q-spaces, $(C(X,\mathbb{R}),\tau_{UC})$ is a subspace of $(D^*(X,\mathbb{R}),\tau_{UC})$, and $(C(X,\mathbb{R}),\tau_{UC})$ is a q-space. By [201, Theorem 3.2], (X,τ) must be hemicompact.

In the following part, we also improve the result concerning complete metrisability of $(D^*(X,\mathbb{R}),\tau_{UC})$ in [102]. To study the complete metrisability of $(D^*(X,\mathbb{R}),\tau_{UC})$, we will deal with the complete metrisability of $(G(X,S(Y)),\tau_{UC})$.

We start with the following useful lemmas. If $G \in F(X,S(Y))$ and $B \subseteq Y$, then we put

$$G^-(B) := \{x \in X : G(x) \cap B \neq \varnothing\} \quad \text{and} \quad G^+(B) := \{x \in X : G(x) \subseteq B\}.$$

Lemma 4.7.7 ([197]). *Let (X,τ) be a topological space, and let (Y,τ') be a locally compact space. If $G \in F(X,S(Y))$, then $G \in G(X,S(Y))$ if and only if for every $K \in K(Y)$, $G^-(K)$ is a closed set in (X,τ).*

Elements G of $F(X,S(Y))$ such that $G^-(K)$ is a closed set in (X,τ) for every $K \subseteq Y$ are called in the literature *upper Fell semicontinuous* [197] or *c-upper semicontinuous* [117] set-valued mappings.

Lemma 4.7.8. *Let (X,τ) be a k-space, and let (Y,τ') be a locally compact space. If $G \in F(X,S(Y))$, then $G \in G(X,S(Y))$ if and only if for every $K \in K(Y)$ and every $C \in K(X)$, $G^-(K) \cap C$ is a closed set in (X,τ).*

Proof. It trivially follows from Lemma 4.7.7 and the definition of a k-space. □

Theorem 4.7.9 ([114]). *Let (X,τ) be a topological space, and let (Y,d) be a locally compact complete metric space. Then the following are equivalent:*
(i) *The uniformity $\mathfrak{U}_{UC}$ on $G(X,S(Y))$ is induced by a complete metric;*
(ii) *(X,τ) is a hemicompact k-space.*

Proof. (ii) ⇒ (i). By Theorem 4.7.2 even the hemicompactness of (X,τ) guarantees that the uniformity $\mathfrak{U}_{UC}$ on $F(X,S(Y))$ is induced by a complete metric. Thus it is sufficient to prove that $G(X,S(Y))$ is a closed set in $(F(X,S(Y)),\tau_{UC})$.

Let $G \in F(X,S(Y))$ be in the closure of $G(X,S(Y))$ in $(F(X,S(Y)),\tau_{UC})$. Let $(G_n : n \in \mathbb{N})$ be a sequence in $G(X,S(Y))$ converging to G in $(F(X,S(Y)),\tau_{UC})$. To prove that $G \in G(X,S(Y))$, we use Lemma 4.7.8. Suppose that there are $K \in K(Y)$ and $C \in K(X)$ such that $G^-(K) \cap C$ is not closed in (X,τ). Let

$$x \in \overline{G^-(K) \cap C} \setminus G^-(K).$$

Then $K \cap G(x) = \emptyset$ and the local compactness of (Y,d) imply that there is $\varepsilon > 0$ with $B(K,\varepsilon) \cap G(x) = \emptyset$ and $D := \overline{B(K,\frac{\varepsilon}{4})}$ is compact. Consider $W(G,C,\frac{\varepsilon}{4})$. There is $n_0 \in \mathbb{N}$ such that $G_n \in W(G,C,\frac{\varepsilon}{4})$ for every $n \geq n_0$. Thus

$$G_n(x) \subseteq B\left(G(x),\frac{\varepsilon}{4}\right) \subseteq Y \setminus D \quad \text{and} \quad x \in \overline{G_n^-(D) \cap C}$$

for every $n \geq n_0$, a contradiction since every $G_n \in G(X,S(Y))$.

(i) ⇒ (ii). By Theorem 4.7.3, (i) implies the hemicompactness of (X,τ). Let $\{K_n : n \in \mathbb{N}\}$ be a countable cofinal family in $K(X)$ with respect to set-inclusion. Suppose that (X,τ) is not a k-space. Then there is a subset A of X that is not closed in (X,τ) but whose intersection with each K_n is closed. Let $x \in \overline{A} \setminus A$. Without loss of generality, we may assume that $x \in K_1$ and that each K_n is contained in K_{n+1}. Let a,b be two distinct points in Y. For each $n \in \mathbb{N}$, let $G_n \in G(X,S(Y))$ be defined as follows:

$$G_n(z) := \begin{cases} \{a\} & \text{if } z \in A \cap K_n, \\ \{b\} & \text{if } z = x, \\ \emptyset & \text{otherwise.} \end{cases}$$

It is easy to verify that $(G_n : n \in \mathbb{N})$ is a $\mathfrak{U}_{UC}$-Cauchy sequence in $G(X,S(Y))$. The completeness of $\mathfrak{U}_{UC}$ on $G(X,S(Y))$ implies that there must exist $G \in G(X,S(Y))$ such that $(G_n : n \in \mathbb{N})$ converges to $G \in (G(X,S(Y)),\mathfrak{U}_{UC})$, a contradiction since $G(x) = \{b\}$ and $G(z) = \{a\}$ for every $z \in A$ (if $z \in A$, then there is $n \in \mathbb{N}$ with $z \in K_n$, that is, $G_m(z) = \{a\}$ for every $m \geq n$). □

We should remark that if a topological space (X,τ) is a hemicompact k-space and if (Y,τ') is a locally compact metrisable space, then since (Y,τ') admits a complete metric [74], $(G(X,S(Y)),\tau_{UC})$ is completely metrisable with the induced Hausdorff metric on $S(Y)$.

In the last part of this section, we will mention also the topology of uniform convergence on compact sets on the space $U(X,Y)$ of all usco set-valued mappings from a topological space (X,τ) to a metric space (Y,d). It is known that $U(X,Y) \subseteq G(X,S(Y))$ [59, 73] and that $\Phi \in F(X,S(Y))$ is upper semicontinuous if and only if $\Phi^-(B)$ is a closed set for every $B \in CL(Y)$; see Exercise 1.1.2.

It should be noticed that in [113] the Hausdorff metric topology on $U(X,Y)$ was studied. It was proved in [113] that if (X,τ) is a locally compact metrisable space and (Y,d) is completely metrisable, then $X\times Y$ admits a complete metric that makes $U(X,Y)$ completely metrisable with the induced Hausdorff metric.

We have the following result concerning the complete metrisability of $(U(X,Y),\tau_{UC})$, the proof of which is similar to that of Theorem 4.7.9.

Theorem 4.7.10 ([114]). *Let (X,τ) be a hemicompact k-space, and let (Y,d) be a complete metric space. Then $(U(X,Y),\tau_{UC})$ is completely metrisable.*

Proof. By Theorem 4.7.2, $(F(X,S(Y)),\tau_{UC})$ is completely metrisable. We prove that $U(X,Y)$ is a closed set in $(F(X,S(Y)),\tau_{UC})$. Let $G \in F(X,S(Y))$ be in the closure of $U(X,Y)$ in $(F(X,S(Y)),\tau_{UC})$. Let $(G_n : n \in \mathbb{N})$ be a sequence in $U(X,Y)$ converging to $G \in (F(X,S(Y)),\mathfrak{U}_{UC})$. It is easy to verify that $G(x)$ must be a nonempty compact set for every $x \in X$ (since $K(Y)$ is a closed set in $(S(Y),H_d)$ for a complete metric space (Y,d)). To prove that $G \in U(X,Y)$, it is sufficient to verify that $G^-(B) \cap C$ is a closed set in (X,τ) for all $C \in K(X)$ and $B \in CL(Y)$. The proof is similar to that of Theorem 4.7.9 (ii) $\Rightarrow$ (i). □

Remarks 4.7.11. We should note that if (X,τ) is a topological space and (Y,d) is a complete metric space with at least three distinct points, then by an easy modification of the proof of Theorem 4.7.9 we can show that the uniformity $\mathfrak{U}_{UC}$ on $U(X,Y)$ is induced by a complete metric if and only if (X,τ) is a hemicompact k-space.

In the next part, we will also mention a result concerning complete metrisability of the space of all minimal usco mappings.

Proposition 4.7.12 ([114]). *Let (X,τ) be a locally compact topological space, and let (Y,d) be a metric space. Then $MU(X,Y)$ is a closed set in $(U(X,Y),\tau_{UC})$.*

Proof. Let Φ be in the closure of $MU(X,Y)$ in $(U(X,Y),\tau_{UC})$. Let Ψ be a minimal usco mapping contained in Φ. We claim that $\Phi = \Psi$. Suppose that there is $(z,y) \in \mathrm{Gr}(\Phi) \setminus \mathrm{Gr}(\Psi)$. There are open sets $U \subseteq X$, $V \subseteq Y$ such that $z \in U$, $y \in V$ and $(U\times V) \cap \mathrm{Gr}(\Psi) = \emptyset$. Let $\varepsilon > 0$ be such that $B(y,\varepsilon) \subseteq V$. Let K be a compact neighbourhood of z with $K \subseteq U$. Let $\Lambda \in W(\Phi,K,\frac{\varepsilon}{4}) \cap MU(X,Y)$. There must exist $v \in \Lambda(z)$ with $v \in B(y,\frac{\varepsilon}{4})$. By Corollary 1.5.18 there is a residual set E in (X,τ) such that Λ is single-valued at each

$x \in E$. Now let g be any selection of Λ. By Theorem 4.1.3, g is quasicontinuous, and $\Lambda = \mathrm{USC}(g)$. The quasicontinuity of g implies that $\mathrm{USC}(g) = \mathrm{USC}(g|_E)$. Thus there is $a \in E \cap \mathrm{int}(K)$ with $g(a) \in B(v, \frac{\varepsilon}{4})$. We must have $\Psi(a) \subseteq \Phi(a) \subseteq B(g(a), \frac{\varepsilon}{4}) \subseteq B(v, \frac{\varepsilon}{2}) \subseteq B(y, \frac{3\varepsilon}{4})$, a contradiction. □

Theorem 4.7.13 ([114]). *Let (X,τ) be a locally compact hemicompact space, and let (Y,d) be a complete metric space. Then $(MU(X,Y),\tau_{UC})$ is completely metrisable.*

Proof. By Theorem 4.7.10, $(U(X,Y),\tau_{UC})$, is completely metrisable. By Proposition 4.7.12, $MU(X,Y)$ is a closed set in $(U(X,Y),\tau_{UC})$, and thus $(MU(X,Y),\tau_{UC})$ is completely metrisable. □

Remarks 4.7.14. If (X,τ) is a Baire space, then by Remark 4.2.6 we have $D^*(X,\mathbb{R}) := MU(X,\mathbb{R})$. Thus if (X,τ) a Baire space, then we can apply to $(MU(X,\mathbb{R}),\tau_{UC})$ all results valid for $(D^*(X,\mathbb{R}),\tau_{UC})$ from [199]; particularly, if (X,τ) is a locally compact space, then $(MU(X,\mathbb{R}),\tau_{UC})$ is a locally convex linear topological space.

Using the previous remark, we have the following corollaries:

Corollary 4.7.15 ([114]). *Let (X,τ) be a locally compact topological space. Then $D^*(X,\mathbb{R})$ is a closed set in $(U(X,\mathbb{R}),\tau_{UC})$.*

Corollary 4.7.16 ([114, 199]). *Let (X,τ) be a locally compact hemicompact space. Then $(D^*(X,\mathbb{R}),\tau_{UC})$ is completely metrisable.*

The following example shows that the condition of local compactness of (X,τ) in Proposition 4.7.12 and Corollary 4.7.15 is essential.

Example 4.7.17. Consider the topological space (X,τ) and the sequence $(f_n : n \in \mathbb{N})$ from Example 4.6.15. For every $n \in \mathbb{N}$, we have $f_n \in DC(X,\mathbb{R})$, since $C(f_n) = X \setminus \{\omega_1\}$. Put $F_n := \mathrm{USC}(f_n|_{C(f_n)})$ for every $n \in \mathbb{N}$. Then $F_n \in D^*(X,\mathbb{R})$. Of course, $D^*(X,\mathbb{R}) = MU(X,\mathbb{R})$. Now define $F : X \to 2^{\mathbb{R}}$ by

$$F(x) := \begin{cases} \{0,1\} & \text{if } x = \omega_1, \\ \{1\} & \text{otherwise.} \end{cases}$$

It is easy to verify that the sequence $(F_n : n \in \mathbb{N})$ converges to F in $(U(X,\mathbb{R}),\tau_{UC})$. However, F cannot belong to $D^*(X,\mathbb{R})$. (Suppose there is a function $f \in DC(X,\mathbb{R})$ with $\mathrm{USC}(f|_{C(f)}) = F$. Then, of course, $X \setminus \{\omega_1\} \subseteq C(f)$, and $f(x)$ has to be 1 for every $x \neq \omega_1$, that is, $F(\omega_1)$ must be $\{1\}$, a contradiction.)

Note that whereas the first countability of $(D^*(X,\mathbb{R}),\tau_{UC})$ is equivalent to the first countability of $(C(X,\mathbb{R}),\tau_{UC})$, the second countability of $(D^*(X,\mathbb{R}),\tau_{UC})$ is very far from being equivalent to the second countability of $(C(X,\mathbb{R}),\tau_{UC})$. For example, $(C([0,1],\mathbb{R}),\tau_{UC})$ is known to have a countable base, whereas the weights of both $(D^*([0,1],\mathbb{R}),\tau_{UC})$ and $(MU([0,1],\mathbb{R}),\tau_{UC})$ are $\mathfrak{c}$, as we can deduce from results of

[130], where the cardinal functions like the weight, density, cellularity and network weight applied to $(D^*(X,\mathbb{R}),\tau_{UC})$ are studied.

Exercises

1. Prove the following characterisation. Let (X,τ) be a topological space. Then the following are equivalent:
   (i) $(MU(X,\mathbb{R}),\tau_{UC})$ is metrisable;
   (ii) $(MU(X,\mathbb{R}),\tau_{UC})$ is first countable;
   (iii) $(MU(X,\mathbb{R}),\tau_{UC})$ is of pointwise countable type;
   (iv) $(MU(X,\mathbb{R}),\tau_{UC})$ is a q-space;
   (v) (X,τ) is hemicompact.
2. Prove the following characterisation. Let (X,τ) be a topological space, and let (Y,d) be a metric space. Then the following are equivalent:
   (i) $(MU(X,Y),\tau_{UC})$ is metrisable;
   (ii) $(MU(X,Y),\tau_{UC})$ is first countable;
   (iii) (X,τ) is hemicompact.

4.8 Ascoli theorem for minimal usco and minimal cusco mappings

In this part, we prove Ascoli-type theorems for minimal usco and minimal cusco mappings. Our results concerning compact subsets of minimal cusco mappings equipped with the topology of uniform convergence on compact sets are entirely new in the literature. Note that some relative results using quasicontinuous and subcontinuous selections can be found in [139]. Ascoli-type theorems for so-called densely continuous forms and locally bounded densely continuous forms were proved in [102] and [199].

Some topological properties of the space $(MU(X,Y),\tau_{UC})$ were mentioned in the previous section. Cardinal invariants of the space $(MU(X,\mathbb{R}),\tau_{UC})$ are studied in [130].

In what follows, let (X,τ) be a Hausdorff topological space, and let (Y,d) be a metric space.

We say that a subset $\mathcal{E}$ of $MU(X,Y)$ is *pointwise bounded* if for every $x\in X$, $\overline{\bigcup\{L(x):L\in\mathcal{E}\}}$ is compact in (Y,d).

We say that a subset $\mathcal{E}$ of $F(X,CL(Y))$ is *densely equicontinuous* [199] if for every $x\in X$ and every $\varepsilon>0$, there exists a finite family $\mathcal{U}$ of open subsets of (X,τ) such that $\overline{\bigcup\mathcal{U}}$ is a neighbourhood of x and such that for all $L\in\mathcal{E}$ and $U\in\mathcal{U}$, $\operatorname{diam}[L(U)]\le\varepsilon$.

Remarks 4.8.1. Note that if a densely equicontinuous family $\mathcal{E}$ is a subset of $C(X,Y)$, then the space of continuous functions from (X,τ) to (Y,d) is equicontinuous.

Lemma 4.8.2 ([124]). *Let (X,τ) be a locally compact topological space, and let (Y,d) be a metric space. Then the set $U(X,Y)$, of all usco mappings from (X,τ) to 2^Y is a closed subset of $(F(X,K(Y)),\tau_{UC})$.*

Proof. Let $L \in F(X, K(Y))$, and let $(L_\sigma : \sigma \in \Sigma)$ be a net of usco mappings from (X, τ) to 2^Y that converges to L in $(F(X, K(Y)), \tau_{UC})$. We prove that L is upper semicontinuous. Suppose there is a point $x \in X$ in which L is not upper semicontinuous. There is an open set U in Y with $L(x) \subseteq U$ and nets $(x_a : a \in A)$ and $(y_a : a \in A)$ such that $y_a \in L(x_a) \setminus U$. There is $\varepsilon > 0$ such that $\{y \in Y : d(y, L(x)) \leq \varepsilon\} \subseteq U$. Let G be an open set such that $x \in G$ and $\overline{G}$ is compact. The convergence of $(L_\sigma : \sigma \in \Sigma)$ to L in $(F(X, K(Y)), \tau_{UC})$ implies that there is $\sigma_0 \in \Sigma$ such that

$$H_d(L_\sigma(s), L(s)) < \varepsilon/4 \quad \text{for all } s \in \overline{G} \text{ and } \sigma \geq \sigma_0.$$

Let $\sigma \geq \sigma_0$. Put $H := \{y \in Y : d(y, L(x)) < \varepsilon/4\}$. Then $L_\sigma(x) \subseteq H$, and the upper semicontinuity of L_σ in x implies that there is an open set V in (X, τ) such that $x \in V$ and $L_\sigma(z) \subseteq H$ for all $z \in V$. Let $a \in A$ be such that $x_a \in V \cap G$. Since $H_d(L_\sigma(x_a), L(x_a)) < \varepsilon/4$, we have that $L(x_a) \subseteq B(H, \frac{\varepsilon}{4}) \subseteq U$, a contradiction, since $y_a \in L(x_a) \setminus U$. □

Lemma 4.8.3. *Let (X, τ) be a locally compact space, and let (Y, d) be a metric space. Then $MU(X, Y)$ is a closed subset of $(F(X, K(Y)), \tau_{UC})$.*

Proof. By Proposition 4.7.12, $MU(X, Y)$ is a closed subset of $(U(X, Y), \tau_{UC})$. Thus, by Lemma 4.8.2, $MU(X, Y)$ is a closed subset of $(F(X, K(Y)), \tau_{UC})$. □

Theorem 4.8.4 ([124]). *Let (X, τ) be a locally compact topological space, and let (Y, d) be a metric space. A subset $\mathcal{E} \subseteq MU(X, Y)$ is compact in $(F(X, K(Y)), \tau_{UC})$ if and only if $\mathcal{E}$ is closed in $(MU(X, Y), \tau_{UC})$, densely equicontinuous and pointwise bounded.*

Proof. Let $\mathcal{E}$ be compact in $(F(X, K(Y)), \tau_{UC})$. Let $x \in X$. The mapping

$$H_x : (MU(X, Y), \tau_{UC}) \to (K(Y), H_d)$$

defined by $H_x(F) := F(x)$ for $F \in MU(X, Y)$ is continuous. Thus $H_x(\mathcal{E})$ is compact in $(K(Y), H_d)$. Then $\bigcup_{F \in \mathcal{E}} F(x)$ is compact in (Y, d) [206]. (The Vietoris topology and the Hausdorff metric topology coincide on $K(Y)$.)

We now show that $\mathcal{E}$ is densely equicontinuous. Let $x \in X$ and $\varepsilon > 0$. There is an open set $U(x)$ containing x such that $A := \overline{U(x)}$ is compact. There are $F_1, F_2, \ldots, F_n \in \mathcal{E}$ such that

$$\mathcal{E} \subseteq \bigcup_{1 \leq i \leq n} W(A, \varepsilon/3)[F_i].$$

Since every F_i is an usco for $1 \leq i \leq n$, $F_i(A)$ is compact for $1 \leq i \leq n$. Let $\mathcal{V} = \{V_1, \ldots, V_m\}$ be an open finite cover of $\bigcup_{1 \leq i \leq n} F_i(A)$ such that each V_i has diameter less than $\epsilon/3$. Let $\pi : \{1, 2, \ldots, n\} \to \{1, 2, \ldots, m\}$. Then define

$$W_\pi := \bigcup \{y \in U(x) : F_i(y) \subseteq V_{\pi(i)} \text{ for every } 1 \leq i \leq n\}.$$

Clearly, each W_π is open, and by applying Proposition 1.3.5 several times we see that

$$\bigcup \{W_\pi : \pi \in \{1, 2, \ldots, m\}^{\{1,2,\ldots,n\}}\} \quad \text{is dense in } U(x).$$

It is easy to verify that for every $F \in \mathcal{E}$ and every $\pi \in \{1,2,\dots,m\}^{\{1,2,\dots,n\}}$, we have $\operatorname{diam}[F(W_\pi)] < \varepsilon$.

Let $\mathcal{E} \subseteq (MU(X,Y), \tau_{UC})$ be closed, densely equicontinuous and pointwise bounded. Using Lemma 4.8.3, we see that $\mathcal{E}$ is closed in $(F(X,K(Y)), \tau_{UC})$.

For every $x \in X$, put $\mathcal{E}_x := \overline{\bigcup\{L(x) : L \in \mathcal{E}\}}$. Then $\mathcal{E}_x$ is a compact subset of (Y,d), and so the space $(K(\mathcal{E}_x), H_d)$ is a compact metric space. We claim that $(\mathcal{E}, \mathfrak{U}_{UC}|_{\mathcal{E}})$ is a complete uniform space. Let $(L_\sigma : \sigma \in \Sigma)$ be a Cauchy net in $(\mathcal{E}, \mathfrak{U}_{UC}|_{\mathcal{E}})$, and let L be the pointwise limit of $(L_\sigma : \sigma \in \Sigma)$, which exists since for each $x \in X$, $(L_\sigma(x) : \sigma \in \Sigma)$ is a Cauchy net in $(K(\mathcal{E}_x), H_d)$. It is now routine to show that $(L_\sigma : \sigma \in \Sigma)$ converges to L in $(F(X,K(Y)), \tau_{UC})$. In particular, $L \in \mathcal{E}$. This shows that $(\mathcal{E}, \mathfrak{U}_{UC}|_{\mathcal{E}})$ is a complete uniform space.

So we need only show that $\mathcal{E}$ is totally bounded. To this end, let $K \in K(X)$ and $\varepsilon > 0$. The local compactness of (X,τ) implies that there is an open set G such that $K \subseteq G$ and $\overline{G}$ is compact. Since $\overline{G}$ is compact, there exists a finite collection $\{V_1, V_2, \dots, V_n\}$ of nonempty open subsets of G such that $\bigcup_{1\le j\le n} V_j$ is dense in G and H_d-$\operatorname{diam}[F(V_j)] < \varepsilon/4$ for all $F \in \mathcal{E}$. For each $1 \le j \le n$, choose $x_j \in V_j$. Now $\{F(x_j) : 1 \le j \le n \text{ and } F \in \mathcal{E}\}$ is totally bounded in $(K(Y), H_d)$. Therefore there exist elements $\{A_1, A_2, \dots, A_m\}$ in $K(Y)$ such that

$$\{F(x_j) : 1 \le j \le n \text{ and } F \in \mathcal{E}\} \subseteq B(A_1, \varepsilon/4) \cup B(A_2, \varepsilon/4) \cup \dots \cup B(A_m, \varepsilon/4).$$

Let $\pi : \{1,2,\dots,n\} \to \{1,2,\dots,m\}$. Then define

$$S_\pi := \{F \in \mathcal{E} : F(x_i) \in B(A_{\pi(i)}, \varepsilon/4) \text{ for all } 1 \le i \le n\}.$$

Next, let $F_1, F_2 \in S_\pi$ and $x \in \bigcup_{1\le j\le n} V_j$. Then there exists $1 \le i \le n$ such that $x \in V_i$. Then

$$\begin{aligned} H_d(F_1(x), F_2(x)) &\le H_d(F_1(x), F_1(x_i)) + H_d(F_1(x_i), A_{\pi(i)}) \\ &\quad + H_d(A_{\pi(i)}, F_2(x_i)) + H_d(F_2(x_i), F_2(x)) < \varepsilon, \end{aligned}$$

that is, $H_d(F_1(x), F_2(x)) < \varepsilon$ for all $x \in \bigcup_{1\le j\le n} V_j$. Since $F_1|_G, F_2|_G$ are minimal usco mappings on G, by Lemma 4.6.4, $H_d(F_1(x), F_2(x)) \le \varepsilon$ for all $x \in G$ and thus also for all $x \in K$. Since $\mathcal{E} \subseteq \bigcup\{S_\pi : \pi \in \{1,2,\dots,m\}^{\{1,2,\dots,n\}}\}$, $\mathcal{E}$ is totally bounded and hence compact. □

Corollary 4.8.5 (Arzelà–Ascoli theorem in [159]). *Let (X,τ) be a locally compact space, and let (Y,d) be a metric space. A subset $\mathcal{E} \subseteq C(X,Y)$ is compact in $(C(X,Y), \tau_{UC})$ if and only if $\mathcal{E}$ is closed in $(C(X,Y), \tau_{UC})$, equicontinuous and pointwise bounded.*

Proof. The proof uses Remark 4.8.1 and the fact that $C(X,Y)$ is closed in $(MU(X,Y), \tau_p)$ and thus also in $(MU(X,Y), \tau_{UC})$. □

Theorem 4.8.6 ([124]). *Let (X,τ) be a locally compact space, and let (Y,d) be a metric space. If $\mathcal{E} \subseteq MU(X,Y)$ is densely equicontinuous, then the topology τ_p of point-*

wise convergence and the topology τ_{UC} of uniform convergence on compact sets coincide on $\mathcal{E}$.

Proof. Let $(L_\sigma : \sigma \in \Sigma)$ be a net in $\mathcal{E}$ that converges to L in $(\mathcal{E}, \tau_p)$. We will show that $(L_\sigma : \sigma \in \Sigma)$ converges to L in $(\mathcal{E}, \tau_{UC})$. Let $K \in K(X)$ and $\varepsilon > 0$. The local compactness of (X, τ) implies that there is an open set G such that $K \subseteq G$ and $\overline{G}$ is compact. Since $\overline{G}$ is compact, there exists a finite collection $\{V_1, V_2, \ldots, V_n\}$ of nonempty open subsets of G such that $\bigcup_{1\leq j\leq n} V_j$ is dense in G and H_d-diam$[F(V_j)] < \varepsilon/3$ for all $F \in \mathcal{E}$. For each $1 \leq j \leq n$, choose $x_j \in V_j$. For each $1 \leq j \leq n$, there is $\sigma_j \in \Sigma$ such that $H_d(L_\sigma(x_j), L(x_j)) < \varepsilon/3$ for all $\sigma \geq \sigma_j$. Let $\sigma_0 \in \Sigma$ be such that $\sigma_0 \geq \sigma_j$ for all $1 \leq j \leq n$. Then $H_d(L_\sigma(x), L(x)) \leq \varepsilon$ for all $x \in \bigcup_{1\leq j\leq n} V_j$ and all $\sigma \geq \sigma_0$. Since $L|_G, L_\sigma|_G$ $(\sigma_0 \leq \sigma)$ are minimal usco mappings on G, by Lemma 4.6.4, $H_d(L(x), L_\sigma(x)) \leq \varepsilon$ for all $x \in G$ and so also for all $x \in K$. □

Let (X, τ) be a topological space, and let $(Y, \|\cdot\|)$ be a normed linear space. Recall that the function $\varphi : MU(X, Y) \to F(X, CL(Y))$ defined by

$$\varphi(F)(x) := \overline{\text{co}}(F(x)) \quad \text{for } x \in X$$

is a homeomorphism betwen $(MU(X, Y), \tau_{UC})$ and $(MC(X, Y), \tau_{UC})$, if (X, τ) is a locally compact space and $(Y, \|\cdot\|)$ is a Banach space by Theorem 4.6.5.

Theorem 4.8.7 ([124]). *Let (X, τ) be a locally compact topological space, and let $(Y, \|\cdot\|)$ be a Banach space. A subset $\mathcal{E} \subseteq MC(X, Y)$ is compact in $(F(X, K(Y)), \tau_{UC})$ if and only if $\mathcal{E}$ is closed in $(MC(X, Y), \tau_{UC})$, densely equicontinuous and pointwise bounded.*

Proof. Let $\mathcal{E} \subseteq MC(X, Y)$ be compact in $(F(X, K(Y)), \tau_{UC})$. Then, of course, $\mathcal{E}$ must be a closed set in $(MC(X, Y), \tau_{UC})$. Let $x \in X$. The mapping

$$H_x : (MC(X, Y), \tau_{UC}) \to (K(Y), H_d)$$

defined by $H_x(F) := F(x)$ for $F \in MC(X, Y)$, is continuous. Thus $H_x(\mathcal{E})$ is compact in $(K(Y), H_d)$. Then $\bigcup_{F\in\mathcal{E}} F(x)$ is compact in $(Y, \|\cdot\|)$ [206]. (The Vietoris topology and the Hausdorff metric topology coincide on $K(Y)$.)

Since

$$\varphi : (MU(X, Y), \tau_{UC}) \to (MC(X, Y), \tau_{UC})$$

is a homeomorphism, $\varphi^{-1}(\mathcal{E})$ is a compact set in $(MU(X, Y), \tau_{UC})$ and thus also in $(F(X, K(Y)), \tau_{UC})$. By Theorem 4.8.4 the family $\varphi^{-1}(\mathcal{E})$ must be densely equicontinuous. It is easy to verify that $\mathcal{E}$ is also densely equicontinuous.

Conversely, suppose that $\mathcal{E}$ is densely equicontinuous, pointwise bounded and closed in $(MC(X, Y), \tau_{UC})$. It is easy to verify that $\varphi^{-1}(\mathcal{E})$ is densely equicontinuous, pointwise bounded and closed in $(MU(X, Y), \tau_{UC})$. Furthermore, by Theorem 4.8.4, $\varphi^{-1}(\mathcal{E})$ is compact in $(F(X, K(Y)), \tau_{UC})$ and thus also in $(MU(X, Y), \tau_{UC})$. Therefore $\mathcal{E}$ is compact in $(MC(X, Y), \tau_{UC})$ since φ is a homeomorphism. □

4.9 Hausdorff metric on the space of usco mappings

Let (X, d_X) and (Y, d_Y) be metric spaces. Consider the product $X \times Y$ metrised in the following way:

$$\rho((x_1, y_1), (x_2, y_2)) := \max\{d_X(x_1, x_2), d_Y(y_1, y_2)\} \quad \text{for } (x_1, y_1), (x_2, y_2) \in X \times Y.$$

As before, we write $U(X, Y)$ for the space of all usco set-valued mappings from (X, d_X) to $2^{(Y,d_Y)}$. If we identify the elements of $U(X, Y)$ with their graphs, then the Hausdorff distance H_ρ defines a metric on $U(X, Y)$ (if $F \in U(X, Y)$, then $\mathrm{Gr}(F)$ is a closed set in $(X \times Y, \rho)$ by Proposition 1.1.8).

For a nonempty subset R of $X \times Y$, we will use the following notation for the vertical section at x of R: $R(x) := \{y : (x, y) \in R\}$. Then we may define a set-valued mapping $H_R : X \to 2^Y$ (induced by R) by $H_R(x) := R(x)$ for $x \in X$. Then $\mathrm{Gr}(H_R) = R$.

Let (Y, d_Y) be a metric space, and let χ be a functional defined on 2^Y as follows: $\chi(\varnothing) = 0$, and if A is nonempty subset of Y, then

$$\chi(A) := \inf\{\varepsilon > 0 : A \text{ is a finite } \varepsilon\text{-dense subset}\}.$$

In the literature, χ has been called the *Hausdorff measure of noncompactness functional*. Basic facts about this functional and its relatives can be found in [16, 146]. We record the following easily verified facts as a lemma.

Lemma 4.9.1. *Let χ be the Hausdorff measure of noncompactness functional defined on 2^Y. Then:*
(a) *$\chi(A) = \infty$ if and only if A is unbounded;*
(b) *$\chi(A) = 0$ if and only if A is totally bounded;*
(c) *if $A \subseteq B$, then $\chi(A) \le 2\chi(B)$;*
(d) *if A is totally bounded, then $\chi(B(A, \varepsilon)) \le \varepsilon$ for each $\varepsilon > 0$;*
(e) *$\chi(\overline{A}) = \chi(A)$.*

Theorem 4.9.2 ([113]). *Let (X, d_X) be a metric space. The following are equivalent.*
(i) *(X, d_X) is locally compact,*
(ii) *For each complete metric space (Y, d_Y), the space $(U(X, Y), H_\rho)$ is a closed subspace of $(CL(X \times Y), H_\rho)$.*

Proof. (ii) $\Rightarrow$ (i). The proof in this direction uses some ideas of the proof of [18, Theorem 7]. Suppose that (X, d_X) is not locally compact. Let $x_0 \in X$ fail to have a local base of compact sets. Let $\delta_1 := 1$. There is a sequence $(x_i^1 : i \in \mathbb{N})$ of distinct points of $\{z \in X : 0 < d_X(x_0, z) < \delta_1\}$ with no cluster points in (X, d_X). There exists $\varepsilon_1 > 0$ such that $\varepsilon_1 < d_X(x_0, x_i^1)$ for every $i \in \mathbb{N}$. Next, let $\delta_2 := \frac{\varepsilon_1}{2}$, and let $(x_i^2 : i \in \mathbb{N})$ be a sequence of distinct points of $\{z \in X : 0 < d_X(x_0, z) < \delta_2\}$ with no cluster points in (X, d_X). Choose $\varepsilon_2 > 0$ such that $\varepsilon_2 < d_X(x_0, x_i^2)$ for every $i \in \mathbb{N}$ and set $\delta_3 := \frac{\varepsilon_2}{2}$. Continuing, we can

produce for each $n \in \mathbb{N}$ a sequence $(x_i^n : i \in \mathbb{N})$ of distinct points with no cluster points in (X, d_X) such that:
(a) $\delta_n := \frac{\varepsilon_{n-1}}{2}$;
(b) $0 < \varepsilon_n < \delta_n$;
(c) $\{x_i^n : i \in \mathbb{N}\} \subseteq \{z \in X : \varepsilon_n < d_X(x_0, z) < \delta_n\}$.

For each $n \in \mathbb{N}$, let $(\lambda_i^n : i \in \mathbb{N})$ be a sequence of positive integers such that:
$(*)$ $\lambda_i^n < 1/i$ for every $i \in \mathbb{N}$;
$(**)$ the family $\{\overline{B(x_i^n, \lambda_i^n)} : i \in \mathbb{N}\}$ is pairwise disjoint;
$(***)$ $\overline{B(x_i^n, \lambda_i^n)} \subseteq \{z \in X : \varepsilon_n < d(x_0, z) < \delta_n\}$ for each $i \in \mathbb{N}$.

Now let (Y, d_Y) be an arbitrary noncompact complete metric space. Let $(y_n : n \in \mathbb{N})$ be a sequence in (Y, d_Y) of distinct terms with no convergent subsequences. For each $n \in \mathbb{N}$, let g_n be a bijection from the set $\{x_i^n : i \in \mathbb{N}\}$ to the set $\{y_m : n \le m\}$. Define the set-valued mappings F_n $(n = 1, 2, \ldots)$ as follows:

$$F_n(x) := \begin{cases} \{y_1, g_n(x_i^n)\} & \text{if } x \in \overline{B(x_i^n, \lambda_i^n)}, i = 1, 2, \ldots, \\ \{y_1, y_m\} & \text{if } x \in \overline{B(x_i^m, \lambda_i^m)}, m < n \text{ and } i = 1, 2, \ldots, \\ \{y_1\} & \text{otherwise.} \end{cases}$$

Since the union of any subfamily $\{\overline{B(x_i^n, \lambda_i^n)} : i \in \mathbb{N}\}$ is a closed subset of (X, d_X) for each $n \in \mathbb{N}$, it is easy to check that $F_n \in U(X, Y)$ for every $n \in \mathbb{N}$. Now define the set-valued mapping F from (X, d_X) to 2^Y as follows:

$$F(x) := \begin{cases} \{y_1, y_m\} & \text{if } x \in \overline{B(x_i^m, \lambda_i^m)} \text{ for some } m \text{ and } i, \\ \{y_1\} & \text{otherwise.} \end{cases}$$

Then $\mathrm{Gr}(F)$ is a closed set in $X \times Y$, and F is not upper semicontinuous at x_0. Since the sequence $(\mathrm{Gr}(F_n) : n \in \mathbb{N})$ converges in the Hausdorff metric to $\mathrm{Gr}(F)$, $(U(X, Y), H_\rho)$ is not a closed subspace of $(CL(X \times Y), H_\rho)$.

(ii) $\Rightarrow$ (i). Let (X, d_X) be a locally compact metric space, and let (Y, d_Y) be a complete metric space. We show that $(U(X, Y), H_\rho)$ is a closed subspace of $(CL(X \times Y), H_\rho)$. Let $(F_n : n \in \mathbb{N})$ be a sequence from $U(X, Y)$ such that the sequence $(\mathrm{Gr}(F_n) : n \in \mathbb{N})$ converges in the Hausdorff metric to a closed subset R of $X \times Y$. We prove that the set-valued mapping H_R induced by R belongs to $U(X, Y)$. We will first show that H_R is densely defined. Put $A := \{x \in X : H_R(x) \neq \emptyset\}$. To see that A is dense in (X, d_X), let V be a nonempty open set in (X, d_X). Let $a \in V$. There exists $\delta > 0$ such that $B(a, \delta) \subseteq V$. There is $k \in \mathbb{N}$ such that $H_\rho(R, \mathrm{Gr}(F_k)) < \delta$. Let $b \in F_k(a)$. There must exist $(x, y) \in R$ such that $\rho((a, b), (x, y)) < \delta$, that is, $x \in V$ and $H_R(x) \neq \emptyset$.

Now, by Exercise 1.1.17, to show that $H_R \in U(X, Y)$, we need only show that, locally, H_R has the compact range. Let $x \in X$ and choose $\delta > 0$ such that $\overline{B(x, \delta)}$ is compact. Put $B := \bigcup\{R(a) : a \in B(x, \frac{\delta}{2})\}$. We show that $\chi(B) = 0$, where χ is the Hausdorff measure of noncompactness functional. Let $\varepsilon < 0$ and put $\alpha := \min\{\frac{\varepsilon}{2}, \frac{\delta}{2}\}$. There is $j \in \mathbb{N}$ such that $H_\rho(\mathrm{Gr}(F_n), R) < \alpha$ for every $j \le n$. Let $j \le n$ be fixed. Then $B \subseteq B(F_n(B(x,\delta)), \alpha)$. By Proposition 1.1.7, $F_n(\overline{B(x,\delta)})$ is a compact subset of (Y, d_Y), and so by (d) of Lemma 4.9.1 we have $\chi(B(F_n(\overline{B(x,\delta)}), \alpha)) \le \alpha \le \frac{\varepsilon}{2}$. The inclusion $B \subseteq B(F_n(\overline{B(x,\delta)}), \alpha)$ and (c) of Lemma 4.9.1 imply that $\chi(B) \le \varepsilon$. Since $\chi(B) \le \varepsilon$ for any $\varepsilon > 0$, we have $\chi(B) = 0$, so that $\chi(\overline{B}) = 0$. By (b) of Lemma 4.9.1, $\overline{B}$ is a totally bounded set, and the completeness of (Y, d_Y) implies that $\overline{B}$ is compact. □

Theorem 4.9.2 fails if we consider the broader class of upper semicontinuous set-valued mappings, rather than the usco ones.

Example 4.9.3. We produce a sequence of upper semicontinuous set-valued mappings with nonempty closed values from the real line to itself that converges in the Hausdorff metric to a set-valued mapping with nonempty closed values that is not upper semicontinuous at the origin. Define the set-valued mappings F_n $(n = 1, 2, \ldots)$ as follows:

$$F_n(x) := \begin{cases} [\frac{1}{x^2}, \infty) & \text{if } \frac{1}{n} \le |x|, \\ [n^2, \infty) \cup \{0\} & \text{if } x = 0, \\ [n^2, \infty) & \text{otherwise.} \end{cases}$$

Evidently, $(\mathrm{Gr}(F_n) : n \in \mathbb{N})$ converges in the Hausdorff metric to $\mathrm{Gr}(F)$, where F is defined by

$$F(x) := \begin{cases} [\frac{1}{x^2}, \infty) & \text{if } x \ne 0, \\ \{0\} & \text{if } x = 0. \end{cases}$$

Theorem 4.9.4 ([113]). *Let (X, d_X) and (Y, d_Y) be complete metric spaces. If (X, d_X) is a locally compact space, then $(U(X, Y), H_\rho)$ is complete.*

Proof. $(CL(X \times Y), H_\rho)$ is a complete metric space, and so, by Theorem 4.9.2, $(U(X, Y), H_\rho)$, as a closed subspace of a complete metric space, is complete. □

We should remark that if (X, τ) is a locally compact metrisable space and (Y, τ') is completely metrisable (i. e. there is a complete metric d_Y that induces τ'), then since (X, τ) admits a metric that is complete [74], $X \times Y$ admits a metric that makes $U(X, Y)$ completely metrisable with the induced Hausdorff metric.

4.10 Applications

Let (X, d_X) and (Y, d_Y) be metric spaces. Let d_1 be the usual metric of uniform convergence on $C(X, Y)$, that is,

$$d_1(f, g) := \sup\{d_Y(f(x), g(x)) : x \in X\} \quad \text{for } f, g \in C(X, Y).$$

If we identify members of $C(X,Y)$ with their graphs, then H_ρ defines a metric on $C(X,Y)$, which we denote by d_2. It is easy to see that $d_2(f,g) \le d_1(f,g)$ (for all $f,g \in C(X,Y)$). If (X,d_X) is a compact metric space, then d_1 and d_2 are equivalent [240]; more generally, if (X,d_X) is an *Atsuji space* [11] (a metric space on which each continuous function into a metric space is uniformly continuous), then d_1 and d_2 are also equivalent [19]. Thus if (X,d_X) is an Atsuji space, then a subset Ω of $C(X,Y)$ is d_1-compact if and only if Ω is d_2-compact.

We have the following variant of the Arzelà–Ascoli theorem.

Theorem 4.10.1 ([113]). *Let (X,d_X) be a locally compact Atsuji space, and let (Y,d_Y) be a complete metric space. If Ω is a subset of $C(X,Y)$, then Ω is compact in the topology of uniform convergence if and only if*

(i) *Ω is d_1-closed, and whenever $(f_n : n \in \mathbb{N})$ is a sequence in Ω convergent in the Hausdorff metric to a closed subset E of $X \times Y$, E is the graph of a function, and*

(ii) *Ω is d_2-totally bounded.*

Proof. Let Ω be d_1-compact. Then Ω is d_2-compact, which means that Ω is d_2-complete and d_2-totally bounded. Thus (i) and (ii) are satisfied. It is sufficient to prove that Ω is d_2-complete. Let $(f_n : n \in \mathbb{N})$ be d_2-Cauchy. Since Atsuji spaces are complete, $(CL(X \times Y), H_\rho)$ is complete, and there is a closed subset E of $X \times Y$ to which $(f_n : n \in \mathbb{N})$ is H_ρ-convergent. By condition (i), E is the graph of a function f, and by Theorem 4.9.2, f is continuous. Since (X,d_X) is an Atsuji space, the sequence $(f_n : n \in \mathbb{N})$ d_1-converges to f, and thus the d_1-closedness of Ω implies that $f \in \Omega$. Hence Ω is d_2-complete. □

Theorem 4.10.1 is proved for a compact metric space (X,d_X) in [20]. It also follows from [21, Theorem 3]. If (X,d_X) is a locally compact Atsuji space and (Y,d_Y) is a complete metric space, then the pointwise total boundedness and equicontinuity of Ω do not ensure the compactness of Ω in the topology of uniform convergence.

Example 4.10.2. Put $X := [0,\frac{1}{2}] \cup \mathbb{N}$ with the usual metric, and let Y be the real line with the usual metric. For each $n \in \mathbb{N}$, define $f_n \in C(X,Y)$ by

$$f_n(x) := \begin{cases} n & \text{if } x = n, \\ 0 & \text{otherwise.} \end{cases}$$

Then $\Omega := \{f_n : n \in \mathbb{N}\}$ is pointwise totally bounded and equicontinuous, but Ω is not compact in the topology of uniform convergence.

It is very easy to see that $(C(X,Y),d_2)$ need not be complete. For example, for $n \ge 1$, let $f_n \in C([0,1],\mathbb{R})$ be the piecewise linear function whose graph connects the following points in succession:

$$(0,1), \left(\frac{1}{n},0\right) \text{ and } (1,0).$$

Then $(f_n : n \in \mathbb{N})$ is a d_2-Cauchy sequence without a cluster point in $C(X, Y)$. Some completeness criteria for closed subsets of $(C(X, Y), d_2)$ can be found in [21]. The last results give some sufficient conditions for the complete metrisability of $(C(X, Y), d_2)$. If (X, d_X) and (Y, d_Y) are complete metric spaces and (X, d_X) is locally compact, then by Theorem 4.9.4, $(U(X, Y), H_\rho)$ is complete. Thus by a theorem of Alexandroff the complete metrisability of $(C(X, Y), d_2)$ will be deduced if we show that $C(X, Y)$ is a G_δ-subset of $(U(X, Y), H_\rho)$.

We need the following definition. A metric space (X, d) is called *boundedly Atsuji* [25] if each closed and bounded subset of (X, d) is Atsuji.

The following result [25] will be further useful.

Lemma 4.10.3. *Let (X, d) be a metric space. The following are equivalent:*
(i) *(X, d) is boundedly Atsuji;*
(ii) *Whenever B is a closed and bounded subset of (X, d) and $\{V_i : i \in I\}$ is a collection of open subsets of (X, d) with $B \subseteq \bigcup\{V_i : i \in I\}$, there exists $\delta > 0$ such that each subset of (X, d) of diameter less than δ that meets B lies entirely within some V_i.*

Theorem 4.10.4 ([113]). *Let (X, d_X) be a boundedly Atsuji space, and let (Y, d_Y) be a metric space. Then $C(X, Y)$ is a G_δ-subset of $(U(X, Y), H_\rho)$.*

Proof. Clearly, $C(X, Y) = \{F \in U(X, Y) : F(x)$ is a singleton for all $x \in X\}$. We use some ideas of the proof of [23, Lemma 3.3]. Let $(H_k : k \in \mathbb{N})$ be the sets

$$H_k := \left\{F \in U(X, Y) : \text{for all } x \in X,\ \operatorname{diam}[F(x)] \le \frac{1}{k}\right\}.$$

Clearly, $C(X, Y) = \bigcap\{H_k : k \in \mathbb{N}\}$. It is very easy to see that $X = \bigcup\{B_n : n \in \mathbb{N}\}$, where each B_n is a closed bounded set, that is, B_n is Atsuji for every $n \in \mathbb{N}$.

Let $k \in \mathbb{N}$. For every $j \in \mathbb{N}$, put

$$H_k^j := \left\{F \in U(X, Y) : \text{for all } x \in B_j,\ \operatorname{diam}[F(x)] \le \frac{1}{k}\right\}.$$

Clearly, $H_k = \bigcap\{H_k^j : j \in \mathbb{N}\}$. We show that H_k^j is a G_δ-subset of $(U(X, Y), H_\rho)$ for every $j \in \mathbb{N}$. Let $j \in \mathbb{N}$. Let $0 < \varepsilon < 1$. First, we prove that there is an open set G_ε in $(U(X, Y), H_\rho)$ such that

$$H_k^j \subseteq G_\varepsilon \subseteq \left\{F \in U(X, Y) : \text{for all } x \in B_j,\ \operatorname{diam}[F(x)] \le \frac{1}{k} + \varepsilon\right\}.$$

Let $F \in H_k^j$. The upper semicontinuity of F implies that for every $x \in B_j$, there is a neighbourhood O_x of x such that

$$F(z) \subseteq B\left(F(x), \frac{\varepsilon}{4}\right) \quad \text{for every } z \in O_x. \qquad (*)$$

Since (X, d_X) is a boundedly Atsuji space and B_j is a closed and bounded set, by Lemma 4.10.3 there is $\alpha > 0$ such that the family $\{B(x, \alpha) : x \in B_j\}$ is a refinement of $\{O_x : x \in B_j\}$. Put $\eta := \min\{\frac{\varepsilon}{4}, \alpha\}$ and

$$O_F^\varepsilon := \{R \in U(X, Y) : H_\rho(\mathrm{Gr}(R), \mathrm{Gr}(F)) < \eta\}.$$

Then O_F^ε is contained in the set

$$\left\{R \in U(X, Y) : \text{for all } x \in B_j, \operatorname{diam}[R(x)] \leq \frac{1}{k} + \varepsilon\right\}.$$

Let $R \in O_F^\varepsilon$. Let $x \in B_j$, and let y_1 and y_2 be two different points from $R(x)$. There are points (x_1, z_1) and $(x_2, z_2) \in \mathrm{Gr}(F)$ such that $\rho((x, y_1), (x_1, z_1)) < \eta$ and also $\rho((x, y_2), (x_2, z_2)) < \eta$. Clearly, $B(x, \alpha)$ containsthe points x_1 and x_2. There is $u \in B_j$ such that $B(x, \alpha) \subseteq O_u$. By $(*)$, $\{z_1, z_2\} \subseteq B(F(u), \frac{\varepsilon}{4})$, that is, there are $v_1, v_2 \in F(u)$ such that $d_Y(z_1, v_1) \leq \frac{\varepsilon}{4}$ and $d_Y(z_2, v_2) \leq \frac{\varepsilon}{4}$, that is, $d_Y(z_1, z_2) \leq \frac{1}{k} + \frac{\varepsilon}{2}$. Thus we have

$$d_Y(y_1, y_2) \leq d_Y(y_1, z_1) + d_Y(z_1, z_2) + d_Y(z_1, y_2) \leq \frac{1}{k} + \varepsilon,$$

that is, $\operatorname{diam}[R(x)] \leq \frac{1}{k} + \varepsilon$ for every $x \in B_j$. Put $G_\varepsilon := \bigcup\{O_F^\varepsilon : F \in H_k^j\}$. It is very easy to see that

$$H_k^j \subseteq \bigcap_{n\in\mathbb{N}} G_{\frac{1}{n}} \subseteq \bigcap_{n\in\mathbb{N}} \left\{F \in U(X, Y) : \text{for all } x \in B_j, \operatorname{diam}[F(x)] \leq \frac{1}{k} + \frac{1}{n}\right\} \subseteq H_k^j.$$

Thus $H_k^j = \bigcap\{G_{\frac{1}{n}} : n \in \mathbb{N}\}$, that is, H_k is a G_δ-subset of $(U(X, Y), H_\rho)$. Since $k \in \mathbb{N}$ was arbitrary, $C(X, Y)$ is a G_δ-subset of $(U(X, Y), H_\rho)$. ◻

Theorem 4.10.5 ([113]). *Let (X, d_X) be a locally compact boundedly Atsuji space, and let (Y, d_Y) be a complete metric space. Then $(C(X, Y), d_2)$ is completely metrisable.*

Proof. Since boundedly Atsuji spaces are complete [25], by Theorem 4.9.4, $(U(X, Y), H_\rho)$ is complete. By Theorem 4.10.4, $C(X, Y)$ is a G_δ-subset of $(U(X, Y), H_\rho)$. By the well-known theorem of Alexandroff, $(C(X, Y), d_2)$ is completely metrisable. ◻

Another sufficient condition for the complete metrisability of $(C(X, Y), d_2)$ can be found in [23].

Exercises

1. Let (X, d_X) and (Y, d_Y) be metric spaces. Let $Un(X, Y)$ be the space of uniformly continuous mappings from (X, d_X) to (Y, d_Y). Prove that the metrics d_1 and d_2 are equivalent on $Un(X, Y)$.

4.11 Topological properties of the space of minimal cusco mappings

Cusco mappings are interesting because they describe common features of maximal monotone operators, of the convex subdifferential and of the Clarke generalised gradient. Minimal cusco mappings are very important in functional analysis [37, 257], where differentiability of Lipschitz functions is deduced by their Clarke subdifferentials being minimal cuscos. Minimal cusco mappings also appear in the study of weak Asplund spaces [81, 234], optimisation [63] and in the study of differentiability of Lipschitz functions [34, 216].

As for topologies on spaces of set-valued mappings, there are mainly two approaches in the literature, function space topologies [102, 114, 130] and hyperspace topologies [113, 126, 140, 198, 240, 241], in which set-valued mappings are identified with their graphs and are considered as elements of a hyperspace.

In this part, we study the (upper) Vietoris topology on the space of minimal cusco mappings. Most of the results in this part are contained in the paper of Holá and Novotný [135]. The main motivation for our study is Theorem 4.13.5, which claims that if (X, τ) is a normal topological space, then the space $C(X, \mathbb{R})$ of continuous real-valued functions is dense in the space $MC(X, \mathbb{R})$ of real-valued minimal cusco mappings equipped with the Vietoris topology. Then we could expect that the Vietoris topology on $MC(X, \mathbb{R})$ shares some nice properties with the Vietoris topology on $C(X, \mathbb{R})$, for example, that the sum of minimal cusco mappings is continuous in the Vietoris topology and that the Vietoris topology coincides with the upper Vietoris topology on the space of minimal cusco mappings. Unfortunately, as Examples 4.13.9 and 4.13.11 show, this is not the case.

The main result of this part is a full characterisation of the Polishness of $MC(X, \mathbb{R})$ with the upper Vietoris topology, obtained by utilizing techniques from topological games. We also complete a list of countability properties of the upper Vietoris topology (also called graph topology; see [27, 240]) on $C(X, \mathbb{R})$ from [137]. Using our mentioned results, we improve those from [126] concerning the (upper) Vietoris topology on the space $L(X, \mathbb{R})$ of all cusco mappings.

Concerning (further) results on approximations of relations by continuous functions in hyperspace topologies, we also mention papers [9, 22, 57, 109, 129, 131, 132].

Let (X, τ) be a Hausdorff topological space. Recall that $CL(X)$ denotes the space of all closed nonempty subsets of (X, τ). On this space, we will consider the *Vietoris topology*, denoted by τ_V. Recall that τ_V is generated by the subbase

$$\{V^+ : V \text{ is an open subset of } X\} \cup \{V^- : V \text{ is an open subset of } X\},$$

where

$$V^+ := \{A \in CL(X) : A \subseteq V\} \quad \text{and} \quad V^- := \{A \in CL(X) : A \cap V \neq \emptyset\}.$$

The first part, namely $\{V^+ : V \text{ is an open subset of } X\}$, is the base of a topology called the *upper Vietoris topology*, denoted by τ_V^+, and it will be of the main interest for us as explained later. Analogously, the other part generates the *lower Vietoris topology* τ_V^-. The usual notation for the base sets of the Vietoris topology is

$$[W_1, \dots, W_n] := W_1^- \cap \dots \cap W_n^- \cap \left(\bigcup_{k=1}^{n} W_k \right)^+ .$$

We will not distinguish set-valued mappings (multifunctions) and their graphs. Let (X, τ) and (Y, τ') be Hausdorff topological spaces. We will be interested in the following spaces:

$$\begin{aligned} L(X, Y) &:= \{F \subseteq X \times Y : F \text{ is cusco}\}, \\ L_0(X, Y) &:= \{F \subseteq X \times Y : F \text{ is cusco and for every isolated } x,\ F(x) \text{ is singleton}\}, \\ MC(X, Y) &:= \{F \subseteq X \times Y : F \text{ is minimal cusco}\}, \\ C(X, Y) &:= \{f : X \to Y : f \text{ is continuous}\}. \end{aligned}$$

Observe that

$$C(X, Y) \subseteq MC(X, Y) \subseteq L_0(X, Y) \subseteq L(X, Y) \subseteq CL(X \times Y),$$

so all these spaces can inherit both Vietoris and upper Vietoris topologies from $CL(X \times Y)$.

Remarks 4.11.1. Note that when we will work with the space, for example, $L_0(X, \mathbb{R})$, then instead of writing $W^+ \cap L_0(X, \mathbb{R})$ or $[W_1, W_2] \cap L_0(X, \mathbb{R})$ we will write only W^+ or $[W_1, W_2]$, respectively. The reader can easily figure out the complete expression from the context.

Note that if (X, τ) is regular, then $X \times \mathbb{R}$ is also regular with the product topology. Thus we have that $(CL(X \times \mathbb{R}), \tau_V)$ is Hausdorff and so are $L(X, \mathbb{R})$, $L_0(X, \mathbb{R})$, $MC(X, \mathbb{R})$ and $C(X, \mathbb{R})$ with τ_V. The situation for τ_V^+ is more complicated in general, and we will address this later.

We are interested in completeness and countability properties of these spaces. For the cardinal invariants that are needed, we refer to the book of Juhász [154]. Denote by ω the first infinite cardinal, that is, the set $\{0, 1, 2, 3, \dots\}$, and by ω_1 the first uncountable cardinal.

Since we will also work with normal and perfectly normal spaces, we will use the following results.

Proposition 4.11.2 (1.7.15 (b) in [78]). *A T_1 space (X, τ) is normal if and only if for every $f, h : X \to \mathbb{R}$ such that $f \leq h$, f is upper semicontinuous and h is lower semicontinuous, there is $g \in C(X, \mathbb{R})$ such that $f \leq g \leq h$.*

Proposition 4.11.3 (1.7.15 (c) in [78]). *A T_1 space (X, τ) is perfectly normal if and only if for every lower semi continuous (upper semicontinuous) function $f : X \to \mathbb{R}$, there is a sequence $(f_n : n \in \mathbb{N})$ such that $f_n \in C(X, \mathbb{R})$ and $f_n \nearrow f$ ($f_n \searrow f$).*

Theorem 4.11.4 (Theorems 2.4 and 2.6 in [122]). *Let (X, τ) be a topological space, and let $F \subseteq X \times \mathbb{R}$ be a set-valued mapping. The following are equivalent:*
(i) *F is minimal cusco;*
(ii) *F has a quasicontinuous subcontinuous selection $f : X \to \mathbb{R}$ such that for every $x \in X$, $\overline{\mathrm{co}}(\mathrm{USC}(f)(x)) = F(x)$;*
(iii) *there is a dense subset A of X and a quasicontinuous (on A) and subcontinuous (on X) function $f : A \to \mathbb{R}$ such that $\overline{\mathrm{co}}(\mathrm{USC}(f)(x)) = F(x)$ for every $x \in X$.*

Recall that for a set-valued mapping $F \subseteq X \times Y$, we denote

$$S(F) := \{x \in X : F(x) \text{ is a singleton}\},$$

and for a function $f : X \to Y$, we denote

$$C(f) := \{x \in X : f \text{ is continuous at } x\}.$$

If (X, τ) is a Baire space and $F \in MC(X, \mathbb{R})$, then by Corollary 3.3.24, $S(F)$ is a dense G_δ-subset of (X, τ).

4.12 Countability properties of $C(X, \mathbb{R})$

In this section, we mostly collect some useful facts about the space $(C(X, \mathbb{R}), \tau_V)$ needed for the study of the space $MC(X, \mathbb{R})$. Denote by τ_U the *topology of uniform convergence* on $C(X, \mathbb{R})$, that is, the topology generated by the supremum metric

$$d_1(f, g) := \sup_{x \in X} |f(x) - g(x)| \quad \text{for } f, g \in C(X, \mathbb{R}).$$

Note that on $C(X, \mathbb{R})$, we have $\tau_U \subseteq \tau_V$. The following theorem is a well-known fact; see, for example, [137].

Theorem 4.12.1. *The topologies τ_V^+ and τ_V coincide on $C(X, \mathbb{R})$.*

Theorem 4.12.2 (Theorem 5.1 in [137]). *Let (X, τ) be a Tychonoff space. Then the following are equivalent:*
(i) *$(C(X, \mathbb{R}), \tau_V)$ is metrisable;*
(ii) *$(C(X, \mathbb{R}), \tau_V)$ is first countable;*
(iii) *$(C(X, \mathbb{R}), \tau_V)$ has a countable π-character;*
(iv) *$(C(X, \mathbb{R}), \tau_V)$ is a Fréchet space;*
(v) *$(C(X, \mathbb{R}), \tau_V)$ is sequential;*
(vi) *$(C(X, \mathbb{R}), \tau_V)$ has a countable tightness;*
(vii) *(X, τ) is countably compact;*
(viii) *$\tau_V = \tau_U$.*

Proof. In addition to [137, Theorem 5.1,] we have (ii) and (viii). The equivalence of (ii) and (iii) follows from the fact that character and π-character coincide for topological groups and $(C(X,\mathbb{R}),\tau_V)$ is a topological group with respect to addition. The equivalence of (vii) and (viii) is proven in [103]. □

The following theorem is a generalisation of [121, Proposition 1.2].

Theorem 4.12.3. *Let (X,τ) be a Tychonoff space. Then the following are equivalent:*
(i) $(C(X,\mathbb{R}),\tau_V)$ *is second countable;*
(ii) $(C(X,\mathbb{R}),\tau_V)$ *is separable;*
(iii) $(C(X,\mathbb{R}),\tau_V)$ *satisfies the countable chain condition;*
(iv) $(C(X,\mathbb{R}),\tau_V)$ *is Lindelöff;*
(v) $(C(X,\mathbb{R}),\tau_V)$ *has a countable π-base;*
(vi) $(C(X,\mathbb{R}),\tau_V)$ *has a countable network;*
(vii) $(C(X,\mathbb{R}),\tau_V)$ *has a countable spread;*
(viii) $(C(X,\mathbb{R}),\tau_V)$ *has a countable extent;*
(ix) X *is compact and metrisable.*

Proof. Recall that all considered cardinal invariants are less than or equal to the weight of τ_V and greater than or equal to either cellularity or extent. Therefore we need to prove only (ix) ⇒ (i), (iii) ⇒ (ix) and (viii) ⇒ (ix). The first two implications are proven in [121, Proposition 1.2]. For the last, suppose that $(C(X,\mathbb{R}),\tau_V)$ has a countable extent. Then $(C(X,\mathbb{R}),\tau_U)$ has a countable extent, and therefore (X,τ) is compact and metrisable. □

Theorem 4.12.4 (Corollary 3.3 in [137]). *The space $(C(X,\mathbb{R}),\tau_V)$ is Baire.*

4.13 Relationship of $C(X,\mathbb{R})$ and $MC(X,\mathbb{R})$

We need the following useful lemma.

Lemma 4.13.1 (Lemma 4.1 in [126]). *Let $F \in L(X,\mathbb{R})$, and let $W \subseteq X \times \mathbb{R}$ be an open set containing F. Then there exists an open set $G \subseteq X \times \mathbb{R}$ such that $G(x)$ is connected for each $x \in X$ and $F \subseteq G \subseteq W$.*

Note that in Lemma 4.13.1, $F \in [G] \subseteq [W]$ and $[G],[W] \in \tau_V^+$. The following result improves that from [129, Lemma 4.1], where (X,τ) is assumed to be countably paracompact and normal. We use some ideas from its proof.

Theorem 4.13.2 ([135]). *Let (X,τ) be a normal space, let $F \in L(X,\mathbb{R})$, and let $F \in [W] \in \tau_V^+$. Then $[W] \cap C(X,\mathbb{R}) \neq \emptyset$.*

Proof. By Lemma 4.13.1 we can assume that $W(x)$ is connected for every $x \in X$. Let $\varphi : X \times (-\pi/2,\pi/2) \to X \times \mathbb{R}$ be the homeomorphism defined by $\varphi(x,t) := (x,\tan(t))$, and let $F_0 := \varphi^{-1}(F)$ and $W_0 := \varphi^{-1}(W)$. Define $f_1, f_2, h_1, h_2 : X \to \mathbb{R}$ by

$$\begin{array}{ll} h_1(x) := \inf F_0(x), & f_1(x) := \inf W_0(x), \\ h_2(x) := \sup W_0(x), & f_2(x) := \sup F_0(x), \end{array} \quad \text{for } x \in X.$$

We can verify that f_1, f_2 are upper semicontinuous, h_1, h_2 are lower semicontinuous and $f_1 < h_1 \le f_2 < h_2$. By Proposition 4.11.2 there are $g_1, g_2 \in C(X,\mathbb{R})$ such that

$$f_1 \le g_1 \le h_1 \le f_2 \le g_2 \le h_2.$$

Put $g_0 := \frac{g_1+g_2}{2}$ and observe that

$$-\frac{\pi}{2} \le f_1 < g_0 < h_2 \le \frac{\pi}{2}.$$

Therefore $g := \tan \circ g_0 \in C(X,\mathbb{R})$ and $g(x) \in W(x)$ for all $x \in X$. □

Corollary 4.13.3 ([135]). *Let (X,τ) be a normal space. Then $C(X,\mathbb{R})$ is dense in $(L(X,\mathbb{R}), \tau_V^+)$ and, consequently, also in $(L_0(X,\mathbb{R}), \tau_V^+)$ and $(MC(X,\mathbb{R}), \tau_V^+)$.*

From Theorems 4.12.4 and 4.12.1 we know that for a Tychonoff space X, the space $(C(X,\mathbb{R}), \tau_V^+)$ is a Baire space, and from this we immediately have the following:

Corollary 4.13.4 ([135]). *Let (X,τ) be a normal space. Then $(L(X,\mathbb{R}), \tau_V^+)$, $(L_0(X,\mathbb{R}), \tau_V^+)$ and $(MC(X,\mathbb{R}), \tau_V^+)$ are Baire spaces.*

Theorem 4.13.5 ([135]). *Let (X,τ) be a normal space, let $F \in L_0(X,\mathbb{R})$, and let $F \in [W_1,\dots,W_n] \in \tau_V$. Then $[W_1,\dots,W_n] \cap C(X,\mathbb{R}) \ne \emptyset$.*

Proof. By Lemma 4.13.1 we can choose open $W_0 \subseteq \bigcup_{n=1}^{n} W_k$ such that $F \in W_0^+$ and $W_0(x)$ is connected for all $x \in X$. By Theorem 4.13.2 there is $f \in C(X,\mathbb{R}) \cap W_0^+$. Now for $k = 1,\dots,n$, choose $(x_k, t_k) \in W_0 \cap W_k$. Since $F(x)$ is a singleton for every isolated x, we may assume that all x_k are distinct. Now choose open intervals V_k for $k = 1,\dots,n$ such that $\{t_k, f(x_k)\} \subseteq V_k \subseteq \overline{V}_k \subseteq W_0(x_k)$ and choose open sets U_k such that $f(U_k) \subseteq V_k$, $U_k \times \overline{V}_k \subseteq W_0$ and U_k are pairwise disjoint. Define the functions $h_1, h_2 : X \to \mathbb{R}$ such that $h_1(x) := h_2(x) := f(x)$ whenever $x \in X \setminus (U_1 \cup \dots \cup U_k)$, $h_1(x_k) := h_2(x_k) := t_k$ whenever $1 \le k \le n$, and for all $1 \le k \le n$ and $x \in U_k \setminus \{x_k\}$, $h_1(x) := \min V_k$ and $h_2(x) := \max V_k$. Observe that h_1 is upper semicontinuous, h_2 is lower semicontinuous, and $h_1 \le h_2$, so Proposition 4.11.2, yields $g \in C(X)$ with $h_1 \le g \le h_2$. It is easy to see that g has the following properties: f and g agree on $X \setminus (U_1 \cup \dots \cup U_n)$, $g(x_k) = t_k$ and $g(U_k) \subseteq \overline{V}_k$. Therefore $g \in [W_1,\dots,W_n] \cap C(X,\mathbb{R})$. □

Since minimal cusco mappings are single valued at isolated points of (X,τ), we have the following corollary.

Corollary 4.13.6 ([135]). *Let (X,τ) be a normal space. Then $C(X,\mathbb{R})$ is dense in $(L_0(X,\mathbb{R}), \tau_V)$ and, consequently, also in $(MC(X,\mathbb{R}), \tau_V)$.*

Similarly to Corollary 4.13.4, we have the following.

Corollary 4.13.7 ([135]). *Let (X,τ) be a normal space. Then $(L_0(X,\mathbb{R}),\tau_V)$ and $(MC(X,\mathbb{R}),\tau_V)$ are Baire spaces.*

The following example shows that the normality of (X,τ) is essential in Theorems 4.13.2 and 4.13.5. It is similar to Example 9 in [9].

Example 4.13.8. Let (X,τ) be the space of non-limit ordinals less than ω_1, together with ω_1, equipped with the order topology. Let $X := X_0 \cup X_1 \cup \{\omega_1\}$, where X_0, X_1 are disjoint uncountable sets not containing ω_1. Let $Y := [0,\omega]$ with the order topology and $Y_0 := [0,\omega)$. Let $Z := X \times Y \setminus \{(\omega_1,\omega)\}$. Let $F \subseteq Z \times [0,1]$ be a mapping defined by

$$F(x,y) := \begin{cases} \{0\} & \text{if } x \in X_0 \text{ and } y \in Y, \\ \{1\} & \text{if } x \in X_1 \text{ and } y \in Y, \\ [0,1] & \text{if } x = \omega_1 \text{ and } y \in Y_0. \end{cases}$$

We can verify that Z is not normal and F is a cusco. We will show that there is an open set $W \subseteq Z \times \mathbb{R}$ such that $F \subseteq W$ and $W^+ \cap C(Z) = \emptyset$. Put

$$W := (X_0 \times Y \times (-1,1/3)) \cup (X_1 \times Y \times (2/3,2)) \cup (X \times Y_0 \times (-1,2)).$$

Suppose that there is $f \in C(Z) \cap W^+$. Since f is continuous, then for every fixed $y < \omega$, we have that $f(x,y)$ is eventually constant, that is, there is x_y such that for every $x > x_y$, $f(x,y) = f(\omega_1,y)$. Since $cf(\omega_1) = \omega_1$, we have that $\hat{x} = \sup\{x_y : y < \omega\} < \omega_1$. Since $f \in W^+$, for all $x_0 \in X_0$ and $x_1 \in X_1$, we have $f(x_0,\omega) < 1/3$ and $f(x_1,\omega) > 2/3$. Since $\hat{x}$ is countable, we can choose $\hat{x}_0 \in X_0$, $\hat{x}_0 > \hat{x}$, and $\hat{x}_1 \in X_1$, $\hat{x}_1 > \hat{x}$. Since f is continuous, there is $n_0 < \omega$ such that for every $y > n_0$, we have $f(\hat{x}_0,y) < 1/3$ and $f(\hat{x}_1,y) > 2/3$, which contradicts the fact that both values are supposed to be equal to $f(\omega_1,y)$.

Note that in the above proofs (especially in the proof of Theorem 4.13.5) the "nice" part of τ_V is actually τ_V^+, and τ_V^- is a nuisance that we have to deal with. From Theorem 4.12.1 we know that $\tau_V = \tau_V^+$ for $C(X,\mathbb{R})$. Since for a Baire space (X,τ), spaces $MC(X,\mathbb{R})$ and $C(X,\mathbb{R})$ are quite close in the sense that any $F \in MC(X,\mathbb{R})$ is single-valued on a residual set, it is natural to ask if this is true also for $MC(X,\mathbb{R})$. Unfortunately, the answer is negative as the following example shows.

Example 4.13.9. Let $X := \mathbb{R} \setminus \{\frac{1}{n} : n \in \mathbb{N}\}$, and let $F, f_n \subseteq X \times \mathbb{R}$ for $n \in \mathbb{N}$ be defined by

$$F(x) := \begin{cases} \{0\} & \text{if } x < 0, \\ [0,1] & \text{if } x = 0, \\ \{1\} & \text{if } x > 0, \end{cases} \quad \text{and} \quad f_n(x) := \begin{cases} 0 & \text{if } x < \frac{1}{n}, \\ 1 & \text{if } x > \frac{1}{n}. \end{cases}$$

We can easily check that (X,τ) is Baire, $F, f_n, (n \in \mathbb{N}) \in MC(X,\mathbb{R})$ and $f_n \to F$ in τ_V^+. Now put $W := X \times (0,1)$. Since $F \in W^-$ and $f_n \notin W^-$ for $n \in \mathbb{N}$, we have that $f_n \not\to F$ in τ_V^-.

On the other hand, there is a trivial example of a space X where $\tau_V = \tau_V^+$ on $MC(X,\mathbb{R})$. Namely, any discrete space (X,τ), since then we have $C(X,\mathbb{R}) = MC(X,\mathbb{R}) = L_0(X,\mathbb{R})$. It could be interesting to investigate for which spaces (X,τ) we have $\tau_V = \tau_V^+$ on $MC(X,\mathbb{R})$. Nevertheless, in general, the topology τ_V^+ may be better suited for the study of the space $MC(X,\mathbb{R})$.

Proposition 4.13.10 ([135]). *For a Baire space (X,τ), the space $(MC(X,\mathbb{R}),\tau_V^+)$ is Hausdorff.*

Proof. Let $F, G \in MC(X,\mathbb{R})$ be such that $F \neq G$. From Corollary 3.3.24 we have that $S(F)\cap S(G)$ is a dense G_δ-set. In particular, there is $x \in S(F)\cap S(G)$ such that $F(x) \neq G(x)$; otherwise, they would agree on a dense subset, and by Theorem 4.11.4 they would be equal. Choose open disjoint $U, V \subseteq \mathbb{R}$ such that $F(x) \in U$ and $G(x) \in V$. Define open sets $U_0 := [X \times U] \cup [(X \setminus \{x\}) \times \mathbb{R}]$ and $V_0 := [X \times V] \cup [(X \setminus \{x\}) \times \mathbb{R}]$. Then the sets U_0^+ and V_0^+ are open and disjoint, $F \in U_0^+$, and $G \in V_0^+$. □

Suppose again that (X,τ) is a Baire space. The fact that minimal cusco mappings are single-valued on a residual set allows us to define an addition on the set $MC(X,\mathbb{R})$ in a natural way. Notice that for $F, G \in MC(X,\mathbb{R})$, we have that $S(F) \cap S(G)$ is a dense G_δ-set, and

$$f := F|_{S(F)\cap S(G)}, \quad g := G|_{S(F)\cap S(G)}$$

are continuous functions, which are also subcontinuous (by Proposition 4.1.1 every selection of an usco mapping is subcontinuous, and therefore its densely defined restriction is also subcontinuous). We can define

$$(F + G)(x) := \overline{\mathrm{co}}(\mathrm{USC}(f + g)(x)) \quad \text{for } x \in X.$$

From Theorem 4.11.4 we have that $F + G \in MC(X,\mathbb{R})$. However, this addition is continuous neither in τ_V^+ nor in τ_V, as the following example shows.

Example 4.13.11. Let $F, G, f_n, g_n (n \in \mathbb{N}) \subseteq \mathbb{R} \times \mathbb{R}$ be defined by

$$F(x) := \begin{cases} \{0\} & \text{if } x < 0, \\ [0,1] & \text{if } x = 0, \\ \{1\} & \text{if } x > 0, \end{cases} \quad G(x) := \begin{cases} \{0\} & \text{if } x < 0, \\ [-1,0] & \text{if } x = 0, \\ \{-1\} & \text{if } x > 0, \end{cases}$$

$$f_n(x) := \begin{cases} 0 & \text{if } x < -\frac{1}{n}, \\ nx + 1 & \text{if } x \in [-\frac{1}{n}, 0], \\ 1 & \text{if } x > 0, \end{cases} \quad g_n(x) := \begin{cases} 0 & \text{if } x < 0, \\ -nx & \text{if } x \in [0, \frac{1}{n}], \\ -1 & \text{if } x > \frac{1}{n}. \end{cases}$$

We can easily see that $F + G = 0$, $f_n \to F$, $g_n \to G$ in τ_V, and obviously they converge also in τ_V^+. But $f_n + g_n \nrightarrow 0 = F + G$ in τ_V^+ and therefore in τ_V.

4.14 Countability and completeness-like properties of cusco mappings

Theorem 4.14.1 ([135]). *Let (X,τ) be a Tychonoff space. If any of the spaces $(MC(X,\mathbb{R}), \tau_V^+)$, $(L_0(X,\mathbb{R}),\tau_V^+)$, $(L(X,\mathbb{R}),\tau_V^+)$, $(MC(X,\mathbb{R}),\tau_V)$, $(L_0(X,\mathbb{R}),\tau_V)$ or $(L(X,\mathbb{R}),\tau_V)$ is first countable, then (X,τ) is countably compact.*

Proof. First, the countability of any of these spaces implies the first countability of $(C(X,\mathbb{R}),\tau_V^+)$, and from Theorem 4.12.2 we have that (X,τ) has to be countably compact. □

Proposition 4.14.2 ([135]). *Let (X,τ) be a normal space. Then the following are equivalent:*
(i) *$(L(X,\mathbb{R}),\tau_V^+)$ is first countable,*
(ii) *(X,τ) is countably compact and perfectly normal.*

Proof. We start with (ii) ⇒ (i). This is in fact stated in [126, Theorem 4.7] without proof, so for the reader's convenience, we provide one. Fix an arbitrary $F \in L(X,\mathbb{R})$ and put $f(x) := \inf F(x)$ and $g(x) := \sup F(x)$. Observe that f is lower semicontinuous and g is upper semicontinuous. Since (X,τ) is perfectly normal, there are by Proposition 4.11.3 continuous functions $f_n \nearrow f$ and $g_n \searrow g$. We can suppose that for every $n \in \mathbb{N}$ and every $x \in X$, $f_n(x) < f(x)$ and $g(x) < g_n(x)$. Defining open sets $W_n \subseteq X \times \mathbb{R}$ by $W_n(x) := (f_n(x), g_n(x))$, we have that $F = \bigcap_{n\in\mathbb{N}} W_n$. Since $X \times \mathbb{R}$ is normal, for every $n \in \mathbb{N}$, there is an open V_n such that $F \subseteq V_n \subseteq \overline{V}_n \subseteq W_n$. We will prove that $\{V_n : n \in \mathbb{N}\}$ is a local base at F by showing that for every open W containing F, there is $n \in \mathbb{N}$ such that $V_n \subseteq W$. Suppose it is not. Then there is an open W containing F such that for every $n \in \mathbb{N}$, there is $(x_n, y_n) \in V_n \setminus W$. Since (X,τ) is countably compact, there are $\alpha_n, \beta_n \in \mathbb{R}$ such that $\overline{V}_n \subseteq X \times [\alpha_n, \beta_n]$, that is, $\overline{V}_n \setminus W$ is countably compact, and therefore $((x_n, y_n) : n \in \mathbb{N})$ has a cluster point $(x,y) \in \bigcap_{n\in\mathbb{N}} \overline{V}_n \setminus W = \varnothing$, contrary to supposition.

Now we prove (i) ⇒ (ii). From Theorem 4.14.1 we have that (X,τ) is countably compact. To prove that it is also perfectly normal, we will use Proposition 4.11.3. Take an arbitrary upper semicontinuous function $f : X \to \mathbb{R}$. For every $n \in \mathbb{N}$, put $f_n(x) := \max\{f(x), -n\}$ and $F_n(x) := [-n, f_n(x)]$. Observe that $F_n \in L(X,\mathbb{R})$, f_n is upper semicontinuous and $f_n \to f$ as $n \to \infty$. There is a local base $\{W_{n,m}^+ : m \in \mathbb{N}\}$ at F_n. Put $g_{n,m}(x) := \sup W_{n,m}(x)$. Without loss of generality we can suppose that $g_{n,m}$ is a real-valued function. Since $g_{n,m}$ is lower semicontinuous, there is by Proposition 4.11.2 a function $f_{n,m} \in C(X,\mathbb{R})$ such that $f_n \le f_{n,m} \le g_{n,m}$. It is easy to see that $f_{n,m} \to f_n$ as $m \to \infty$. Put $g_n := \min\{f_{p,q} : p, q = 1, \ldots, n\}$. We can verify that $g_n \searrow f$ as $n \to \infty$. □

Corollary 4.14.3 ([135]). *Let (X,τ) be a countably compact and perfectly normal space. Then $(L_0(X,\mathbb{R}),\tau_V^+)$ and $(MC(X,\mathbb{R}),\tau_V^+)$ are first countable.*

Note that by [126, Theorem 5.2] $(L(X,\mathbb{R}),\tau_V)$ is first countable if and only if (X,τ) is countably compact, perfectly normal and hereditarily separable space.

Theorem 4.14.4 ([135]). *Let (X,τ) be a Tychonoff space, and let Z be any of the following spaces $(MC(X,\mathbb{R}),\tau_V^+)$, $(L_0(X,\mathbb{R}),\tau_V^+)$, $(L(X,\mathbb{R}),\tau_V^+)$, $(MC(X,\mathbb{R}),\tau_V)$, $(L_0(X,\mathbb{R}),\tau_V)$ or $(L(X,\mathbb{R}),\tau_V)$. The following are equivalent:*
(i) *Z is second countable;*
(ii) *Z has a countable network;*
(iii) *Z is hereditarily Lindelöff;*
(iv) *Z has a countable spread;*
(v) *(X,τ) is compact and metrisable.*

Proof. (i) $\Rightarrow$ (ii), (ii) $\Rightarrow$ (iii), (iii) $\Rightarrow$ (iv) are clear. To prove (iv) $\Rightarrow$ (v), suppose Z has a countable extent. Then since $(C(X,\mathbb{R}),\tau_V^+)$ is embedded in Z, it has also a countable spread, and by Theorem 4.12.3 the space (X,τ) must be compact and metrisable. Finally, to prove (v) $\Rightarrow$ (i), assume that (X,τ) is compact and metrisable. Then (X,τ) has a countable base. It is easy to verify that $L(X,\mathbb{R}) \subseteq K(X\times\mathbb{R})$, where $K(X\times\mathbb{R})$ is the space of all nonempty compact subsets of $X\times\mathbb{R}$. Since $X\times\mathbb{R}$ has a countable base, we know that $(K(X\times\mathbb{R}),\tau_V)$ has also a countable base; see [206]. The proof that also $(K(X\times\mathbb{R}),\tau_V^+)$ has a countable base is straightforward. The second countability of the considered spaces follows from these two cases. □

Using Corollary 4.13.3, we give a simpler proof of a slight generalisation of [126, Proposition 5.5].

Theorem 4.14.5 ([135]). *Let (X,τ) be a normal space. If any of the spaces $(L(X,\mathbb{R}),\tau_V)$, $(L(X,\mathbb{R}),\tau_V^+)$, $(L_0(X,\mathbb{R}),\tau_V)$, $(L_0(X,\mathbb{R}),\tau_V^+)$, $(MC(X,\mathbb{R}),\tau_V)$ or $(MC(X,\mathbb{R}),\tau_V^+)$ satisfies the countable chain condition, then (X,τ) is compact and metrisable.*

Proof. First, note that if $(L(X,\mathbb{R}),\tau_V)$ satisfies the countable chain condition, then so does $(L(X,\mathbb{R}),\tau_V^+)$. Since $C(X,\mathbb{R})$ is dense in all the mentioned spaces (except for $(L(X,\mathbb{R}),\tau_V)$), then also $(C(X,\mathbb{R}),\tau_V)$ satisfies the countable chain condition. From Theorem 4.12.3 we have that (X,τ) must be compact and metrisable. □

The following theorem extends the results of Theorem 4.14.4 in the case of normal X.

Theorem 4.14.6 ([135]). *Let (X,τ) be a normal space, and let Z be any of the following spaces $(MC(X,\mathbb{R}),\tau_V^+)$, $(L_0(X,\mathbb{R}),\tau_V^+)$, $(L(X,\mathbb{R}),\tau_V^+)$, $(MC(X,\mathbb{R}),\tau_V)$, $(L_0(X,\mathbb{R}),\tau_V)$ or $(L(X,\mathbb{R}),\tau_V)$. The following are equivalent:*
(i) *Z has a countable π-base;*
(ii) *Z is separable;*
(iii) *Z satisfies the countable chain condition;*
(iv) *(X,τ) is compact and metrisable.*

Proof. (i) $\Rightarrow$ (ii) and (ii) $\Rightarrow$ (iii) are clear. (iii) $\Rightarrow$ (iv) follows from Theorem 4.14.5, and (iv) $\Rightarrow$ (i) follows from Theorem 4.14.4. □

For $f, g : X \to \mathbb{R}$ such that $f < g$, denote

$$M_{f,g} := \{(x, y) \in X \times \mathbb{R} : f(x) < y < g(x)\}.$$

Proposition 4.14.7 (Proposition 4.4 in [126]). *Let (X, τ) be a countably paracompact normal space, and let D be a dense subset of $(C(X, \mathbb{R}), \tau_V^+)$. Then*

$$\mathcal{B} := \{M_{f,g}^+ : f, g \in D, f < g\}$$

is a base of $(L(X, \mathbb{R}), \tau_V^+)$.

Lemma 4.14.8 ([135]). *Let (X, τ) be a space, and let $f, g \in C(X, \mathbb{R})$ be such that $f < g$. Then in $(MC(X, \mathbb{R}), \tau_V^+)$, we have that $\operatorname{cl}(M_{f,g}^+) \subseteq \overline{M}_{f,g}^+$, where $\operatorname{cl}(M_{f,g}^+)$ is the closure of the set $M_{f,g}^+$ in the topology τ_V^+.*

Proof. Put $W := M_{f,g}$. Observe that $\overline{W}(x) = \overline{W(x)} = [f(x), g(x)]$. Choose an arbitrary $F \in \operatorname{cl}(W^+)$. We will prove by contradiction that for every $x \in X$, $F(x) \cap \overline{W}(x) \neq \emptyset$. Suppose that there is $x_0 \in X$ such that $F(x_0) \subseteq \mathbb{R} \setminus \overline{W}(x_0)$. Then $F \subseteq V = (X \times \mathbb{R}) \setminus (\{x_0\} \times \overline{W}(x_0))$, where V is an open subset of $X \times \mathbb{R}$. Since $F \in V^+ \cap \operatorname{cl}(W^+)$, $V^+ \cap W^+ \neq \emptyset$, a contradiction.

We have proved that $F \cap \overline{W}$ has nonempty values, and since $F \in MC(X, \mathbb{R})$ and $\overline{W}$ is closed and convex-valued upper semicontinuous mapping, we have that $F \cap \overline{W}$ is cusco. From the minimality of F we have that $F \cap \overline{W} = F$, and thus $F \subseteq \overline{W}$, that is, $F \in \overline{W}^+$. □

Theorem 4.14.9 ([135]). *Let (X, τ) be a countably paracompact normal space. Then the space $(MC(X, \mathbb{R}), \tau_V^+)$ is regular.*

Proof. Choose arbitrary $F \in MC(X, \mathbb{R})$ and open $W \subseteq X \times \mathbb{R}$ such that $F \in W^+$. Since X is countably paracompact and normal, $X \times \mathbb{R}$ is normal (see [200]), and there is an open set W_0 such that $F \subseteq W_0 \subseteq \overline{W}_0 \subseteq W$. From Proposition 4.14.7 we have that $\mathcal{B}$ is a base of τ_V^+, so there is $M_{f,g}$ such that $F \in M_{f,g}^+ \subseteq W_0^+$, and from Lemma 4.14.8 it follows that $F \in M_{f,g}^+ \subseteq \operatorname{cl}(M_{f,g}^+) \subseteq \overline{M}_{f,g}^+ \subseteq \overline{W}_0^+ \subseteq W^+$. □

Let (X, τ) be a topological space. The *Choquet game* G_X on (X, τ) is defined as follows. Players I and II take turns in playing open sets. Player I starts with U_0, Player II follows with $V_0 \subseteq U_0$, and then again Player I plays $U_1 \subseteq V_0$ and so on, creating an infinite sequence called a *run of the game*. Player I *wins* if $\bigcap_{n\in\omega} U_n = \emptyset$; otherwise, Player II *wins*. A *strategy* for Player I is a rule that assigns exactly one U_n to any $U_0, V_0, \ldots, V_{n-1}$. A strategy is a *winning strategy* if Player I wins every run consistent with this strategy. A *tactic* for Player I is a rule that assigns exactly one U_n for every V_{n-1}. A *winning tactic* is defined analogously to a wining strategy. Also a (winning) strategy and tactic for Player II are defined analogously.

A space (X,τ) is called a *Choquet space* if Player II has a winning strategy. (A more rigorous definition can be found in [158, Definition 8.10]).

Note that every winning tactic defines a winning strategy, but the converse is not true.

Let (X,τ) be a topological space. The *strong Choquet game* G_X^s on (X,τ) is defined similarly to G_X, but Player I plays a pair $(x_n, U_n) \in X \times \tau$ with $x_n \in U_n$ instead of just U_n, and Player II has to choose V_n so that $x_n \in V_n \subseteq U_n$.

A space (X,τ) is called *strong Choquet* if Player II has a winning strategy [158, Definition 8.14].

Note that every strong Choquet space is Choquet, but the converse is not true.

Theorem 4.14.10 ([135]). *Let (X,τ) be a countably paracompact normal space. Then $(MC(X,\mathbb{R}), \tau_V^+)$ is a strong Choquet space.*

Proof. We will present a winning tactic for Player II in the strong Choquet game. Suppose that Player I has chosen $F_k \in MC(X,\mathbb{R})$ and $\mathcal{U}_k \in \tau_V^+$ such that $F_k \in \mathcal{U}_k \subseteq \mathcal{V}_{k-1}$, where $\mathcal{V}_{k-1}$ is the choice of Player II from the previous turn. By the proof of Theorem 4.14.9 there are $f_k, g_k \in C(X,\mathbb{R})$ such that

$$F_k \in M_{f_k,g_k}^+ \subseteq \overline{M}_{f_k,g_k}^+ \subseteq \mathcal{U}_k.$$

Player II chooses $\mathcal{V}_k := M_{f_k,g_k}^+$. Now we have

$$f_0 < f_1 < \cdots < f_k < \cdots < g_k < \cdots < g_1 < g_0.$$

Put $F(x) := \bigcap_{k\in\omega} [f_k(x), g_k(x)] \neq \varnothing$. Since F is the intersection of a decreasing system of cusco mappings, we have that F is cusco, and therefore it contains a minimal cusco mapping F^*. Since $F^* \in \bigcap_{k\in\omega} \mathcal{U}_k$, we have described a wining tactic. □

Recall that a second countable and completely metrisable topological space is called *Polish*.

Theorem 4.14.11 ([135]). *Let (X,τ) be a Tychonoff space. Then $(MC(X,\mathbb{R}), \tau_V^+)$ is Polish if and only if the space (X,τ) is compact and metrisable.*

Proof. If $(MC(X,\mathbb{R}), \tau_V^+)$ is Polish, then it is second countable, and from Theorem 4.14.4 we have that (X,τ) must be compact and metrisable. By [158, Theorem 8.18] a nonempty second countable topological space is Polish if and only if it is T_1, regular and strong Choquet. By Theorem 4.14.4 the space $(MC(X,\mathbb{R}), \tau_V^+)$ is second countable, by Theorem 4.14.9 it is regular, by Theorem 4.14.10 it is strong Choquet, and by Proposition 4.13.10 it is Hausdorff. □

Lemma 4.14.12 ([135]). *Let (X,τ) be a countably paracompact normal space, and let $\mathcal{U}$ be an open set in the space $(L(X,\mathbb{R}), \tau_V)$ such that there is $h \in \mathcal{U} \cap C(X,\mathbb{R})$. Then there are $f, g \in C(X,\mathbb{R})$ such that $h \in M_{f,g}^+ \subseteq \overline{M}_{f,g}^+ \subseteq \mathcal{U}$.*

Proof. Without loss of generality, we can suppose that $\mathcal{U} = W_0^+ \cap W_1^- \cap \cdots \cap W_n^-$, where W_k are open subsets of $X \times \mathbb{R}$. Choose $x_1, \ldots, x_n$ so that $(x_k, h(x_k)) \in W_k$. Let $x_1, \ldots, x_{n'}$ be all x that are distinct (rearranged if necessary). For $k = 1, \ldots, n'$, we can choose open sets U_k and open intervals J_k so that U_k are pairwise disjoint and $(x_k, h(x_k)) \in U_k \times J_k \subseteq W_k$. Put $\mathcal{U}' := W_0^+ \cap (U_1 \times J_1)^- \cap \cdots \cap (U_{n'} \times J_{n'})^-$ and observe that $h \in \mathcal{U}' \subseteq \mathcal{U}$. Since by Theorem 4.12.1 we have that $\tau_V = \tau_V^+$ on $C(X, \mathbb{R})$, by Proposition 4.14.7 there are $f', g' \in C(X, \mathbb{R})$ such that

$$h \in M_{f',g'}^+ \cap C(X, \mathbb{R}) \subseteq \mathcal{U}' \cap C(X, \mathbb{R}).$$

We can verify that there are $f, g \in C(X, \mathbb{R})$ such that $f' < f < h < g < g'$, and we have that $h \in M_{f,g}^+ \subseteq \overline{M_{f,g}^+} \subseteq \mathcal{U}'$. □

Theorem 4.14.13 ([135]). *Let (X, τ) be a countably paracompact normal space. Then $(L_0(X, \mathbb{R}), \tau_V)$ and $(MC(X, \mathbb{R}), \tau_V)$ are Choquet spaces.*

Proof. We will present a winning tactic for Player II in the Choquet game on the space $(MC(X, \mathbb{R}), \tau_V)$. Suppose that Player I has chosen $\mathcal{U}_k \in \tau_V$ such that $\mathcal{U}_k \subseteq \mathcal{V}_{k-1}$, where $\mathcal{V}_{k-1}$ is the choice of Player II from the previous turn. Since by Corollary 4.13.6 the set $C(X, \mathbb{R})$ is dense in $(MC(X, \mathbb{R}), \tau_V)$, there is $h_k \in C(X, \mathbb{R}) \cap \mathcal{U}_k$. By Lemma 4.14.12 there are $f_k, g_k \in C(X, \mathbb{R})$ such that

$$h_k \in M_{f_k,g_k}^+ \subseteq \overline{M_{f_k,g_k}^+} \subseteq \mathcal{U}_k.$$

Player II chooses $\mathcal{V}_k := M_{f_k,g_k}$. We have constructed

$$f_0 < f_1 < \cdots < f_k < \cdots < g_k < \cdots < g_1 < g_0.$$

Similarly, as in the proof of Theorem 4.14.10, put $F(x) := \bigcap_{k \in \omega} [f_k(x), g_k(x)] \neq \emptyset$. Since F is an intersection of a decreasing system of cusco mappings, we have that F is cusco, and therefore it contains a minimal cusco mapping $F^* \in \bigcap_{k \in \omega} \mathcal{U}_k$, which concludes the proof for the space $(MC(X, \mathbb{R}), \tau_V)$. The proof for the space $(L_0(X, \mathbb{R}), \tau_V)$ is analogous. □

Even though Theorem 4.14.11 is a result about metrisability of the space $(MC(X, \mathbb{R}), \tau_V^+)$, we have no answer to the following problem.

Question 4.14.14. Characterise those spaces (X, τ) for which the space $(MC(X, \mathbb{R}), \tau_V^+)$ or $(MC(X, \mathbb{R}), \tau_V)$ is metrisable.

In the rest of this section, we collect some miscellaneous results concerning the above question.

Let us define a (possibly infinite-valued) metric L on $MC(X, \mathbb{R})$ by

$$L(F, G) := \sup\{H_d(F(x), G(x)) : x \in X\} \quad \text{for } F, G \in MC(X, \mathbb{R}),$$

where H_d is the Hausdorff metric generated by the usual metric d on $\mathbb{R}$. Denote by τ_L the topology generated by L.

Theorem 4.14.15 ([135]). *Let (X, τ) be a countably compact space. Then $\tau_V \subsetneq \tau_L$ on $L(X, \mathbb{R})$.*

Proof. For $F \in L(X, \mathbb{R})$, we denote by $B(F, \varepsilon)$ the open ball with respect to metric L with centre F and radius ε. It is easy to prove that whenever F is an usco and $\mu > 0$, the multifunction F_μ defined as $F_\mu(x) := \{y \in \mathbb{R} : d(F(x), y) \leq \mu\}$ is an usco.

We will prove by contradiction that for every open $W \subseteq X \times \mathbb{R}$ and every $F \in L(X, \mathbb{R})$ such that $F \in W^+$, there is $\varepsilon > 0$ such that $B(F, \varepsilon) \subseteq W^+$. Suppose this is not true. Then there are an open $W \subseteq X \times \mathbb{R}$ and $F \in W^+$ such that for every $\varepsilon > 0$, there is $G \in B(F, \varepsilon) \setminus W^+$. Thus there is $(x, y) \in G \setminus W$ such that $y \in F_\varepsilon(x)$. Taking $\varepsilon := \frac{1}{n}$, we construct a sequence $(x_n, y_n) \in (X \times \mathbb{R}) \setminus W$ such that $y_n \in F_{1/n}(x_n) \subseteq F_1(x_n)$.

The space (X, τ) being countably compact implies that the sequence $(x_n : n \in \mathbb{N})$ has a subnet $(x_{n_\alpha} : \alpha \in A)$ such that $x^* := \lim_{\alpha \in A} x_{n_\alpha}$ for some $x^* \in X$. Since the multifunction F_1 is an usco at x^*, by Proposition 4.1.1 it is subcontinuous at x^*, and thus $(y_{n_\alpha} : \alpha \in A)$ has a cluster point $y^* \in \mathbb{R}$. Since $(X \times \mathbb{R}) \setminus W$ is closed, $(x^*, y^*) \in (X \times \mathbb{R}) \setminus W$, but from the continuity of the distance we have $y^* \in F(x^*)$, which is a contradiction. □

A topological space (Y, τ') is said to be *weakly π-metrisable* if it has a σ-disjoint π-base [295]. Recall that a σ-disjoint system is a countable union of families consisting of pairwise disjoint sets.

Theorem 4.14.16 ([135]). *Let (X, τ) be a countably compact perfectly normal space. Then $(MC(X, \mathbb{R}), \tau_V^+)$ is weakly π-metrisable.*

Proof. Since (X, τ) is countably compact, $(C(X, \mathbb{R}), \tau_V) = (C(X, \mathbb{R}), \tau_V^+)$ is metrisable. Since (X, τ) is also perfectly normal, by Corollary 4.14.3 the space $(MC(X, \mathbb{R}), \tau_V^+)$ is a first countable Hausdorff space, which by Corollary 4.13.3 contains a dense metrisable subspace. Thus by [312, Theorem 2.6] the space $(MC(X, \mathbb{R}), \tau_V^+)$ has a σ-disjoint π-base, that is, it is weakly π-metrisable. □

Exercise

1. Let (X, τ) be a topological space, and let (Y, d) be a boundedly compact metric space, that is, every closed bounded set in (Y, d) is compact. Let $F : X \to 2^Y$ be an usco. Prove that for every $\mu > 0$, the set-valued mapping F_μ defined by

$$F_\mu(x) := \{y \in Y : d(F(x), y) \leq \mu\} \quad \text{for } x \in X$$

is an usco on (X, τ).

Bibliography

[1] L. Alaoglu. Weak topologies of normed linear spaces. *Ann. Math. (2)*, 41:252–267, 1940.

[2] P. S. Alexandroff. *Einführung in die Mengenlehre und die Theorie der reellen Funktionen*. VEB Deutscher Verlag der Wissenschaften, Berlin, 1956.

[3] C. Aliprantis and K. C. Border. *Infinite dimensional analysis*. Springer, Berlin, 3rd edition, 2006.

[4] T. M. Apostol. *Calculus. Vol. II: Multi-variable calculus and linear algebra, with applications to differential equations and probability*. Blaisdell Publishing Co. Ginn and Co., Waltham, Mass.-Toronto, Ont.-London, 2nd edition, 1969.

[5] A. V. Arhangelskiĭ. On a class containing all metric and all locally bicompact spaces. *Dokl. Akad. Nauk SSSR*, 151:751–754, 1963.

[6] A. V. Arhangelskiĭ. Some topological spaces that arise in functional analysis. *Usp. Mat. Nauk*, 31(5(191)):17–32, 1976.

[7] A. V. Arhangelskiĭ. *Topological invariants in algebraic environment. North-Holland, Recent Progress in General Topology II edition*, 2002.

[8] K. J. Arrow and G. Debreu. Existence of an equilibrium for a competitive economy. *Econometrica*, 22:265–290, 1954.

[9] G. Artico, Ľ. Holá, U. Marconi, and R. Moresco. Approximation by continuous functions in the Fell topology. *Topol. Appl.*, 155(17–18):2150–2157, 2008. Proceedings of the 2006 International Conference on Topology and its Applications.

[10] E. Asplund. Fréchet differentiability of convex functions. *Acta Math.*, 121:31–47, 1968.

[11] M. Atsuji. Uniform continuity of continuous functions of metric spaces. *Pac. J. Math.*, 8(1):11–16, 1958.

[12] R. Baire. Sur les fonctions de variables réelles. *Ann. Mat.*, 3(3):1–123, 1899.

[13] R. Baire. *Leçons sur les fonctions discontinues. Les Grands Classiques Gauthier-Villars. [Gauthier-Villars Great Classics]*. Éditions Jacques Gabay, Sceaux, 1995. Reprint of the 1905 original.

[14] V. S. Balaganskiĭ and L. P. Vlasov. The problem of the convexity of Chebyshev sets. *Usp. Mat. Nauk*, 51(6(312)):125–188, 1996.

[15] T. Banakh. Quasicontinuous functions with values in Piotrowski spaces. *Real Anal. Exch.*, 43(1):77–104, 2017.

[16] J. Banas and K. Goebel. *Measure of Noncompactness in Banach Spaces*. Dekker, New York, 1980.

[17] R. G. Bartle and L. M. Graves. Mappings between function spaces. *Trans. Am. Math. Soc.*, 72:400–413, 1952.

[18] G. Beer. The approximation of real functions in the Hausdorff metric. *Houst. J. Math.*, 10:325–338, 1984.

[19] G. Beer. Metric spaces on which continuous functions are uniformly continuous and Hausdorff distance. *Proc. Am. Math. Soc.*, 95(4):653–658, 1985.

[20] G. Beer. Hausdorff distance and a compactness criterion for continuous functions. *Can. Math. Bull.*, 29(4):463–468, 1986.

[21] G. Beer. Complete subsets of $C(X, Y)$ with respect to Hausdorff distance. *Math. Balk.*, 2:78–84, 1988.

[22] G. Beer. On a theorem of Cellina for set-valued functions. *Rocky Mt. J. Math.*, 18(1):37–48, 1988.

[23] G. Beer. Topological completeness of function spaces arising in the Hausdorff approximation of functions. *Can. Math. Bull.*, 35(4):439–448, 1992.

[24] G. Beer. *Topologies on closed and closed convex sets*, volume 268 of *Mathematics and its Applications*. Kluwer Academic Publishers Group, Dordrecht, 1993.

https://doi.org/10.1515/9783110750188-005

[25] G. Beer and A. DiConcilio. A generalization of boundedly compact metric spaces. *Comment. Math. Univ. Carol.*, 32(2):361–367, 1991.

[26] G. Beer and S. Levi. Strong uniform continuity. *J. Math. Anal. Appl.*, 156:1271–1288, 2009.

[27] G. Beer and S. Naimpally. Graphical convergence of continuous functions. *Acta Math. Hung.*, 140(4):305–315, 2013.

[28] C. Berge. *Topological spaces*. Dover Publications, Inc., Mineola, NY, 1997. Including a treatment of multi-valued functions, vector spaces and convexity, Translated from the French original by E. M. Patterson, Reprint of the 1963 translation.

[29] E. Bishop and R. R. Phelps. A proof that every Banach space is subreflexive. *Bull. Am. Math. Soc.*, 67:97–98, 1961.

[30] W. W. Bledsoe. Neighbourly functions. *Proc. Am. Math. Soc.*, 3:114–115, 1952.

[31] J. Borsík. Quasiuniform limits of quasicontinuous functions. *Math. Slovaca*, 42:269–274, 1992.

[32] J. Borsík. Points of continuity, quasicontinuity and cliquishness. *Rend. Ist. Mat. Univ. Trieste*, 26(1–2):5–20, 1995.

[33] J. Borsík, Ľ. Holá, and D. Holý. Baire spaces and quasicontinuous mappings. *Filomat*, 25:69–83, 2011.

[34] J. M. Borwein. Minimal cuscos and subgradients of Lipschitz function. In *Fixed Point Theory and its Applications*, volume 252 of *Pitman Research Notes*, pages 57–81. Longman, Harlow, 1991.

[35] J. M. Borwein and A. S. Lewis. *Convex analysis and nonlinear optimization*, volume 3 of *CMS Books in Mathematics/Ouvrages de Mathématiques de la SMC*. Springer, New York, 2nd edition, 2006. Theory and examples.

[36] J. M. Borwein and W. B. Moors. Essentially strictly differentiable Lipschitz functions. *CECM preprint*, 95:029:1–50, 1995.

[37] J. M. Borwein and W. B. Moors. Essentially smooth Lipschitz functions. *J. Funct. Anal.*, 149:305–351, 1997.

[38] J. M. Borwein, W. B. Moors, and X. Wang. Lipschitz functions with prescribed derivatives and subderivatives. *Nonlinear Anal.*, 29(1):53–63, 1997.

[39] J. M. Borwein, W. B. Moors, and X. Wang. Generalized subdifferentials: a Baire categorical approach. *Trans. Am. Math. Soc.*, 353(10):3875–3893, 2001.

[40] J. M. Borwein and D. Preiss. A smooth variational principle with applications to subdifferentiability and to differentiability of convex functions. *Trans. Am. Math. Soc.*, 303(2):517–527, 1987.

[41] J. M. Borwein and Q. J. Zhu. Multifunctional and functional analytic techniques in nonsmooth analysis. In F. H. Clarke and R. J. Stern, editors, *Nonlinear Analysis, Differential Equations and Control*, pages 61–157. Kluwer Academic Publishers, 1999.

[42] J. M. Borwein and Q. J. Zhu. *Techniques of variational analysis*, volume 20 of *CMS Books in Mathematics/Ouvrages de Mathématiques de la SMC*. Springer-Verlag, New York, 2005.

[43] J. M. Borwein and Q. J. Zhu. *Techniques of variational analysis*. Springer, 2005.

[44] R. D. Bourgin. *Geometric aspects of convex sets with the Radon–Nikodým property*, volume 993 of *Lecture Notes in Mathematics*. Springer-Verlag, Berlin, 1983.

[45] A. Bouziad. Continuity of separately continuous group actions in p-spaces. *Topol. Appl.*, 71:119–124, 1996.

[46] A. Bouziad. Every Čech-analytic Baire semitopological group is a topological group. *Proc. Am. Math. Soc.*, 124:953–959, 1996.

[47] A. Bouziad and J. -P. Troallic. Lower quasicontinuity, joint continuity and related topics. *Topol. Appl.*, 157:2889–2894, 2010.

[48] A. Brøndsted and R. T. Rockafellar. On the subdifferentiability of convex functions. *Proc. Am. Math. Soc.*, 16:605–611, 1965.

[49] L. E. J. Brouwer. Über Abbildung von Mannigfaltigkeiten. *Math. Ann.*, 71(1):97–115, 1911.
[50] D. Burke, D. Lutzer, and S. Levi. Functional characterisations of certain p-spaces. *Topol. Appl.*, 20:161–165, 1985.
[51] T. Byczkowski and R. Pol. On the closed graph and open mapping theorems. *Bull. Acad. Polon. Sci. Ser. Sci. Math. Astronom. Phys.*, 24:723–726, 1976.
[52] J. Cao and W. B. Moors. Separate and joint continuity of homomorphisms defined on topological groups. *N.Z. J. Math.*, 33:41–45, 2004.
[53] J. Cao and W. B. Moors. Quasicontinuous selections of upper continuous set-valued mappings. *Real Anal. Exch.*, 31(1):63–71, 2005/06.
[54] B. Cascales, V. Kadets, and J. Rodríguez. Measurability and selections of multi-functions in Banach spaces. *J. Convex Anal.*, 17(1):229–240, 2010.
[55] E. Čech. On bicompact spaces. *Ann. Math. (2)*, 38(4):823–844, 1937.
[56] A. Cellina. Approximation of set valued functions and fixed point theorems. *Ann. Mat. Pura Appl. (4)*, 82:17–24, 1969.
[57] A. Cellina. A further result on the approximation of set valued mappings. *Rend. Accad. Naz. Lincei*, 48:412–416, 1970.
[58] G. Choquet. Convergences. *Ann. Univ. Grenoble. Sect. Sci. Math. Phys. (N. S.)*, 23:57–112, 1948.
[59] J. P. R. Christensen. Theorems of Namioka and R. E. Johnson type for upper semicontinuous and compact valued set-valued mappings. *Proc. Am. Math. Soc.*, 86:3649–655, 1982.
[60] F. H. Clarke. Generalized gradients and applications. *Trans. Am. Math. Soc.*, 205:247–262, 1975.
[61] F. H. Clarke. *Optimization and nonsmooth analysis. Canadian Mathematical Society Series of Monographs and Advanced Texts*. John Wiley & Sons, Inc., New York, 1983. A Wiley-Interscience Publication.
[62] M. M. Čoban, P. S. Kenderov, and J. P. Revalski. Generic well-posedness of optimization problems in topological spaces. *Mathematika*, 36(2):301–324, 1990.
[63] M. M. Čoban, P. S. Kenderov, and J. P. Revalski. Densely defined selections of multivalued mappings. *Trans. Am. Math. Soc.*, 344(2):533–552, 1994.
[64] E. F. Collingwood and A. J. Lohwater. *The theory of cluster sets*, volume 56 of *Cambridge Tracts in Mathematics and Mathematical Physics*. Cambridge University Press, Cambridge, 1966.
[65] M. D. Contreras and R. Payá. On upper semicontinuity of duality mappings. *Proc. Am. Math. Soc.*, 121(2):451–459, 1994.
[66] A. Crannell, M. Frantz, and M. LeMasurier. Closed relations and equivalence classes of quasicontinuous functions. *Real Anal. Exch.*, 31:409–424, 2006.
[67] G. Debs and J. Saint-Raymond. Topological games and optimization problems. *Mathematika*, 41(1):117–132, 1994.
[68] F. Deutsch. *Best approximation in inner product spaces*, volume 7 of *CMS Books in Mathematics/Ouvrages de Mathématiques de la SMC*. Springer-Verlag, New York, 2001.
[69] R. Deville, G. Godefroy, and V. Zizler. A smooth variational principle with applications to Hamilton–Jacobi equations in infinite dimensions. *J. Funct. Anal.*, 111(1):197–212, 1993.
[70] R. Deville, G. Godefroy, and V. Zizler. *Smoothness and renormings in Banach spaces*, volume 64 of *Pitman Monographs and Surveys in Pure and Applied Mathematics*. Longman Scientific & Technical, Harlow, 1993; copublished in the United States with, John Wiley & Sons, Inc., New York.
[71] J. Dieudonné. Une généralisation des espaces compacts. *J. Math. Pures Appl.*, 9(23):65–76, 1944.
[72] J. Doboš. Some generalizations of the notion of continuity and quasi-uniform convergence. *Časopis Pěst. Mat.*, 106(4):431–434, 437, 1981. With a loose Russian summary.

[73] L. Drewnowski and I. Labuda. On minimal upper semicontinuous compact-valued maps. *Rocky Mt. J. Math.*, 20(3):737–752, 1990.

[74] J. Dugundji. *Topology*. Allyn and Bacon, Boston, 1966.

[75] N. Dunford and J. T. Schwartz. *Linear Operators. I. General Theory*, with the assistance of W. G. Bade and R. G. Bartle, volume 7 *Pure and Applied Mathematics*. Interscience Publishers, Inc., New York/London, 1958.

[76] I. Ekeland. On the variational principle. *J. Math. Anal. Appl.*, 47:324–353, 1974.

[77] R. Engelking. *Outline of General Topology. North-Holland Publishing Co./PWN-Polish Scientific Publishers/Interscience Publishers Division*. John Wiley & Sons, Inc., Amsterdam/Warsaw/New York, 1968. Translated from the Polish by K. Sieklucki.

[78] R. Engelking. *General Topology*, volume 60 of *Monografie Matematyczne [Mathematical Monographs]*. PWN—Polish Scientific Publishers, Warsaw, 1977. Translated from the Polish by the author.

[79] J. Ewert and T. Lipski. Lower and upper quasicontinuous functions. *Demonstr. Math.*, 16:85–93, 1983.

[80] M. Fabian, P. Habala, P. Hájek, V. Montesinos, and V. Zizler. *Banach space theory. CMS Books in Mathematics/Ouvrages de Mathématiques de la SMC*. Springer, New York, 2011. The basis for linear and nonlinear analysis.

[81] M. J. Fabian. *Gâteaux differentiability of convex functions and topology. Canadian Mathematical Society Series of Monographs and Advanced Texts*. John Wiley & Sons, Inc., New York, 1997. Weak Asplund spaces, A Wiley-Interscience Publication.

[82] K. Fan. Fixed-point and minimax theorems in locally convex topological linear spaces. *Proc. Natl. Acad. Sci. USA*, 38:121–126, 1952.

[83] J. Fletcher and W. B. Moors. Chebyshev sets. *J. Aust. Math. Soc.*, 98(2):161–231, 2015.

[84] V. P. Fonf and J. Lindenstrauss. Boundaries and generation of convex sets. *Isr. J. Math.*, 136:157–172, 2003.

[85] M. K. Fort. Points of continuity of semi-continuous functions. *Publ. Math. (Debr.)*, 2:100–102, 1951.

[86] Z. Frolík. Remarks concerning invariance of Baire spaces under mappings. *Czech. Math. J.*, 86:381–385, 1961.

[87] R. Fuller. Relations among continuous and various non-continuous functions. *Pac. J. Math.*, 25(3):495–509, 1968.

[88] F. García, L. Oncina, J. Orihuela, and S. Troyanski. Kuratowski's index of non-compactness and renorming in Banach spaces. *J. Convex Anal.*, 11(2):477–494, 2004.

[89] K. R. Gentry and H. B. Hoyle. C-continuous functions. *Yokohama Math. J.*, 73:18–21, 1970.

[90] K. R. Gentry and H. B. Hoyle. Somewhat continuous functions. *Czech. Math. J.*, 21(96):6–12, 1971.

[91] J. R. Giles. *Convex analysis with application in the differentiation of convex functions*, volume 58 of *Research Notes in Mathematics*. Pitman (Advanced Publishing Program), Boston, Mass.-London, 1982.

[92] J. R. Giles and M. O. Bartlett. Modified continuity and a generalization of Michael's selection theorem. *Set-Valued Anal.*, 1:365–378, 1993.

[93] J. R. Giles, P. S. Kenderov, W. B. Moors, and S. D. Sciffer. Generic differentiability of convex functions on the dual of a Banach space. *Pac. J. Math.*, 172(2):413–431, 1996.

[94] J. R. Giles and W. B. Moors. A continuity property related to Kuratowski's index of noncompactness, its relevance to the drop property, and its implications for differentiability theory. *J. Math. Anal. Appl.*, 178(1):247–268, 1993.

[95] J. R. Giles and W. B. Moors. A selection theorem for quasi-lower semicontinuous set-valued mappings. *J. Nonlinear Convex Anal.*, 2(3):345–350, 2001.

[96] I. Glicksberg. Uniform boundedness for groups. *Can. J. Math.*, 14:269–276, 1962.

[97] I. L. Glicksberg. A further generalization of the Kakutani fixed theorem, with application to Nash equilibrium points. *Proc. Am. Math. Soc.*, 3:170–174, 1952.
[98] G. Godefroy. Boundaries of a convex set and interpolation sets. *Math. Ann.*, 277(2):173–184, 1987.
[99] Z. Grande. Sur le quasi-continuité approximative. *Fundam. Math.*, 129:167–172, 1988.
[100] G. Gruenhage. Generalized metric spaces. In K. Kunen and J. E. Vaughan, editors, *Handbook of Set-Theoretic topology*, pages 423–501. North-Holland, 1984.
[101] H. Hahn. Über Funktionen mehrerer Veränderlicher, die nach jeder einzelnen Veränderlichen steting sind. *Mat. Z.*, 306–313, 1919.
[102] S. T. Hammer and R. A. McCoy. Spaces of densely continuous forms. *Set-Valued Anal.*, 5:247–266, 1997.
[103] J. D. Hansard. Function space topologies. *Pac. J. Math.*, 35(2):381–388, 1970.
[104] G. Hansel and J. P. Troallic. Quasicontinuity and Namioka theorem. *Topol. Appl.*, 46:135–149, 1992.
[105] R. W. Hansell. On Borel mappings and Baire functions. *Trans. Am. Math. Soc.*, 194:195–210, 1974.
[106] R. W. Hansell, J. E. Jayne, and M. Talagrand. First class selectors for weakly upper semicontinuous multivalued maps in Banach spaces. *J. Reine Angew. Math.*, 361:201–220, 1985.
[107] E. Hewitt and K. A. Ross. *Abstract harmonic analysis. Vol. I: Structure of topological groups. Integration theory, group representations*, volume 115 of *Die Grundlehren der mathematischen Wissenschaften*. Academic Press, Inc., Publishers/Springer-Verlag, New York/Berlin-Göttingen-Heidelberg, 1963.
[108] Ľ. Holá. There are $2^{\mathfrak{c}}$ quasicontinuous non Borel functions on uncountable Polish space. *Results Math.*, 76:126, 2021.
[109] Ľ. Holá. On relations approximated by continuous functions. *Acta Univ. Carol., Math. Phys.*, 28:67–72, 1987.
[110] Ľ. Holá. Some conditions that imply continuity of almost continuous multifunctions. *Acta Math. Univ. Comen.*, LII–LIII:159–163, 1987.
[111] Ľ. Holá. An extension theorem for continuous functions. *Czech. Math. J.*, 38:398–404, 1988.
[112] Ľ. Holá. Functional characterization of Baire spaces. *Proc. Am. Math. Soc.*, 109:1121–1124, 1990.
[113] Ľ. Holá. Hausdorff metric on the space of upper semicontinuous multifunctions. *Rocky Mt. J. Math.*, 22(2):601–610, 1992.
[114] Ľ. Holá. Spaces of densely continuous forms, usco and minimal usco maps. *Set-Valued Anal.*, 11:133–151, 2003.
[115] Ľ. Holá. Spaces of densely continuous forms. *Topol. Proc.*, 31:145–150, 2007.
[116] Ľ. Holá. Functional characterisations of p-spaces. *Cent. Eur. J. Math.*, 11(12):12197–2202, 2013.
[117] Ľ. Holá, V. Baláž, and T. Neubrunn. Remarks on c-continuous multifunctions. *Acta Math. Univ. Comen.*, L–LI:51–59, 1987.
[118] Ľ. Holá and D. Holý. Minimal usco maps, densely continuous forms and upper semicontinuous functions. *Rocky Mt. J. Math.*, 39:545–562, 2009.
[119] Ľ. Holá and D. Holý. Pointwise convergence of quasicontinuous mappings and Baire spaces. *Rocky Mt. J. Math.*, 41:1883–1894, 2011.
[120] Ľ. Holá and D. Holý. Relation between minimal usco and minimal cusco maps. *Port. Math.*, 70(3):211–224, 2013.
[121] Ľ. Holá and D. Holý. Spaces of lower semicontinuous set-valued maps. *Math. Slovaca*, 63(4):863–870, 2013.

[122] Ľ. Holá and D. Holý. New characterizations of minimal cusco maps. *Rocky Mt. J. Math.*, 44(6):1851–1866, 2014.
[123] Ľ. Holá and D. Holý. Minimal usco and minimal cusco maps. *Khayyam J. Math.*, 1:126–152, 2015.
[124] Ľ. Holá and D. Holý. Minimal usco and minimal cusco maps and compactness. *J. Math. Anal. Appl.*, 439(2):737–744, 2016.
[125] Ľ. Holá and D. Holý. Quasicontinuous subcontinuous functions and compactness. *Mediterr. J. Math.*, 13(6):4509–4518, 2016.
[126] Ľ. Holá, T. Jain, and R. A. McCoy. Topological properties of the multifunction space $L(X)$ of cusco maps. *Math. Slovaca*, 58:763–780, 2008.
[127] Ľ. Holá and G. Kwiecinska. Pointwise topological convergence and topological graph convergence of set-valued maps. *Filomat*, 31:2779–2785, 2017.
[128] Ľ. Holá and G. Kwiecińska. On the Borel classes of set-valued maps of two variables. *Ann. Math. Sil.*, 34(1):81–95, 2020.
[129] Ľ. Holá and R. A. McCoy. Relations approximated by continuous functions. *Proc. Am. Math. Soc.*, 133(7):2173–2182, 2005.
[130] Ľ. Holá and R. A. McCoy. Cardinal invariants of the topology of uniform convergence on compact sets on the space of minimal usco maps. *Rocky Mt. J. Math.*, 37:229–246, 2007.
[131] Ľ. Holá and R. A. McCoy. Relations approximated by continuous functions in the Vietoris topology. *Fundam. Math.*, 195(3):205–219, 2007.
[132] Ľ. Holá, R. A. McCoy, and J. Pelant. Approximations of relations by continuous functions. *Topol. Appl.*, 154(11):2241–2247, 2007.
[133] Ľ. Holá and A. K. Mirmostafaee. Norm continuity of quasicontinuous mappings into $C_p(Y)$. *Colloq. Math.*, 154(1):97–105, 2018.
[134] Ľ. Holá and B. Novotný. Subcontinuity of multifunctions. *Math. Slovaca*, 62:345–362, 2012.
[135] Ľ. Holá and B. Novotný. Topological properties of the space of convex minimal usco maps. *Set-Valued Var. Anal.*, 28(2):287–300, 2020.
[136] Ľ. Holá and Z. Piotrowski. Set of continuity points of functions with values in generalized metric spaces. *Tatra Mt. Math. Publ.*, 42:148–160, 2009.
[137] Ľ. Holá and L. Zsilinszky. Completeness and related properties of the graph topology. *Topol. Proc.*, 46:1–14, 2015.
[138] D. Holý. Uniform convergence on spaces of multifunctions. *Math. Slovaca*, 57:561–570, 2007.
[139] D. Holý. Ascoli-type theorems for locally bounded quasicontinuous functions, minimal usco and minimal cusco maps. *Ann. Funct. Anal.*, 6:29–41, 2015.
[140] D. Holý and P. Vadovič. Hausdorff graph topology, proximal graph topology and the uniform topology for densely continuous forms and minimal usco maps. *Acta Math. Hung.*, 116:133–144, 2007.
[141] D. Holý and P. Vadovič. Densely continuous forms, pointwise topology and cardinal functions. *Czech. Math. J.*, 58(133):79–92, 2008.
[142] R. Hrycay. Noncontinuous multifunctions. *Pac. J. Math.*, 35(1):141–154, 1970.
[143] S. Hu and N. S. Papageorgiou. *Handbook of Multivalued Analysis, volume I*. Kluwer Academic Publishers, Dordrecht-Boston-London, 1997.
[144] Z. Hu and B. -L. Lin. Smoothness and the asymptotic-norming properties of Banach spaces. *Bull. Aust. Math. Soc.*, 45(2):285–296, 1992.
[145] T. Husain. Almost continuous mappings. *Prace Mathematyczne*, 10:1–7, 1966.
[146] V. Istratescu. *Fixed Point Theory*. Reidel, Dordrecht, 1981.
[147] R. C. James. Reflexivity and the supremum of linear functionals. *Ann. Math. (2)*, 66:159–169, 1957.
[148] R. C. James. Weakly compact sets. *Trans. Am. Math. Soc.*, 113:129–140, 1964.

[149] J. E. Jayne, J. Orihuela, A. J. Pallarés, and G. Vera. σ-fragmentability of multivalued maps and selection theorems. *J. Funct. Anal.*, 117(2):243–273, 1993.

[150] J. E. Jayne and C. A. Rogers. Borel selectors for upper semicontinuous set-valued maps. *Acta Math.*, 155(1–2):41–79, 1985.

[151] J. E. Jayne and C. A. Rogers. *Selectors*. Princeton University Press, Princeton, NJ, 2002.

[152] L. Jokl. Minimal convex-valued weak* USCO correspondences and the Radon–Nikodým property. *Comment. Math. Univ. Carol.*, 28(2):353–376, 1987.

[153] J. E. Joseph. Multifunctions and graphs. *Pac. J. Math.*, 79(2):509–529, 1978.

[154] I. Juhász. *Cardinal Functions in Topology – Ten Years Later*. Matematisch Centrum, Amsterdam, 1980.

[155] S. Kakutani. A generalization of Brouwer's fixed point theorem. *Duke Math. J.*, 8:457–459, 1941.

[156] O. F. K. Kalenda. Weak Stegall spaces. Unpublished, 3 pages, 1997.

[157] O. F. K. Kalenda. Stegall spaces that are not fragmentable. *Topol. Appl.*, 96:121–132, 1999.

[158] A. S. Kechris. *Classical Descriptive Set Theory*. Springer-Verlag, New York, 1995.

[159] J. L. Kelley. *General Topology*. Van Nostrand, New York, 1955.

[160] J. L. Kelley and I. Namioka. *Linear topological spaces*. With the collaboration of W. F. Donoghue, Jr., Kenneth R. Lucas, B. J. Pettis, Ebbe Thue Poulsen, G. Baley Price, Wendy Robertson, W. R. Scott, Kennan T. Smith. *The University Series in Higher Mathematics*. D. Van Nostrand Co., Inc., Princeton, N.J., 1963.

[161] S. Kempisty. Sur les fonctions quasi-continues. *Fundam. Math.*, 19:184–197, 1932.

[162] P. S. Kenderov. The set-valued monotone mappings are almost everywhere single-valued. *C. R. Acad. Bulgare Sci.*, 27:1173–1175, 1974.

[163] P. S. Kenderov. Most of the optimization problems have unique solution. In *Parametric optimization and approximation (Oberwolfach, 1983)*, volume 72 of *Internat. Schriftenreihe Numer. Math.*, pages 203–216. Birkhäuser, Basel, 1985.

[164] P. S. Kenderov, I. S. Kortezov, and W. B. Moors. Continuity points of quasi-continuous mappings. *Topol. Appl.*, 109:321–346, 2001.

[165] P. S. Kenderov, I. S. Kortezov, and W. B. Moors. Topological games and topological groups. *Topol. Appl.*, 109:157–165, 2001.

[166] P. S. Kenderov, I. S. Kortezov, and W. B. Moors. Norm continuity of weakly continuous mappings into Banach spaces. *Topol. Appl.*, 153(14):2745–2759, 2006.

[167] P. S. Kenderov and W. B. Moors. Fragmentability of Banach spaces. *C. R. Acad. Bulgare Sci.*, 49(2):9–12, 1996.

[168] P. S. Kenderov and W. B. Moors. Game characterization of fragmentability of topological spaces. In *Mathem. and Education in Mathem*, pages 8–18, *Proceedings of 25th*. Spring Union of Bulgar. Mathematician, Kazanlak, 1996.

[169] P. S. Kenderov and W. B. Moors. Separate continuity, joint continuity and the Lindelöf property. *Proc. Am. Math. Soc.*, 134(5):1503–1512, 2006.

[170] P. S. Kenderov, W. B. Moors, and J. P. Revalski. Dense continuity and selections of set-valued mappings. *Serdica Math. J.*, 24(1):49–72, 1998.

[171] P. S. Kenderov and J. P. Revalski. Variational principles in non-metrizable spaces. *Top*, 20(2):467–474, 2012.

[172] P. S. Kenderov and J. P. Revalski. Fragmentability and selections of set-valued mappings. *J. Nonlinear Convex Anal.*, 16(6):1069–1082, 2015.

[173] P. Kostyrko. On convergence of transfinite sequences. *Mat. Čas. Slov. Akad. Vied*, 21:233–239, 1971.

[174] M. Krein. Sur quelques questions de la géométrie des ensembles convexes situes dans un espace linéaire norme et complet. *C. R. (Dokl.) Acad. Sci. URSS, n. Ser.*, 14:5–7, 1937.

[175] W. Kubiś, O. Okunev, and P. J. Szeptycki. On some classes of Lindelöf Σ-spaces. *Topol. Appl.*, 153(14):2574–2590, 2006.
[176] C. Kuratowski. Sur les espaces complets. *Fundam. Math.*, 15:301–309, 1930.
[177] C. Kuratowski. Les fonctions semi-continues dans l'espace des ensembles fermés. *Fundam. Math.*, 18(1):148–159, 1932.
[178] K. Kuratowski and C. Ryll-Nardzewski. A general theorem on selectors. *Bull. Acad. Polon. Sci. Sér. Sci. Math. Astronom. Phys.*, 13:397–403, 1965.
[179] M. Lavrentiev. Contribution á la théorie des ensembles homéomorphes. *Fundam. Math.*, 6:149–160, 1924.
[180] P. D. Lax. Change of variables in multiple integrals. *Am. Math. Mon.*, 106(6):497–501, 1999.
[181] H. Lebesgue. *Oeuvres scientifiques (en cinq volumes). Vol. III.* Institut de Mathématiques de l'Université de Genève, Geneva, 1972. Sous la rédaction de François Châtelet et Gustave Choquet.
[182] G. Lebourg. Valeur moyenne pour gradient généralisé. *C. R. Acad. Sci. Paris Sér. A-B*, 281(19):Ai:A795–A797, 1975.
[183] A. Lechicki. On bounded and subcontinuous multifunctions. *Pac. J. Math.*, 75(1):191–197, 1978.
[184] A. Lechicki and S. Levi. Extensions of semicontinuous multifunctions. *Forum Math.*, 2:341–360, 1990.
[185] N. Levine. Semi-open sets and semicontinuity in topological spaces. *Am. Math. Mon.*, 70:36–41, 1963.
[186] J. Lindenstrauss, D. Preiss, and J. Tišer. *Fréchet differentiability of Lipschitz functions and porous sets in Banach spaces*, volume 179 of *Annals of Mathematics Studies*. Princeton University Press, Princeton, NJ, 2012.
[187] S. Marcus. Sur les fonctions quasi-continues au sense Kempisty. *Colloq. Math.*, 8:45–53, 1961.
[188] N. F. G. Martin. Quasi-continuous functions on product spaces. *Duke Math. J.*, 28:39–44, 1961.
[189] M. Maschler, E. Solan, and S. Zamir. *Game theory*. Cambridge University Press, Cambridge, 2013. Translated from Hebrew by Ziv Hellman and edited by Mike Borns.
[190] M. Matejdes. Sur les sélecteurs des multifonctions. *Math. Slovaca*, 37(1):111–124, 1987.
[191] M. Matejdes. Minimality of multifunctions. *Real Anal. Exch.*, 32:519–526, 2007.
[192] M. Matejdes. Upper quasicontinuous maps and quasicontinuous selections. *Czechoslov. Math. J.*, 60:517–525, 2010.
[193] M. Matejdes. Selection theorems and minimal mappings in a cluster setting. *Rocky Mt. J. Math.*, 41:851–857, 2011.
[194] J. P. May. *A concise course in algebraic topology. Chicago Lectures in Mathematics*. University of Chicago Press, Chicago, IL, 1999.
[195] S. Mazur. Über die kleinste konvexe Menge, die eine gegebene kompakte Menge enthält. *Stud. Math.*, 2:7–9, 1930.
[196] S. Mazur. Über konvexe Mengen in linearen normierten Räumen. *Stud. Math.*, 4:70–84, 1933.
[197] R. A. McCoy. Fell topology and uniform topology on compacta on spaces of multifunctions. *Rostock. Math. Kolloqu.*, 51:127–136, 1997.
[198] R. A. McCoy. Recent progress in function spaces. In *Quad. Mat. 3, chapter Comparison of hyperspace and function space topologies*, pages 241–258. Aracne, Rome, 1998.
[199] R. A. McCoy. Spaces of semicontinuous forms. *Topol. Proc.*, 23:249–275, 1998.
[200] R. A. McCoy. Densely continuous forms in Vietoris hyperspaces. *Set-Valued Anal.*, 8:267–271, 2000.
[201] R. A. McCoy and I. Ntantu. *Topological properties of spaces of continuous functions*, volume 1315 of *Lecture Notes in Mathematics*. Springer-Verlag, 1988.

[202] L. McKenzie. On equilibrium in Graham's model of world trade and other competitive systems. *Econometrica*, 22:147–161, 1954.
[203] R. E. Megginson. *An introduction to Banach space theory*. Springer, 1998.
[204] M. G. Megrelishvili. Fragmentability and continuity of semigroup actions. *Semigroup Forum*, 57(1):101–126, 1998.
[205] Y. Mibu. On quasi-continuous mappings defined on a product space. *Proc. Jpn. Acad.*, 34:189–192, 1958.
[206] E. Michael. Topologies on spaces of subsets. *Trans. Am. Math. Soc.*, 71:152–182, 1951.
[207] E. Michael. Continuous selections. I. *Ann. Math. (2)*, 63:361–382, 1956.
[208] E. Michael. Continuous selections. II. *Ann. Math. (2)*, 64:562–580, 1956.
[209] E. Michael. Selected Selection Theorems. *Am. Math. Mon.*, 63(4):233–238, 1956.
[210] E. Michael. Continuous selections. III. *Ann. Math. (2)*, 65:375–390, 1957.
[211] E. Michael. A note on closed maps and compact sets. *Isr. J. Math.*, 2:173–176, 1964.
[212] D. P. Milman. Characteristics of extremal points of regularly convex sets. *Dokl. Akad. Nauk*, 57:119–122, 1947.
[213] A. K. Mirmostafaee. Continuity of separately continuous mappings. *Math. Slovaca*, 64(4):1019–1026, 2014.
[214] W. B. Moors. Continuity conditions defined by indices of non-compactness and applications to differentiability theory. PhD thesis, The University of Newcastle, 1992.
[215] W. B. Moors. On a weak* drop property. *Bull. Aust. Math. Soc.*, 47(3):407–414, 1993.
[216] W. B. Moors. A characterization of minimal subdifferential mappings of locally Lipschitz functions. *Set-Valued Anal.*, 3:129–141, 1995.
[217] W. B. Moors. A selection theorem for weak upper semi-continuous set-valued mappings. *Bull. Aust. Math. Soc.*, 53(2):213–227, 1996.
[218] W. B. Moors. Some more recent results concerning weak Asplund spaces. *Abstr. Appl. Anal.*, 3:307–318, 2005.
[219] W. B. Moors. An elementary proof of James' characterization of weak compactness. *Bull. Aust. Math. Soc.*, 84(1):98–102, 2011.
[220] W. B. Moors. Any semitopological group that is homeomorphic to a product of Čech-complete spaces is a topological group. *Set-Valued Var. Anal.*, 21:627–633, 2013.
[221] W. B. Moors. A note on Fort's theorem. *Topol. Appl.*, 160(2):305–308, 2013.
[222] W. B. Moors. Semitopological groups, Bouziad spaces and topological groups. *Topol. Appl.*, 160:2038–2048, 2013.
[223] W. B. Moors. Fragmentability by the discrete metric. *Bull. Aust. Math. Soc.*, 91(2):303–310, 2015.
[224] W. B. Moors. Invariant means on chart groups. *Khayyam J. Math.*, 1(1):36–44, 2015.
[225] W. B. Moors. Fragmentable mappings and CHART groups. *Fundam. Math.*, 234:191–200, 2016.
[226] W. B. Moors. Some Baire semitopological groups that are topological groups. *Topol. Appl.*, 230:381–392, 2017.
[227] W. B. Moors. Nearly Chebyshev sets are almost convex. *Set-Valued Var. Anal.*, 26(1):67–76, 2018.
[228] W. B. Moors. A gentle introduction to James' weak compactness theorem and beyond. *Methods Funct. Anal. Topol.*, 25(1):35–83, 2019.
[229] W. B. Moors and I. Namioka. Furstenberg's structure theorem via CHART groups. *Ergod. Theory Dyn. Syst.*, 33(3):954–968, 2013.
[230] W. B. Moors and S. Sciffer. Sigma-fragmentable spaces that are not countable unions of fragmentable subspaces. *Topol. Appl.*, 119(3):279–286, 2002.
[231] W. B. Moors and S. Somasundaram. Some recent results concerning weak Asplund spaces. *Acta Univ. Carol., Math. Phys.*, 43(2):67–86, 2002. 30th Winter School on Abstract Analysis (Lhota nade Rohanovem/Litice u České Lípy, 2002).

[232] W. B. Moors and S. Somasundaram. USCO selections of densely defined set-valued mappings. *Bull. Aust. Math. Soc.*, 65(2):307–313, 2002.

[233] W. B. Moors and S. Somasundaram. A weakly Stegall space that is not a Stegall space. *Proc. Am. Math. Soc.*, 131:647–654, 2003.

[234] W. B. Moors and S. Somasundaram. A Gâteaux differentiability space that is not weak Asplund. *Proc. Am. Math. Soc.*, 134(9):2745–2754, 2006.

[235] W. B. Moors and N. O. Tan. An abstract variational theorem. *J. Convex Anal.*, 26(4):1123–1142, 2019.

[236] W. B. Moors and N. O. Tan. Dual differentiation spaces. *Bull. Aust. Math. Soc.*, 99(3):467–472, 2019.

[237] W. B. Moors and S. J. White. An elementary proof of James' characterisation of weak compactness. II. *Bull. Aust. Math. Soc.*, 95(1):133–137, 2017.

[238] S. A. Morris. *Pontryagin duality and the structure of locally compact abelian groups*, volume 29 of *London Mathematical Society Lecture Note Series*. Cambridge University Press, Cambridge-New York-Melbourne, 1977.

[239] K. Nagami. Σ-spaces. *Fundam. Math.*, 65:169–192, 1969.

[240] S. Naimpally. Graph topology for function spaces. *Trans. Am. Math. Soc.*, 123(1):267–272, 1966.

[241] S. Naimpally. Multivalued function spaces and Atsuji spaces. *Appl. Gen. Topol.*, 4(2):201–209, 2003.

[242] I. Namioka. Right topological groups, distal flows, and a fixed-point theorem. *Math. Syst. Theory*, 6:193–209, 1972.

[243] I. Namioka. On class $\mathcal{S}$ of Stegall, October 1983.

[244] I. Namioka. Kakutani-type fixed point theorems: a survey. *J. Fixed Point Theory Appl.*, 9(1):1–23, 2011.

[245] I. Namioka and R. R. Phelps. Banach spaces which are Asplund spaces. *Duke Math. J.*, 42(4):735–750, 1975.

[246] J. Nash. Non-cooperative games. *Ann. Math. (2)*, 54:286–295, 1951.

[247] J. F. Nash Jr. Equilibrium points in n-person games. *Proc. Natl. Acad. Sci. USA*, 36:48–49, 1950.

[248] T. Neubrunn. A generalized continuity and product spaces. *Math. Slovaca*, 26:97–99, 1976.

[249] T. Neubrunn. Quasi-continuity. *Real Anal. Exch.*, 14:259–306, 1988.

[250] T. Neubrunn, T. Šalát, and J. Smítal. On certain properties characterizing locally separable metric spaces. *Časopis Pěst. Mat.*, 92:157–161, 1967.

[251] T. Neubrunn, T. Šalát, and M. Švec. *Matematická analýza*. Veda, 1990.

[252] A. Neubrunnová. Transfinite convergence and locally separable metric spaces. *Acta Fac. Rerum Nat. Univ. Comen. Math.*, 34:177–181, 1979.

[253] D. Noll. Baire spaces and graph theorems. *Proc. Am. Math. Soc.*, 96:513–522, 1986.

[254] B. Novotný. On subcontinuity. *Real Anal. Exch.*, 31:535–545, 2005.

[255] W. F. Osgood. Non-Uniform Convergence and the Integration of Series Term by Term. *Am. J. Math.*, 19(2):155–190, 1897.

[256] J. C. Oxtoby. Cartesian products of Baire spaces. *Fundam. Math.*, 49:157–166, 1961.

[257] R. R. Phelps. *Convex functions, monotone operators and differentiability*, volume 1364 of *Lecture Notes in Mathematics*. Springer-Verlag, Berlin, 2nd edition, 1993.

[258] R. S. Phillips. On weakly compact subsets of a Banach space. *Am. J. Math.*, 65:108–136, 1943.

[259] Z. Piotrowski. Continuity points in $\{x\} \times Y$. *Bull. Soc. Math. Fr.*, 108:113–115, 1980.

[260] Z. Piotrowski. Quasicontinuity and product spaces. In *Proceedings of the International Conference on Geometric Topology (Warsaw, 1978)*, pages 349–352. PWN, Warsaw, 1980.

[261] Z. Piotrowski. On the theorems of Y. Mibu and G. Debs on separate continuity. *Int. J. Math. Sci.*, 19(3):495–500, 1996.

[262] Z. Piotrowski. Separate and joint continuity in Baire groups. *Tatra Mt. Math. Publ.*, 14:109–116, 1998.

[263] V. I. Ponomarev. A new space of closed sets and many-valued continuous mappings of bicompacts. *Mat. Sb. (N. S.)*, 48(90):191–212, 1959.

[264] L. S. Pontryagin. *Topological groups*. Gordon and Breach Science Publishers, Inc., New York-London-Paris, 1966. Translated from the second Russian edition by Arlen Brown.

[265] D. Repovš and P. V. Semenov. *Continuous selections of multivalued mappings*, volume 455 of *Mathematics and its Applications*. Kluwer Academic Publishers, Dordrecht, 1998.

[266] J. P. Revalski and N. V. Zhivkov. Best approximation problems in compactly uniformly rotund spaces. *J. Convex Anal.*, 19(4):1153–1166, 2012.

[267] E. Reznichenko. The continuity of the inverse. http://2201gtopology.math.msu.su/erezn-ci.

[268] N. K. Ribarska. Internal characterization of fragmentable spaces. *Mathematika*, 34:243–257, 1987.

[269] N. K. Ribarska. Internal characterization of fragmentable spaces. *C. R. Acad. Bulgare Sci.*, 40:9–10, 1987.

[270] C. Richter. Representing cliquish functions as quasiuniform limits of quasicontinuous functions. *Real Anal. Exch.*, 27:209–222, 2001/2002.

[271] M. A. Rieffel. Dentable subsets of Banach spaces, with application to a Radon–Nikodým theorem. In *Functional Analysis (Proc. Conf., Irvine, Calif., 1966)*, pages 71–77. Academic Press/Thompson Book Co., London/Washington, D.C., 1967.

[272] R. T. Rockafellar. *Convex analysis*, volume 28 of *Princeton Mathematical Series*. Princeton University Press, Princeton, N.J., 1970.

[273] R. T. Rockafellar. Favorable classes of Lipschitz-continuous functions in subgradient optimization. In *Progress in nondifferentiable optimization*, volume 8 of *IIASA Collaborative Proc. Ser. CP-82*, pages 125–143. Internat. Inst. Appl. Systems Anal, Laxenburg, 1982.

[274] R. T. Rockafellar and R. J. -B. Wets. *Variational analysis*, volume 317 of *Grundlehren der Mathematischen Wissenschaften [Fundamental Principles of Mathematical Sciences]*. Springer-Verlag, Berlin, 1998.

[275] C. A. Rogers. A less strange version of Milnor's proof of Brouwer's fixed-point theorem. *Am. Math. Mon.*, 87(7):525–527, 1980.

[276] C. A. Rogers and J. E. Jayne. *K-analytic sets. in Analytic Sets*. Academic Press, New York - London, 1980.

[277] S. Rolewicz. On drop property. *Stud. Math.*, 85(1):27–35, 1987.

[278] S. Rolewicz. On Δ-uniform convexity and drop property. *Stud. Math.*, 87(2):181–191, 1987.

[279] M. E. Rudin. A new proof that metric spaces are paracompact. *Proc. Am. Math. Soc.*, 20:603, 1969.

[280] W. Rudin. *Functional analysis. McGraw-Hill Series in Higher Mathematics*. McGraw-Hill Book Co., New York-Düsseldorf-Johannesburg, 1973.

[281] J. Schauder. Der fixpunktsatz in funktionalräumen. *Stud. Math.*, 2:171–180, 1930.

[282] W. Sierpinski. Sur les suites transfinies convergentes de fonctions de Baire. *Fundam. Math.*, 1:132–141, 1920.

[283] I. Singer. *The theory of best approximation and functional analysis*, volume 13 of *Conference Board of the Mathematical Sciences Regional Conference Series in Applied Mathematics*. Society for Industrial and Applied Mathematics, Philadelphia, Pa, 1974.

[284] M. Sion. On general minimax theorems. *Pac. J. Math.*, 8:171–176, 1958.

[285] M. A. Smith. Some examples concerning rotundity in Banach spaces. *Math. Ann.*, 233(2):155–161, 1978.

[286] E. Sperner. Neuer Beweis für die Invarianz der Dimensionszahl und des Gebietes. *Abh. Math. Semin. Univ. Hamb.*, 6(1):265–272, 1928.

[287] V. V. Srivatsa. Baire class 1 selectors for upper semicontinuous set-valued maps. *Trans. Am. Math. Soc.*, 337(2):609–624, 1993.

[288] R. M. Starr. *General equilibrium theory: An Introduction*. Cambridge University Press, Cambridge, 1997.

[289] S. B. Stečkin. Approximation properties of sets in normed linear spaces. *Rev. Math. Pures Appl.*, 8:5–18, 1963.

[290] C. Stegall. The duality between Asplund spaces and spaces with the Radon–Nikodým property. *Isr. J. Math.*, 29(4):408–412, 1978.

[291] C. Stegall. Optimization of functions on certain subsets of Banach spaces. *Math. Ann.*, 236(2):171–176, 1978.

[292] C. Stegall. A class of topological spaces and differentiation of functions on Banach spaces. In *Proceedings of the conferences on vector measures and integral representations of operators, and on functional analysis/Banach space geometry (Essen, 1982)*, volume 10 of *Vorlesungen Fachbereich Math. Univ. Essen*, pages 63–77. Univ. Essen, Essen, 1983.

[293] C. Stegall. Functions of the first Baire class with values in Banach spaces. *Proc. Am. Math. Soc.*, 111(4):981–991, 1991.

[294] A. H. Stone. Paracompactness and product spaces. *Bull. Am. Math. Soc.*, 54:977–982, 1948.

[295] D. Stover. On π-metrizable spaces, their continuous images and products. *Comment. Math. Univ. Carol.*, 50(1):153–162, 2009.

[296] K. R. Stromberg. *Introduction to classical real analysis*. *Wadsworth International Mathematics Series*. Wadsworth International, Belmont, Calif, 1981.

[297] W. L. Strother. Continuous multi-valued functions. *Bol. Soc. Mat. São Paulo*, 10:87–120, 1958.

[298] F. E. Su. Rental harmony: Sperner's lemma in fair division. *Am. Math. Mon.*, 106(10):930–942, 1999.

[299] S. Swaminathan. Normal structure in Banach spaces and its generalisations. In *Fixed points and nonexpansive mappings (Cincinnati, Ohio, 1982)*, volume 18 of *Contemp. Math.*, pages 201–215. Amer. Math. Soc., Providence, RI, 1983.

[300] H. P. Thielman. Types of functions. *Am. Math. Mon.*, 60:156–161, 1953.

[301] M. Tkachenko. Paratopological and semitopological groups versus topological groups. In *Recent Progress in General Topology. III*, pages 825–882. Atlantis Press, Paris, 2014.

[302] S. L. Troyanski. On some generalizations of denting points. *Isr. J. Math.*, 88(1–3):175–188, 1994.

[303] A. Tychonoff. Ein Fixpunktsatz. *Math. Ann.*, 111(1):767–776, 1935.

[304] L. Vašák. On one generalization of weakly compactly generated Banach spaces. *Stud. Math.*, 70(1):11–19, 1981.

[305] H. Vaughan. On locally compact metrizable spaces. *Bull. Am. Math. Soc.*, 43:532–535, 1937.

[306] L. P. Vlasov. Approximative properties of sets in normed linear spaces. *Usp. Mat. Nauk*, 28(6(174)):3–66, 1973.

[307] J. von Neumann. Zur Theorie der Gesellschaftsspiele. *Math. Ann.*, 100(1):295–320, 1928.

[308] J. von Neumann. Über ein ökonomisches Gleichungssystem und eine Verallgemeinerung des Brouwerschen Fixpunktsatzes. *Erg. Math. Kolloqu.*, 8:73–83, 1937.

[309] J. von Neumann and O. Morgenstern. *Theory of Games and Economic Behavior*. Princeton University Press, Princeton, New Jersey, 1944.

[310] C. E. Weil and K. -S. Lau. Differentiability via directional derivates. *Proc. Am. Math. Soc.*, 70:11–17, 1978.

[311] D. Weller. Fair division of a measurable space. *J. Math. Econ.*, 14(1):5–17, 1985.

[312] H. E. White. First countable spaces that have countable pseudo-bases. *Can. Math. Bull.*, 21(1):103–112, 1978.

[313] S. Xiang, W. Jia, J. He, S. Xia, and Z. Chen. Some results concerning the generic continuity of set-valued mappings. *Nonlinear Anal.*, 75(8):3591–3597, 2012.

[314] L. Zsilinszky. Products of Baire spaces revisited. *Fundam. Math.*, 183:115–121, 2004.

Index of notation and conventions

Unless otherwise stated, we will assume that all vector spaces and all normed linear spaces are over the field of real numbers. We use the notation ":=" to mean that the symbol to the left of ":=" is defined to be equal to the object on the right.

- Suppose that X and Y are sets, $f : X \to Y$ is a function and $\Phi : X \to 2^Y$ is a set-valued mapping. Then:
  - If U is a subset of X, then $\Phi(U) := \bigcup\{\Phi(u) : u \in U\}$;
  - If U is a subset of X, then $f(U) := \{f(u) : u \in U\}$, and if V is a subset of Y, then $f^{-1}(V) := \{x \in X : f(x) \in V\}$;
  - $\mathrm{Dom}(\Phi) := \{x \in X : \Phi(x) \neq \varnothing\}$. $\mathrm{Dom}(\Phi)$ is called the *domain of* Φ or the *effective domain of* Φ;
  - If X possesses a topology τ, then we say that Φ is *densely defined* if $\mathrm{Dom}(\Phi)$ is dense in (X, τ);
  - If A is a subset of X and $\Phi : X \to 2^Y$, then $\Phi|_A : A \to 2^Y$ is defined by $\Phi|_A(a) := \Phi(a)$ for all $a \in A$;
  - $\mathrm{Gr}(\Phi) := \{(x, y) \in X \times Y : y \in \Phi(x)\}$, called the *graph of* Φ.
- If (M, d) is a metric space, then:
  - If $0 < r < \infty$ and $x \in M$, then $B[x, r] := \{x \in M : d(x, y) \leq r\}$ is called the *closed ball of radius r with centre x*;
  - If $0 < r < \infty$ and $x \in M$, then $B(x, r) := \{x \in M : d(x, y) < r\}$ is called the *open ball of radius r with centre x*;
  - If $0 < r < \infty$ and $x \in M$, then $S(x, r) := \{x \in M : d(x, y) = r\}$ is called the *sphere of radius r with centre x*;
  - If $0 < r < \infty$ and A is a subset of M, then $B(A, r) := \bigcup_{a \in A} B(a, r)$, and $B[A, r] := \bigcup_{a \in A} B[a, r]$;
  - $S_\varepsilon(M) := \{X \in 2^M : \text{ for every distinct } x, y \in X, \varepsilon \leq d(x, y)\}$, the elements of $S_\varepsilon(M)$ are called *ε-separated sets* or *ε-nets*;
  - For a subset C of M and an element $x \in M$, we define the *distance from x to C* by $d(x, C) := \inf\{d(x, c) : c \in C\}$;
  - $\mathcal{B}^1(X, M)$ denotes the set of all *Baire one* functions from a topological space (X, τ) into M;
  - $\mathcal{B}^1(X)$ denotes the set of all real-valued Baire one functions on a topological space (X, τ).
- The *natural numbers*, $\mathbb{N} := \{1, 2, 3, \ldots\}$.
- The *integers*, $\mathbb{Z} := \{\ldots, -2, -1, 0, 1, 2 \ldots\}$.
- The *rational numbers*, $\mathbb{Q} := \{a/b : a, b \in \mathbb{Z}, b \neq 0\}$.
- The *real numbers*, $\mathbb{R}$.
- The first infinite ordinal, ω.
- The first uncountable ordinal, ω_1.
- For any set X, 2^X is the set of all subsets of X (sometimes, we denote by $\mathcal{P}(X)$ the set of all subsets of X as well). The symbol $\varnothing$ denotes the empty set.

https://doi.org/10.1515/9783110750188-006

- If (X, τ) and (Y, τ') are topological spaces, then the product topology on the set $X \times Y$ generated by (X, τ) and (Y, τ') is denoted by $\tau \times \tau'$.
- Let A be a nonempty subset of a set X. Then we call the weak topology on $\mathbb{R}^X$ generated by the functions $\{\delta_a : a \in A\}$ the *topology of pointwise convergence on* A, where for each $a \in A$, $\delta_a : \mathbb{R}^X \to \mathbb{R}$ is defined by

$$\delta_a(f) := f(a) \quad \text{for } f \in \mathbb{R}^X.$$

  We denote the topology of pointwise convergence on A by $\tau_p(A)$.
- For any subset A of a topological space (X, τ), we define:
  - $\text{int}(A)$, called the *interior* of A, is the union of all open sets contained in A;
  - $\overline{A}$, called the *closure* of A, is the intersection of all closed sets containing A;
  - $\text{Bd}(A)$, called the *boundary* of A, is $\overline{A} \setminus \text{int}(A)$,
- Let (X, τ) and (Y, τ') be topological spaces. For any $f : X \to Y$, we define

$$C(f) := \{x \in X : f \text{ is continuous at } x\} \quad \text{and} \quad D(f) := X \setminus C(f).$$

- Let (X, τ) and (Y, τ') be topological spaces. Then

$$DC(X, Y) := \{f \in Y^X : C(f) \text{ is dense in } (X, \tau)\}.$$

- Let (X, τ) and (Y, τ') be topological spaces. Then $MU(X, Y)$ denotes the set of all minimal usco from (X, τ) into subsets of (Y, τ'). If $(Y, +, \cdot, \tau')$ is a linear topological space, then $MC(X, Y)$ denotes the set of all minimal cusco from (X, τ) into subsets of (Y, τ').
- Let (X, τ) be a topological space, and let (M, d) be a metric space. For any $f : X \to M$, we define

$$A(f) := \{x \in X : f \text{ is cliquish at } x\}.$$

- For any points x and y in a vector space $(X, +, \cdot)$, we define the following intervals:
  - $[x, y] := \{x + \lambda(y - x) : 0 \leq \lambda \leq 1\}$;
  - $(x, y) := \{x + \lambda(y - x) : 0 < \lambda < 1\}$;
  - $[x, y) := \{x + \lambda(y - x) : 0 \leq \lambda < 1\}$;
  - $(x, y] := \{x + \lambda(y - x) : 0 < \lambda \leq 1\}$.
- For any normed linear space $(X, \|\cdot\|)$, we define
  - $B[x, r] := \{y \in X : \|x - y\| \leq r\}$ for any $x \in X$ and $r > 0$;
  - $B(x, r) := \{y \in X : \|x - y\| < r\}$ for any $x \in X$ and $r > 0$;
  - $B_X := B[0, 1]$;
  - $S_X := \{x \in X : \|x\| = 1\}$.
- Given a compact Hausdorff space (K, τ), we write $C(K)$ for the set of all real-valued continuous functions on (K, τ). This is a vector space under the operations of pointwise addition and pointwise scalar multiplication. $C(K)$ becomes a Banach space when equipped with the uniform norm $\|\cdot\|_\infty$ defined by

$$\|f\|_\infty := \sup_{x \in K} |f(x)| \quad \text{for } f \in C(K).$$

- For a normed linear space $(X, \|\cdot\|)$, we denote by X^* the set of bounded linear maps from X to $\mathbb{R}$. If X^* is equipped with the operator norm given by

$$\|f\| := \sup_{x \in B_X} |f(x)| \quad \text{for } f \in X^*,$$

then $(X^*, \|\cdot\|)$ is a Banach space and is called the *dual space* of $(X, \|\cdot\|)$. For each $x \in X$, we denote by $\widehat{x} : X^* \to \mathbb{R}$ the continuous linear functional on X^* defined by $\widehat{x}(x^*) := x^*(x)$ for $x^* \in X^*$. Then $\widehat{X} := \{\widehat{x} : x \in X\} \subseteq (X^*)^* =: X^{**}$.
- If (X, τ) and (Y, τ') are topological spaces and $\Phi : X \to 2^Y$, then we define $\mathrm{USC}(\Phi) : X \to 2^Y$ by

$$\mathrm{USC}(\Phi)(x) := \bigcap\{\overline{\Phi(U)} : U \text{ is an open neighbourhood of } x\}.$$

- If (X, τ) is a topological space, $(Y, +, \cdot, \tau')$ is a topological vector space and $\Phi : X \to 2^Y$, then we define $\mathrm{CSC}(\Phi) : X \to 2^Y$ by

$$\mathrm{CSC}(\Phi)(x) := \bigcap\{\overline{\mathrm{co}}\, \Phi(U) : U \text{ is an open neighbourhood of } x\}.$$

- Let (X, τ) and (Y, τ') be topological spaces. Then

$$D(X, Y) := \{\mathrm{USC}(f|_{C(f)}) : f \in DC(X, Y)\}.$$

- Let X be a set, and let $(Y, \leq)$ be a totally ordered set. For any function $f : X \to Y$, we define

$$\mathrm{argmax}(f) := \{x \in X : f(y) \leq f(x) \text{ for all } y \in X\},$$
$$\mathrm{argmin}(f) := \{x \in X : f(x) \leq f(y) \text{ for all } y \in X\}.$$

- Let A be a subset of a vector space $(X, +, \cdot)$. Then the *convex hull* of A, denoted by either $\mathrm{co}(A)$ or $\mathrm{co}\, A$, is defined as the intersection of all convex subsets of $(X, +, \cdot)$ that contain A. If $(X, +, \cdot, \tau)$ is a topological vector space, then the *closed convex hull* of A, denoted by either, $\overline{\mathrm{co}}(A)$ or $\overline{\mathrm{co}}\, A$, is defined as the intersection of all closed and convex subsets of $(X, +, \cdot, \tau)$ that contain A.
- Let X be a set, and let $f : X \to [-\infty, \infty]$. Then

$$\mathrm{Dom}(f) := \{x \in X : f(x) \in \mathbb{R}\}.$$

We say that f is a *proper function* if $\mathrm{Dom}(f) \neq \varnothing$.
- Let $(X, \|\cdot\|)$ be a normed linear space, and let $f : X \to [-\infty, \infty]$. Then the *Fenchel conjugate* of f is the function $f^* : X^* \to [-\infty, \infty]$ defined by

$$f^*(x^*) := \sup_{x \in X}(x^* - f)(x) \quad \text{for } x^* \in X^*.$$

The function f^* is convex, and if f is a proper function, then f^* never takes the value $-\infty$.

- If f is a convex function defined on a nonempty convex subset K of a normed linear space $(X, \|\cdot\|)$ and $x \in K$, then we define the *subdifferential of f at x* as the set $\partial f(x)$ of all $x^* \in X^*$ satisfying

$$x^*(y - x) \leq f(y) - f(x) \quad \text{for all } y \in K.$$

- We assume that the reader has a basic working knowledge of metric spaces, normed linear spaces and even basic general topology. In particular, we assume that the reader is familiar with Tychonoff's theorem, the Banach–Alaoglu theorem and the separation theorem.

Theorem (Tychonoff's theorem [78]). *The Cartesian product $\prod_{s \in S} X_s$, where $X_s \neq \emptyset$ for all $s \in S$, is compact if and only if all spaces X_s are compact.*

Theorem (Banach–Alaoglu theorem [1]). *Let $(X, \|\cdot\|)$ be a normed linear space. Then $(B_{X^*}, weak^*)$ is compact.*

Theorem (Separation theorem [75, p. 418]). *Let $(X, +, \cdot, \tau)$ be a locally convex space over $\mathbb{R}$, and let C be a nonempty closed convex subset of $(X, +, \cdot, \tau)$. If $x_0 \notin C$, then there exists a continuous linear functional x^* on $(X, +, \cdot, \tau)$ such that*

$$\sup\{x^*(c) : c \in C\} < x^*(x_0).$$

Index

De Gruyter Studies in Mathematics

Volume 76
Steven Lord, Fedor Sukochev, Dmitriy Zanin
Singular Traces. Volume 1: Theory, 2021
ISBN 978-3-11-037779-8, e-ISBN 978-3-11-037805-4, e-ISBN (ePUB) 978-3-11-039231-9

Volume 80
Xiao-Xiong Gan
Formal Analysis. An Introduction, 2021
ISBN 978-3-11-059785-1, e-ISBN 978-3-11-059945-9, e-ISBN (ePUB) 978-3-11-059860-5

Volume 79
Rajendra V. Gurjar, Kayo Masuda, Masayoshi Miyanishi
Affine Space Fibrations, 2021
ISBN 978-3-11-057736-5, e-ISBN 978-3-11-057756-3, e-ISBN (ePUB) 978-3-11-057742-6

Volume 78
Kai Liu, Ilpo Laine, Lianzhong Yang
Complex Delay-Differential Equations, 2021
ISBN 978-3-11-056016-9, e-ISBN 978-3-11-056056-5, e-ISBN (ePUB) 978-3-11-056040-4

Volume 77
Laura Costa, Rosa Maria Miró-Roig, Joan Pons-Llopis
Ulrich Bundles. From Commutative Algebra to Algebraic Geometry, 2021
ISBN 978-3-11-064540-8, e-ISBN 978-3-11-064768-6, e-ISBN (ePUB) 978-3-11-064580-4

Volume 76
Anton Zettl
Recent Developments in Sturm-Liouville Theory, 2021
ISBN 978-3-11-071884-3, e-ISBN 978-3-11-071900-0, e-ISBN (ePUB) 978-3-11-071938-3

Volume 75
Elena Cordero, Luigi Rodino
Time-Frequency Analysis of Operators, 2020
ISBN 978-3-11-053035-3, e-ISBN 978-3-11-053245-6, e-ISBN (ePUB) 978-3-11-053060-5

Volume 25
Karl H. Hofmann, Sidney A. Morris
The Structure of Compact Groups, 4th Edition, 2020
ISBN 978-3-11-069595-3, e-ISBN 978-3-11-069599-1, e-ISBN (ePUB) 978-3-11-069601-1